IS NATURE SUPERNATURAL?

IS NATURE SUPERNATURAL?

A Philosophical
Exploration
of Science and Nature

SIMON L. ALTMANN

Prometheus Books
59 John Glenn Drive
Amherst, New York 14228-2197

Published 2002 by Prometheus Books

Inquiries should be addressed to
Prometheus Books
59 John Glenn Drive
Amherst, New York 14228–2197
VOICE: 716–691–0133, ext. 207
FAX: 716–564–2711
WWW.PROMETHEUSBOOKS.COM

06 05 04 03 02   5 4 3 2 1

Library of Congress Cataloging-in-Publication Data

Altmann, Simon L., 1924–
    Is nature supernatural? : a philosophical exploration of science and nature / Simon L. Altmann.
        p. cm.
    Includes bibliographical references and index.
    ISBN 1–57392–916–6 (alk. paper)
    1. Science—Philosophy. I. Title.

Q175 .A53 2001
501—dc21

Q
175
A53
2002   2001048278

Printed in the United States of America on acid-free paper

To the memory of my mother,
Matilde Branover de Altmann,
from whom I learnt to work.

*Das Denken gehört zu den größten Vergnügungen der menschlichen
Rasse.*

Thinking is one of the greatest pleasures of the human race.
Bertolt Brecht (1955), Scene 3, p. 40.

*Nel suo profondo vidi che s'interna
    Legato con amore in un volume
    Ciò che per l'universo si squaderna:
Sustanzia ed accidente e lor costume,
    Tutti conflati insieme per tal modo,
    Che ciò che io dico è un semplice lume.
La forma universal di questo nodo
    Credo ch'io vidi, perchè più di largo,
    Dicendo questo, mi sento che io godo.*

*I saw in its depth the contents
    Bound with love in one volume
    of that which along the universe is dispersed:
Substance and accident and their behaviour,
    All conflated together in such a manner
    that what I can say is but a gleam.
The universal shape of this knot I think that I saw,
    because on saying this
    I feel that I enjoy myself even more.*
Dante Alighieri (1317), Canto 33, 85–93.

All references, including the epigraphs, may be identified by author and date in the Bibliography at the end of the book.

CONTENTS

18 IS NATURE SUPERNATURAL?

PREFACE

There are dozens of popular science books, mainly written by scientists, which touch on philosophical aspects of the subject and numerous books on the philosophy of science written by professional philosophers. This book attempts to bridge the gap between these two sets, while addressing readers with no previous experience either of philosophy or of science. Although I do not follow the wise injunction of Stephen Hawking's editor, but use a few formulas and, horror of horrors, even some tables, I have planned this carefully, so that the reader needs no previous knowledge to use them. Even more, whenever I could, I have written the formulas in English rather than in mathematical or logical notation. Many of these formulas and tables cover very elementary ideas of logic, and they are given in order to fulfill one of the purposes of this book, which is to allow people to get a feeling about how one works and how one thinks in logic and philosophy.

In my cheeky moments I have thought, in fact, that a suitable subtitle of this book would have been *What Sophie Was Not Told*; this is how to face philosophical problems and how to try to provide solutions. If philosophy does not teach people to think, what is the good of it? Merely being given a list of thoughts regurgitated from past philosophers is not a way to improve understanding, and I do not propose to present a collection either of physical facts or of philosophical ideas. Of course, I have to cover some fundamental philosophical questions, like necessity and causality, and I have to review some of the ideas of relativity and quantum mechanics—including things like Schrödinger's cat and the Einstein, Podolsky, and Rosen paradox—but I shall do this constantly from the point of view of what these things teach us about nature. At a more pedestrian but very useful level, I have also provided some examples to show how logic may help in the correct enunciation of physical laws.

I expect this book to be most useful for scientists who want to penetrate the mysteries of philosophy, and for others who want to get started on both subjects. I imagine, however, that it might be read by some philosophers, and if this were the case, a few remarks might be appropriate. Although Hume has been under attack of late, I lean strongly on his ideas. I have attempted to read him not so much as a philosopher, but as a natural scientist. Scientists appeal cautiously to causality but, when they do so, they use it in a way that does not satisfy the requirements of most modern philosophers. Hume's definitions, instead, are easily adapted to current scientific use. Even more, I shall claim that Hume was a proto-Darwinian, and that his concept of *custom* or *habit* can easily be grafted into an evolutionary interpretation—evolutionary theory being in a way the backbone of the book. I have also extensively used Nelson Goodman's concept of *entrenchment*, not so far developed in the philosophy of science.

Although circularities are wrong in logic, the situation is different in natural science. Some property of nature must be "fitted" by a method of successive approximations that entails unavoidable circularity, but which is finally justified by its empirical adequacy. I claim, for instance, that this is the only way in which the definition of *state* can be made, and I give as an example a reconstruction of Newton's second law, which I hope will be found useful, even by some scientists. (I include in this treatment an informal account of Newton's invention of the calculus, the relevance of which becomes far clearer if its importance for the definition of state is understood.)

One way of writing philosophy is to start each subject with a critical review of other people's ideas, an approach which might mislead the general public into believing that philosophy is more concerned with what philosophers write than with anything else. Over four hundred years ago, the French essayist Michel de Montaigne already had reason to complain: "there are more books on books than on any other subject. We do nothing but write comments on one another" (*Essays*, Book 3, chap. 13). I have therefore avoided critical reviews and used quotations in the text only when absolutely necessary. I have relegated such material to notes that follow each chapter: they are not just an adjunct for the scholar, but a complement to the text which will often help clear up the reader's doubts, for which purpose the glossary at the end of the book will also be useful.

The reader may notice that there are certain subjects that recur, and are even repeated to some extent, in different parts of the book. This is intentional, first, because writing for what I expect might be a not very experienced audience, I felt that in some cases, rather than referring readers back to a previous section in order to refresh their memories, it was worthwhile providing a brief summary. Second, this has made each chapter reasonably self-contained, thus allowing readers to read the book in the order that better reflects their interests.

The book is basically divided into two parts. The first ten chapters cover the main ideas of logic and philosophy necessary for the philosophy of science. In these chapters, I also treat the basic questions involved in the classical approach to time, space, and mechanics, ending with a discussion of natural laws and of the scientific method. In between the last two subjects, I have discussed the problems that arise when space and time are chopped up into indefinitely smaller intervals. This is, of course, the subject of the Zeno paradoxes. I hope that I might not be unduly optimistic in expecting that my solution to the infamous paradox of Achilles and the Tortoise might clear up this problem once and for all for many people. In discussing the scientific method in the last of the first ten chapters, I had to violate my rule of avoiding reviews of other people's ideas, since the latter are all incorporated in one way or the other in my final synthesis of the subject. Many practicing scientists are still too firmly attached to Popperian modes of thinking, and I hope I might help them to wake up from such dogmatic slumbers.

The second part of the book starts with a review of the implications of relativity for the concepts of space and time, and continues with a chapter on the arrow of time that shows how the concept of randomness had to be brought into science. C. P. Snow maintained that no man or woman could be held to be attuned to our culture who did not understand the second principle of thermodynamics. Although I do not entirely agree with this dictum, I hope I have provided here an entry for the many into his academy. What really matters is not some more or less detailed knowledge of specific scientific results but, rather, the flexibility of mind necessary to understand the way in which scientists think and operate. For the real problem is that of two subcultures separated by a common language, which is not here a mere witticism—serious misunderstandings occur by a lack of appreciation that some common words, such as "causes" and "states," are used in science in an unfamiliar way.

Two chapters follow dealing with probability and mathematics, respectively. My main concern here is the possibility for the world to be, or not, some sort of an adjunct of a Platonic, mathematical superworld— a view that appears to have some following these days. I discuss within this framework the Gödel theorems and their implications. I hope that my treatment of mathematics here will help bring it firmly back to this sublunar world. The last few chapters are devoted to the microworld of quantum mechanics, which provides a strong test for the ideas developed in the first part of the book. Revered principles, such as that of causality, I have shown, cannot be expected to be valid outside macroscopic environments. This carefully prepared philosophical background helps to demythologize quantum mechanics and bring back some of the rationality lost in many popular expositions on the subject.

I include a discussion of the Aspect experiments and the Bell inequalities which, despite a large variety of so-called popular presentations, are still difficult to grasp. Of course, I discuss the consequences of these results on action at a distance, which I consider one of the major conundrums of our time. The last chapter, which bears the book's title, is a review of some ontological problems, mainly related to idealism and physicalism. I also deal with the effects of cultural relativism and social constructivism on the public's perception of science. About this last question, I provide a strong critique of Thomas Kuhn's ideas which, in my view, have created serious misunderstandings. The book ends with a dialogue on science and belief.

Readers will understand, I hope, that I had to impose two major constraints on my writing. I avoided in many cases details that, however important to the specialist, would have obscured the subject for the beginner. For the same reason, I did not always follow a strict historical treatment, unless doing so was important in understanding the subject. Despite these two perilous constraints, I have tried to maintain a modicum of honesty in my arguments.

I should perhaps say a few words about how this book originated: the story goes back to 1947 or 1948, when I gave the opening lecture of a newly founded seminar on the philosophy of science at the University of Buenos Aires with a (much too long) lecture on causality. At the end of the lecture, my mathematics teacher, Professor Julio Rey Pastor, who had strongly encouraged the creation of the seminar, challenged my statements about Hume's views on the subject. I was fortunately armed with my copy of the *Enquiries*, and, to his surprise, my quotation had been precise. This experience has been repeated more than once, which encouraged me to believe that the time had come to publish my material, most of which has not appeared in print before. However, parts, especially my chapter on *grue*, have been much circulated.

It is a pleasure to thank the many friends who have helped me in writing this book. Mrs. Julia Turton bravely accepted the role of a lay reader, and her suggestions will undoubtedly help many others in their understanding of the book. I have had numerous discussions with Dan and Gerry Altmann, from whose expertise in philosophy and psycholinguistics, respectively, I benefited greatly. (I must not forget Paul Altmann, whose medical skills saved my computer more than once from ignominious death.) I have had useful correspondence on the question of identity and indistinguishability of elementary particles with Professors Peter

Higgs and Sir Michael Berry, to whom I am grateful. Professor Ian Shanker and Dr. Alan Weir most kindly read and commented on chapter 14, particularly on the Gödel theorem. Professor Ian Aitchison read the whole manuscript, and I am much indebted for his remarks. I owe a great deal to Professor Paul Horwich for encouraging comments and for detailed and very useful criticism of most of the philosophical sections of the book. I am also grateful to my philosophy colleague at Brasenose, Marianne Talbot, for useful discussions. I could never have written this book without the support of my wife Bocha, who not only had to put up with an almost virtual husband for many years, but who even volunteered to check some difficult passages for readability.

In writing about matters philosophical, there are two constraints that an author must accept. The first one is that whatever you say has already been said before by somebody else, often much better. The second is that, as a consequence of the first, whatever you say has already been demonstrated by other writers to be wrong or incomplete or somehow otherwise defective. In this play, all you can hope for is to swell a progress, start a scene or two, knowing full well that Prince Hamlet can never be got hold of to demonstrate the truth or at least the consistency of your representation. With this sobering thought, I leave this book in the reader's hand.

Simon L. Altmann
Brasenose College, Oxford
October 2000.

ACKNOWLEDGMENTS

Thanks are due to the following for permission to quote extracts from works under copyright:

Wynstan H. Auden, *Collected Shorter Poems 1930–1944*. Faber & Faber, London, 1950. Reprinted by permision of Faber & Faber. "Shorts," copyright © 1974 by The Estate of W. H. Auden, from W. H. AUDEN: COLLECTED POEMS by W. H. Auden. Used by permission of Random House, Inc.

John S. Bell, *Speakable and unspeakable in quantum mechanics*. Cambridge University Press, 1987. Reprinted by kind permission of Cambridge University Press.

Anthony Burgess, *Byrne: A novel*, published by Hutchinson. Used by permission of the Estate of Anthony Burgess and by kind permision of The Random House Group Limited.

Jean-Pierre Changeux and Alain Connes, *Conversations on mind, matter, and mathematics*. Copyright © 1995 by Princeton University Press. Reprinted by kind permission of Princeton University Press.

Albert Einstein, *The meaning of relativity*. Methuen, London (1950). Copyright © 1950 by Princeton University Press. Reprinted by kind permission of Princeton University Press.

Richard P. Feynman, *The character of physical law*. Penguin, London, 1992, and MIT Press, Cambridge, Mass., 1967, © Richard P. Feynman, 1965. Reproduced by kind permission of Penguin Books Limited and The MIT Press.

Murray Gell-Mann, *The quark and the jaguar: Adventures in the simple and the complex*. Abacus, London, 1995; Little Brown and Company, London, 1994. Reproduced by kind permission of Little Brown and Company, and of W. H. Freeman and Company/Worth.

Salman Rushdie, *The Moor's last sigh*, published by Jonathan Cape, London, 1995. Used by kind permission of The Random House Group Limited.

1

WHAT THIS BOOK IS ALL ABOUT

Autant peut fair le sot celui qui dit vrai que celui qui dit faux, car nous sommes sur la manière, non sur la matière du dire.

He who says what is true may be as foolish as another who utters falsities, for we are concerned with the manner, not with the matter, of discourse.
Michel de Montaigne (1588), Book 3, chap. 8, p. 192.

There is a wonderful short story by Anton Chekhov,[1] in which a young man, Kovrin, believes that he sees from time to time a black monk; but he fears that he is in fact hallucinating. One day, on meeting the monk again, Kovrin confronts him: " 'You are a mirage,' he says, upon which the monk answered in a low voice, not immediately turning his face towards him, 'The legend, the mirage, and I are all products of your excited imagination. I am a phantom.'

" 'Then you don't exist?' said Kovrin.

" 'You can think as you like,' said the monk, with a faint smile. 'I exist in your imagination, and your imagination is part of nature, so I exist in nature.' "

This story will serve as a metaphor for the whole of this book. People often ask such questions as: why does the language of mathematics fit reality? But, just as for Kovrin, our imagination, and thus our language, is part of nature.[2] If natural language, let alone the language of mathematics, had been created by a disembodied intelligence unchallenged by the flow of natural phenomena, it would have never fit anything. On the other hand, it is a matter of observation that all natural phenomena entail a balance or harmony, of which ecosystems are an example: language, be it that of science and mathematics or not, harmonizes likewise with phenomena, of which, of course, it is itself a part. People are not surprised if vegetables exist that vegetarian animals might eat, yet they are often amazed at the uncanny validity of the languages of science and mathematics. I shall, in fact, argue that the main purpose of science and mathematics is that of *setting up* a language with which natural phenomena may be fruitfully discussed. "Setting up" here entails the long process of natural selection by which language is tested and discarded or accepted.

Another way of describing what I have been saying is by stressing that the most remarkable property of our world is that it is self-referential; through us, the world talks to itself, in a beautiful mirror image of Kovrin's self conversation: whatever is in our minds is also in nature.[3] This is very important: as we shall see later, some physicists tend to make heavy weather about the difference between what we know (what the philosophers call *epistemology*) and what there is (what philosophers and theologians call *ontology*). Kovrin's monk or phantom is not very worried about such fine nuances; and although people often insist on neatly dividing things in pigeonholes, there is a lot to be said for taking a wider and more relaxed attitude: pigeonholes are the coffins of ideas, and it is with the latter and not with their corpses that we must engage.[4]

Most people will not find it difficult to accept that our minds are part of nature, and that unless we understand how this partnership works, we shall understand very little about nature itself. It is easy to believe that what the scientific mind does is deal with facts and with the words that neatly describe them. And that, eventually, the mastery of the words required to deal with the facts, leads to the mastery of the facts themselves: it appears to be undeniable that it is in this way, for instance, that humanity (aka NASA) managed to put a man on the moon. (Remember all those messages ending "roger"!)

Yes, of course, facts and words will be important for us in our quest, but we face here a serious problem, for these two things are not quite what the woman in the street thinks they are. Facts are not nice little packages that nature throws at us for us to inspect: obviously this cannot be so because if it were so, there would be a sharp boundary separating us from the rest of nature, whereas we are nothing else than an integral part of it. Naked facts hardly exist at all: they are processed by us through a network of theoretical constructs. Luigi Pirandello[5] put this very neatly: "a fact is like a sack, empty, it does not stand." Conversely, just like a sack collapses without its contents, our minds would collapse without the input that fills them up and gives them the characteristics that they have. Thus, the boundary between the facts that create this input and the mind is fuzzy, to say the least.[6]

If facts are not the reliable, hard, and eminently external data that we might think they are, at least a rose is a rose is a rose: nobody, we perhaps believe, can take words from us: they are our property, our exclusive invention. Ludwig Wittgenstein[7] (1889–1951), however, recognized that the real situation is much more involved than that when he said, "a great deal of stage-setting in the language is presupposed," in the mere act of naming. And we shall have to understand how this stage-setting works.

Of course, despite these worries, we shall undoubtedly deal with facts and with words in this book: I do not claim that I shall be able to disentangle even most of the problems involved in their use, but I hope that when readers finally close this book, they will be a bit wiser about them.

IS SCIENCE IN CRISIS?

Most readers will have read articles in the press, or will have seen television programs that purport to show that physics is in a crisis, that quantum mechanics has brought us near to magic, or at least that it has gotten stuck in apparent mysteries. Science, almost by definition, has always been in crisis. Like the stock exchange, science goes through bull as well as bear periods; but that the wonderfully reassuring bull periods were much longer a couple of centuries ago, is only a reflection of the fact that the science markets were moving much more slowly than now. No, a lot of the present feeling of crisis is largely a reflection of the sense of insecurity that has descended upon us in the last two or three decades: our worries about the environment, weapons, and the social consequences of advanced technologies, have resulted in a somewhat antiscientific, if not even antirational mood, which sees a crisis or a mystery where there is a mere lack of understanding—the latter being a normal part of the human condition.

That there are right now some apparent contradictions in physics is nothing new, neither, for that matter, is this a situation that has afflicted science alone. The contradictions in theology brought about by Darwin and Huxley last century were far deeper and more dolorous, and yet, theology is all the better for having gone through them.

Of course, because of the nature of science, it would be absurd to expect a book like this to solve any problems: this must be left to new experiments and new theory. But I am afraid that, at the risk of disappointing the reader, I shall go even further and say that I shall not be primarily concerned with discovering what is true and what is false. My purpose is to explore the manner in which one can fruitfully discourse about nature, a modest aim in which I follow the wise dictum of Montaigne quoted in the epigraph to this chapter.

As for the present so-called contradictions of physics, I shall likewise say that it is not my business to resolve them,[8] but to get a clear view of whatever it is that troubles us in this subject; that is, to describe reasonably accurately the state of affairs before they are resolved.

METAPHYSICS AND META-PHYSICS

In order to sketch some of the major problems discussed in this book, I shall now introduce in a rough way concepts that will be discussed later in more detail; so, the reader should not expect these ideas to be fully comprehensible until then. Imagine that we are in the middle of the nineteenth century, and, therefore, the possibility of observing the far side of the moon is inexistent. It could even be argued not only that this is not possible, but that it will *never* be possible: I do not commend this latter assertion, but even distinguished scientists are wont to make statements for all time. We have three lines of action as regards discussing the properties of the other side of the moon. (1) The entirely skeptical approach will urge us to remain silent. (2) We could invoke some sort of principle of uniformity or regularity of nature[9] and suggest that one might therefore expect the far side of the moon to be similar to the near side. (3) We could say that the other side of the moon is covered in blue cheese or, as Fontenelle[10] more elegantly did, we could place there a new and strange civilization.

The approach exemplified in the first option was important in the nineteenth century and in the first half of the twentieth century, during which period several philosophical schools adopted as a program the abolition of *metaphysics*: that is, roughly, of propositions that cannot be verified by observation. In the nineteenth century, when atoms were as unobservable as the other side of the moon, Ernst Mach (1838–1916), a highly influential German physicist, refused to accept their use in physical theory.[11] *Logical positivism*, a school that rejected the need for metaphysics, took in our century as its banner the last sentence of Wittgenstein's first book: "Whereof we cannot speak, thereof we must remain silent."[12]

This is not the way science works. An approach such as that proposed in (2) above is much nearer the mark. Undoubtedly, the *principle of uniformity or regularity of nature* invoked is not one that can be *proved* in any way, by observation or otherwise,[13] although it is known to be applicable to a variety of systems. We shall see that science requires the use of certain *normative principles* that have a much greater generality than *physical laws* and which are so named because they entail norms that organize scientific discourse. However well-grounded they are, they are not subject to confirmation in the same way as physical laws are, and because of this, I shall call such principles *meta-physical*. All this requires, of course, almost immediate clarification.

Before I do that, however,[14] let me say that the approach exemplified in option (3) above entails an *ontological* statement; that is, a statement about the *existence* of something. Moreover, if it is accepted (we are in the nineteenth century!) that the alleged existence cannot, now or ever, be open to confirmation by any procedure grounded in natural science, I shall call those propositions *metaphysical*.[15] Such statements have their place in poetry, ethics, and religion,[16] but they are even used in science during periods when things are so obscure that any speculation is better than nothing. It is precisely because the use of the word *metaphysics* is so varied that I shall qualify as *meta-physical* statements that are directly wedded to experience without being, nevertheless, derivable from it. I shall later discuss this in detail,[17] when we shall see that our propensity to use such meta-physical statements is grounded upon a well-established adaptive process entailing our interaction with the environment, and thus on *objective* experience. I shall, instead, qualify as *metaphysical* statements that cannot be so grounded.

I now come to the promised clarification of the very important concepts that I have named above. The idea of a *physical law* is not one that should worry many people even at this stage, and it will be discussed in detail in chapter 8. In order to get some idea about how they are distinguished from the *normative principles* that I have mentioned, the following short account should suffice for the time being. Physical laws are applied to individual experiments, and the collection of all such results and theories forms the sum total of scientific knowledge embodied in what later in this book will be called the *scientific mesh*. The principles that guide the construction of that mesh are called *normative principles* because they give the norms or standards that may be followed in scientific argument. One of the major normative principles, for instance, is the *principle of causality*, and we shall see in chapter 4 that although this principle is closely related to experience, it cannot be derived from it, which is why I call it a *meta-physical (normative) principle*. All the normative principles that I shall suggest as guiding the construction of the scientific mesh are in the same manner meta-physical.

It will be useful to discuss in a little more detail the role of the normative meta-physical principles, which, as should follow from the above, are clearly distinct from physical laws. Whereas a scientific law can be proposed in direct relation to a specific set of observations, and will in general provide *predictions* as to future events (which can then be either verified or falsified), this is not exactly the case for the normative meta-physical principles. Rather than predicting experience, they are used to organize it, if necessary by appeal to theoretical nonobservable constructs. Although they do not entail direct predictions as to the course of events, they direct our attention to ways in which *relations* between events may be described, as for instance when distinguishing between *causes* and their *effects*. And readers should try to adapt themselves to the idea that such relations are theoretical, rather than observable, constructs. We shall discuss this at length in chapter 4 as regards the causal relation, when the significance of what a meta-physical principle is will become much clearer.

It might be useful to illustrate the relation between a normative principle such as causality and a physical law such as Newton's second law, which deals with motion. In formulating this law, the events that are observed are forces, positions, velocities, and accelerations (which are the rates of change of the velocities).[18] What the principle of causality does, is to direct our mind to try to establish a connection between some of these events; and indeed, Newton's second law identifies the *forces* as the causes of the *accelerations*, as we shall see in chapter 6. Once this law is formulated, the acceleration of a body at a given time can be predicted by that law.

One of the major ideas in this book is that scientific statements do not stand or fall on their own but only as members of the scientific mesh, and that the latter grows by a rather organic process of natural evolution, certain ideas surviving (perhaps after some modification) at any one time and others being discarded. The *principle of natural selection* will thus recur throughout all our work, so that I shall discuss now a very simple biological illustration, although I shall present natural selection in chapter 8 as an application of a much wider, normative principle. Let us suggest that a trait is preserved during evolution if it serves a useful purpose for the individual members of the species. Whatever you may think about Darwinian theory, it is a fact that in applying a principle such as the one stated, the rigorous methods of science must be used. If, for example, we ask the question: "Why do horses have long tails?" we can answer: "Because it allows the animal to get rid of flies." But for this to be acceptable, it has to be proved *experimentally*, for instance, that flies adversely affect the development of the animal and that horses and flies coexisted for a sufficiently long period so that natural selection could operate. Of course, other questions remain, especially how it is that different individuals of the species, at any one time, may appear with diverse traits; but this is a question of practical genetics with which we need not be involved. Having discovered that the tail of the horse has a purpose, however, it is not unnatural that people tried to find the purpose of the horse itself, for instance as providing transport to the "favored" human species. This is a purely *metaphysical* statement, and it cannot be grounded by anything except revelation and faith.

To summarize a little, normative principles are required in order to organize experience, and they are meta-physical because although they have become a part of our mental structure through a learning process based on experience, they cannot be *derived* from experience. Metaphysics, instead, adds to and is beyond experience. The problem is that meta-physical and metaphysical propositions can very easily be mixed up. Even distinguished scientists and mathematicians jump with little compunction from one to the other. I do not believe that metaphysical statements can be abolished (nor for that matter that they need be abolished), but I take the view that it is the moral duty of a scientist to recognize when he or she puts forward a metaphysical argument and to own that this is so. Metaphysics, however, has a role in science, and some uses of it, very often in a way that is very nearly meta-physical, cannot be avoided: naked metaphysics is sometimes used in science in the context of discovery, to keep work going at a period when the experimental situation is very confusing. It is extremely dangerous, however, to bring in metaphysical arguments through the back door in order to sustain theories that cannot be properly validated by the methods of physical science.

TWO VIEWS OF NATURE

It will be useful at this stage to comment briefly about two entirely antithetic views of nature. The earliest one (at least within our Western tradition), probably goes back to Gen. 2:7: "And the Lord God formed man of the dust of the ground, and breathed into his nostrils the breath of life; and man became a living soul."[19] And in John 3:8, the Spirit is the wind that we cannot tell "whence it cometh and whither it goeth." This belief that there are two entirely distinct constituents of nature, the material substance and the spiritual substance, is called the *dualistic* view of nature. Among the Greek philosophers, Plato (428–348 B.C.E) was the most influential in postulating not only a nonmaterial and eternal soul but also an entirely independent, nonphysical, world of ideas. In modern times the most notable champion of dualism was René Descartes (1596–1650). At the other end of the scale, there is the *materialistic-empiricist* view that everything is physical and that everything known is known only by observation through the five senses, a view vigorously asserted by Thomas Hobbes (1588–1679) and much developed during the French Enlightenment: it is enough to quote the title of J. O. de La Mettrie's book, published in 1748, *L'homme machine* (*Man A Machine*). The heirs of this philosophical tradition in our own time supported in one form or other a *physicalist* doctrine, in which all statements about matters of fact or existence were supposed to be capable of formulation in terms of physical objects or actions.

Of course, I shall try and find myself a place between these two extremes. To start with, Darwin and modern cosmology have forced us to read Genesis as a metaphor, and most people accept that the creator (be it the God of Genesis or the *big bang*) allowed his work to be driven by natural laws and the process of natural selection. Having built walls, the creator did not write graffiti on them to advertise himself: thus, room was left for free will in accepting or rejecting faith.[20] Although I am sure that a substantial body of people do largely share this view, scientists and mathematicians often appear to look for such graffiti; and a favorite case for this purpose is mathematics. Mathematics is so beautiful, and mathematicians are so very clever that they often claim that mathematics has a life of its own, so to speak, outside the normal domain of nature. This is what must be meant when they express a belief in mathematical truth forming a reservoir of verities that are outside human sensorial experience, but to which gifted mathematicians have direct intuitive access. (Mathematical ideas are not invented but rather *discovered* by tapping this reservoir, they claim).[21] From this, to recognizing this Platonic world as a graffito scratched on the walls of creation by its maker, is one easy and obvious step.[22]

Far be it from me to decry any belief that any human being has ever held. This is not the purpose of this book. What I want to do is to see whether there is any need for us to postulate such graffiti or, more properly, whether there is any evidence for them in studying the natural world; those who want to see them round the corner, however, are welcome, but there will not be much dialogue with them in these pages. This does not mean that such dialogue is meaningless and should not be conducted, but I hold that it should take place in its proper way, within well-grounded theological discourse. What I shall try to avoid is bringing in such discourse by the back door when discussing the natural world, but I shall likewise shun an extreme materialist or physicalist position: there is a lot more to a book than being a stack of sheets of paper with ink marks on them, a simple fact about the world which I shall carefully keep in mind.

A good deal of the rest of this work will be dedicated to showing that everything we do in describing the physical world is contingent, that is, that in so doing—except for some linguistic conventions—we never assert necessary truths (which the creator's graffiti or a Platonic world of ideas would entail); but propositions entirely built *through* (but not necessarily *from*) experience. All this will be much clearer when we progress along the book, but readers must make a mental note that an acid test of a dualistic view of nature is the concept of *necessity*. If you believe that part of our nature is a reflection of a world of immutable ideas, then some statements, be they theological or mathematical or even physical, must be *necessary* truths valid for all time. This is why I shall try to discuss in some detail in the next two or three chapters a few very simple examples where the concepts of necessity and contingency can be illustrated with a modicum of precision.

We shall start our stage-setting of the scientific language in the next chapter and you will see how even the most innocent sentences such as "Snow is white," "Swans are white," "Gold is yellow," entail a great many problems which we must address. What will be very useful is that we shall have a good opportunity to investigate the meaning of necessity and contingency, concepts that, as I have argued, lie at the root of our understanding of a dualist view of nature.

NOTES

1. Tchehov (1962), pp 327–28.
2. My parable about young Kovrin and his imagination must not be taken to mean that we shall start our exploration of nature from our mental states, à la Descartes. Not at all: I shall be concerned only with such properties of the mind as are significant in order to understand how these mental processes are wedded to nature.
3. This must not be taken to entail that the black monk and my writing table have the same form of 'existence': one of the things we have to learn is how to discriminate in this respect.
4. Pigeonholing has nevertheless its place. Philosophy contains a huge corpus of ideas that have already been subject to exhaustive scrutiny. Sometimes, by identifying the pigeon-hole to which an idea pertains, one can quickly discover its weaknesses and its strengths.
5. Pirandello (1958), p. 55: "un fatto è come un sacco, vuoto non si regge."
6. See Achinstein (1970) for the impossibility of a complete separation between facts and theories. This problem will be further discussed in chapter 20.
7. Wittgenstein (1967b), p. 257.
8. I am here paraphrasing Wittgenstein (1967b), p. 125.
9. Such a principle was in fact used by John Stuart Mill (1806–1873), who claimed (wrongly) that it was confirmed by experience.
10. Fontenelle (1955). Strictly speaking, Fontenelle's new world covered the whole of the moon, not just the far side of it.
11. See for instance Nyhof (1988), but Mach will be a frequent visitor to these pages.
12. "Wovon man nicht sprechen kann, darüber muß man schweigen" Wittgenstein (1961), 7, p. 150. Although Wittgenstein had close contacts with the most influential group of logical positivists, the so-called Vienna Circle, he was not one himself, and his book contains strong echoes of Arthur Schopenhauer (1788–1860), who did openly embrace some form of transcendental or metaphysical idealism. Ironically, it appears that these echoes were not recognized by the logical positivists (see Hacker 1972, p. 188), who made the sentence that I have quoted a manifesto for a whole generation of philosophers. This radical program had to be abandoned in midcentury, Wittgenstein himself leading the withdrawal.
13. It is, in fact, known not to be universally applicable.
14. Remember that, in any case, the discussion in this chapter is purely preliminary, and that these ideas will not be fully tackled until chapters 4 and 8. The reader will find there specific examples that are essential to grasp these concepts, but which would be premature to treat at this stage.
15. Metaphysics is often taken in a much more general sense than the one described here. Also, the border line between it and meta-physics can be a bit blurred.
16. Kant himself took the view that metaphysics should deal only with "God, freedom, and immortality." See Kant (1781), p. 325. (Added in the second edition.) It is in order to establish a clear difference between propositions involving concepts such as these, and

the great normative principles mentioned, that I have introduced the expression *meta-physics*. The word *metaphysics*, on the other hand, if not cast into utter darkness as a nonsense by some philosophers like the early A. J. Ayer (1910–1989), has been used in a wide variety of ways, from a guide to constructing 'reality' (Hamlyn 1984, p. 8), to a guide to morals (Murdoch 1992).

17. See chapters 4 and 8.

18. For the sake of this example I disregard the problems that arise when, for instance, the meaning of the concept of force is critically analyzed.

19. This had a profound effect in Christian theology after the second ecumenical council at Constantinople in 381 complemented the Nicenean Creed by introducing the Holy Spirit as the giver of life. Although this is the life of faith and the resurrection of the death, it also carries the sense of Gen. 2:7.

20. In theology, the concept of the hidden God, *Deus absconditus*, is well known if not, of course, universally accepted. On the contrary, the medieval God was manifest: it was widely believed that movement, of the planets for instance, was a demonstration of God's activity. It was only when the principle of inertia showed in the seventeenth century that movement as such required no cause (because an object in motion would persist in it without any agency that sustained the motion), that the idea of a continuous activity of God on nature was dropped. (Hans Blumenberg, quoted by Pannenberg 1993, p. 19.)

21. A famous example is Hardy (1992), pp. 123–24: "I believe that mathematical reality lies outside us, that our function is to discover or to *observe* it, and that the theorems which we prove, and which we describe grandiloquently as our 'creations', are simply notes of our observations. This view has been held, in one form or another, by many philosophers of high reputation from Plato onwards." Half a century later (Hardy's book was first published in 1940) this dogma is still endorsed by Penrose (1989), p. 428: "I imagine that whenever the mind perceives a mathematical idea, it makes contact with Plato's world of mathematical concepts." (See also Penrose 1994, p. 412.) I quote these statements in the spirit of Wittgenstein's acute remark: "what a mathematician is inclined to say about the objectivity and reality of mathematical facts, is not a philosophy of mathematics, but something for philosophical *treatment*" (Wittgenstein 1967b, 254).

22. This, however, could hardly be said of Hardy, a well-known atheist for whom C. P. Snow wrote, "God was his personal enemy" (Hardy 1992, p. 20).

2

WORDS AND NECESSITY

And now to work. Work, the what's-its-name of the thingummy and the thing-um-a-bob of the what-d'you call it.

P. G. Wodehouse (1979), p. 152.

Diz que la luna es luna
diz que la mar es mar.
Diz que lo que dicen
se lo pueden bien guardar.

They say that the moon is moon
they say that the sea is sea.
They say that what they say
is worth nothing to me.

Abel Martin (1940), p. 19.

One of the major tasks of science and mathematics, as I have already said, is that of setting up a language with which we can discourse about nature. In trying to do this, alas, I shall have a problem: I cannot use philosophical or scientific language in full strength, and even less can I indulge in the language of the marketplace. So, I shall have to compromise with a language just adequate for our purposes.

Everything that is the case, meaning, anything that can be construed as an element of a fact, is *contingent*, in the sense that it could just as well be or not be the case. If I say, *There is a table in my room*, there may or may not be a table in my room. And I shall claim that such *necessities* as we shall meet stand themselves on the shoulders of contingencies. Humankind, however, seems to long for a certain amount of necessity, and the search for necessarily true propositions is one that people do not give up readily. The safest bet in this respect seems to be the tautology.

A ROSE IS A ROSE: TAUTOLOGY

A rose is a rose is an archetype of a tautology, a proposition that is said to be necessarily true by virtue of its form. Naturally, this type of statement does not advance knowledge very much, if at all, but tautology is to logic as the number zero is to mathematics. Without the zero we would never know that two numbers a and b are identical (they must differ by zero). Likewise, as we shall see later, tautologies can reveal when two propositions are logically the same.

The first point I want to make is that not every sentence that has the clear form of a tautology is *necessarily* true. Take *Scott is Scott*. We appear to say here something like *The author of Waverley is the author of Waverley*, which is necessarily true. On the other hand, the word *Scott* is a variant spelling of Scot[1] (a native of Scotland), in which case, if the second entry in my example is used in this sense, the statement is clearly contingent (since Sir Walter Scott could have been born outside Scotland). There are of course many other possible examples of this situation, like *Blue is blue* (the second entry in the sense of *dismal*). Thus, the existence of homonyms nullifies the idea that we can decide that a sentence is tautological (and thus *necessarily* true) by merely inspecting its *form*: we must also investigate the *meaning* (also called in philosophy the *intension*) of the words used. We thus find the paradoxical result that the property of a sentence to be a tautology is itself *contingent*! So, even if you think that tautologies are useful (which they can be), and that they entail necessity (which they do), their existence does not do much for the view that there are propositions, entailed by tautological sentences, that are *genuinely* necessary, because their necessity depends on the contingent nature of the corresponding tautologies.

However, true tautologies can be produced, and they do have the property of being unconditionally true by virtue of their logical form. I have to agree first that the letter p will stand for one and only one name (that is I shall not accept homonyms), and I can then write p is p. That this is necessarily true we can accept as cast-iron, but the *referential* content of this proposition to anything that might be the case is null. If I say *The sea is the sea* or *The moon is the moon*, I am saying in both cases exactly one and the same thing about the world, namely nothing.[2] The reader should not be surprised about this: $5 - 5$, and $1 - 1$, denote exactly the same number, *zero*.[3]

ANALYTIC PROPOSITIONS

Having failed to establish tautologies to be so by considerations that are entirely free from contingencies, that is, independent of what the world is like, we pass on to another candidate for necessity, the *analytic proposi-tions*. If I say *A bachelor is an unmarried man*, I might consider this sentence as a definition, or as the specification of a synonymy, and therefore to be *independent of fact*. All that that sentence does, we could argue, is to license us to replace every use of the word *bachelor* by the words *unmarried man*. I do not have to know what these words *mean* in order to understand this rule, and therefore the sentence may be *necessarily true* but entirely devoid of any factual content.

This approach is good if this rule were given to us by, let us say, a lex-icographer we are prepared to trust. But imagine that I am the lexicogra-pher and that I am trying to invent the rule. If this is so, I am immediately involved in a lot of contingencies. Suppose I were a member of a society in which the concept of marriage does not exist: I would not know what a *married* man is, and even less would I understand the meaning of *unmarried man*. Suppose now that we understand marriage: are we going to call widowers and divorced men *unmarried*? Obviously they would not do as *bachelors*. So, the next thing I do is add a note to the effect that, for the purpose of this definition, *unmarried man* means *a man who has not pre-viously been married*. My worries do not end here: my friend Pepino has married in Italy under canon (church) law, and has had his marriage annulled under the same law. Since annulment by the Vatican is not divorce but rather a wiping off of the previous marriage, is Pepino *unmar-ried*? (I confess that I do not know the answer to this question.) To com-plicate matters further, in the United Kingdom, Mary and George living together, never having gone through a form of marriage, can be said to be *married in common law*. It would be difficult to call George an unmarried man and therefore a bachelor. Yet, the famous nineteenth-century Cam-bridge mathematician George Green never married Mary Smith, mother of his five children, because under the statutes of his college, he would have lost his fellowship unless he remained a *bachelor*.

Let us see what we have so far. We have a sentence of the form *p is q*, where the *meaning* of *q* is wholly contained in the meaning of *p*. If you believe, for instance, that it is not possible for snow to be other than white (about which more later), then the sentence *Snow is white* is true inde-pendently of fact, and it is an *analytic proposition*.[4] However, and most importantly, once I want to use this type of proposition in any non-purely abstract way, its *meaning* has to be established contingently. Such necessi-ties as analytic propositions might entail are no more than lexicographic conventions (see note 8). Having reached this sober conclusion, the reader may rightly feel that we have been wasting our time. We have gained two things, however: we have understood that formulating defi-nitions entails setting up a language, and we have opened our eyes to the risk of having illusions that we can formulate *necessary* statements that are entirely independent of all experience.

SYNTHETIC PROPOSITIONS

Synthetic propositions have the same form, *p* is *q*, as analytic propositions, but the property denoted by *q* is not one that is deemed to be inherent to *p*. Thus, the relation entailed by the proposition is not *necessary*, as when I say: *The cover of this book is blue.* This is clearly contingent, since it could have been red. The traditional distinction between analytic and synthetic propositions, however, is one that scientists cannot always countenance, although it was one of the cornerstones of the philosophical system created by the great German philosopher Immanuel Kant (1724–1804). Kant[5] gave *Gold is a yellow metal* as an example of an analytic proposition because, he said, the concept of gold necessarily entails that it is yellow. But the concept of what is what changes with time, even within the same culture: any solid-state scientist would now say that the concept of gold is that of being a cubic close-packed crystal with atoms of atomic weight 79 (disregarding isotopes for simplicity) at the lattice sites.[6] Given this formulation, she would input the data in her computer and get in a few minutes, from the basic principles of quantum mechanics, the so-called band structure of this material and find, in fact, that there is an absorption band that gives the material its yellow characteristic color: thus this property is entirely contingent (because it depends on properties of the crystal and the electronic structure of the material) and it is not inherent to the concept of *goldness*. Contrary to Kant, therefore, *Gold is a yellow metal* must now be considered as synthetic.

Most people, on the other hand, would reckon *Snow is white* as analytic, on the cosy grounds that nothing that is not white would be recognized as snow, but this is a purely psychological trait of our minds because snow happens to be a substance that entails emotional reactions, dramatic as Thomas Mann imposed on Hans Castorp, or sentimental as in *White Christmas*. The true fact, however, is that, as for gold, the (hexagonal) crystal structure of the water crystal is responsible for its color. If by snow we mean whatever falls from the sky when it is very cold, it could just as well have been brown, and therefore the sentence in question must be considered synthetic.

I shall discuss one more example to show that the distinction between analytic and synthetic propositions in scientific use is arbitrary (and thus contingent!). If by *swan* I mean the common or mute swan, *Cygnus mansuetus*, the statement *Swans are white* would be analytic if ornithologists took color as an essential *differentia* to determine the species within the genus: in this case the Australian black swans, *Chenopsis atratus*, would not be swans at all. Suppose that it were true (which it is not) that ornithologists use color in the way stated, that animal activists discovered that black swans were pigmentally discriminated, and that they forced ornithologists to treat such birds as humans are, disregarding their color: the proposition *Swans are white* with reference to the mute swan would become synthetic and not analytic, since the color of a swan would not necessarily be a characteristic of its species.[7]

NECESSITY

I must now come clean: I have so far talked a lot about *necessity* without ever defining what I mean by it. I shall now try to bridge this gap and, in so doing, we shall see that a few things already accepted as clearly established will have to be modified a little. In order to do this I shall introduce a useful fiction. We can imagine that this world in which we live is not the only possible one, and that many other possible universes might also exist now or at other times. These universes will be perhaps very different from ours. In some, life will not exist at all; in others, the laws of nature might be so different that even the elementary particles that make up our world (electrons, protons, and the like) do not exist, let alone more complex things like gold atoms. Once this scenario is accepted, I shall say that a proposition is *necessary* if (and only if) it is *true in all possible worlds*, including our own, of course. In order to give an example, let me go back to gold. We had agreed that *Gold is yellow*, contrary to Kant, is not analytic. We have replaced this proposition by *Gold is a cubic close-packed crystal with atoms of atomic weight 79 at the lattice sites*. I shall assert that this proposition is true in our world, and it is also true in *all possible worlds*. This is surprising because in some worlds crystals or even solids might not exist at all, and, in others, atoms as we know them do not exist. Suppose, however, that I am a transworld traveler, sent by my masters to find and report worlds that do contain gold. I shall send back reports such as *Gold does not exist in World Number 3459* or, more happily: *I have found gold in World 100345*. And, unless I have gone deranged, in both cases I mean exactly the same thing with the word *gold* namely *a cubic closed-packed crystal with atoms of atomic number 79*. Thus if it is true *in our world* (and therefore *contingently true*) that *Gold is a cubic closed-packed crystal with atoms of atomic number 79*, then the noun "gold" is a *rigid designator* that designates such crystals *in all possible worlds* and the proposition in question is necessary and thus *analytic*. But remember that this concept of analyticity and necessity hangs on the far more robust requirement that the proposition in question be true *in our world* and thus only *contingenly* true.

I am sorry if many readers might feel by now that I am squaring circles or gilding lilies, but what we are trying to establish is very important, namely that even when the concept of necessity is carefully grounded, thus allowing us to speak of *analyticity* with some degree of verisimilitude, contingency is always lurking in the background through the requirement that truth, in *our* world, must be established. (The defining sentence for gold when uttered in his own language by a native of World Number 3459 is untrue or meaningless![8])

THE A PRIORI

An *a priori truth* is one that can be *known* independently of all experience.[9] Notice the emphasis on *known*: the concept of a priori is fundamentally epistemological (that is, related to our knowledge of facts rather than to the facts themselves). Many philosophers nowadays would confine the use of *a priori* to whatever is *true by convention* or *true by definition*, but even then caution is required.[10]

The history of the *a priori* as an epistemological concept, that is as a concept useful in organizing our knowledge, deserves a little thought. Remember that Kant thought that he could make a clear distinction between analytic and synthetic propositions, the truth of the latter being an empirical matter. To refresh our minds: he said that *Some bodies are heavy* is *synthetic* because, he alleged, the concept of *heaviness* is not necessarily contained in that of *body*, in which contrary case he would have called the proposition *analytic*. I have already argued that whatever is or is not an essential constituent of a concept is not a matter that can safely be left to philosophical speculation alone, but let us play Kant's game for the time being.

What is quite remarkable is that Kant had the courage, for better or for worse, to argue that there are *a priori synthetic propositions*, and that they do indeed play a fundamental part in organizing our knowledge. At first sight, this appears to be a contradictory nonsense: the a priori is by definition nonempirical and *necessary*, whereas the synthetic is based on experience and is *contingent*. If the reader is unhappy, so am I, but life is all about resolving conflict, so let us be a bit patient and try to understand a little what Kant had in mind, because although it might not be exactly what he thought it was, it nevertheless contains the germ of something which is very important indeed.

As an example of an *a priori synthetic* proposition, Kant gave the following one: *Everything which happens has its cause.* The concept of *cause* does not necessarily belong to *everything which happens* and the proposition is therefore considered by Kant to be synthetic. On the other hand, it is not the result of experience, and it is therefore a priori. (Here I am jumping over a big hurdle because, although practically all philosophers after Hume agree that the concept of cause is not obtained from observation alone—a point that for the time being we must accept—this is not by any means a trivial matter, and it will require far more serious analysis in chapter 4.)[11]

So, what we must remember for the moment without getting too agitated about it (because the apparent paradoxes here will be resolved in chapter 4, as far as I can claim to resolve anything!) is that, for Kant, the statement *Everything which happens has its cause* is, amazingly, both *a priori* and *synthetic*. What he claimed was, roughly, that the concept of *causality* was an essential and preordained feature of our mind (what he called a *category*) and thus a priori, although it was applied to *contingent* relations between cause and effect. One of the troubles, of course, is that how such a convenient category was prefabricated in our mind is not a problem that Kant would consider: "But it is there," his answer would have been. Even if we were not to disagree with the guts of this analysis, not everything that Kant has said along this line can be accepted outright.[12]

What is most important is Kant's claim that natural science uses *principles* that are synthetic a priori. On the other hand, even he can be demonstrably wrong, thus providing absolute proof of the serious dangers of unguided a priori thinking. Another example that he gives of the synthetic a priori (Kant 1781, *Introduction*, V, p. 54) now known, after Einstein, to be totally false: "in all changes of the material world the quantity of matter remains unchanged." Modern science, of course, must be far more cautious than Kant in using principles that assume a priori contingent relations, but the important idea is that we do need principles that, although incapable of being established experimentally (that is contingently), are necessary in organizing experience, and they are the normative principles which I called in chapter 1 *meta-physical*. These principles are very much related to Kant's a priori synthetic propositions, but I prefer not to use this terminology because the very expression *a priori* is one that carries with it a history of unfortunate preconceptions. We shall discuss these important meta-physical principles in chapter 4, where all this will become, I hope, far more clear. Meanwhile, we shall make one last attempt at finding necessity.

POSITIONAL PREDICATES

When we predicate of an object that it is *white*, we cannot possibly claim any necessity for this property. More than that, what appears to us as white might appear to an Inuit to be of a different color.[13] If, on the other hand, I see an object above the table, although there is nothing necessary here, it appears to be reasonable that the quality of *aboveness* must be a sort of invariant of language, that is, whatever the word used in any different language, it must *necessarily* express the same relation. (We tend to believe that the concept of *aboveness* must survive translation, although such survival is by no means universally valid for other concepts.) Predicates with this alleged property were called *positional* and enjoyed a rightly brief vogue twenty or thirty years ago. It is not worthwhile trying to produce a watertight definition, and it will be sufficient for our purposes here to define them as requiring a relation between two objects, such relation being thought to be *transcultural*, that is, *language-invariant*. On the other hand, a predicate that can be predicated of a single object, as color is, was called *qualitative*.

Within this mode of thinking one could not find a pair of more clearly positional predicates than *above* and *below*: the inhabitants of Ka-'u in Hawaii, however, appeared to use them as qualitative properties. Objects were *qualified* as belonging to *above* and *below*, and once these qualities were assigned, they were never exchanged, whatever the positions of the respective objects, with disagreeable consequences for economically minded strangers attempting to rearrange their bedding: once a top sheet, always a top sheet![14] So, it is far from clear that the *relations* entailed by positional predicates are *necessarily* invariant. If our transworld traveler, for instance, found herself in a flat world of only two dimensions, she would not be able to use the words *above* and *below* at all. (Whereas it might be a contingent fact that gold does not exist in that world, those words *must* be synonymous in it!)

CODA

The import of our discussion so far is that necessities, if they are logical, are entirely devoid of empirical content, and if they are not, they either are a matter of linguistic convention (which must be very carefully formulated) or they must be regarded with great suspicion: the scientist will do well in avoiding them as far as possible, since very often the alleged necessities of a generation are shown to be false in the next one. The only true analytic propositions that we have found are those that entail synonymy, and, as such, they are nothing more than mere rules for the use of a contingent language. Other so-called analytic propositions, such as *Snow is white* and *Gold is yellow*, we have recognized as undoubtedly synthetic. The reader will have found here, I hope, more than one example of what *setting up a language*, as was mentioned in chapter 1, means. On the other hand, we began to notice the possibility of synthetic propositions that are nevertheless nonempirical. These will be recognized later as important meta-physical principles.

NOTES

1. Sir Walter Scott himself seems to have favored that spelling, at least in his letters, of which there are three examples in the Oxford English Dictionary.

2. A true tautology cannot *determine reality in any way*. Wittgenstein (1961), 4.463. It might be adduced, on the other hand, that some tautologies, like *Atlantis is Atlantis*, may fail to refer and that they cannot be considered true. I shall not use this approach because I shall accept (with Wittgenstein) that the meaning of a word is its use, and that as long as the use of a word is understood, that is enough. Referents, however, are important in trying to understand *natural kinds*, and are discussed in note 8.

3. It is useful to remember that names (like *Scot*) must be distinguished from their *meanings* or *intensions* (*natives of Scotland*, for instance), and their *extensions* (the *class* of all natives of Scotland). The two different names *animal with a heart* and *animal with a kidney* have different intensions but the same extension (that is, they apply to exactly the same aggregate of referents, because all animals with a heart also have a kidney and vice versa).

4. Notice that this definition is less restrictive than that entailed in *A bachelor is an unmarried man*. This analytic proposition establishes a *synonymy* and thus the *extensions*, as well as the *intensions* (meanings) of p and q must be identical: it would not do for some bachelors not to be members of the class of unmarried men or for some unmarried men not to be bachelors. On the other hand, *Snow is white* can be construed to be analytic, in the sense that it can be taken to be a necessary constituent of the meaning of *snow* to be white: anything not white cannot be snow. (This point of view will be criticized later on in the section on synthetic propositions.) This is the reason behind the definition in the text.

5. Kant (1783), p. 17.

6. The crystal structure must be stated. The element carbon, atomic weight 12, has six *allotropes*, all with the same atomic number of course, but different crystal structures and thus different properties. (The three most commonly known are: diamond, graphite, and fullerene.)

7. Despite various precedents for the distinction between analytic and synthetic statements, there is no doubt that it was Kant (1781), pp. 48–50, and Kant (1783), pp. 16–17, that persuaded philosophers for a couple of centuries to take this distinction seriously. It was only after the last war that philosophers started to express serious doubts about this distinction, and the crucial piece of work that cast the concept into utter darkness was Quine (1951). Hempel (1965) pp. 132–33, was also very influential in this respect. Perhaps the most forceful position as regards necessities was that of Goodman (1965), pp. 19, 60, who entirely discarded the distinction between the analytic and the synthetic. Goodman, in fact (in a letter to the author), took the view that *all* propositions are synthetic. For a conditional revival of the concept of analyticity see note 8 below.

8. Much of this section is based on ideas vigorously presented in the seventies by the American philosophers Saul Kripke (1940–), Keith Donnellan, Hilary Putnam (1926–), and various followers. Notice that the many-worlds device used by them is metaphysical, and it was through their efforts, and especially also those of the Oxford philosopher Sir Peter Strawson (1919–), that this word became once more respectable in philosophy. The fundamental papers are Kripke (1971) and Kripke (1972), and a very good review is the editor's

introduction in Schwartz (1977). The important idea of a rigid designator is due to Kripke (1971), (1972), and (1980). Sellars (1948), p. 292, was the first to introduce the many-worlds idea in this context, going back to proposals originally made by Leibniz.

The last sentence in this paragraph, about the goldless native of World 3459, requires a little discussion. When uttering the definition of *gold*, he or she is in the same situation as when any of us says, *A unicorn is a horselike animal with one horn*. For many purposes, such an utterance would be considered as true, for instance when considered as a purely lexicographic statement (in which case a definition entailing two horns would be *untrue*). For my purposes here, however, I use the word *true* for the definition of a name if its extension (that is the aggregate of objects to which the name applies) is, is believed to be, or was, not null in our world. This is sufficient for scientific purposes, in which case one avoids defining inexistent things, an activity which, alas, is not always exorcised out of science, as such words as *phlogiston* and *ether* testify. The truth is that the concept of *truth* is too delicate for me to get into it in any detail here. A good discussion may be found in Horwich (1990).

That my discussion is somewhat oversimplified can be seen from the following. If I admit that the definition of a unicorn as a one-horned horse is lexicographically true in our world, then the word unicorn is a *rigid designator* in the sense that our world traveler, when finding a one-horned donkey in World 2001, would not report back home that he had found a unicorn. So, in this sense, the definition of a unicorn is both analytic and it does not require its contingent verification in our world. I have disregarded such cases as likely to be of no interest in our study of nature.

Some philosophers, following Kripke (1980), would object to my contention that, however much the proposed definition of gold is a rigid designator, it still depends on its *contingent* validity in our world. Kripke (1980) claims that the fact that we have discovered the physical constitution of gold, say, does not make it contingent. But the *existence* of such a substance in our world *is* contingent. Fullerene, an allotropic form of carbon (and thus a solid with the same status as gold), did not exist until 1985, when it was chemically made by Professor Kroto and his collaborators. To say that its existence is not contingent is to claim that such a discovery was *predetermined*, at least since the big bang. Of course, the same could be claimed for Michelangelo's David.

9. I am quoting Kripke (1971), p. 84 almost verbatim.

10. See for instance Horwich (1987), p. 132.

11. For the discussion about *bodies are heavy* see Kant (1783), p. 16. His concept of *cause* is discussed in Kant (1781), *Introduction*, IV, p. 50.

12. Kant's view of mathematics, especially geometry, is not acceptable in modern science. He maintains that all "*mathematical propositions . . . are always judgments a priori, not empirical; because they carry with them necessity, which cannot be derived from experience.*" (Kant 1781, *Introduction*, V, p. 52; see also part I, sec. I, §3, p. 70 as regards geometry.) This question is treated in more detail in chapter 14.

13. I am not subscribing here to the view that Eskimos can discern a large number of different snow hues, which has now been debunked (Pullum 1991).

14. The story about Ka-'u is quoted in Lévi-Strauss (1962), p. 190. A discussion of positional predicates and their alleged transcultural properties can be found in Swinburne (1973), pp. 80, 106. See also Goodman (1965), p. 78, and Goodman (1960).

HOW TO DEAL WITH PROPERTIES BY LOGIC

Life's nothing but words, words, words; and how are we to know when words are true?

P. G. Wodehouse (1958), p. 174.

> **Health warning.** Some readers may have a natural reluctance, if not aversion, to dealing with anything that looks like formulas or tables, and if this is the case they might find, at first sight, some parts of this chapter abhorrent. I hope very much, however, that this does not put them off reading the rest of the book, which will be much less formal. This chapter can be read at first with almost cynical disregard for its more formal aspects: as long as readers get some form of a handle on the concepts introduced, even if their manipulation is skipped, they can proceed to the next chapters, and then return to this one when they have acquired more confidence about the significance of the work introduced here.

Most readers who have heard anything at all about logic must have been exposed to the ancient argument:

All men are mortal; Socrates is a man; therefore Socrates is mortal.

For many people, this type of argument, which is called a *syllogism*, is a model of rational discourse, and indeed it was taken to be so from Aristotle through the medieval schools and well into our century. And yet, although the shell is there, so to speak, there is hardly any really useful content in this syllogism. Let me try another one of exactly the same form, but which is a lot more satisfactory (assuming that you get satisfaction by getting things right even when they are pretty useless):

All objects in this box are white; this object is in this box; therefore this object is white.

Let us compare these two syllogisms. In the second one, assuming that we understand the words *object* and *white*, the first sentence (which is called the *major premise)* is clear: the box will contain, say, six objects and *all* means that the six objects are white. If we now look back at the major premise of the Socrates syllogism, we immediately realize that we are in a mess. There are two serious problems. First, what is the meaning of *all* here? Do we have to inspect all six billion or so human beings to use it? Do we have to wait another million years, say, and examine those future generations, before we can license ourselves to use *all*? And secondly, when we examine humans, how do we convince ourselves that they are mortal? Do we wait 120 years and see whether they qualify for the coffin, or do we have to shoot bullets into their hearts? If you now compare this dismal situation with the major premise of the box syllogism, you can see that, in this second case, nothing is seriously in doubt.

We can have a look at the second leg of our syllogisms (called the *minor premise*). In the box case, *object* and *white* can easily be accepted to be known, without interfering with the flow of the syllogism. But in the Socrates case, because the whole thing hinges on an alleged property of something called *man*, we must be sure that we understand what we mean by this word. The major problem is that, if we run into a person who looks like a man, but happens to be immortal (fictitious as this is, if we don't admit this possibility there is no point whatsoever in enunciating the syllogism), do we then say that he is not a man? Of course, we could adopt this convention, but then the whole syllogism falls to the ground because its enunciation appears to follow some rational mode of discourse, whereas all that we would really be saying is that, by a language convention, nobody who is immortal can be called a man, very much like in chapter 2, where we had agreed that no one previously married could be called a bachelor.

The result of all this is somewhat sad. The box syllogism is good both in form and in content. But its content is useless. Anyone looking at our box with six white objects and articulating the syllogism either is a pedant or profoundly disadvantaged: you do not need the syllogism at all to know its conclusion. As for the Socrates syllogism, it means absolutely nothing unless we accept blindly the use of words that really require a lot of discussion, which shows straightaway that rational discourse is not and cannot be a mere exercise in logic.

The reader will think, at this stage, that I am cutting my own throat: obviously, because of the nature of this chapter, I am trying to persuade people that a little bit of logic, painful as it must be to learn it, is a good thing. On the other hand, it appears to be the case that logic is all form with no content. Then, why do we want it? Think, however, about arithmetic: a knowledge of sums will not (necessarily) improve your bank balance, but it will be mightily helpful in checking your bank statement. Logic, likewise, will help you organize rational discourse and discover what you can and cannot do with it. And it will help us understand some of the formalities of formulating laws of nature.

TOWARD LAWS OF NATURE: PROPERTIES

We all know what we mean when we say *My shoes are black* or *My sugar cube has dissolved in water*, but anyone who has done any science must have met statements that appear to claim some general validity for properties which are posited for a whole set of objects. Consider for example the sentences *Ravens are black* or *Sugar is soluble*. When such statements are used, the set of objects to which the predicate (*black* or *soluble*) refers is for all practical purposes infinite (in the sense that enumeration of its members can never terminate), comprising *all* ravens, *all* sugar samples, and so forth. At this level of generality we are getting near to what one understands as a *law of nature*, but, first, laws of nature are delicate things with which we shall have to deal in some detail later; and, second, infinite sets entail serious problems of their own which have to be properly tackled. So, for the time being, I shall leave conveniently vague the *extensions* (that is, the sets) to which I refer with such words as *ravens*, *sugar*, and so on.

What I really want to do to start with is to negotiate the logical structure of a proposition like *Ravens are black*. The first problem is this: everyone has experienced the fact that if one wants to solve some simple mathematical questions, such as *What is the price of one apple if three apples cost 90 cents*, a little intuition will yield the result with no need for the algebraic equation $3x = 90$. The moment we want to analyze more involved questions, however, some formal notation and formal rules of operation become necessary. What is true for algebra is also true for logic: unless we work out some simple formal rules, no amount of arguments in English will allow us to get to the bottom of the logical structure of propositions. So, for the first part of this chapter, we shall deal with a few of the aforementioned dreaded formalities. I do not need to stress the importance of what we are trying to do: you will soon see that positing a property, such as *black* or *soluble*, is not by any means as simple as it looks, and once we acquire a salubrious respect for properties, we shall realize that talking about the far more involved problem of laws of nature must be very delicate, indeed. But first of all, we have to learn a little about the basis of logic; and I shall start by considering a practical example.

RAVENS ARE BLACK: MATERIAL IMPLICATION

Some philosophers appear to have the amazing gift of knowing by direct intuition what such things as ravens or gold are. Scientists (perhaps I should say good scientists) do not have such immediate insight into nature, and they have to learn by hard work how to use such words. I do not want to get involved in this problem at this stage, so I shall assume that if I want to talk about birds, I shall have a well-trained ornithologist next to me. I see this person picking up specimens of birds and labeling them with such words as *raven, finch, magpie,* and the like. After a bit of looking over his or her shoulder, I form in my mind the hypothesis: *Ravens are black.* If I want to state rigorously what I mean by this proposition, I am afraid that I have to use the joker x, that has already reared its probably unwelcome head in the previous section. Let me use x as a shorthand for *a bird.* Easy enough: whenever you read x, you translate in your mind *a bird.* The statement I want to establish (that is, I want to show whether it is true) is:

If x *is a raven,* **then** x *is black.*

This is still too long. Notice that, besides the two words in **bold**, the above sentence contains two propositions, for which we shall introduce a further shorthand. Call:

x *is a raven* $= p.$ x *is black* $= q.$

So, we now have:

If $p,$ **then** $q.$

The two propositions here are connected by the logical relation given by the words **if . . . then**. This logical relation is called *material implication,* but it is also called the conditional. In this game, propositions like p and q play a similar role to the variables $x, y,$ and so forth in algebra, and logical relations such as **if . . . then** act like the operations of addition (+), multiplication (×), and so on. Just as for the latter, it is useful to shorten the logical relation by means of a symbol. We shall use:

Material implication: **If . . . then** $= \Rightarrow.$

Let us now display all the steps in our progress, which I shall do for clarity by enclosing each step in parentheses.

(**If** x *is a raven,* **then** x *is black*) = (**If** $p,$ **then** q) = ($p \Rightarrow q$).

Of course, we have done nothing so far about testing our hypothesis: we have only restated it in a convenient notation. We now look over our ornithologist's shoulder and note the results of four pieces of labeling that he has done, which are given in the four rows a to d in table 1.

In column 1, we write down the name assigned by the ornithologist to each bird, and in column 2 its color. (Do not worry about the green raven: it may have been a moldy specimen!) In columns 3 to 5 we write

Table 1. Material implication.

The letters T and F mean *true* and *false* respectively.

	Bird	Color	p (bird is raven)	q (color is black)	$p \Rightarrow q$
a	raven	black	T	T	T
b	raven	green	T	F	F
c	blackbird	black	F	T	T
d	finch	not black	F	F	T
	1	2	3	4	5

down the true (T) or false (F) value of the proposition at the head of the corresponding column. Columns 3 and 4 should entail no problems, but the main results in column 5 must be discussed. The entries in a and b are obvious: the first "experiment" verifies the implication that if a bird is a raven, then its color is black; and the second falsifies it. You may be a bit surprised by the remaining two entries but what they mean is that in both cases the implication is *not false*, but because we have only two truth values, *not false* must mean *true*. You might not be very happy about this, and you might prefer a logic with three truth values, T, F, and NF (not false), say. Such logics have been developed,[1] but they are not as attractive as it might appear from the problem in question, and they will not be used in this book (in fact, they are seldom used).

The reader must appreciate that the material implication, although defined in terms of the *conditional* **if . . . then**, works in a somewhat different way: you would not normally use a conditional such as *If Paris is in Germany, then Rome is the capital of Italy*, which in accordance to row c of the table, is true. This is why the word *material* is adjoined to *implication* to stress the technical sense of its use, but it is important to remember that a false proposition implies not only false propositions but also *all* true ones. Our conditional is thus a little unorthodox in comparison with the vernacular, but we shall see in the next section that it is very good for establishing *sufficient conditions*, which are, of course, very important.

TRUTH TABLES AND THE LOGICAL FUNCTIONS

This section might be unduly boring, but take heart: it is short and its main value is that it provides a compact and easy reference for various important practical applications that will appear later. You may simply return to it if and when required.

In order to summarize the results for the truth values of the material implication $p \Rightarrow q$, I collect them in table 2. Such a table is called a *truth table* and it *fully describes* the properties of the logical function given in it.

Table 2. Truth table for the material implication (conditional)

The letters T and F mean *true* and *false* respectively.

p	q	$p \Rightarrow q$
T	T	T
T	F	F
F	T	T
F	F	T

As it might be expected, there are other logical functions, the description and symbols[2] of which are given in table 3 where, for shortness, material implication is entered as *implication*.

Table 3. The logical functions

Operation	Symbol	Example	Description
Negation	~	$\sim p$	Not p
Conjunction	·	$p \cdot q$	p **and** q
Disjunction	∨	$p \vee q$	p **or** q
Implication (*conditional*)	$\Rightarrow$	$p \Rightarrow q$	p implies q = **if** p **then** q
Equivalence (*biconditional*)	$\Leftrightarrow$	$p \Leftrightarrow q$	p implies q = **if** p **then** q = **if and only if** p **then** q

The truth table for the negation is fairly obvious, as shown in table 4.

Table 4. Truth table for the negation

The letters T and F mean *true* and *false* respectively.

p	$\sim p$
T	F
F	T

The truth tables of the remaining functions in our table are collated in table 5, although the reader need not worry too much about them at this stage, but rather return to them when needed.[3]

Table 5. Truth tables for the logical functions

The letters T and F mean *true* and *false* respectively.

	p	q	$p \cdot q$	$p \vee q$	$p \Rightarrow q$	$p \Leftrightarrow q$
a	T	T	T	T	T	T
b	T	F	F	T	F	F
c	F	T	F	T	T	F
d	F	F	F	F	T	T
	1	2	3	4	5	6

Columns 1 and 2 give all the four possible combinations of the truth values of p and q, and the rules given for the logical functions in the remaining columns are fairly straightforward. The conjunction $p \cdot q$ (*and*) is true only when its two prongs are true, whereas the disjunction $p \vee q$ (*or*) is true whenever either prong is true. The material implication $p \Rightarrow q$, as we have seen, is false only when the first prong is true and the second is false. The entries in column 6 for the equivalence $p \Leftrightarrow q$ (*if and only if*) are fairly clear, since the biconditional must be false, not only when the first prong is true and the second false, but, conversely, when the second is true and the first is false. It can thus be understood that the biconditional entails the equivalence of its two prongs in the sense that it is only true when both prongs have the same truth value.

It is possible for a logical proposition to have truth values identically equal to T for all permitted values of p and q, in which case the proposition is a *tautology* (Although none of the entries in table 5 have this property, an example may be seen in table 6 in note 5.)

Notice that each logical function is *fully characterized* by its (distinct) truth table; two functions that have identical truth tables must be regarded as being *equivalent*, that is as being the same function. (Do not confuse the use of the word *equivalent* here with the function called *equivalence* as in $p \Leftrightarrow q$.[4]) For example, we should expect $p \Leftrightarrow q$ to be the same function as $\{(p \Rightarrow q) \cdot (q \Rightarrow p)\}$, and it can easily be verified that they have the same truth tables. On the other hand, it can be proven (see notes[5]) that

$$p \Rightarrow q \text{ is equivalent to the } \textbf{\textit{contrapositive}} \; (\sim q) \Rightarrow (\sim p),$$

a result, called the *law of contraposition*, which will have remarkable consequences later, and which is also valid when the conditionals in it are replaced by biconditionals.

To finish with these formalities, let us discuss the question of sufficient and necessary conditions with which most readers will to some extent be familiar. When q is implied by p, we say that p is a *sufficient* condition for q if, whenever p is true, then q is also true; and we say that it is *necessary* if whenever p is false, q is also false. These conditions can be easily related to the conditional and biconditional defined in table 5. Consider in the table the three cases when $p \Rightarrow q$ is true: we immediately see (row *a*) that this condition is sufficient but (row *c*) not necessary. Thus, $p \Rightarrow q$ being true entails a sufficient but not necessary condition. For example, the numbers m and n being both positive (say p true) entail their product being positive (q true), which satisfies row *a* of the table. This condition, however, is not necessary, since the product would also be positive in case both m and n were negative.

As regards $p \Leftrightarrow q$, the only relevant, true, rows are *a* and *d*; the first one shows that the condition is sufficient and the second that it is necessary. Thus $p \Leftrightarrow q$ being true entails a *necessary and sufficient* condition, as its description in table 3 by the **if and only if . . . then** entailment signifies. For example: if and only if the numbers m and n are both of the same sign (call this proposition p), then their product is positive (proposition q). Clearly, if p is true q is true, and if p is false q is false, in which two cases the implication is true, in agreement with rows *a* and *d* of column 6 of the table. We have now done all we need about the formalities of logic, and we can begin our real work.

THE UNIVERSAL QUANTIFIER: INDUCTION

Let us go back to the Socrates syllogism:

> *All men are mortal; Socrates is a man; therefore Socrates is mortal.*

Undoubtedly, as we have seen, the conclusion is correctly derived in logic from the premises, but we also know that the use of the word *all* in the major premise is far from transparent: thus the expectation that we have *demonstrated beyond any possible doubt* that Socrates is mortal is an ungrounded illusion. And we had better get used to the idea that, as I have warned before, pure logic says nothing whatsoever about the world.[6]

The infamous word *all*, or more precisely the expression *for all*, is appropriately called a *universal quantifier*; and we shall have to talk a little about it, with the proviso that if we were able to ground its use on cast-iron logical premises, then we would be in line not for one but for a whole string of Nobel Prizes in philosophy, if such prizes were awarded (which very sensibly they aren't).

Suppose that we want to establish the proposition *All ravens are black*. Let us accept that color is not an essential *differentia* that *defines* the species

Corvus corax, the common raven, so that this proposition is not analytic. Thus, there might be albino ravens, and we may assume that the proposition under discussion is directed at proving that they do not exist. Basically, the method used for accepting such a proposition is the method of *induction*, whereby, on examining *n* instances of the proposed *fact*, all of which confirm the hypothesis, we may accept the hypothesis with a higher degree of confidence the larger *n* is, which means that we would expect that the next instance after *n* would also confirm the hypothesis.

This looks plausible, but if you unreservedly believe in this, you could just as well believe in fairies. Suppose that we have a bag with 10000 balls, all black except one that is white. Consider these two situations: (1) We know nothing about the contents of the bag, in which case after drawing 9900 consecutive black balls, our confidence that the next ball will be black inductively *increases*. (2) We know that *one* of the balls is white, in which case after drawing 9900 consecutive black balls, our confidence that the next one will be white *increases*. (It would be a certainty after 9999 consecutive black balls!)

The problem of induction has been at the center of philosophy for centuries, the Elizabethan courtier Francis Bacon (1561–1626) being its first major advocate in scientific study. It was the great Scottish philosopher David Hume (1711–1776) who firmly debunked the view that there could be a logical or even a rational argument for believing in the method of induction,[7] as the example above shows, given that we reached two *opposite* conclusions at the end of apparently the same set of events. This question has been exhaustively studied by philosophers ever since Hume, but purely philosophical enquiry is just as likely to build a brick wall as to solve this problem. This is so because induction, essential as it is in scientific method,[8] obtains such precarious rationality as it has only insofar as it is an *informed* guess. If your information were nil, as in case 1 above, and your life depended on the outcome, would you not guess that the ball number 9901 is black? Such dramatic situations, however, are seldom the case in science,[9] and any scientist worth her salt would always inform herself of any feature that could possibly be relevant before she uses induction. The whole art of induction depends on a careful choice of the finite sample used to make a prediction for one of infinite (or very large) size. We shall see in the next section that any inductive procedure depends crucially on the nature of this sample; and further discussion of Hume's ideas will be deferred to chapter 4.[10]

I must remark for the moment that the question of the use of induction in scientific practice remains one of our major problems. I shall later propose a solution based on the principle that propositions in science never stand or fall on their own; that they must be closely knitted within what I shall call the scientific mesh of facts and theories, and that the use of induction for a proposition can only be legitimized when the proposition is integrated (or, as I shall call it, *entrenched*) as a part of this scientific mesh. (See chapters 7 and 8.)

RAVENS ARE BLACK: PARADOXES OF CONFIRMATION

Whatever laws of nature are, they certainly must be general rather than particular propositions: *My left shoe is black* does not even begin to look like a law of nature but *All ravens are black* appears to fit the bill.[11] (Although I prefer to consider this type of proposition as merely stating a *property*, rather than stating a law.) Let us now make our *variable x* more general by defining

$$x = some\ arbitrary\ object.$$

As before, we shall also define

$$p = x\ is\ a\ raven\ ;\qquad q = x\ is\ black.$$

We would like to confirm by induction the proposition:[12]

$$For\ all\ x\ (p \Rightarrow q),$$

which means:

$$For\ all\ x,\ if\ x\ is\ a\ raven,\ then\ x\ is\ black.$$

We now know that we cannot ask for pie in the sky: all that we can modestly aim at is to establish a stepwise procedure whereby our *confidence* in the validity of the above proposition is gradually increased. Let us try for this purpose two criteria, called *Nicod's rules*, which appear to be eminently sensible:[13]

Rule 1. *Every instance that confirms $p \Rightarrow q$ increases our confidence about the inductive generalization above.*

Rule 2. *Confirmation of any proposition logically equivalent to $p \Rightarrow q$ increases our confidence about the inductive generalization above.*

(An example will soon reinforce our knowledge of *logical equivalence*.) Good as this stuff appears to be, blunt application of these rules may lead into arrant nonsense, and we must learn how to avoid silly mistakes. Let us see how these rules work for the ravens. Rule 1 means that if I pick up an object x and find that it is a raven and that it is black, then, because the truth values of p and q are both T, and thus, from table 5, $p \Rightarrow q$ is also T, this instance increases our confidence to assert the induction for all x. This seems to be unassailable, but wait a little: you will see that it is not *always* as good as it sounds in this case. (Watch for grasshoppers!)

Where the trouble is indeed serious is when we use rule 2. We have shown previously that $p \Rightarrow q$ is logically equivalent to $(\sim q) \Rightarrow (\sim p)$, in the sense that both functions have identically the same truth table. (This is the *law of contraposition* proved in table 6, note 5 of this chapter.) In plain English, what the contrapositive, quantified for all x, means is (with q black and p raven):

For all x, if x is not black, then x is not a raven.

Let us now try to apply rule 2, in accordance to which an instance that confirms $(\sim q) \Rightarrow (\sim p)$ also confirms $p \Rightarrow q$ and thus adds to our confidence of the induction proposed that all ravens are black.

What we are saying is that to establish that all ravens are black is the same as to establish that all nonblack objects, $(\sim q)$, are nonravens, $(\sim p)$. So, we pick up a piece of paper, observe that it is white, and of course conclude that it is not black. With equal perspicacity we also observe that the piece of paper is not a raven. Result: this comfortable experiment confirms (that is, adds confidence to) the induction that *all ravens are black*. People who doubt the uses of philosophy have here an example to the contrary: the advantages of doing ornithology in your own sitting room while having tea and toast are obvious.

That all this is a load of nonsense is also obvious. The reason for our confusion is this: although logic is a marvelous tool, it, like all tools, has to be used with care. Even more, as we have seen, induction stands on the shoulders of logic, but *it is not a logical process*. We have already insisted that the choice of the sample that is used in an induction process is crucial: this is the source of the paradox, as we shall now discuss.

Remember that induction consists of having an implication, such as $p \Rightarrow q$, where p and q state properties of an object x. The set of all x for which $p \Rightarrow q$ is true, is called the *domain of confirmation* or *evidence class* of the implication. The problem is that this domain of confirmation is in practice so huge (all black ravens, for instance), that all that we can do is *sample* this domain, from the positive results for a comparatively small set of x (any ravens, for example), to try to induce the proposition to the whole domain of confirmation. Let us call a *confidence index* the ratio between the size of the sample of all objects x that are tested and belong to the domain of confirmation, to that of the *whole* domain of confirmation. For the induction process to be sensible, it is necessary that, as more instances are taken from the sample chosen, this index increases *significantly*. (I say *significantly* here because we shall soon see that it is easy to be mesmerized by *nonsignificant* increases of this ratio.)

We illustrate this problem in figure 1. First of all, you must notice that this figure is hugely out of scale: the ellipse that contains all black objects should be minute compared with the whole world, and, likewise, the circle that is the set of all ravens should be minute compared with the ellipse. If we use rule 1, the domain of confirmation is just that part of the circle that is inside the ellipse (black ravens), and the induction process, if successful, must result in this circle being entirely inside the ellipse, that is, no instance should be found in the *domain of falsification*, which is the hatched area outside the ellipse (nonblack ravens).

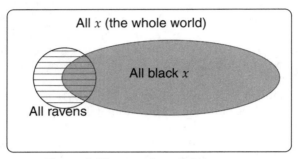

Figure 1. The paradox of the ravens.

The circle is the set of all ravens. The domain of confirmation (black ravens) is the intersection of this circle with the ellipse. The domain of falsification (nonblack ravens) is the hatched area outside the ellipse.

We now realize that we have two conditions for sampling: the first is the one already stated that, on increasing the size of the sample, the proportion of the domain of confirmation covered must increase *significantly*. The second is that the sampling must be done so that the chances of hitting the domain of falsification are reasonable. This is important because the most profitable event one can encounter in the process of induction is just *one* hit at the domain of falsification. Whereas thousands of positive results in the domain of confirmation do not *prove* that the induction is true (they merely increase confidence in it), just *one instance* from the domain of falsification *proves* that the induction is false.

We can now see in figure 1 how, *in this case* (see later), rule 1 works well. If we choose as our sample a set (of gradually increasing size) of ravens, we do two good things. First, because this set is within the comparatively small circle, and because the domain of confirmation is wholly included in the latter, when we increase the size of the sample, we increase reasonably the ratio between this set and the domain of confirmation. Second, this sample allows us to hit the domain of falsification, if it exists. So, we have done a good job.

Consider now the use of rule 2 by dealing with the logically equivalent proposition $(\sim q) \Rightarrow (\sim p)$ (not black entails not raven). The domain of confirmation is huge: it is the intersection between the part of the whole world which lies outside the ellipse (nonblack objects) and that part of the whole world that lies outside the circle (nonravens). This intersection is, therefore, that part of the whole world which lies outside both the circle and the ellipse in the figure (nonblack objects that are nonravens). It is easy to see that the domain of falsification is the same as before. Samples for the induction must be obtained from sets of any x, that is, of any objects in the world. Suppose I take two samples, call them a and b, of 1000 and 2000 objects, respectively, and that both of them lie within the domain of confirmation. So far so good. You can even rejoice because, whatever the value of the ratio, call it r, between the size of sample a and the size of the domain of confirmation, this ratio has surely *doubled* for sample b, thus apparently increasing the confidence index. This is utterly wrong, however: because the domain of confirmation is so huge (almost the whole world!), the ratio r (which is a divided by an exceedingly large number) is for all uses and purposes *nothing*, and double nothing is still nothing. So, however much we increase the size of the sample, the confidence index for the process remains nil *for all practical purposes*. (I am not interested in the possibility of a billion people spending a billion years inspecting all grains of sand in the Sahara!) Moreover, by the same token, the chances of hitting the domain of falsification in this sampling are negligible. So, in this case, rule 1 must be used, because it is the only one that provides us with sensible samples. This, alas, will not always be the case, as we shall now see when we turn our attention to grasshoppers.[14]

Consider the proposition:

No grasshoppers live on the Pitcairn Islands.

We shall rewrite this as follows:

x = *any insect,*	u = *x is a grasshopper,*
v = *x lives outside Pitcairn,*	$\sim v$ = *x lives on Pitcairn.*

We want to establish the proposition:

For all x, (u $\Rightarrow$ v),

which, from the law of contraposition (table 6 in the notes), is logically equivalent to:

For all x, ($\sim v \Rightarrow \sim u$).

Suppose that we use rule 1, which means that we look outside the Pitcairn Islands: every time we find a grasshopper, we verify $u \Rightarrow v$ and thus increase our confidence for the desired induction. This appears to be unimpeachable, but wait a moment, and you will run into delicious trouble. To see how this arises, let us consider one other proposition:

w = *x lives anywhere in the world,* and: *for all x, (u $\Rightarrow$ w).*

It should be abundantly clear that all the instances that confirmed for us the proposition $u \Rightarrow v$ equally well confirm the proposition $u \Rightarrow w$, and thus also increase confidence in the induction that grasshoppers live all over the world, *including* the Pitcairn Islands! So, rule 1 is defective in this case. On the other hand, if we use rule 2 and consider the proposition

$\sim v \Rightarrow \sim u$, this means that we must prove that any insect that lives on Pitcairn is not a grasshopper. Here we choose a sample from all insects on Pitcairn (notice that this sample is far more compact than the previously unwieldy one of rule 1), and an observation that they are not grasshoppers confirms only the proposition wanted and no other.

WHAT HAVE WE LEARNED ABOUT INDUCTION?

A material implication $p \Rightarrow q$ is a logical proposition and it can be confirmed or falsified from its truth table. On the other hand, confirmation of the quantified proposition *for all* x, $(p \Rightarrow q)$ can never be absolute. Instances (values of x) for which $p \Rightarrow q$ is true or false belong to the domains of confirmation and falsification, respectively. On sampling x, a *single* instance belonging to the domain of falsification stops the induction and *proves* the falsity of the quantified proposition. The larger the number of instances that belong to the domain of confirmation, the greater the confidence that can be placed on the truth of the quantified proposition, but this truth cannot ever be absolutely established. Moreover, in sampling during the process of induction, the two Nicod rules must *sensibly* be used in order to satisfy the following three conditions. First, the ratio between the size of the sample and the size of the domain of confirmation (*confidence index*) must increase *significantly* as the size of the sample increases. (If the confidence ratio is infinitesimally small, doubling it, for example, will not lead to a *significant* increase of it.) Second, the chances of hitting the domain of falsification must be reasonable. Third, care must be taken to avoid the possibility that the domain of confirmation of the implication selected be contained in that of a different implication that might contradict the desired one or which might confirm independent but wrong projections.[15] Naturally, these requirements are the mere bones of an induction process. In scientific practice, a great deal of work must be done in order to choose the sample to be tested, so as to provide a good degree of confidence with the smallest possible size, for which purpose specific statistical methods are used.[16]

One final and very important point. In describing the process of induction, I have made no attempt whatsoever at justifying it: I have simply described what appears to be reasonable practice, if various pitfalls are avoided. Any justification of the induction process cannot be logical, we know. Then, why do we believe in induction? This will be discussed in chapter 4.

PROPERTIES

I shall now try to use our logical knowledge (I am afraid with not very splendid results) in order to define a property S (solubility in water, for example) of a substance (sugar, say). Let us use the following notation:

x = *a sample of sugar.*
S = *x is soluble in water.*
P = *x is placed in water.*
D = *x is dissolved in water.*

It appears plausible to try the definition:
$$S = \text{for all } x, (P \Rightarrow D).$$

This means in plain English: x is soluble in water if, for all x, when x is placed in water, x is dissolved in water. To make our life less miserable, I shall not bother about the induction implied by *for all x*: we shall see this that is the least of our troubles. The above definition means whenever $P \Rightarrow D$, then S is also true. But as we know from the properties of the material implication (see table 5), $P \Rightarrow D$ is true *whenever the antecedent P is false.* This leads to the absurd conclusion that I do not need to wet my hands to know that sugar is soluble: for all samples *not placed in water* (for which P is false), S is true! This is a great weakness of the operation of *material implication,* and a great deal of (not very satisfactory) work is required in order to improve the situation.

It might be thought that the way out is to substitute the biconditional $\Leftrightarrow$ (if and only if) for the conditional $\Rightarrow$ (if):
$$S = \text{for all } x, (P \Leftrightarrow D).$$

This is wrong for the following reason: It means that S is true if both implications $P \Rightarrow D$ and $D \Rightarrow P$ are true. But the second prong of this conjunction is wrong, because saying that x *is dissolved in water* implies that x *is placed (or has been placed) in water.* But salt, for instance, can be dissolved in water never having been placed in water; it could have been produced there by a chemical reaction.

One way out is to use the following tentative definition, which for simplicity we shall do leaving out the universal quantifier:

(1) If x is placed in water, then x is soluble if and only if it dissolves in water:
$$P \Rightarrow (S \Leftrightarrow D).$$
(2) The above implication should only be used when the antecedent P is true.

Because of this second condition, this definition is called a *partial definition.* It specifies the meaning of S not for all cases, but only for the cases when the test condition P has been satisfied.[17] This gets nearer scientific practice but not quite. When the scientist says, "Sodium chloride is soluble in water," he is not necessarily referring to samples of sodium chloride that *have been placed* in water. (New ideas will be required in order to cope with this.)

The important point to notice so far is that logic alone is unable to do anything as simple as provide a cast-iron definition of a property! There was a time, however, when people expected that laws of nature, which must be much more difficult to handle, could be recognized by their logical forms, a chimera that by now is safely forgotten.[18]

Before we move on, I must say that I have left out a number of very important points that arise in dealing with this problem, in particular the use of the *counterfactual conditional*, which is very important in science in relation to the so-called *dispositions*. The latter are important in discussing properties such as solubility and will be treated in chapter 4, whereas we shall now have a brief look at counterfactuals.

USEFUL UNTRUTHS: COUNTERFACTUALS

Let us have another go at the generalized statement "Sodium chloride is soluble in water," which is typical of the form of a law of nature (although as I have already said, it is more like the enunciation of a *property* since it applies to only one substance). We notice first of all that there is implicit in this statement a universal quantifier: what we really mean by it is: *For all x, if x is a sample of sodium chloride, then x is soluble in water*. So, to start with, we have all the problems associated with the universal quantifier and the use of induction, which we already discussed. On top of that, we have the problem of the meaning of the proposition *soluble in water*. We have seen that this cannot be defined by the material implication *x is placed in water* $\Rightarrow$ *x dissolves*, because whenever the antecedent is untrue, that is for all samples not placed in water, the implication is true and, therefore, the sample satisfies the definition of solubility.

In order to get a better definition of *x is soluble* let us try:

If x were placed in water, then x would dissolve.

This sentence is called a *conditional subjunctive* because the verb (*to be*) is in the subjunctive mood (*were*). You may think that this sentence is a pretty obvious thing to say, but reflect upon the extraordinary fact that we are making a positive scientific statement about something that is a non-fact: *x* is not placed in water. Thus, we are saying something about *x*, although nothing whatsoever happens to *x*. If we are not discussing a fact, what are we doing? Are we merely expressing an opinion? Obviously, we must be trying to do more than that because we seem to want to endow the statement with some sort of a lawlike quality. How can we justify this?

In order to make some progress I shall jump into the deep end by considering a conditional that is even more distant from reality than the above one, because rather than referring to a *nonfact* it refers to something that is *contrary to fact*:

If Mr. Bloggs had not tried to cross the road, he would not have been killed by the bus.

This type of statement is called a *counterfactual conditional*, and we shall see that not only is it important in trying to understand what a law of nature is, but also that its use (or lack of it) plays a significant part in the philosophy of quantum mechanics.

I have taken, on purpose, an example that has legal implications because we shall have to use a level of discourse that is accountable, as any legal judgement is. So, we can assume that Mr. Bloggs's widow is suing the driver of the bus in a civil court for damages, that the judge has to apportion the blame as between the parties, and that the lawyer for the defendant (the driver) has just made the statement above. Of course, if the statement were "true," then the deceased would have had a measure of responsibility for his own death. Our problem is this: how does the judge decide that the statement is "true" or not? I use the word "true" in quotes because the judge cannot order, of course, that an experiment be conducted in which poor Mr. Bloggs does not attempt to cross the road, with a view to verifying that, in that situation, the bus did not actually hit him. This cannot be done, so that such an element of truth that can be assigned to the statement in question cannot be established by experience. Yet, the judge must decide.

A very superficial examination of the defense plea can give the impression that it is manifestly true. Yet, this is far from being the case. Our judge, fortunately, is philosophically inclined and he devises the following method. He invents different histories about Mr. Bloggs's life, or more properly different world histories, in which everything is very much as we know, except that events connected with the last minutes of Mr. Bloggs are changed. In one history, say H(1), Mr. Bloggs does not cross the road, and he is not killed. In H(2), he is drunk, he crosses the road and is killed. In H(3), Mr. Bloggs is as sober as a bishop, but he crosses the road and is killed. In H(4), Mr. Bloggs is carefully waiting at the pavement, but the bus is speeding, and when the driver applies the brakes, the vehicle loses control, mounts the pavement, and kills Mr. Bloggs. H(1) corroborates the statement for the defense above, whereas H(4) proves it wrong. What the judge must do is to assess all available evidence to give weights to all these histories. Even if H(4) is not what actually happened but, say, the driver as a point of fact was drunk, and his vehicle's brakes defective, then H(4) becomes sufficiently weighty as to require the judge to assign a level of responsibility to the driver. The larger the number of histories that entail the killing of Mr. Bloggs when he has not attempted to cross the road, and the more weighty they are, the larger the responsibility of the driver. We begin to see, I hope, how the "truthlikeness" of the counterfactual conditional can be assessed. But before we do that, we shall rewrite the counterfactual as a conditional subjunctive:

> *If it were the case that Mr. Bloggs had not crossed the road, then it would be the case that he would not have been killed.*

This conditional subjunctive is of the general form:

> *If it were the case that p, then it would be the case that q,*

where p and q correspond in the example to the two italicized propositions respectively. In order to understand how we use this conditional, we have to construct an ensemble of possible histories, compatible with the laws of nature and with all the information we have about the situation in question. If the implication stated is true in all possible histories, then the counterfactual expresses a *necessary* condition. If the implication is true in some of these histories only, then the counterfactual is *contingently possible*. If is not true in any, it is *impossible*. These qualities of necessity, possibility, or impossibility are called *modalities*. Clearly, because these modalities must be analyzed with reference to world histories that are not the true world history, modal analysis, however practically important it is, is metaphysical.[19]

We can now see how amazingly complicated it is to say "Sugar is soluble." We must first define x = sample of sugar and then state:

> *For all x, if it were the case that x were placed or produced in water, then it would be the case that x would dissolve.*

Except for the quantifier, this proposition is of the same form as the conditional subjunctive last displayed above, and it would have to be treated just as that one was. Clearly, it would not be acceptable to claim that the statement must be true in all possible histories: we do have here a process of induction, and the truthlikeness of the proposition can only be regarded as *possible* rather than *necessary*, but the degree of confidence in this possibility can be increased by increasing the number of samples considered, with all the provisos we have already entered as regards the choice of samples and the experimental protocol.[20]

Coda

I hope that the reader is not profoundly despondent by now because so many times, at important points in this chapter, we have come unstuck. We have introduced some formidable tools, that for a good part of last century created the hope that one could start from observational propositions, thus known to be true, and construct from them all manner of rigorous results by the methods of logic. Although the tools that we have created are useful, as we shall soon see, it is quite clear that the just-stated program was an illusion: a great deal more is needed in order to do science than just collect results, use perhaps the method of induction in order to make generalized statements, and then turn the handle of logic. In studying nature, however, one must always be grateful when nature refuses to be abused: this is the way one learns.

One important idea that we have obtained is that the method of induction cannot be either logically, or indeed empirically, justified (if by empirically we mean observationally). That it is used, however, cannot be denied, and we shall learn more about the grounds for so doing in chapters 4 and 10. For the time being, the examples discussed in this chapter, both famous and infamous, should have opened the reader's eyes about the dangers of misguided induction and the need to pay a great deal of attention when attempting to use the method.

The sentence used by the chemist "Sodium chloride is soluble in water" defines a *property* of sodium chloride. Scientists would prefer to say that we have here the definition of a property rather than to say that we have a law of nature, the latter being expected to have a wider domain of application than just a single substance—sodium chloride in this case. We have seen that even the definition of this simple property cannot easily be given, if given at all, in the language of logic: syntax is not sufficient to decide on what a property is. If this is so difficult, it can hardly be expected that what constitutes a law of nature could be formulated by examining the syntax of sentences that are used as such laws.

Another important result is that a more detailed analysis of properties requires the concepts of conditional subjunctives or of counterfactuals (the two being intimately related), with the surprising consequence that in order to understand the way in which these strategies are used, one must accept the use of multiple world histories (or even multiple worlds). We shall see later that the treatment of counterfactuals in quantum mechanics is both important and delicate.

We are now ready to embark into the study of one of the most important concepts for our discussion, that of causality, which will allow us to begin to understand how the human mind works and *why* it works as it does.

NOTES

1. Rescher (1969).

2. I am afraid that the notation I use is not entirely standard. The implication sign $\Rightarrow$ is more commonly given as $\supset$, and the biconditional $\Leftrightarrow$ is more commonly $\equiv$. I have used the symbols given in the text because they make clearer the relation between the conditional and the biconditional, and because they agree with usage in science and mathematics. The arrows used follow the notation of, for instance, Lemmon (1965), except that I have doubled them in order not to clash with the standard symbol for mapping in mathematics.

3. A fuller discussion of truth tables may be found in Strawson (1952).

4. Equivalent propositions, however, are related by the operation of equivalence (biconditional) as follows. From table 5, if the antecedent and the consequent have equal truth values, their biconditional is always T. Thus if two propositions are equivalent, their equivalence (biconditional) is identically true, that is, it is a tautology.

5. The corresponding truth tables are constructed in table 6. Columns 1 to 3 are copied from the truth table for $p \Rightarrow q$ (table 2) and columns 4 and 5 are the negatives of 2 and 1, respectively. Column 6 is obtained from the two previous columns from the truth table for the implication, and it agrees as expected with column 3. Column 7 is obtained from the rules for the biconditional in table 5 and, as expected, it is identically true, indicating that the equivalence of the two propositions is tautological, that is, that the two propositions are equivalent. The similar result for the biconditional is obtained in the same manner. The required equivalence in table 6 can be more quickly obtained from the equality of the corresponding truth tables, as it follows from comparison of columns 3 and 6.

Table 6. Equivalence of $p \Rightarrow q$ and $(\sim q) \Rightarrow (\sim p)$

(Law of contraposition)

p	q	$p \Rightarrow q$	$\sim q$	$\sim p$	$(\sim q) \Rightarrow (\sim p)$	$(p \Rightarrow q) \Leftrightarrow \{(\sim q) \Rightarrow (\sim p)\}$
T	T	T	F	F	T	T
T	F	F	T	F	F	T
F	T	T	F	T	T	T
F	F	T	T	T	T	T
1	2	3	4	5	6	7

6. A way to license the use of the word "all" in this case is to claim "mortality" as an essential feature of the definition of "man." If I met a presumptive "man" who happens to be immortal, I would, within this convention, call him a "god," for instance. Thus, *all* men would have to be mortal, whether I actually inspect them or not. In discussing the use of "all," however, I shall rule out such linguistic necessities as being somewhat trivial from the point of view of natural science.

7. There are many statements in Hume's books where the question is broached. For instance: "From the mere repetition of any past impression, even to infinity, there will never arise any new original idea, such as that of a necessary connexion; and the number of impressions has in this case no more effect than if we confin'd ourselves to one only" (Hume 1739, Bk I, Part III, § VI, p. 88). See also chapter 4.

8. Popper (1969), p. 53 hotly denies that induction is ever used as part of the scientific method. This point of view flies in the face of the evidence of scientific practice. First,

whenever scientists repeat an experiment (and a scientist who does not do so is not worth this name), he or she is implicitly using induction. To give just one historical example: it was only because Ørsted repeated the experiment that established the existence of electromagnetic interaction dozens of time between (probably) February and July 1820, always obtaining the same, for him, totally unexpected effect, that he was prepared to stick his neck out and announce the result to the world in July 1820. This is not a matter of conjecture but of fact, because his notebooks are extant. (See Altmann 1992. I shall also return to Ørsted in chapters 8 and 10.) Second, the whole of statistics, which is at the root of scientific method, is intimately connected with induction. Third, it would be almost impossible to do any taxonomy without using, at least implicitly, induction. The so-called logic of scientific discovery of Karl Popper will be discussed later in this book.

9. One case where we have similar trouble, however, is the following one (d'Espagnat 1989, p. 43). If we were to examine 100000 atoms of ^{238}U for one hour, none would be found to decay, and we could conclude that no atom of ^{238}U decays within an hour. For such an event to occur, however, a much larger number of atoms would be required. The correct induction would be that the probability of an atom of ^{238}U decaying in one hour is less than $1/100000$.

10. A comprehensive discussion of induction may be found in Braithwaite (1953), chapter 8, and in Reichenbach (1938), pp. 339–63.

11. The example of the ravens, also known as the paradox of the ravens, is due to Hempel (1965), p. 15. There are dozens of treatments of the paradoxes of confirmation: see, for example, Mackie (1963); Swinburne (1971), 1973; Horwich (1982).

12. The inductive passage from the singular proposition $p \Rightarrow q$ to the quantified one for all x is called a *projection*.

13. See the references in note 11 for the Nicod rules.

14. The grasshopper paradox is due to Swinburne (1971).

15. An example of the first difficulty (contradiction of the desired projection) was given in the text in the case of the grasshoppers. An example of the second possibility, the evidence class supporting an independent but wrong projection, is as follows. Consider, for the case of the ravens, the evidence class as made up of nonblack objects. Examination of this evidence class might reveal that it does not contain any polar bears (which would almost certainly be the case if our diligent search for nonblack objects were carried out in London's Hyde Park). Thus, this evidence class, which supports the projection"all ravens are black," also supports the projection "all polar bears are black." This is another reason against the use of the contrapositive in the case of the ravens.

16. The reader must appreciate that the discussion of this section can in principle be affected by the infirmity I revealed in the example of the bag with 9999 black balls and one white. There are, nevertheless, two redeeming features in the practical application of the discussion of this section. In the case of the bag, two alternative inductions were possible, as long as it was known that a white ball existed among the all-black aggregate. Induction as I have described, instead, is based on *ignorance* of any prior information. On the basis of such ignorance, induction might be used as a *better-than-nothing* procedure. Second, I assume that we draw samples out of very large aggregates; therefore, however many instances are collected, the aggregates left unsampled are still large enough so that, even if you knew that there is one white ball in the bag, the probability of hitting it would still be negligible. As the reader must realize, although induction might usefully be employed in some cases, it can never be considered as a rigorous procedure and must be subject to as much scrutiny as possible in any practical case.

17. Partial definitions, as particular examples of what he called *reduction sentences*, were introduced by Carnap (1937), (1956) as part of a rather quixotic program directed at establishing such things as properties and laws by means of purely logical rules. For a

discussion see Hempel (1965), p. 129, and Brown (1979), p. 43. For the relation of Carnap's partial definition to Bohr's definition of physical properties, see d'Espagnat (1983), chap. 3.

18. Carnap was the leading protagonist of this venture. See note 17.

19. Counterfactuals were brought to the attention of philosophers mainly through the work of Nelson Goodman, (1906–1998), professor of philosophy at Harvard, whose book, Goodman (1965), is a classic. The very important idea of discussing necessity through the concept of many worlds and world histories was introduced by Wilfred Sellars (1912–1989) although it really goes back to Leibniz. (See Sellars 1948.) After that, the most important work is by the Princeton philosopher David K. Lewis (1941–2001), especially Lewis (1973). I have followed Lewis's approach, with one important change: instead of the world histories that I have used, as done by Sellars, he introduces a multiplicity of worlds. We have already met this idea in chapter 2 in discussing linguistic necessities, where different worlds must be conceived because they must be allowed to be entirely different from our own. This is not so much the case when dealing with necessity as regards counterfactuals. Moreover, the idea of a multiplicity of world histories is important in quantum mechanics, where it was introduced by Richard Feynman. On the other hand, some cosmologists and quantum theorists now believe in the *real* existence of a multiplicity of worlds, and Lewis would certainly disagree with my view that the treatment of counterfactuals via world histories is necessarily metaphysical, since he professes to believe in his alternative multiplicity of worlds as being part of *reality*. Thus, in his last book, Lewis (1986a), he declares himself a modal realist. A very good discussion of counterfactuals appears in Horwich (1987), chap. 10.

20. The logic of counterfactuals is not without problems. The reader should consult the references in note 19.

4

OF HUME AND POACHED EGGS: CAUSALITY

It is . . . important to discover whether there is any answer to Hume within the framework of a philosophy that is wholly or mainly empirical. If not, there is no intellectual difference between sanity and insanity. The lunatic who believes that he is a poached egg is to be condemned solely on the ground that he is in a minority.

Bertrand Russell (1946), p. 699.

One of the great triumphs and one of the great disasters of intellectual history was the axiomatization of geometry by Euclid. It was a triumph because it showed how, starting from a few axioms or allegedly self-evident truths, one could proceed to the discovery of numerous consequences by rigorous deductive methods. But it was a disaster because Euclid had been so successful in providing a model for what appeared to be pure thought (that is independent of experience, or a priori), that everyone who could tried to jump onto the bandwagon. Archimedes himself presented mechanics as if it were a branch of geometry, studded with 'self–evident truths,' and this disease lasted a very long time indeed. If we come to the time of John Locke (1632–1704), who was one of the first to make an important contribution to the understanding of causality, we should not be surprised if he had taken as a self-evident truth that when one hits a billiard ball with a cue, it is the *action* of the cue that is the *cause* of the motion of the ball (which, otherwise, by the principle of inertia would have remained still). There is nothing too bad about this idea, except that, contaminated as they were by the alleged *necessary* character of geometrical (and thus, via Archimedes, mechanical) truth, it would also have been claimed that the causal relation invoked was a *necessary* one.[1] It is only against this background that the gigantic achievement of David Hume (1711–1776) in debunking necessity as a modality of causal relations can be judged. (The reader may reflect upon the fact that one-by-one the bastions of necessity are falling down, such modality being so far confined to the humble umbrella of linguistic conventions.)

Why are we so concerned about *necessity*? The man in the street uses this word quite happily, but we have hijacked it into a much more technical meaning. The question is that, if we want to understand anything about our world, we must understand what *necessity* means. For the medieval man, for instance, it was *necessary* that men and women had to have two arms because they were made in the image of God. If we assert instead that this is a *contingent* fact, that it could just as well have been that men had four arms, say, we are at the same time claiming something very important indeed about creation: it is as it is, but it might have been otherwise, that is, it was not preordained. You need not accept this, but you must agree that there is a substantial difference between the two views, and that this difference hinges on what we consider necessary and what we consider contingent. Notice, by the way, that neither point of view commits us to accept or deny creation as the work of the deity, but that the alternative chosen determines, at the very least, our views as regards the type of intervention (preordained or otherwise) that was involved in creation.

Before I start, I had better warn the reader that my treatment of causality here will not be entirely in agreement with the views of the subject discussed in many philosophical treatments. The problem is that the word *cause* is used in *different* ways in science, in philosophy, and in law; and if the reader ends this chapter fully aware of this important and dangerous fact, I shall have done my job.

Philosophers would be quite happy to consider a sentence such as *bacteria cause disease* as an example of a causal relation, whereas scientists would regard it as too vague to be worthy of consideration: not only do many bacteria not cause disease at all but, contrariwise, some diseases are organic rather than produced by bacterial infection. Likewise, one could spend an entertaining, but not very productive, time in discussing such questions as, "What is the cause of a car's motion?" Is it the engine? But the engine does not run without petrol. Is it the petrol? But petrol would not burn without air, and so on. Although I shall have to address some such questions from time to time, they will be entirely marginal to our inquiry. I shall occasionally refer to causal examples in law, but I shall not want to discuss examples such as the following one, which, although exciting to the imagination of philosophers, are of no interest in science. Bill is swimming in the sea, gets a cramp, is in serious difficulty, and shouts for help. A very competent lifeguard runs in for what for her would be a trivial rescue: but she is deliberately tripped up by Bill's nephew, Jack, eager for his inheritance. A court of law might find Jack the *cause* of Bill's death, but this is not for us a causal statement with which we need be concerned. [2]

It is most important, on the other hand, to realize that for lawyers, a statement such as "no causal relation has been established between *A* (say, John pushing Jill) and *B* (say, the death of the latter)," terminates the story for them: John has to be found not guilty, and the lawyers can happily go home. For scientists, instead, such a statement may mean that a vast program of research is required in order to prove or disprove the alleged causal relation. Misunderstanding of this distinction can cause huge problems, especially when politicians (who more often than not have legal rather than scientific training) seek advice from scientists.

Despite these difficulties, I shall not go as far as Bertrand Russell (1872–1970), for whom the law of causality was "a relic of a bygone age, surviving, like the monarchy, only because it is erroneously supposed to do no harm." Causality, as we shall see, has an important role to play in the understanding of science, as long as we keep a view of it firmly based on David Hume's ideas, despite the numerous and strong philosophical arguments that have undermined the Humean tradition in the last thirty years or so.[3]

There is an even more important reason why a discussion of causality is necessary. If we want to study the early history of living forms, we need palaeontology, the study of fossils, which supports Darwin's theory of evolution. But the way in which the brain works is in itself a testimony to the evolution of our species, and to that of our earliest ancestors, spanning millions of years. The main property of the brain is that it is a learning machine, and learning, as we shall see, and it is in any case self-evident, is almost wholly dependent on the apprehension of causal relations. Thus, an understanding of causality will allow us to understand one of the major mechanisms by which our intercourse with nature is conducted.

All these ideas, amazingly, stem from the program of work started in Edinburgh in the eighteenth century by David Hume, well over a hundred years before evolution was even conceived. Thus, not only does his work fit the needs of modern science in addressing causality, but he should also be considered as a proto-Darwinist, his ideas being central to the view of nature that I shall develop in this book.

Before we start on this enterprise, let me repeat that there are many different ways in which the word *cause* is used in learned discourse: people spend a lot of time arguing against each other's ideas when, in fact, they are using different words that are mere homonyms. Be warned.

THE SUN HEATS THE STONE

Rather than follow a historical account, let me illustrate the major aspects of Hume's analysis of causality by means of an example. When we say "the sun heats the stone," we are implying a causal relation. We could just as well have said: "The sun *causes* the stone to warm up." If what I am now going to say surprises you, it is all to the good, because you will then acquire a healthy respect for the novelty of Hume's ideas: for he claimed that the alleged causal relation was *neither necessary nor observable*. And I had better warn readers that a great deal of what I shall say will be so apparently counterintuitive, that unless they happen to have a robust philosophical bent, my discussion will meet with strong rejection. So, try to read what follows in a relaxed manner because it will take quite a bit until all this falls into place. And it is very easy to produce false counter-arguments based simply on misconceptions about what Hume really meant, of which the literature is studded with examples.

Let me first deal with the question of *necessity*, a concept that we have met twice before. In chapter 3, when dealing with counterfactuals, and because they establish in a way a bifurcation between possible world histories, we took *necessary* to mean true in all such possible histories. In chapter 2, instead, because we were dealing with linguistic necessities, we had to impose a much stronger criterion of necessity, namely one that requires a truth value in *all possible worlds*, and it is in this latter sense, which I shall call *universal necessity*, that necessity must now be understood. When we do this, we are bound to recognize that it is perfectly possible to imagine a world in which, whenever the sun comes out, stones cool down. It is thus clear that, whatever the relation is between the antecedent (the sun coming out) and the consequent (stone warming up), this relation is purely contingent and not universally necessary: it might at most be a property of our world.

As regards observability, I am sure that the reader will at least initially scorn my contention that *the causal relation is not observable*. We have all experienced the fact that before dawn the stones on the ground are cold, but that after a few minutes of the sun shining on them, they become warm: we can touch them with our own hands. What Hume wants of us is to recognize that, truly enough, we see the sun coming out, *and* we feel the stone getting warmer: but all that we experience is the *conjunction* of these two successive events and nothing else: any causal relation that we invoke is of our own making.[4] The fact remains that nothing is *observed* further than a conjunction, and that the *perception* of causal relations, when analyzed, is far from being properly established.[5] Whatever the causal relation is, it is in the nature of a theoretical construction rather than a straightforward percept.[6]

HUME THE SKEPTICAL

The reader must appreciate that Hume, in propounding the views described, had to fight a serious battle, and it is not unnatural that in order to do this from a position of strength, he concerned himself with fundamentals and disregarded inessential detail. People who read him, not keeping this in mind, and that includes Bertrand Russell, present a view of Hume that is probably not true to the man. Take the problem of conjunction: much has been made of the fact that mere conjunction can lead to arrant nonsense. In our previous example, for instance, we considered the conjunction of two events, dawn (and thus the sun coming out) and the stone warming up. Consider now this: I have an exceptionally punctual news carrier who delivers my newspaper to my front door every morning precisely at dawn. Do we then surmise that, had Hume been a habitual resident in my household, he would have inferred that the delivery of the newspaper was the cause of the stone warming up? On the contrary, Hume fully recognized the danger of picking an arbitrary pair of events as candidates for a causal conjunction, because conjunctions, he said, "may be arbitrary and casual."[7] In the next section, we shall see how his definitions can rule out the possibility of the allegedly calent news carrier.

Thus, contrary to popular belief, Hume did temper his skepticism in dealing with the conjunction of cause and effect, recognizing that the arbitrary and casual had to be avoided: more than once he states that there must be some *power* by which the cause leads to the effect, although he, quite rightly, warns us against the loose use of this concept,[8] about which, more later. So, we have here a problem, namely how the significant constant conjunctions can be separated from the arbitrary and casual ones, a question that will have to be discussed later.

What most people remember of Hume is that he, sword in hand, so to speak, stated the skeptical view that the invocation of a causal relation is nothing more than a *custom* or *habit*: since we cannot observe the causal relation between *A* and *B*, it is a mere custom or habit that the appearance of the first will bring to our mind the idea of the second. Thus, the cause and effect appear to be merely related by an association of ideas. When Hume affirmed this, however, he also made it quite clear that in asserting this fact, he was merely describing human nature.[9] As Stuart Hampshire has put it, his was a study of "the anthropology of knowledge." If one forgets this, it is very easy to think of Hume as a total agnostic as far as causality goes, and thus as one who would cheerfully go into a McDonald's hoping to find that a hamburger tastes of oysters. I shall discuss later, in much more detail, this idea of the causal relation as a custom or habit, but, before, we must understand better Hume's view of causal relations and learn about some very important aspects of them.

One of the problems is that Hume wrote two books, first the *Treatise* and then the *Enquiries*,[10] and although Hume himself considered the latter the better account of his ideas, many philosophers have placed a greater weight on the earlier work, as being more scholarly in comparison with the more popular later book. The *Enquiries*, however, differ from the earlier book not only in presentation but also, as I shall show later, very deeply in substance. In order to have a quick definition in Hume's words of the causal relation, I shall leave the *Treatise* until later and quote from the *Enquiries*:

> we may define a cause to be an *object, followed by another, and where all the objects similar to the first are followed by objects similar to the second*. Or in other words *where, if the first object had not been, the second never had existed*.[11]

That the effect cannot exist without the cause, gives an urgency to the causal relation that those who overemphasize Hume's skepticism might forget, although it must be clearly remembered that any such urgency is not and cannot be a (universal) *necessity*. The requirement that if *A* causes *B* and *A* does not exist, so must be the case for *B* (which is what would make the cause *A* *necessary*), is one, however, that no respectable philosopher would swallow, so Hume has been much castigated for the quoted definition. At first sight, in fact, it appears to be demonstrably false: the sun, truly enough, heats the stone, but, the sun being absent, the stone heats just as well because it has an electrical element embedded in it. I shall show later, nevertheless, that such causal statements, as are used in science, are made to obey precisely Hume's condition.

This question of necessity is thus a very delicate one, and we shall come back in detail to it later on in this chaper, but the most important thing for us at the moment is the nature of that curious custom or habit, which in accordance to Hume, gives origin to the idea of causal relation. Before we look into that, we must explore in more detail the nature of the causal relation itself.

POWERS, EFFICIENT CAUSES, AND OTHER PRETTY IDEAS

Let us review what we have so far. *A* is the cause of *B*, if whenever we observe *A*, we observe *B*; and if *B* is observed, then *A* must have preceded it. Further than this we cannot go: we cannot say that the relation between *A* and *B* is universally necessary, neither can we observe the causal relation itself; all that we observe is the conjunction. So far, this is orthodox Hume, but it appears to be so negative as to hardly account for what one would like to call causation. The worst thing we can do, however, is to become impatient: good old Hume can go quite a long way, but we must not rush him. Such advice, alas, has not always been followed. One of the major problems is that the observation of constant conjunction is in itself a poor beginning, as I noted before about the news carrier being the *cause* of the stone been heated. Even in this case, however, Hume can set us on the right path, because the day there is a newspaper strike, we still observe that the stone heats, that is that *not A is followed by B*, whence *A* cannot be the cause of *B*. This is already a fairly powerful tool to discriminate against casual conjunctions, as we needed. More about this problem will come later, but one must guard oneself against other proposed solutions of this question. The fact is that philosophers have long been adept to endow the concept of cause with something like *power*; and words like *efficient cause* and the like have been used in an attempt to identify the 'true' cause, thus automatically guarding against the possibility of a spurious cause resulting from an arbitrary conjunction.

It is tempting to do this, for instance, when discussing a running car's engine as the *cause* of the movement of a car. We know only too well that it does not satisfy Hume's criterion as a cause because the engine may be running and the car may be stationary (declutched, for instance). Some philosophers, however, will insist that the engine must be considered as *the* cause because it alone has the 'power' to move the car, whereas the clutch must be considered as a mere *condition* for the cause to become efficient.[12] Many people, on the other hand, have recognized that any attempt to separate causes from conditions is essentially arbitrary,[13] a view which is probably held by most scientists.

The idea that one can associate power with the concept of cause is tempting, largely because in doing so, one provides a semblance of an explanation for the causal relation. (It is agreeable to hear that the sun has the power to heat the stone because it sends off energetic photons that stir the stone into warmth: that is, the sun has the *power* to heat because it produces photons that have the *power* to heat...) Now, I have nothing against explaining, except that this is an occupation that has to be relegated to its correct place: a great deal of confusion can be caused by putting too much emphasis on explanation before one is ready for it. (Charlatans and con men, for instance, thrive on explanation.) Moreover, however important explanation may be, it is often that the explanations of one generation are the nonsense of the next, a reason why scientists use explanation with caution. The example of the car engine, however, is too attractive for anyone who likes the powers of explanation of the idea of power. Let me kill this idea stone-dead with a few examples.

A foundry making wheels for trains has a problem because some of the wheels, when put into use, soon crack. So, they engage a metallurgist with the instructions to tap each newly made wheel with a hammer and, should it crack, to report as to the cause of the crack. At the end of the first day of work, the metallurgist reports as follows: "I have today examined 112 wheels, of which three cracked. In each case, the cause of the crack was the impact of my hammer on the wheels." Obviously, the hapless man was a bad philosopher and a worse metallurgist. No one in his company was interested in anything as elusive as the *power* that caused the cracks observed. From their point of view, the causes that would have satisfied them were, for instance, microcavities in the castings of the cracked wheels that would *explain* the cracks. (Notice that the employers were prepared to accept an *explanation* if provided by means of a micrograph, for instance, whereas the *explanation* provided by the metallurgist was premature and thus useless.)

Many examples about the absurdity of associating powers with causes can be found in legal practice, where the concept of causation is of vital importance. There was a case in the sixties in England in which a young man (*A*) pushed, in the way in which youngsters are wont, another young man (*B*), who fell on the pavement, cracked his skull, and died. You have here two obvious candidates as causes. One (which the judge would not have liked, even as a joke, but which nevertheless is a very serious one) is the law of gravity. The other is the push given by *A*, which the philosopher eager for *powers* would seize as the *efficient cause*. If the jury had found that this was so, *A* would have been guilty of manslaughter. As it happened, he was acquitted. Why? The answer is that the autopsy showed that *B* had an unusually thin and brittle skull that would have broken with an impact that no reasonable person would have expected to be dangerous. So that the *cause* of his death was neither the "power-endowed" agency of gravity, nor the "power-endowed" blow, but the condition of his skull.

Even more dramatic examples of the impossibility of requiring a cause to be endowed of power are common in the law of torts (breaches of duty): in torts of negligence the *causes* are nonevents, and nothing can be more innocent of power than a nonevent. As an example: "The signalman's failure to pull the lever was the cause of the accident."[14]

A more satisfactory concept than power, perhaps, is the akin one of *capacity*, eloquently urged by Nancy Cartwright. It is, however, safer in natural science to keep away from its use.[15]

CAUSAL CHAINS

It is possible to misread Hume in two ways. One is to forget that unless he says anything to the contrary, he assumes that the cause and effect are *contiguous*,[16] that is next to each other or in actual contact (see note 11). The other is to read some of his examples in the light of modern scientific knowledge, and thus to imagine that he was disregarding that condition. Despite this requirement of contiguity, Hume recognized that the relation between cause and effect could be mediated by a chain of intermediate causal relations, a view which allowed him to eliminate possible actions at a distance between the cause and the effect:

> Tho' distant objects may sometimes seem productive of each other, they are commonly found upon examination to be link'd by a chain of causes, which are contiguous among themselves, and to the distant objects; and when in any particular instance we cannot discover this connexion, we still presume it to exist.[17]

This idea is most important. To start with, it is what distinguishes scientific practice from voodoo: conjunctions such as sticking a needle in a wax effigy and a faraway man dying, are rejected out of hand as candidates for causal relations because the two events cannot be bridged by any plausible causal chain. So, we have here a very powerful tool for discarding arbitrary or casual conjunctions. Thus, in good scientific practice, one has to proceed in successive steps. First, one has to find *conjunctions*. This was the first step that was taken by Professor (now Sir) Richard Doll from Oxford, when he discovered that cigarette smoking causes cancer: he found that a great many people who smoked contracted lung cancer. The situation, however, is not so clear-cut in practice because not every smoker was found to get lung cancer and not every sufferer of lung cancer was found to have been a smoker. However, Professor Doll examined a very large sample of people, and the number of coincidences of the two events was found to be *significant* by the methods of statistics. So, Professor Doll was able to state that the conjunction of smoking with contracting lung cancer was statistically significant. This is basically the same as saying, in Hume's language, that the connection between the two events is neither casual nor arbitrary. Was the scientific world then immediately convinced that there was a causal relation between the too events? Not at all: a causal chain had to be found between smoking and cancer. It was only when numerous experiments showed, first, that cigarette smoke contained substance A, say, and that substance A applied on healthy rats was significantly connected with the rats developing cancer, that the scientific world accepted that smoking causes cancer in the statistical sense (that is that there is a higher probability for the conjunction smoking–cancer than for the conjunction smoking–health). It must be appreciated, however, that the causal chain between smoking and cancer was not completed: work was needed for instance to establish the connection between "applying substance A" and "growth of cancer tumor." In fact, the more science progresses, the more fine-grained causal chains become, and this is what most people describe as "providing an explanation" or "discovering a mechanism."

So, was Hume wrong when saying that all that we perceive is the conjunction between the two events A and B? For we appear to be saying that, in fact, we can also perceive further steps in the causal chain, such as $A'\,B'$, $A''\,B''$, and so on. Nevertheless, if you examine the constant conjunction of the most microscopic step such as A'' and B'', you are still faced with Hume's constraint: however much you may understand or explain or model in any way whatever the relation between A'' and B'', the fact remains that all that you *observe* is the conjunction of these two events. A'', for instance, may be the locking of some chemical from the smoke into some vital part of the DNA of a healthy cell, thus leading to cancerous growth: all that you *observe* is that one event follows the other. Pharmacologists these days can observe how a single molecule of a chemical locks, for instance, an enzyme, thus causing a substantial physiological change. However much you now *understand* what happens, all that you *observe* is that one event, the locking of the enzyme, follows the other, the inactivity of the enzyme. Obviously, what you call *understanding* is in your mind: it is not a fact observed.

We can now see how causality works in science. You observe a conjunction, and you then verify by induction (or the methods of statistics, which are no more than a refined form of induction) that the conjunction is significant. This gives you an inference ticket to find a succession of intermediate causes that leads to a succession of intermediate effects, thus linking the two prongs of the original conjunction. To go from this, to say that you have now *observed* the causal relation linking any two elements in the chain, is as mistaken as saying that a rule of grammar, such as requiring that a noun phrase in plural must be accompanied by a verb in plural, is itself a new part of speech.[18]

Although I hope that the matter should be reasonably clear by now, let me give one illustration of how Hume can be misread in the light of modern knowledge. Considering the vibration of a string as cause of sound, Hume writes:[19]

> We say . . . that the vibration of this string is the cause of this particular sound. . . . We either mean *that this vibration is followed by this sound, and that all similar vibrations have been followed by similar sounds: Or, that this vibration is followed by this sound, and that upon the appearance of one the mind anticipates the senses, and forms immediately an idea of the other.* We may consider the relation of cause and effect in either of these two lights; but beyond these, we have no idea of it.

Of course, with modern knowledge, we understand only too well "the intervening mechanism which links the vibration of the string to the sound we hear."[20] However: what we mean by the intervening mechanism" is nothing else than bridging the causal chain between the string vibrating and the sound being produced; and, as I have stressed, Hume's constraint is equally valid for any intermediate steps in that chain.

I shall give another example. The impact of the cue makes the billiard ball move. Why is this so? This appears to be very clear: when the cue hits the ball, it causes a forward displacement of the atoms of the ball, which then bounce back causing a *forward* recoil of the ball. (As a cannon moves back when the projectile moves forward.) We now have to explain why the atoms move back after compression. Easy: they are repelled by other atoms of the ball to which they got nearer than normal. However: why are they repelled? Easy: the electrons of those atoms, being negative, repel each other. Once more, why is this so? Field theory answers that this is the case because electrons emit virtual photons and in doing so repel each other (as we shall see in chapter 11). However: why do electrons emit virtual photons? And so on and on. Notice that, if we had stopped the constant querying at any one step, we would have been left with an *explanation* of a conjunction. But the moment we dig deeper a further explanation is required. Causal chains, therefore, do not provide a contradiction of Hume's argument that, ultimately, all that there is is conjunction, unless the chains are arbitrarily cut down at whatever step our level of ignorance regards as an *explanation*.

We must still come to what will be perhaps the most important part of this chapter, that is how to answer Bertrand Russell's challenge as stated in the epigraph. Before we do this, it will be useful to discuss one more technical point, which is the concept of disposition.

DISPOSITIONS

The problem with causal statements is that they are in most cases *narratives* of the form: *A* has been observed and *B* has followed. Insistence on this format can lead to nonsense, as in the example of the hapless metallurgist; we have also seen that causes can be nonevents, which obviously do not easily lend themselves to the narrative form. Attempts to cure these problems by taking recourse to elaborate distinctions between causes and conditions are to be firmly avoided as sources of unmitigated confusion.[21] Dispositional statements are extremely useful in the examples just mentioned. Our metallurgist could have said: "Three out of 112 wheels had a *disposition* to crack, this disposition being revealed by the accompanying microphotographs, which show in each of the three cases the existence of microcavities." Counsel for the youth accused of manslaughter could have pleaded: "This was an accident waiting to happen, because the skull of the deceased had a *disposition* to break under a very weak blow."

Dispositional statements are very important in physics, where they have the virtue of saying what is necessary without a multiplicity of accessory concepts. It is very interesting to notice that the same subject may oscillate between a causal and a dispositional description from one generation to another. Gravity is perhaps the best example. Aristotle's theory of gravitation was both dispositional and wrong, but the logic of it was exquisite. For Aristotle there were four material elements (Earth, Water, Air, Fire) each of which had its *proper place* in one of four concentric spheres, with the Earth at their center. In our present terminology, we would say, following Aristotle, that when any element is displaced from its proper sphere, it has a disposition to return to it. So: why does a stone fall? Very simply because its substance belongs to the Earth sphere, and it has a disposition to return to it.[22]

Notice how well this approach allows us to avoid narratives and to deal with nonevents. I can say: "If I were to release this stone, it would fall because it has a disposition to return to the Earth sphere." Thus, the link between dispositions and conditional subjunctives and therefore counterfactuals is, I hope, clear. Wrong as Aristotle's dispositional theory was, it was philosophically much more satisfactory than Newton's theory of gravitation, which was causal. This was the first good theory after Aristotle and, as we all know, it fit an enormous range of facts; but it presented a number of philosophical worries which, rightly and fortunately, did not hamper its success. The first problem with Newton was that he had to introduce a *cause* for the acceleration (that is the change of velocity from zero to a nonnull one) of the stone when released. This *cause* was the gravitational force, which is not an observable (even when you make an experiment to measure it, as between two heavy spheres, say, all that you observe is the *motion* of some pointer, nothing more). Second, since this force is exerted on the stone by Earth, which may be at a considerable distance from it, this *cause* must act at a distance. If you accept this, then you can say that the gravitational force exerted by Earth on the stone is the cause of its falling when released.

Newton's theory was one of the most influential events in the whole history of human thought: until then the *cause* of planetary motion planets had to be a supernatural agency, whereas now it was the gravitational pull of the Sun. The medieval man was finally dead and buried, and modern man was born. Theology would never be the same: the thinking man (or indeed woman) was truly very worried. In 1715, the sixty-nine-year-old Gottfried Willhelm Leibniz (Newton was then seventy-three) wrote to Caroline, princess of Wales, complaining about the possible decline of religion in Britain caused by Newton (oh for the days when princesses of Wales received such letters!), and the famous Leibniz–Clarke correspondence ensued in the last year of the German philosopher's life. (It was not so much Newton's age—he had another dozen years to live—but rather his retiring disposition that made him operate through Dr. Clarke's mouthpiece.) Newton himself was not very happy about action at a distance, but it worked. Leibniz, on the other hand, abhorred it, and he called it "a chimerical thing, a scholastic occult quality."[23] It is not necessary for us to get into Leibniz's arguments, which in fact were not very sound, but his intuition was right, in the sense that the more time passes the less scientists like actions at a distance, although we shall see later that quantum mechanics poses the most dramatic problem of all time in this respect.

More than two hundred years after Newton, Einstein brought us back to a dispositional theory of gravitation, thus recovering the neatness of Aristotle's approach. Einstein introduced two great changes to our concept of space and time. First, the three space dimensions and the time dimension must all be put together in a four-dimensional space, which is called *spacetime*. Second, whereas space had before been nothing more than the stage on which objects move, Einstein's spacetime is so profoundly tied up with the objects in the world that without them, it would not exist at all. Its very existence and shape depends on the masses that are in it, and the bigger the masses in a region of spacetime the more *curved* spacetime is. So, when you release a stone near the surface of Earth, where the mass of Earth makes space curved, the stone follows a curved trajectory in this space, very much as a stone would fall down along the track of a roller coaster, if released at the top. Thus, material objects have a disposition to follow certain lines in the curved spacetime, which is what gravity is. No need to introduce phantom forces acting at a distance: they are replaced by the curvature of space (about which more in chapter 11).

One final remark before I close down this section. You will remember how difficult it was to get a clear picture of what we mean when we enunciate a simple property such as *Sugar is soluble*. Much of the trouble is that, when doing this, one cannot use a narrative approach (such as *This lump of sugar has dissolved in this glass of water*), but one needs a conditional subjunctive (such as *If I were to place this lump of sugar in a glass of water it would dissolve*), or even a counterfactual. It is a lot simpler to say *Sugar has a disposition to dissolve in water*. Not a great triumph, but a modest improvement.[24]

THE POACHED EGG AT LAST:
HUME ON CUSTOM OR HABIT

The causal conjunction of *A* followed by *B* does not entail a (universally) *necessary* relation. Neither can anything such as the *action* of *A* on *B* be observed. (Counterintuitive as this statement is, I hope that by now the reader is coming to terms with it.) Thus, all we have when we say that *A* is the cause of *B* is a repeated conjunction of the two, the repetition of which, however well established in the past, does not guarantee its future. So, a rational man intent on tasting oysters and not wanting to spend much, goes to a McDonald's for a hamburger: lunacy is just as rational as sanity, and there is nothing except the vote of the majority to discriminate against it, as Bertrand Russell lamented.

We really have two problems here: one is to describe what rationality is and the other to provide some foundations for it. The first task is almost anthropological, which is the line that Hume took, and in his usual way, he did not spare us the shock. Read what he says:

> For after a frequent repetition, I find, that upon the appearance of one of the objects, the mind is *determin'd* by custom to consider its usual attendant, and to consider it in a stronger light upon account of its relation to the first object.

This seems to be a very flimsy basis for rationality, as expressed through causal statements and, in particular, it appears to sustain the objection that Russell states, that rational behavior can only be supported by, so to speak, a majority vote based on custom.

Let us set to work to try to understand what Hume is about. First, the word *custom* can be understood either as applying to the individual or to the community. If Hume meant the latter, the referendum approach to causality would have been left open, but it is clear that for Hume, custom (or habit, as he also uses, more frequently in the *Enquiries* than in the *Treatise*) is some feature of the mind, *not* of the community. So, one problem less. We are still affected by the unpleasant feeling that the whole of rationality hangs from the flimsy thread of custom, whatever that be. People who abandon Hume at this point are then left with the impression that his skeptical analysis, however impeccable it might be, leaves us stuck in a position from which there is no moving forward.

Move forward, however, we must because Russell's worry is very serious indeed: but it is Hume himself who shows us the way, since his ideas were far less comprehensively skeptical than it might appear. For Hume intuited that his custom or habit was not just one more custom, like wearing gold earrings or playing golf once a week. Not at all, it was something very specific. Admittedly, he was cautious about this in the

Treatise, although even there he says: "Nay, habit is one of the principles of nature, and derives all its force from that origin."[25] It is in the *Enquiries*, however, that Hume greatly expands on this theme; and I shall quote him at some length, because his remarks will lead us to a resolution of our problems that will play a major part in the rest of this book. So, let us have a careful look at the following three quotations from the *Enquiries*.

> Custom, then, is the great guide of human life. It is that principle alone which renders our experience useful to us, and makes us expect, for the future, a similar train of events with those which have appeared in the past.

> Here, then, is a kind of pre-established harmony between the course of nature and the succession of our ideas; and though the powers and forces, by which the former is governed, be wholly unknown to us; yet our thoughts and conceptions have still, we find, gone in the same train with the other works of nature. Custom is that principle, by which this correspondence has been effected; so *necessary to the subsistence of our species*. [My italics.]

> As nature has taught us the use of our limbs, without giving us the knowledge of the muscles and nerves, by which they are actuated; so has she implanted in us an instinct, which carries forward the thought in a correspondent course to that which she has established among external objects; though we are ignorant of those powers and forces, on which this regular course and succession of objects totally depends.[26]

It is evident from these readings that the *custom* that Hume had in mind was something fundamental to human nature, as a quality that allows human beings to be in harmony with their natural environment. Remember that Hume was writing this in the 1740s (when he was still hardly thirty): he did not have the tools by which these ideas could be developed any further, for it is clear that he is approaching the problem almost as a naturalist; and it was only one hundred years later that humanity acquired the tools by which Hume's extraordinary intuition would make sense. So, in order to complete Hume's sketchy program, we must take stock of what we now know about the world and us.

A BRIEF HISTORY OF THE WORLD

Not many people now believe in Genesis in the literal sense. On the other hand, though cosmologists do not yet have a picture of the beginnings of the universe that can be considered well grounded in all its aspects, they have made enormous progress in the last twenty or thirty years. Such a picture as it emerges must of necessity be speculative, but the fact that this picture leads to an understanding, for instance, of the relative abundance of the chemical elements, gives confidence to the general outline of the theories on which it is based. We do not need to get into details, which are neither necessary for us, nor in any case entirely reliable, but a rough picture can be constructed as follows. We have already said that relativity theory completely changed our idea of space and time. Before it, space was like an empty stage, into which objects can be placed that then evolve in time: we could even imagine that this empty stage has a clock, with reference to which the evolving of the objects later placed on the stage is described. Relativity says, on the contrary, that without objects, neither space nor time can exist: so before the actors come, there can be neither stage, nor waiting. (Saint Augustine (354–430), as Maimonides (1135–1204) later did, had already understood that time did not exist before creation.) If we have no objects, that is, no matter, we do not have an empty stage: we have the *vacuum*, but because space and time are not meaningful without matter, the vacuum has neither position nor duration.

The vacuum is a very important concept in modern physics and, contrary to what one might expect, it is curiously complex: we must imagine it as full of particles or blobs of energy that are constantly created but that perfectly annihilate one another, because they always come in pairs of particles and antiparticles, thus keeping the balance of nothingness that must characterize the vacuum. As follows from this picture, the vacuum is, in a way, always on the boil but always, unless there is some external perturbation that hits it, in such perfect balance that there are no particles, no energy, and no spacetime. This perfect balancing act, however, can fail by a well-known physical process, called a *fluctuation* or a *singularity*. When this fluctuation occurred, the fact that, say, two particles in the vacuum failed to annihilate each other, caused a *real* particle to be created, whereupon a huge and immensely powerful reaction followed, which in a minuscule fraction of a microsecond created a plasma of particles. Correspondingly, because it is the masses that fashion spacetime, spacetime was created. This is roughly what cosmologists describe as the *big bang*. It is possible that in the initial very small fraction of a microsecond the newborn particles were not quite what we now have, or that spacetime and the laws of physics were not the same as now. But, in a way, the die was cast in that minuscule period of time, because from what happened then, the world evolved to what we now have; and if things had gone differently at that very early stage, an entirely different universe might have emerged. Cosmologists can even give a reasonably informed guess as to the date of the big bang: say 10 to 15 billion years ago. (I shall always define a billion as a thousand millions.) You might not think that this is very precise, but it is in fact amazing that one can guess at it at all.[27]

A few seconds after the big bang, the universe began to do its job and, out of the hot mass of primeval matter, particles and eventually stars and galaxies emerged. This process took a very long time indeed: about two–thirds of the total life of the universe passed before the solar system was formed, some 5 billion years ago. In comparison, life was an eager newcomer after this: only 2 billion years after the solar system was formed, life appeared on Earth, some 3 billion years ago. These times are so long that, to make them more meaningful, I will measure them also in human generations, taking one generation as 25 years. Thus, life began 3 billion years or 12 million generations ago, and the first upright humanoid species arrived very much later, about 4 million years ago (160000 generations). This species was *Australopithecus*, whose brain size was still only 450 cc. The first *Homo sapiens* (*Neanderthal*) appeared much later, half a million years ago (20000 generations) with a huge brain size, 1500 cc. This means that in only 3.5 million years (140000 generations) the brain size was tripled, the biggest change in a trait recorded for any species, as pointed out by biologist J. B. S. Haldane. Language started only some 100000-50000 years (say 3000 generations) ago, and our own species, *Homo sapiens sapiens*, appeared 50000 years or 2000 generations ago (*Cro-Magnon*). So, although we are only 2000 generations away from our first direct ancestors, we are 160000 generations away from the *Australopithecus*.

In popular parlance, Neanderthal is the ultimate politically incorrect individual, yet their brain was bigger than ours. In fact, the vital statistics of *Homo sapiens sapiens* are: brain size 1400 cc, brain area 2200 cm^2 (chimpanzees 500 cm^2), brain cortex 3 mm thick, and 30 billion neurons (chimpanzees 6).[28] So far, I have given the major facts of the evolution of the human species, but if we want to make sense of this evolution, we need a new principle. This is the *principle of natural selection*.

EVOLUTION AND THE PRINCIPLE OF NATURAL SELECTION

There are, of course, people who still deny that the human species is the result of an evolutionary process that started with primates and finally led to *Homo sapiens sapiens*. If people want to stick to this denial, I prefer to remain silent because I am aware that diversity of belief is inherent in humankind and that no amount of argument, however rational, can change this. (To believe that this is possible at present, in the face of overwhelming evidence to the contrary, would indeed be irrational!)[29] So, I am afraid that readers who entertain such beliefs will not find much for them here. For most people, however, the theory of evolution of Charles Robert Darwin (1809–1882) is perhaps the greatest intellectual event in the history of humanity. Newton, as I had mentioned, had liberated us from the need to postulate that the movement of the planets was the result of some divine agency, but Darwin's theory went even deeper because it removed at a stroke the need for postulating that the creation of humankind was the result of an instantaneous (or even gradual) divine intervention. Even more, it soon led to a picture of a self-organizing nature, until then impossible to entertain.

A statement such as this can be misunderstood as claiming that the combined efforts of Newton and Darwin killed the deity stone-dead. It is just as easy to argue, on the contrary, that what they did was to remove false attributes from the deity, permitting humans to contemplate the latter, if they wish to do so, in a purer and better light.[30]

The idea of evolution preceded Darwin for quite some time, and the person who did most to push forward this concept was Jean-Baptiste Lamarck (1744–1829), although the interpretation that he propounded was wrong. Perhaps not accidentally, the contrast of a dispositional versus a causal description, which we noticed in going from Aristotle to Newton, was repeated in the progression from Lamarck to Darwin. Lamarck believed that species have a *disposition* to better themselves. If the environment in which the species lives requires an organ to be exercised and thus developed (like the short neck of an ancestor giraffe, stretching high to reach the desirable leaves of tall trees), then this improvement is transmitted to the next generation and so on. A very attractive concept on humanitarian grounds (hard work rewarded!), which it is no wonder was often embraced by Marxists with true but misguided devotion. It is not for me to produce evidence against Lamarckianism, which is overwhelming. For the benefit of the reader who wants a quick answer, however: we now know that inheritance is done through the particles of the DNA that are called *genes*, and there is no evidence whatsoever that acquired traits in an individual can modify its genes (a point which can be settled experimentally quite easily with species such as insects where adaptations of one individual to its environment can be controlled).

I shall now briefly describe Darwinism, for brevity in its post-Darwinian form.[31] There are four basic ideas in this theory: (1) Inheritance of traits is done through the genes of the parents. (2) Owing to external, normally random effects, like cosmic radiation, chemical stimuli and the like, genes can mutate, that is, can alter the message that they transmit to the descendant. (Notice that it is only the production of the mutation that is random: the mutation itself will not result in general in a random message, but in a randomly modified version of the original one.) (3) Different individuals of the same species, but belonging to different mutations, compete for food, survival from predators, and opportunities for reproduction, a process called the *struggle for life*. (4) Most of the mutations produced are neutral or negative from the point of view of the position in the struggle for life of the individual that inherits them. Also, in the population of a given species at any one time, there will be a distribution of individuals with the standard genes of the previous generation and of individuals with mutated genes. The most important postulate of Darwinism concerns this mixed population, and it states that its evolutionary progress is guided by a driving force embodied in the *principle of natural selection*. This principle (see p. 98), favors the individuals better adapted to their environment, marginally increasing their chances of survival and successful reproduction; and therefore of passing on to the next generation through their genes the trait acquired through mutation. Even very small advantages caused by a particular mutated gene accumulate after many generations, thus increasing the proportion of individuals in the species with that gene.

BACK TO HUME'S CUSTOM OR HABIT

We can now understand the important quotations (see notes 25 and 26) of Hume on custom, as one of the "principles of nature," deriving its force from it; as "that principle alone which renders our experience useful to us"; as providing "a kind of pre-established harmony between the course of nature and the succession of our ideas . . . so necessary to the subsistence of our species"; and as an instinct "implanted in us" by nature.

If this is not saying that the *custom* of recognizing significant conjunctions as causal relations is not a trait acquired by evolution, what is it? In fact, one hundred years before the *Origin of Species* the language is surprisingly Darwinian. Let us amplify this, in the light of our little history of the world and our account of evolution. The facts that we must bear in mind are the following: about three-quarters of the total age of the universe passed without any form of life on our planet. As regards the appearance of the first humanoid species, this happened in the last, 0.03% of the total life of the universe! Not only is our species new: it is so new that, even starting with our furthest ancestor species, the length of our lifetime on the planet is insignificant!

The reason why this is important is that, according to the big bang picture, the universe was evolving and changing very rapidly for billions of years. Even after the formation of the solar system, things were very changeable until the first primitive forms of life appeared. Then two things happened more or less simultaneously (talking of course within a very coarse-grained, geological, timescale): species appeared with some form of a central nervous system, and the order of events on Earth reached some sort of steady state, with some regularities in the seasons and in the general patterns of life. Obviously, an ability to react on a causal basis was a very useful trait for survival: if, to paraphrase Russell, a gazelle ancestor had recognized a cheetah ancestor some times as such, and some times as a poached egg, its chances of survival in the struggle for life would have been very limited indeed.

It was thus very important to develop, through evolution, organs that allowed the species to learn *through repeated conjunction*. (This is why Hume's insistence on the latter is important if not premonitory.) The brain was such an organ and humankind the species that made the most of this possibility. Remember that, in an unprecedented evolutionary record, the size of the brain was tripled in only 140000 generations until Neanderthal man appeared.[32] Increasing the size of the cranial cavity was not all gain, however: to allow the human fetus to be viable, it had to be born at a comparatively immature stage, otherwise the head could not have gone through a pelvis of reasonable size. Thus, the human baby takes about a year to reach the stage of development of a newborn chimpanzee. The evolutionary advantage of a large brain must therefore have been enormous to compensate for this impediment.[33] How was this organ adapted at exploiting repeated conjunctions?

The crucial organization in the brain is the network of neurons in the cerebral cortex, of which there are some 100 billion. Each *neuron* is made up of a more or less globular body, with a long filament or *axon* stemming out of one end and thinner filaments, like the thin roots of a tree, called *dendrites*, coming out of the other. The latter communicate with the axons of other neurons through a gap called the *synapse*, where electrochemical processes provide communication from one neuron to the next. This way, an axon can end in some 10000 synapses, and one neuron is linked with some 20000 other neurons on average. Such a structure is wonderfully suited to respond to repeated conjunction. What happens, roughly, is that when the system experiences a conjunction (say tiger seen, tiger attacking the observer), a synapse bridge is triggered on, and, the more such events are repeated, the more the given synapse is reinforced (through increased concentration of chemical neurotransmitters). This indicates that the conjunction has been experienced as significant, at the same time making it easier for that conjunction to be recognized and acted upon. Moreover, repeated stimuli produce a rapid growth of axons and dendrites in the brain cortex, which thus becomes ready to respond to further similar stimuli.

Of course, this picture may be oversimplified, and it might not be any more than a model merely *representing* some of the brain features, but it provides the basic properties of a learning system, that is of a system that learns by experience, as computer simulations (appropriately called *neural networks*) prove. Two things, however, are needed for this to work. One is that we require abundant repetition of stable stimuli, that is, of stimuli that appear similar and are in fact similar (it would not do if a high proportion of the tigers spotted had been philanthropic ones).[34] The other is that the neural network (or at least a similarly acting structure) must be there to take advantage of this. We appear to be in a chicken-and-egg situation, until we realize that the principle of evolution tells us that the neural network would not have evolved if the stable stimuli had not been there to make such a structure fruitful in the struggle for life.[35] This type of situation will lead us to the anthropic argument.

THE ANTHROPIC ARGUMENT (ANTHROPIC PRINCIPLE)

What we have achieved so far is, truly enough, nothing more than getting involved in a circular argument: without a steady state in the world that permits repeated and stable stimuli, the brain's neural network would not have evolved (since the advantage of recognizing tigers as ferocious disappears if half of them are as gentle as lambs), and without a neural network, humankind would not have been able to manipulate significant conjunctions of cause and effect. A pure philosopher might deprecate such a circular argument, or even consider it so defective as to throw it away.[36] A natural scientist, instead, would realize that such circularity is an unavoidable condition of any rational description of nature. The world as we know it is self-referential, which means that the big bang that created it was such that, in a very short time, laws of nature emerged which eventually permitted the existence of at least some regions of spacetime sufficiently stable so as to allow for, not only the creation of life, but also of intelligent life; that is, of organisms which, although being part of nature, are also able to describe it and speculate about it. But these organisms, *Homo sapiens sapiens*, being themselves part of nature and thus contingent, cannot be expected to demonstrate the truth of any statements they make by standing out of nature in order to find a body of necessary truths.[37] All that we can aim at is establishing "a kind of pre-established harmony between the course of nature and the succession of our ideas," to repeat our previous quotation from Hume. In order to study this harmony between nature and our minds, we must use circular arguments which, far from being defective, are just about the best we can do. So, they are virtuous rather than vicious circles, and we shall see over and over again that they appear whenever we cast our gaze over different aspects of nature.

I must come clean and admit that I have assumed in the above that intelligence cannot exist without the ability to learn, the latter being essentially the recognition of significant conjunctions and the development of the ability to manipulate them. I find it impossible to try to imagine, even on a science-fiction basis, an intelligent organism that has been exposed to totally contradictory stimuli throughout its life.[38] The problem is, of course, that a major test of what we call intelligent behavior is the ability to predict with some degree of confidence the course of events; and if the events themselves obey no pattern whatsoever, then it is difficult to understand what intelligent behavior might mean.

Having said all this, we can now construct a virtuous circle in order to answer the ancestral question: why is it that things are as they are? Our answer will be: because if they were otherwise no intelligent beings would exist to ask that question. In other words, we must accept that physical constants and the laws of nature are such that they have created an environment in which, by natural selection, it was possible for intelligent beings to emerge: if those conditions had been different, the result would have been a non-self-referential universe. This is one form of the *anthropic argument*, more often called the *anthropic principle*.[39]

Although cosmologists engage in a very specific use of the anthropic principle, I shall loosely bunch together, in what remains of the book, a number of concepts which I shall refer to as the *anthropic principle* and which I shall now summarize. The basic assumption is that, when intelligence appeared, nature was sufficiently stable to permit neural networks to emerge, as being better adapted for survival from the point of view of the principle of evolution. These neural networks are able to recognize as significant regularities that, if perhaps not an entirely objective part of nature itself, are certainly constructed through the interaction of the latter with the perceptual system emerging. For this process to have taken place, certain conditions of stability in the laws of nature are required which, if not fulfilled, would have precluded the emergence of intelligence and thus the self-referential character of the universe.

The reader should notice that the approach used in stating the anthropic argument, insofar as it allows us to dispense with imagined necessities, is similar to the one that I have used in enunciating the principle of natural selection, also based on the idea that there are self-contained and self-organizing systems, for which necessities merely mean the rules whereby such systems are organized. This brings us to the need of having another look at the concept of necessity.

NECESSITY AGAIN

The question of necessity in dealing with our description of nature is not one that can be swept under the carpet because, in the last resort, the picture of nature that anyone constructs depends on the way in which one chooses to provide an answer to this problem. And, alas, the word *necessity* is one of the bugbears of philosophy, largely because its use easily leads into abhorrent recurrences. I shall try to illustrate this by means of a parable. Every visitor to the *Isle of Alba* is told that it is *necessary* to drive on the left in that country. Should the visitor ask why this is so, he or she may be answered that this is required by the Road Traffic Act 1938 or some such statute. If the visitor is philosophically minded, he or she might ask: "Why is obedience to this act *necessary*?" And this query might recur for different levels of the institutional hierarchy in the island, until a terminus is provided for the chain, by the assertion that the *necessary* power of the government is, in Alba, legitimized by its divine origin. End of conversation.

The anthropic principle is, in a way, an alternative conversation stopper, which brings me to the second part of my parable. The island of *Barataria* is populated by eminently sensible people. When motorcars first appeared on the island, people drove on either side of the road, and they drove carefully and slowly in order to minimize accidents. One day, a queue of several big trucks was accidentally formed, all driving on the left because of difficulties in overtaking; private motorists followed the queue behind, for the same reason. After a few hours, vehicles in both directions traveled on their left of the road and, this being one of the few roads on the island, all traffic on the island at the end of the day was on the left, the islanders thus finding that they had less worries about accidents. Having observed this, they still followed the same rule the next day, and soon the habit of left-hand circulation was established on the island.

If a visitor from Alba asks an islander *why* they drive on the left, he will be told: "This is the way we drive here." End of conversation. But the visitor then writes home to a friend: "When you come to Barataria with your car, please remember that it is necessary to drive on the left." The islanders are not interested, like the visitor, about the question *why*? Their main concern is the question *how*? The visitor, fortunately, is philosophically minded, and he soon realizes that his multilayered idea of necessity is not really significant in Barataria, and in order to tie up with his old habits of thought and old vocabulary, he coins the phrase *empirical necessity* to describe the regulation of traffic on the island. This is a necessity that is not a logical (universal) necessity and that is not justifiable by anything except praxis in a self-organizing structure (as Barataria is).

So, to summarize, in Alba, we have an organization, but this is hierarchical, each layer of the hierarchy acquiring legitimacy from the layer above. A recurrence thus arises that is terminated by invoking an entity external to and above the top hierarchy of the island, the existence of which requires a *metaphysical* hypothesis. Barataria, instead, is a self-contained, self-organizing system, in which nothing can be necessary except as a description of how the system is organized. Thus we have this concept of *empirical necessity*, which depends on the history (see chapter 3, note 19) chosen for the system: in a different history, for instance, the bunch of trucks could have stood on the right and the road-traffic rule would have been different. I hope that it is clear that what I have called an *empirical necessity* is, by my own definitions, *contingent*.[40] Although this empirical necessity may be adduced from observation of the self–organizing system, it is nevertheless not an observable within the system (for all you know all traffic in Barataria will change direction on the morrow) and has to be postulated as a *meta-physical* hypothesis.

Notice that the *metaphysical* hypothesis constructed in Alba entails the postulation of an entity existing *outside* the island, whereas the *meta-physical* hypothesis in Barataria does not depend on the existence of any external reality. This hypothesis is a *normative* principle because it establishes rules by which the system is organized, and it is an example of the *normative meta-physical* principles discussed in chapter 1.

It will be fairly clear, I hope, that such necessities as one might wish to introduce in describing nature, within the ideas of the anthropic principle, must be at most contingent *empirical necessities, normative* rather than prescriptive, and internal to the system rather than external to it. Which brings me to the question of necessity in causality. A little logic might help to sharpen the discussion, as I shall do in the next section, and this will allow me to tighten up somewhat the concept of empirical necessity.

THE LOGIC OF CAUSAL STATEMENTS IN SCIENCE

I shall now go back to the use of causality in science. We all know that if I release a stone it will fall, and we also know that the force of gravity always acts on the stone, so: can we say that forces are the causes of motion? This is a statement that every scientist must reject, and this is done by using Hume's requirement that when there is no cause, no effect must follow. (A requirement that, as we have seen, is not entirely compatible with philosophical usage.) In fact, if I observe a billiard ball moving on the baize, no force acts on it (because the force of gravity is canceled by the table) and yet it moves!

Readers must realize that in considering this motion of the billiard ball, I am asking them to forget all about the billiard cue, that is, all about the *origin* of the motion. For centuries, if not millennia, such a demand would have been absurd, while humanity was still in the grips of Aristotelian philosophy. Aristotle (384–322 B.C.E.), faced with the same problem, would have considered it his responsibility to explain the *origin* of the motion, for which the cue was the relevant feature of the situation. And the explanation of such origin was the search for what he called the *first causes*, the billiard cue in our case.

It was Galileo Galilei (1564–1642), and after him Sir Isaac Newton (1642–1727), who taught us that the primary task of science is to *observe and describe events* rather than explain their origins. Science was thus liberated from having to discuss the history of chains of events in order to find first causes: an event like the state of motion of the billiard ball at one instant of time, isolated from the impact of the cue, is what must firmly be in front of our minds for consideration.

Fortunately, Hume's concept of causality permits us to consider an isolated state of motion (that is, disregarding the chain of states going back to its origin), and, because we observe force-free motion, we must conclude that *forces and motions are not causally related*. This statement may appear to be counterintuitive, but it is of fundamental importance. It allowed Newton, as I shall discuss in chapter 6, to discover the correct effects entailed by forces, which would have been impossible if he had been involved in the fruitless Aristotelian search for first causes.

It must be abundantly clear by now that the logic of causality, as used in science, is extremely specific to it, as I shall now try to analyze. If p is the cause of an effect q, we might wish to say that it entails the latter, at least in the sense that, given the cause, the effect must follow. So, we might try to write $p \Rightarrow q$, but the required relation between cause and effect is stronger than that. Remember, in fact, that the entailment described by the material implication is such that it is true just in case the antecedent is false and the consequent is true (see table 5, chapter 3). That is, if the cause is missing and the effect still appears, the implication is true, which is unacceptable for a causal relation. Hume himself required (see page 74) that if the cause does not exist, the effect must not appear, and this is a condition that every scientist would insist upon if the concept of cause is at all used. So, it appears that a causal relation must be expressed by the biconditional $\Leftrightarrow$, which fails if the cause does not exist but the effect appears (see table 5, chapter 3): $p \Leftrightarrow q$.[41] Although this is nearer the bone, I must warn the reader that the logic of causality cannot be expressed sufficiently closely by any of the logical functions. And naturally, any form of necessity entailed by the causal relation is to be understood in the sense of an *empirical necessity*, as explained above and as it shall be put in a more precise manner a little later.[42]

The problem, despite Hume's conditions, is that the relation between cause and effect is, strictly speaking, *neither necessary nor sufficient*. Consider, for instance, the following causal statement: "Short circuits in houses cause fires." The cause, short circuits, is not necessary because the fire can be caused by other agents such as lightning. Neither is it sufficient because short circuits in nonflammable stone and cement houses do not cause fires.

Was then Hume wrong? This conundrum exercises most philosophers, whereas scientists sail through it with complete abandon. To avoid the difficulty, rather than writing "if and only if p, then q," as we have done in our logical notation, we should say "if and only if p, and *all other conditions being equal*, then q." That is, a short circuit can be considered the cause of fire if no changes in what scientists would call the experimental conditions are permitted. (The house must be of a certain type, and all other external agents must be carefully excluded.) The whole art of the scientist in discovering causal relations consists, in fact, in making sure that there is no possible condition that has been ignored which could interfere with the presumptive relation between cause and effect. This ability is what characterizes a good experimentalist, and, alas, it is not an ability that can be described in logical terms because it depends on intuition and experience as much as on any rational quality.[43] Philosophers, however, do worry a great deal about the *all other conditions being equal* requirement (called *ceteris paribus* in philosophy). These worries are not by any means trivial and show that mere logical notation is not entirely adequate for describing the causal relation.

To summarize a little: in science, the relation between a cause p and its effect q must always be expressed by a biconditional, $p \Leftrightarrow q$, which entails that the effect cannot exist without the cause, but this biconditional *must* always be supplemented by *expert* use of *ceteris paribus* conditions.[44] This is perhaps the most significant feature of the concept of *empirical necessity*, so far left on somewhat general terms.

There is one serious possibility of confusion entailing the use of the word *cause* which scientists need not worry about, but which could perturb the philosophically minded. I hope, in fact, that some of my readers might have already spotted what appears to be a counterexample of my description of the causal relation as requiring: "If no cause then no effect and vice versa." Scientists say "smoking causes cancer," but we all know poor Mr. Bloggs, who was never in contact with tobacco smoke, and nevertheless died of lung cancer at thirty, whereas Ms. Other, who smokes forty a day, still dances the night away in fashionable discotheques at the age of ninety-two. The question here is that, when scientists use causal statements in cases such as this, they do it *statistically*, that is, the referent of such statements is never any single individual but a *sample* of individuals. If samples

A and B are smokers and nonsmokers, respectively, the causal statement that "smoking causes cancer" requires that the proportion of people who acquire cancer in sample A be *significantly higher* (in the statistical sense) than that in sample B. Thus, *statistical causality* properly understood satisfies the conditions that we have required of causal relations.[45]

FACTS AND HOW TO DIGEST THEM

There is a very important point that we must clear up about the treatment so far given in this chapter. We have more or less taken for granted that our sensorial system receives stimuli, and thus bombards our brain with facts from which we eventually learn about nature. The question is this: How do these alleged facts relate to the "world out there"? I have already said in chapter 1 that facts are not the clear-cut things that we often expect them to be, and let me now expand a little on this point.

To start with, we are all familiar with the limitations of our senses. The red color of a tomato cannot be just "out there," because if our optical system were somewhat different and we were sensitive to, say, infrared light, a tomato would not look red at all. We also know that Daltonians perceive colors differently from normal people. There is also a remarkable perceptual phenomenon called *synaesthesia*, which throws even stronger doubts about the straightforward significance of our perceptions. In subjects with this condition, the channels corresponding to one of the senses appear to cross over to those of another. People may hear certain sounds when they see some colors, as Oliver Messiaen famously did, and it is even possible that Bernard Berenson's rather cryptic remarks about the "tactile values" of pictures were due to the same cause.[46] So our first remark is that whatever we observe through our senses is no more than a sort of mapping[47] between the "world out there" and our mind.

Things would not be too bad if this mapping were a decent one, that is, if the mind did not interfere at the point of processing the information received in order to map it. If the mapping were as we have called it *decent*, the mind would have in store maps that are independent of the mind that does the storing. Unfortunately, this is very unlikely to be so: we are like bad geographers who read their instruments and apply some rigorous mathematics to find the correct projection point on the map of the land feature observed, but who then distort the final map in some way to satisfy some political or military preconceptions they have. So our mental maps do not correspond to naked facts, but rather to dressed ones, the dressing being provided by the whole set of ideas or theoretical concepts acquired by the given mind during its previous history.[48]

A very simple demonstration of this situation was given in some old Cambridge experiments, in which the subject had to observe circles at a glancing angle, so that what they perceived were ovals, which they had to compare with those from a printed set. If the observers knew that the circles where in fact pennies or plates, they consistently identified incorrectly the ovals that they saw (which the experimenters could determine from the glancing angle) with rounder ovals: the knowledge that what they were looking at was a circle made them dress up the stimuli with this theoretical preconception. In the history of science, the most famous case where the *facts*, as apparently observed, had to be disregarded, appeared when Copernican theory was propounded. For the majority of the learned people of the early seventeenth century, it was virtually impossible to accept that what was clearly their experience—that the Sun raises in the East, rotates around Earth and then sets in the West—was not a *fact*.[49]

The relation between what we observe and what is "out there" may be far more tenuous than so far described. Plato, in the seventh book of the *Republic*, produced a very good parable for this purpose. Imagine people chained in a cave with their backs to the entrance and facing a blank wall at the back: they see the shades of the objects and people "out there" projected on the blank wall, and they think that that is what "out there" is like, not realizing that they only perceive shades of the *real* things. For Plato, the real things of which we only perceive the shades are *eternal ideas*. So, the idea of a triangle is true because it is part of this ideal world, whereas what we perceive as a triangle is only a reminiscence (or a shadow in his parable) of the real thing. This idea is, of course, very attractive for those who prefer to view nature as having a somewhat dual structure, physical, material, and contingent, but with a nonmaterial supranatural sphere somehow added to it, from which certain things, such as the soul or the mind, partake. (All this will be fully discussed in chapter 14.)

It goes without saying that Hume had no time for the world of Platonic ideas. Nevertheless, the parable of the cave created a serious worry as to the relation between what we perceive and what is perceived, thus leading to a second form of skepticism about our construction of the so–called external world. It was Immanuel Kant (1724–1804), who had been "interrupted" in his "dogmatic slumbers" by Hume, as he said, who combined these two forms of skepticism. Rather than inventing, like Plato, a second and more powerful world to sustain our perceptions, he relegated the "world out there" to be eternally unknown, and he called it the *noumenon* or *the thing-in-itself*.

Stimuli are generated from the noumenon and are picked up by our sensorial system. But this "picking up" is active rather than passive because the sensorial system cannot work without certain theoretical props. Thus, the regularities that we observe may not be "out there" at all, but rather be due to the fact that our mind is like an electronic receiver tuned to pick up signals of a certain type only. Whereas Hume made no assumptions about the nature of our mind (providing, instead a lead as to how some of the mind's functions emerged in the two-way trade with the world of stimuli), Kant postulated that the mind had some preordained functions that he called *categories*, one of which was *causality*. Naturally, like Hume, Kant was not aware of the principle of natural selection and, like Bertrand Russell, he had to provide a justification for rationality: so our minds were born so to speak with mental calipers, the *categories*, which allow us to organize the perception of external stimuli. Space and time do not exist "in themselves" but are only "modes of representation."[50] Thus they are like a stage that we ourselves construct in order to place into it the phenomena that we observe. As we know, this point of view is given the lie by the modern view of space and time given by relativity theory.[51]

TWO GREAT NORMATIVE META-PHYSICAL PRINCIPLES

The concept of causality is one that provides a norm by which to organize experience: it directs our minds to the search of constant significant conjunctions of events; and when these are found, we can distinguish between causes and their effects. We can formulate this normative principle, which I shall call the *principle of causality*, as requiring that *every effect must have its cause, without which the effect cannot appear*. This is of course a very rough statement because it must be complemented with all the caveats which I attached to the use of causal relations. It is nevertheless a useful principle to bear in mind, which I shall relate in chapter 8 to the *principle of sufficient reason*.

Because the relation between cause and effect, as we have seen, is not observable, the principle of causality is not a physical law, and it is an example of what I have called in chapter 1 a *normative meta-physical principle*. Whereas metaphysical principles cannot be grounded in anything like physical experience, we have seen that it was through the accumulated experience of our species, as embodied in the evolutionary process, that the principle of causality became part of the tools by which our minds organize our observations. Thus, although it cannot be validated by experience, this principle is intimately wedded to it, and I accordingly propose for causality a meta-physical rather than a metaphysical status.

I shall now move on toward a second normative meta-physical principle. If you remember the story of Alba and Barataria, their organizations were described in antithetic ways. In Alba, we required a fully metaphysical hypothesis, whereas the structure of Barataria could be described without any reference to anything external to the island. It would be too much, however, to expect that what is metaphysical in Alba becomes at a stroke physical in Barataria. To describe the latter as a self-organizing structure is not to make a statement of anything directly observable, for three reasons. The first and most important is that the organization of the structure requires a continuity which has to be postulated rather than observed (however much it can be grounded on previous experience, as it also happens with inductive processes). The second is that the *driving forces* which determine the direction of change in the structure are, like the relation between cause and effect, not directly observable themselves. Finally, however plausible my description of Barataria as a self-organizing structure is, I could be quite wrong because I can never exclude the possibility of some supernatural agency that acts in an entirely hidden way on the island, so that what appears to be accidental in Barataria is in fact the result of the actions or plans of that agency. Because of these reasons, I describe my statement that Barataria is a self-organizing structure as a meta-physical one.

We are now ready to consider the principle of natural selection which, although it was postulated in order to understand the evolution of living species, has a much wider import. Not only species evolve: theories, words, ideas, customs, political systems, even religions, also evolve; and the principle of natural selection, as applied to our previous history, allows us to understand how it is that things are as they are and not otherwise: natural selection has determined the successful outcome. In its most general form, the principle asserts that nature is a self-organizing system in which the driving force for change is the competition between the structures that emerge in the evolutionary process.

This important principle, however, is not anything that can be directly observed: at no time, for instance, can we determine or measure the driving force for change. Rather, it is a principle that we use in order to organize our experience, in order to satisfy ourselves, as we have done with our description of Barataria, that whatever is there, is there not because of a particular a priori necessity, but because it must have competitively survived the strains imposed by adaptation to the environment, be it a physical or a mental one. Because of this, I consider that the *principle of natural selection*, in asserting that nature is a self-organizing structure, is better described as a meta-physical principle.[52] Notice also that, as we have already indicated, the principle of natural selection is causal rather than dispositional. Even the fact that within this principle, we can refer to a *driving force* for evolution is reminiscent of Newton's causal language.

In conclusion, it is worthwhile recognizing that application of the principle of natural selection to an evolving structure entails a fundamental departure from earlier views. In the classical theological model, change is clearly directed toward a given *end*, so that the driving force for change is taken to be *teleological*. This means that all change is informed by a *purpose*, this being for instance that of creating the *best possible* species; and this purpose can best be described by what Aristotle called the *final cause*. The species or the structure that emerges from an evolutionary process, instead, is not taken to be the best in all possible worlds, but rather the one best adapted to the circumstances prevailing during the changing process. As in Barataria, however, the possibility of *hidden* supernatural intervention can never be ruled out, and a rational use of the principle of evolution must always entail this caveat. The latter, however, does not affect scientific practice, since, when using the principle, it must be accepted that such supernatural intervention as might exist must be hidden and not manifest. Any statement to the contrary is necessarily theological rather than scientific, and in saying this I am trying to establish acceptable boundaries for scientific discourse.

More on meta-physics and metaphysics

Since the distinction between these two concepts is delicate, I shall review it here again, but the reader must be aware that I shall consider only a certain type of metaphysical statement, of which the following is an example: "There is a world of mathematical ideas entirely independent from nature, but which determines the properties of the latter." This is the view of the mathematical Platonists, discussed in chapter 14.

Such a statement entails two things. First, a postulated ontology, namely the existence of the so-called world of mathematical ideas. Second, a concept of a priori or universal necessity, since the rules assumed for this world are deemed to be universally valid, independent of the contingencies of nature as we know it.

The principle of causality, as I have discussed it, may be regarded to be outside nature itself, since it provides the norms or rules which we use in order to describe the latter. Principles such as this, however, differ from metaphysical principles in two ways. They do not postulate any ontology extraneous to nature, and they do not entail universal but empirical necessities; that is, they are themselves subject to the contingencies of the environment in which they are applied. This is very important: because the principle of causality arises from the interaction of *macroscopic* nature with our rational and cognitive system, it cannot be expected to be *necessarily* valid for microscopic phenomena, which we shall find to be the case. This makes it absolutely clear that their validity cannot be based on a universal necessity.

It should be remembered, however, that in negating any ontology extraneous to nature, we are in fact making a metaphysical postulate about nature itself, and however restrained this postulate is, it somehow reflects on the nature of the principles in question, which is the reason why I qualify them as meta-physical.

It could be argued that a principle such as "every event has a cause" could be empirically testable. I shall disregard that this proposal entails the odious problem of induction in order to justify the use of the word "every." Even worse, it would be useless arguing around this difficulty because such a principle is *untrue* when dealing with microscopic bodies, as we shall see later. But there is a far more important problem. If I wear red-tinted spectacles, any statements I make about the colors of objects are open to suspicion. But if I try to make experiments to verify the causal principle stated, I must use my rational and cognitive system, which, as I have argued, has been molded by nature itself to operate in acceptance of the principle. Finally, it is most important, if such an attempt were made, to avoid false *causes*. It has been repeated dozens of times, in illustrating the phenomenon of chaos, that the flapping of a butterfly's wings in the Australian bush can *cause* a hurricane in North Carolina. It must be understood that if the latter event has a *cause* at all, this is given by innumerable initial conditions: a simultaneous sneeze of an Inuit in Newfoundland is just as good as the proverbial butterfly. To try to identify one *cause* is in this case, like in many others, a Byzantine exercise, so that the concept of *cause* in chaotic phenomena is largely irrelevant. As far as science is concerned, the use of the principle of causality is restricted to such cases where experience indicates that the principle can be applied! Not at all the pure clean-cut situation that is philosophically desirable.

CODA

That the relation between cause and effect is not *logically necessary* is a point well made by Hume and generally accepted. Some philosophers, however, either explicitly or implicitly, claim that such a relation is *observable*, thus negating the second prong of Hume's argument. To recognize with Hume, as I have done in this chapter, that the relation between cause and effect is not observable is most important, however, since otherwise we would have no reason whatsoever to inquire into the origin of this relation. If causes were as observable as tables are, we would not need to discuss why we use them; and we would then miss a most important opportunity to examine the way in which our mind has evolved, and the way in which it works.

As regards this second question, how the mind works, Hume's skeptical analysis appeared to have left us with the notion that all that is entailed in causality is the apprehension of a constant conjunction processed by the mind through a habit or custom. Kant attempted to move forward from this apparent cul-de-sac, by proposing that the mind must necessarily operate through certain a priori categories, of undisclosed origin, but regarded as built-in preconditions for rational thinking. His solution was thus entirely metaphysical. Hume, instead, had intuited a far better solution, well within the realm of natural science, and I have attempted to complete his program, by suggesting that the disposition of the mind to use causal statements correlates with physical properties that the brain acquired during the evolution of the species.

The capacity of the mind to apprehend some regular conjunctions as significant must be understood through two steps. First, the history of the species (that is, its *phylogenesis*) favored the emergence of a brain structure (involving *neural networks*) capable of learning, that is of recognizing and reacting to repeated regular stimuli. Genetics determines that this structure is developed mostly after birth, thus permitting the second necessary step, through the development of the individual (that is, *ontogenetically*). It is during this development that the mind learns to recognize significant conjunctions, thus acquiring a disposition to use causal relations.

Having understood this, I hope that the reader will be able to see through the fallacy implicit in some statements that might be adduced against Hume's analysis. A psychologist might find that his subjects, when exposed to certain conjunctions of events on a computer screen *observe* causal relations. A philosopher might find a case where a great scientific discovery was made after a single instance was observed, thus negating the importance of repeated conjunction. What is wrong with these two arguments? The answer is very simple: in both cases the observers are adults, and thus people who have already trained minds, trained by repeated conjunctions so as to acquire a disposition to use causal statements in the way that I have discussed. It is only if one could have observers with perfectly innocent minds, never previously exposed to repeated conjunctions, that experiments on the possibility of the perception of causality would have any significance. Philosophers who claim that a single instance of conjunction may demonstrate a causal connection are looking at a problem which is not the one we have been concerned with. If you say that you understand a spectrograph because it produces spectra, you are using the word *understand* in a very anemic way. We are in business if we open the instrument, find a diffraction grating inside, and work out how it functions. Our mind is, in fact, that diffraction grating that registers and manipulates the regularities of nature; and it was the job of this chapter to understand this.

If we are prepared to accept that an understanding of what we might call the *mind-in-nature* is firmly tied up both to its phylogenesis and its ontogenesis, we must expect trouble when applying such concepts as causality in the study of physical phenomena because we have as a species acquired these concepts through the interaction of macroscopic stimuli with the evolutionary process. Our mind, however, has never had to process microscopic stimuli, such as might result from the direct perception of electrons and other elementary particles. Thus, whereas causality and other related concepts are well-adapted for the description of macroscopic phenomena, we cannot expect that their extension for the understanding of the microscopic world be painless.

Understanding is, in fact, one of the activities helped by the concept of causality. At least in the macroscopic world (in principle, we should always qualify thus what we have so far learned), causal action at a distance is to be regarded with great caution: instances where such causal relations appear are to be regarded as inference tickets for causal explanations, which are then constructed by determining a chain of contiguous elements that causally connect the original cause with the final effect. But the question remains whether *true*, that is not mediated, action at a distance can be countenanced within a physical theory. This does not stop scientists from time to time proposing physical theories where action at a distance is admitted, like Newton pragmatically did with his theory of gravitation, but experience so far shows that when theories become more sophisticated, such actions at a distance are usually eliminated. We shall see later how this is done and also how this, at least at present, cannot always be done.

Although causal connections can help understanding by the construction of causal chains, it must be remembered that science uses causality, in the last analysis, for one and only one purpose: as an *instrument of prediction*. The reader might reflect that such a purpose is precisely the one that makes the ability to recognize causal relations valuable from the evolutionary point of view, as Hume had amazingly intuited. Everything else about causality, such as the seductive ideas of powers and capacities, is entirely discardable in science. Scientists might use them, of course, in order to construct mental models to organize their thoughts, but, whereas a predictive tool will last for ever within some range of accuracy, any other such props as I have mentioned can be jettisoned, and are in fact jettisoned, at a moment's notice. Science is thus not as ephemeral as some people like to think: scientists may throw away unnecessary baggage, but a good well-grounded causal relation that serves as an instrument of prediction in appropriate circumstances is too good to be discarded, as long as those circumstances are of any interest.

In establishing causal connections, science is ruthlessly single-minded, and thus the reading of nature that it effects is not necessarily the same as that of the man in the street, even when the latter happens to be a philosopher. When the scientist sees a billiard ball moving on the table, she does not ask the usual questions that might come to mind: Why is it moving? What is it moving for? Those are the questions that snared Aristotle in his first and final causes respectively. All that the scientist tries to do is predict the position of the billiard ball at some later instant, as we shall see in chapter 6. This does not mean that Aristotle's questions are meaningless. Obviously, they are not. But they are not questions that have or require scientific answers, and this is so crucial that medieval man gave way to modern man when this point was understood.

An important point to remember is that such sort of necessity as might be entailed in a causal statement cannot be a logical one, so that I have introduced for it the notion of *empirical necessity*. We have seen, on the other hand, that Hume's definitions, which appear to regard the causal relation as necessary and sufficient (at least within the *empirical* proviso), cannot be taken to be valid in general; this is nevertheless the way in which causal relations are cautiously used in science, very careful attention having to be paid to what we have called the *ceteris paribus* conditions, including such statements as the timing of events.

It is important to recognize that our work is not yet complete. Whereas we understand on general lines how causal statements come to be made, we do not yet know how they come to be validated. The concept of empirical necessity thus appears in this chapter mainly as a verbal shorthand, but it will be grounded in chapter 7 in terms of the concept of *entrenchment*. This will be central to many of the arguments used in this book, especially when we discuss laws of nature in chapter 8, when the reader will get a clearer picture of the way in which causal statements (and others) are validated. It is also important to recognize that I have used the fact that the mind has a disposition to use causal statements, in order to endow with some objectivity the properties of regularity observed in nature "out there." (This is so, since the anthropic argument suggests that without such regularities our neural network would not have developed as it has done and, therefore, that the mind as we know it would not exist.) Since induction entails the use of a *regularity of nature* rule, its use is now understandable, providing an insight into Hume's original problem.

Newton's theory of classical mechanics is our archetype of a causal theory, but there is an important feature of it that we have so far disregarded. If we think of causality as requiring that the same causes must produce the same effects, we must understand that duration must be considered among the effects: the time a given stone takes to fall to the ground from the top of the tower of Pisa must always be the same. This relation between causality and time is very subtle and will be studied in the following chapter.

NOTES

1. See Warnock (1963), p. 59. The reader must appreciate that the word *necessary* is used here in the sense of *noncontingent* or *universally necessary*. (That is, valid in every possible world.) We are not concerned at this stage with the more limited question of necessity and sufficiency as discussed in chapter 3 after table 5. In most cases in this chapter *necessary* will be understood as stated in this note.

2. Bill's drowning is rehearsed by Edgington (1997), p . 432.

3. Russell's dictum appears in Russell (1912), p. 1. Hume himself gave a warning about the misuse of causality: "Nothing is more requisite for a true philosopher, than to restrain the intemperate desire of searching into causes" (Hume 1739, Bk I, Part I, § IV, p. 13). The most useful and complete modern treatment of causality which I have found is Mackie (1980); Ehring (1997) is a very useful introduction. Nancy Cartwright and Rom Harré are among the most vigorous critics of Hume, as exemplified in Cartwright (1983), (1989); Harré et al. (1975), chap. 3.

4. This is made quite clear in the *Treatise*, Hume (1739) Bk I, Part III, § XIV, p. 166: we must repeat to ourselves "*that* the simple view of any two objects or actions, however related, can never give us any idea of power, or of a connexion betwixt them: *that* this idea arises from the repetition of their union: *that* the repetition neither discovers nor causes any thing in the objects, but has an influence only in the mind."

5. Some psychological experiments were made, in which the subjects observed figures colliding on a screen, and under certain conditions inferred causal relations (see Michotte 1946). This does not mean that causal relations were *observed*, but rather that under certain conditions the subjects experienced associations of ideas between what they saw and the concept of causal relation. Such experiments would be meaningful to reveal the perception of causality if the subjects involved had *never* been exposed to repetitive stimuli, as already recognized by Hume. (See note 6.) Unfortunately, as experience with unstimulated children abundantly shows, such subjects would not only be unable to perceive causal relations, but would hardly be capable of performing the tasks usually associated with intelligent behavior.

6. Some readers may still prefer to think that they *perceive* that the motion of the cue is the cause of the motion of the billiard ball. If they do so, they are in unimpeachable company. The French American philosopher Curt John Ducasse (1881–1969) was probably the most outspoken modern advocate of the view that the causal relation is observable (see, for instance, Ducasse 1966), but critics of Hume along this line are as numerous as they are distinguished. (See the treatment of the Buffon clock in Cartwright 1989, p. 92.) This, however, is not a line of thought that I shall entertain for the following reasons.

First, although we must agree that the act of observing is complex, there is no question that when we observe a red patch, some receptor organ has been stimulated, however much this stimulus might be processed by our minds. On the other hand, no one has ever suggested that there is a receptor organ that is stimulated by the perception of causal relations. What we call *observing* requires the hardware of our perceptual system, whereas the apprehension of a causal relation depends on some sort of software; and it is a virtue of Hume's approach that it leads us into examining the significance and even the origin of the latter.

Second, it is inherent to the position that I am trying to deconstruct that the *perception* of causality takes place within a single event, and although this may be of philosophical interest it is not so in science. You can *perceive* that this brick breaks this pane of glass: however, as a difference with Ducasse and Cartwright, we are not in science interested in single instances but, rather, in what licenses us in making *universally quantified* statements such as "Bricks thrown at glass panes are the cause of the latter breaking." Scientifically, we would have to do some experiments in order to sustain such a statement, and we would most certainly observe some cases where the proposed rule fails because the particular glass pane is unbreakable. Thus, our original *perception* of a causal relation in a single event is not necessarily fruitful within a scientific approach. (Cartwright 1989, p. 92 states however that "Ørsted's single experiment was quite sufficient," but we have already seen on p. 67 that the Dane reproduced his experiment many times before he went on print, and, as shown in Altmann 1992, pp. 13–14, the experiment was *repeated* within weeks in every major European capital before its results were accepted. Why Ørsted himself required not *one* experiment but a series of *systematic* experiments will be discussed in chapter 8.)

I am afraid that I have to say something more about bricks. (Ducasse (1968), p. 8, says: "When any philosophically pure-minded person sees a brick strike a window and the window break, he judges that the impact of the brick was the cause of the breaking, *because* he believes that impact to have been the only change which took place in the immediate environment of the window." Why should *one change* (the window breaking) have to be caused by *another*, rather than by nothing at all, for instance? And this is precisely the point that Hume urges on us as requiring examination, and quite rightly so, because, as we shall see later, events in quantum mechanics do happen that are not related to any other possible *change*.

I should like to stress that those who argue for singular instances as evidentiaries of causal relations (see Ducasse 1968; Mackie 1980, p. 135; Cartwright 1989, chap. 3) entirely miss the point of Hume's arguments. For, even if that were true, it would still be us, it would still be our minds, that do this *perceiving*, and Hume's question was, essentially, why has the human mind a disposition to use causal statements at all? To hammer the point down: for the single-event argument to have any value, it would have to refer to a subject who perceives a single event *never before* having experienced repetition of any events. Hume himself saw this very clearly (Hume 1739, Bk I, Part III, § VIII, p. 105): we must consider "that tho' we are suppos'd to have had only one experiment of a particular effect, yet we have many millions to convince us of this principle; *that like objects, plac'd in like circumstances, will always produce like effects.*"

As it some times happens in philosophy, those who find exceptions to Hume do so by addressing a *different* problem than the one that preoccupied him. Hume's concept of causality encapsulates within it the thorny problem of induction, as shown by his insistence on *repeated* conjunction. If you consider single instances instead, you may be philosophically purer, but you throw away the very problem that most concerns us, namely how can universally quantified statements be made. The American philosopher Donald Davidson (1917–) put this very sharply: "Ducasse is right that singular causal statements entail no law; Hume is right that they entail there is a law." (Davidson 1980, p. 160).

There is another reason why the *perception* of causal relations must be taken with a large pinch of salt, and it is that nonscientific cultures tend to be more prone to causal pronouncements, which they are often persuaded are based on objective causal perception. Anyone with a little knowledge of the Neapolitan hinterland is bound to have met people who *experience* the effects of the evil eye. The Azande rationalized so profoundly the causal relation they postulated in witchcraft, that they believed that the latter is a substance housed in the witch's body, and they *perceived* flashes of light indicating its passage toward the victim (Evans-Pritchard 1937, pp. 33–34). Cultural differences must of course be accepted in their proper context, but a theory of causality in modern science that does not exclude such cases would be totally defective. It is thus prudent to be skeptic as to the possibility of unmediated *perception* of causal relations, as good old Hume was.

7. Hume (1748), Sect. V, Part I, 35, p. 42: it is not "reasonable to conclude, merely

because one event, in one instance, precedes another, that therefore one is the cause, the other the effect. Their conjunction may be arbitrary and casual." There is a *propensity* to infer the existence of one object from the appearance of the other. Despite this disclaimer by Hume, the requirement of repeated conjunction of a causal relation has often been criticized on the grounds that, for instance, we would conclude that "day is the cause of night." The Finnish philosopher Georg Henrik von Wright (1916–), who for a time succeeded Wittgenstein at Cambridge, proposed a solution to this conundrum (see von Wright 1974, 1993) by specifying that such conjunctions had to be such that they can be made to happen "at will" (thus requiring laboratory conditions). This solution, however, would still permit the following proposition to be causal: "Life is the cause of death," since, unfortunately, any number of live animals could be killed in experimental conditions.

This type of problem shows another instance where the approach to causality by scientists widely differs from that of the philosophers. The day-night conjunction exercises philosophers because for them the concept of powers or *efficient causes* is still significant. Scientists, in the few cases where they use causality, do not necessarily worry about such concepts. One field where causality is both important and legitimate in science is classical mechanics, which will be discussed in chapter 6, but consider this example. When a billiard ball moves, its *state* at any time, which is carefully defined in that chapter, *determines* its state at any later time. Thus, the *first state* is the *cause* and the *second* is the *effect*. And this relation is valid even if the ball is stationary, in which case the connection between cause and effect is well beyond the concept of an efficient cause. Yet, it is essential in science to treat causality in this way. Thus, all you want of a cause is that the effect will follow, that states will follow states in a predictable way. Whatever other attributes may be assigned to the causal relation, it stands or falls in science by one and only one criterion: that it must be an instrument of prediction; and the more accurate this prediction is, the sounder the corresponding causal statement will be.

Night can only happen on planets exposed to light periodically, as death can only happen to beings previously alive. In either case, this imposes a condition on the accessibility of certain states. (Just as, if the billiard ball is stationary, the only state accessible from it will be a stationary state.) But the predictions involved in either of the two causal statements, "Day cause of night" and "Life cause of death," are very poor, and they are superseded by much better ones. The starting moment of nighttime can be calculated to the nearest second on using as *causes* the position of Earth on its orbit and that of the observer on Earth's surface. And thousands of medical data will predict death far better than the mere presence of life does.

8. When in all instances two objects have always been conjoined Hume says (1748, Sect. VII, Part II, 59, p. 75): "We then call the one object, *Cause*; the other, *Effect*. We suppose that there is some connexion between them; some power in the one, by which it infallibly produces the other." Hume (1739), Book I, Part I, § IV, p.12: "two objects are connected by the relation of cause and effect, when the one produces a motion or any action in the other, but also when it has a power of producing it." See however Hume (1748), Sect. VII, Part II, 60, footnote to p. 77, and Hume (1739), Book I, Part III, § XIV, pp. 157–161, where he strongly criticizes the concepts of power, force, efficacy, connection and so on.

9. Hume (1748), Sect. V, Part I, 36, p. 43: "this propensity is the effect of *Custom*. By employing that word, we pretend not to have given the ultimate reason of such a propensity. We only point out a principle of human nature . . ." "All inferences from experience, therefore, are effects of custom, not of reasoning." The reference to Stuart Hampshire is from Hampshire (1963), p. 4. Even the title of one of his books, *The Natural History of Religion*, confirms Hume's views in his own writings.

10. These are, respectively, Hume (1739), and Hume (1748).

11. Hume (1748), Sect. VII, Part II, 60, p. 76. Notice that the last sentence is not a paraphrase of the first, a question that baffles some people although, in fact, it agrees with the

conditions given in the *Treatise*: Hume (1739), Bk I, Part III, § XV, p. 173 gives rules for the cause-effect relationship. The most important ones are: (1) Contiguity in space and time. (2) The cause must be prior to the effect. (3) There must be a constant union between cause and effect. (4) "The same cause always produces the same effect, and the same effect never arises but from the same cause."

12. See Harré (1964).

13. See Collingwood (1938). Is the fuel a mere condition or is *it* that it has the power, by combusting, to move the engine and ultimately the car? A causal analysis of this type plays no part whatever in genuine scientific practice.

14. Hart et al. (1985), p. 16. It cannot be overemphasized that the way in which the concept of causality is applied in law can be almost diametrically opposed to that in science, as discussed earlier in this chapter.

15. Cartwright (1989). I shall give three examples, first on using causal statements and then using capacities.

(1) Forces cause accelerations. (See chapter 6 below.) (2) Electric charges cause electric fields. (3) Electrical currents cause magnetic fields. All three statements, with a bit of care, can be used in science without serious trouble. (They are at least useful slogans easy to memorize: many scientists would hardly give them a higher status.) We now assume that the causes do what they do by virtue of possessing certain capacities:

(1') Forces have the capacity to accelerate bodies. (2') Electrical charges have the capacity to create electrical fields. (3') Electrical currents have the capacities to create magnetic fields.

However: forces, despite their general acceptance by the public, are considered in physics as largely fictitious (see chapter 8, p. 195) and thus poor candidates to be possessed of anything as objective as a capacity. Also, the only difference between a charge and a current is that the latter entails moving charges. Since in relativity theory the state of rest is indistinguishable from that of uniform motion, charges and currents are physically interchangeable. Thus, if we were to endow currents, for instance, with the *capacity* to generate a magnetic field, this *capacity* cannot have any physical meaning, since in a frame in which the charges carried by the current are at rest such *capacity* would disappear.

16. In fact, like Leibniz, Hume rejects the possibility of action at a distance.

17. Hume (1739), Bk I, Part I, § II, p. 75.

18. I lean heavily here on Ryle (1949), pp. 121–22.

19. Hume (1748), Sect. VII, Part II, 60, p. 77.

20. Harré (1964), p. 356. However: Harré ignores that Hume was writing in the 1740s, and that not much was known about vibration and sound until Lord Rayleigh published *The Theory of Sound* in 1877. So, for Hume, the relation between the vibration of the string and the production of sound was as immediate as that between the impact of the cue and the moving of the billiard ball had been at the time of Locke.

21. This does not mean that the question of *conditions* can be entirely disregarded, since it is important in controlled experimental work. (See note 44 and related text.)

22. For an account of Aristotle's ideas see Hesse (1961), p. 66.

23. See Alexander (1956) for this correspondence. The quotation given is from p. 94.

24. It is important to realize, however, that not all counterfactuals can be reworded as causal statements.

25. This quotation and the previous one are from Hume (1739), Bk I, Part III, § XVI, pp. 156 and 179, respectively.

26. The last three quotations are from Hume (1748), Sect. V, Part I, 36, p. 44; Sect. V, Part II, 44, pp. 54–55, and Sect. V, Part II, 45, p. 55, respectively. I hope that these quotations will persuade the reader that Hume is not "the most ruthless and skeptical of the empiricists," as Edelman (1994), reflecting popular opinion, asserts.

27. For alternative views of the origin of the universe see Hawking (1993), (1988). Maimonides (1904), Part II, Ch. XIII, p. 63 rejects time's existence before creation.

28. The subjects given in this section are well discussed, with somewhat varying details, in a number of popular expositions. The big bang is treated in Weinberg (1993). For the evolution picture see, for instance, Ornstein (1991), Dennett (1995).

29. The more serious manifestations of irrationality are very well analyzed in Sutherland (1992), chapter 23, where, indeed, an evolutionary explanation is given for their survival in modern society. Superstitious belief, such as fear of snakes, helps survival: it is more beneficial unnecessarily to run away from snakes a hundred times than to be killed once by a viper. (See Ornstein 1991, p. 262.) I shall later argue that irrationality is essential in order to boot up rational behavior.

30. Even theological language cannot remain frozen, and just as science must strive for new and better forms of expression, so does theology, as the scriptures aver. "These things have I spoken unto you in proverbs: but the time cometh, when I shall no more speak unto you in proverbs," (John 16: 25); and, even more explicitly: "When I was a child, I spake as a child, I understood as a child, I thought as a child: but when I became a man, I put away childish things." (1 Cor. 13:11).

31. Almost simultaneously with Darwin, Russell Wallace (1823–1912) also proposed an evolutionary theory, but his work did not catch the public eye in the same way as Darwin's, partly because of the immense amount of data with which Darwin supported his proposals, and partly because Darwin, who was too ill to present his ideas in public, was defended by Thomas H. Huxley (1825–1895), a wonderful public speaker and debater. Charles Darwin published *Origin of Species* (Darwin 1859) six years before the Moravian priest Gregor Mendel started the science of genetics. Even then, Mendel's work was ignored until 1900, when it was rediscovered, thus leading to the modern science of genetics and a reformulation of Darwinism.

32. Possible explanations of this remarkable development are that Neanderthal engaged in complex activities, like the throwing of hand axes (sharpened flattened stones). Although this task appears simple enough, it entails very complex mental processes and the development of an accurate perceptual system requiring a very large neural structure. (In note 1 of chapter 6, in fact, I quote evidence to show how complex catching a cricket ball is.) Simple decision taking, which even at an elementary level must have separated sharply early humans from other primates, requires already an ability to perform in one way or another a number of logical functions. It is probable, however, that brain development was driven by factors not directly related to intelligence: Barton (1998) suggests that the size of the brain is correlated to the increasingly sophisticated processing of visual stimuli. Intelligence, thus, might have been a mere by-product of other mental features, capable of producing evolutionary advantages more rapidly manifested than in the case of intelligence.

33. The relative helplessness of the human infant had of course profound implications in the development of social forms: biological evolution had to go hand in hand with all other aspects of human life. Some form of social structure was essential, even at the earliest stages of our species, in order to permit infants to reach adulthood. It must also be remembered that most of the complex structure of the brain is developed after birth, and that this development is profoundly sensitive to the stimuli received by the young organ. Thus in mice, who receive a great deal of their physical stimuli through their whiskers, removing a single one in a newborn animal shows a substantial later difference in brain structure. The development of the human infant has surely changed along the generations, owing to the very rapid changes (in comparison with the evolutionary timescale) in the quality of the stimuli to which the infant is subjected, and much of this stimuli (such as speech sounds) crucially depends on the social environment in which the baby grows. It is therefore possible that the comparative immaturity of the human newborn was advantageous in

requiring a longer period of development outside the womb, which made the human baby more able, along successive generations, to profit by the rapidly increasing richness of the stimuli provided to it by the society around. Social evolution is, in fact, thousands of times faster than the biological one. (See note 35.)

It is clear, I hope, that the brain's neural network is admirably adapted to processing repetitive stimuli, which thus gives it a disposition to the use of causal arguments. But there is more to it because learning depends crucially on making some predictions as to the course of events, and then testing them. (This is evident in the computer neural networks developed by Jeffrey Elman as well as in psycholinguistic experiments, a remark that I owe to Gerry Altmann.)

34. The effect of repeated stimuli on the growth of the neuronal system has been physiologically corroborated, as for instance in the case of mice deprived of one whisker (note 33), or by such experiments as those of Colin Blakemore on kittens raised under unusual optical stimuli.

Notice that the human baby is born with very few instincts, which means that its brain is only lightly prewired. An instinctive brain, in fact, although it can react efficiently to certain stimuli, is hardly capable of learning, whereas the human brain is a most efficient learning machine.

35. By the time *Cro-Magnon* appeared some 2000 generations ago, the human brain must have been capable of performing any of the tasks to which today it is put to, like doing higher mathematics. This is so because 2000 generations is too little in evolutionary terms to allow for major changes. If this is so, the evolutionary driving forces created an instrument well beyond the needs of the time, since the intellectual stimuli 50000 years ago were insignificant compared with the tasks that the brain was later required to perform. This is, perhaps, a case of a grossly overdeveloped organ, although it is quite possible that we badly underestimate the difficulty of the tasks that the early hunters and agriculturists had to perform. (See note 32.) It is also important to recognize that the process of evolution, guided by the same principle of natural selection, also took place as between different social, religious, and intellectual structures. Although you do not have here genes that mutate, you have different habits, beliefs, or ideas that are put to trial and either succeed or fail in the struggle for life. This psychosocial evolution is many times faster than the biological one, as the last 1000 years or so have witnessed.

36. See Achinstein (1968) for a criticism of circularity in arguments to support induction. A remarkable defense of circular arguments, coming from one of the most distinguished Oxford philosophers of our time, Sir Peter F. Strawson (1919–), recognizes that in elaborate networks such arguments must not only be necessary but also positively useful. (See Strawson 1992, pp. 18–19.) This entirely agrees with the view I sustain throughout this book that if one wants to study nature, than which no more complex system can be imagined, circular arguments are actually virtuous. One famous form of a circular argument in physics is the *bootstrap principle* used in elementary particle theory. (See Gell-Mann 1995, p. 128.)

37. The reader is invited to disagree violently with the author over this statement. Many people, of course, accept sacred texts or the pronouncement of their religious leaders as necessary truths, but even serious believers are nowadays less adept to use such sources in relation to natural science.

38. Unless the intelligence of the organism had been "inserted" into it by an intelligent supranatural being.

39. I prefer to avoid the use of the word *principle* in this context, since the anthropic argument is not a principle in the same sense as the principle of sufficient reason, for example, is a principle. (See p. 183.) As discussed in the text, the anthropic argument does not really organize our ideas, like principles do, but it is rather a sort of warning that "further than this we should not go," or words to that effect. As David Lewis puts it, the anthropic argument is not in itself an explanation. "Rather it is a reason why we may be

content, if need be, to do without one." (Lewis 1986a, p. 132). Having said this, I neverthe-less reserve the right to continue using the expression *anthropic principle*, as more likely to be familiar to the reader.

A stronger form of the anthropic principle is propounded by those cosmologists and philosophers (like David Lewis) who believe that ours is only one of many universes, most of which would not be self-referential. In those that are so, the physical constants and the laws of nature are what they are because otherwise that universe would not be self-refer-ential, that is, it would not admit of intelligent life. See Hawking (1988), p. 124; Lewis (1986a), p. 132; Barrow et al. (1985); Bertola et al. (1989). The book by Barrow et al. (1986a) contains a wealth of information on all aspects of the anthropic principle.

The first enunciation of the principle seems to have been made by Brandon Carter in 1974 in the *International Astronomical Union Symposium* no. 63, where the anthropic principle was stated as a requirement "that what we can expect to observe must be restricted by the condi-tions necessary for our presence as observers." The principle, however, was implicitly used before, as in an article by G. J. Whitrow (1956), in which he presented reasons for space being three-dimensional because only two or three dimensions would lead to Earth orbits compat-ible with evolution, whereas two dimensions were incompatible with higher forms of life.

40. The need to consider nonlogical necessities was conceived by Leibniz (1697), p. 339, who introduced *hypothetical necessities*: "we must go beyond the physical or hypothet-ical necessity, according to which the later things of the world are determined by the earlier. . . . For the present world is necessary physically or hypothetically, but not absolutely or metaphysically. That is to say, the nature of the world being such as it is, it follows that things must happen in it just as they do." (See also Mates 1986, p. 118.)

41. There is a serious problem here, discussed by Strawson (1992), p. 129. The relations $p \Leftrightarrow q$ and $q \Leftrightarrow p$ are equivalent, which means that they both have the same truth table (use table 5, chapter 3). Therefore, the blunt use of the biconditional would make causes and effects interchangeable, that is, symmetrical, which contradicts the usual interpretation of the concept of cause. Thus, we never expect the effect to precede the cause, despite attempts to show that this is possible, which are dismissed by Strawson. In note 43 I shall make a pro-posal to solve this difficulty without having to recourse to breaking down the apparent symmetry of cause and effect just revealed by invoking concepts such as *causal efficacy*.

42. As explained in the text, such necessities are contingent. Contingent necessities, ("which, *given the other circumstances*," are "physically necessary for the outcome") as well as contingent sufficiencies, were first introduced by Scriven (1964), p. 408. Mackie (1980), chap. 1, classifies several varieties of necessity used in discussing causality. My empirical necessity is akin to Mackie's *necessity*. See his p. 12 and also his *natural necessity* on p. 194 of the same reference.

43. Herein lies the solution of the apparent symmetry between cause and effect dis-cussed in note 41. The conditions that must be used in science to validate causal relations must always entail time determinations for both cause and effect, and such time determi-nations automatically distinguish the cause from the effect as always preceding it. I shall discuss in chapter 5 this deep relation between causality and the timescale.

44. A major problem is that, if the necessary conditions entailed by *ceteris paribus* are represented with c, the relation $p \Leftrightarrow q$ would have to be replaced by $(p \cdot c) \Leftrightarrow q$. However it can be proved that $(p \Leftrightarrow q) \Leftrightarrow \{(p \cdot c) \Leftrightarrow q\}$ is a tautology, which means that c could be any-thing on earth! This result is called the *Law of weakening*, and it throws invincible doubts on the use of the logic biconditional to represent causal relations. This is a point very well made by van Fraassen (1980), p. 114. The same problem arises when rewording the relevant propositions as counterfactuals: in order to assert them, it is not sufficient to require the truth of the proposition *if* the relevant conditions were obtained. Rather, we must commit ourselves to the *actual* truth of those conditions before asserting the counterfactual

(Goodman 1965, p. 8). This is entirely in keeping with scientific practice, but it much restricts the applicability of logic in these cases.

The precise nature of the relation between cause and effect is extremely carefully analyzed in Mackie (1980), chap. 2 and p. 62, resulting in what he calls the *inus* conditions, which are also reviewed by Cartwright (1989), p. 25. Bas van Fraassen (1980), p. 114, however, produces a cogent criticism of them. The fact remains that in scientific practice, if a conjunction is established as causal, the *ceteris paribus* conditions are so well defined and checked as to make the relation both necessary and sufficient. This means that such instances of the conjunction as are observed satisfy necessary and sufficient conditions, but it does not mean, as a difference with the biconditional in logic, that this relation can be universally quantified. That the causal relation must be treated in science as necessary and sufficient was always recognized by those philosophers of science who actually practised the latter. An early statement to this effect was given by Herschel (1830), p. 151, art. 145.

Given all the stated provisos, I shall often use in the book both the conditional and the biconditional in order to discuss relations of the type described, but it must be understood that this notation is no more than a convenient shorthand, subject to all the caveats discussed. This will allow us to differentiate quickly full causal relations, which must be represented, with these provisos, by biconditionals from other relations that can only be represented by conditionals.

45. The reader should note that the use of the word *significant* in discussing statistical proportions or averages is not the one attached to it in ordinary discourse. This word entails quantitative requirements defined by very specific statistical techniques.

46. See Berenson (1952), p. 94. It used to be thought that *synaesthesia* was merely due to association of ideas, but it has recently been found that the capacity of some subjects to associate a color with some other percept is so specific and it is so consistently repeated after long periods of time, that their claims that they actually *see* colors must be accepted. Likewise, synaesthetic experiences of textures and shapes arising from color stimuli are now well attested.

47. In order to understand what a mapping is, we must have two aggregates, for instance two groups of objects (one group of numbered apples and the other of numbered coins, say). The mapping is a rule that establishes a correspondence between one (or more) objects of one aggregate and one (or more) objects of the second aggregate. A geographical map itself is a mapping of a given terrain, and the menu in a restaurant is a mapping of the food prepared in the kitchen. The important concept for us to understand here is that the aggregate and its map are, in general, entirely different objects. A serious philosophical mistake is to take the map for the object: when you go to a restaurant, you must not eat the menu!

48. That pure facts undressed by theory do not exist was already recognized by the great Cambridge *scientist* (he invented this word) William Whewell (1794–1866). See Whewell (1837), (1858–60).

49. For the penny experiments see Thouless (1931*a*), (1931*b*), cited by Gombrich (1962), p. 255. The problems with the acceptance of the Copernican theory are discussed in detail in Feyerabend (1993), pp. 61 ff. See also Glymour (1996) for *relative truth*.

50. The quotations given in these two paragraphs are from Kant (1783), p. 9 and § 52c, p. 106, respectively. *Categories* had been used, in a different way, by Aristotle.

51. It is therefore rather strange that the only modern theoretical physicist who embraced a Kantian position was Sir Arthur Stanley Eddington (1882–1944), who was one of the pioneers of relativity theory. Eddington (1939), p. 16 propounded an epistemological approach which he exemplified with the parable of the ichthyologist, a man who studies fish in a lake, and who after much research concludes that all the fish in the lake are larger than two inches. An epistemologist comes and claims that those experiments are a waste of time: it is sufficient to look at the net used by the ichthyologist to be able to predict the results without doing any fishing at all, because the net is a two-inch one. The example of the ichthy-

ologist is a very good one, but the problem is that we cannot, as he did, look at the net, and neither could Eddington: but he attempted, by pure a priori thought à la Kant, to discover its characteristics. We now know that Kant went wrong in this endeavor, and there is no reason to believe that Eddington really did better, but his arguments are often very persuasive and may contain some element of truth: even wrong methods can yield useful results.

52. This must in no way be interpreted as reducing the status of Darwinian theory. All physical theories make use of some meta-physical principles, as is the case with Newton's laws, which depend on the principle of causality. Notice also that the principle of natural selection discussed here is far more general than the one required in biological evolution.

5

VIRTUOUS CIRCLES: LENGTH AND TIME

quid est ergo tempus? si nemo ex me quaerat scio; si quaerenti explicare velim, nescio

what is time, therefore? if you do not ask me I know; but if you want an explanation, I don't

Augustine (397), Book 11, chap. 14.

¿Qué me aprouecha á mí que dé doze horas el relox de hierro, si no las ha dado el del cielo?

What is the good for me that the iron clock struck twelve, if the heavenly one has not done so?

Fernando de Rojas (1499), Aucto XIV, pp. 300–301.

The problems discussed in the last chapter are very central, not only as regards their subject matter but, most importantly, as regards the strategy adopted. Hume's skeptical criticism of the causal relation left us hanging on the borderline of sanity: Bertrand Russell was certainly right in demanding a rational way out of this *post-Humean problem*. The overwhelming question here is what we mean by *rational* because different people have different views in this respect. Many philosophers for many generations have taken the view that a rational argument must be one obtained by *pure thought*. But to require this when we are discoursing about nature (or of our construction of nature, which is the same thing since we are ourselves part of it), is tantamount to requiring us to pull ourselves up from our own bootstraps out of nature in order to argue about it, but independently of it. I admit that it might be possible to attempt such an amazing feat, but you require for this purpose some form of revelation or some sort of insight into a world of pure ideas, as mystics and some mathematicians claim. That this can be done, or at least claimed to be done, is undeniable, but one problem about this approach is that, even if one seriously believes in it, it can be premature, thus endangering the understanding of the very nature of the supernatural world that it is desired to embrace: the concept of the deity, for instance, that was forced upon us before the Copernican or Darwinian revolutions, is not one that many modern theologians would identify with.[1]

The strategy that was adopted in the last chapter in order to deal with the post-Humean problem was unashamedly circular. We accepted as a fact Hume's philosophical discovery that the mind has a *custom* to reason on causal lines; and we explicated this *custom* not quite as philosophers but as natural scientists: for the *custom* to exist, nature must present the mind with repeatedly consistent stimuli and, at the same time, it must favor, through the principle of natural selection, organisms whose central nervous systems are able to learn from this repeated stimuli. Thus, the construction of our mental map of nature involves the processing of repeated stimuli, the existence of which must be postulated as some sort of intrinsic property of nature (or at least of our perception of it), without which our mind as we know it would not exist. A vicious circle for some philosophers, but I commend it to the reader as a virtuous one. More than that, I consider this as an exemplar of *rational* thinking: to explicate nature by pure thought is not just impossible but ill-conceived. The most that we can achieve is to establish a *harmony* (to borrow a word used by Hume), or an *internal consistency* between our description of nature and the mental events whereby that description is constructed, and it is only after this approach has been followed to the limit that any attempt at the supernatural might be fruitful. My job, however, is not theological, so my purpose in this chapter and the next is to show how the construction of virtuous circles is at the very core of science.[2]

One final set of remarks before we embark on the work of the chapter. I shall be interested here in the ways in which we measure space and time, rather than in trying to understand what they are; that is, we shall be concerned with the *epistemology* rather than the *ontology* of the subject, but I shall nevertheless say now a few words about the ontology of time.

THE ONTOLOGY OF TIME

There is a tendency for people to shoot from the hip when they engage in ontology, never more so than in relation to time. Before we embark into a review of such activities, therefore, let me warn you straightaway about a very important result of modern physics: neither space nor time are, as such, *elements of reality*.[3] I shall tell you more about this later in this chapter, but I shall first review briefly what time was thought to be before relativity theory opened our eyes.

Time for Newton was as much of an absolute as space:

> Absolute, true, and mathematical time, of itself, and from its own nature, flows equably without relation to anything external.[4]

Newton's absolute time is therefore like a river, on which a cork floats and thus moves in synchronization with the time: although we do not see the river, it is the motion of the cork that we read as *time*. This picture entails of course some worrying problems. The instant that we call *now* is mapped by the position of the cork. But, if Newton's river flows, then there is river ahead of the cork: is there a future already there, waiting to be reached by the moving cork? Not only have we this disturbing feature, but the very postulation of the unobservable river is dubious: cannot we just manage only with the observable cork? (This is very much what I shall do later, that is, merely relate the time to *events*.)

Already in 1908, the Cambridge philosopher John McTaggart Ellis McTaggart (1866–1925) faced these problems by simply denying the existence of time. He argued that in the above discussion we have two series, the *time series* which entails *past, present,* and *future,* and the *events series* in which we qualify events as *precedent, simultaneous,* and *following.* It is clear that the *events* series presupposes the *time* one. But if I say that the event *x* which I *now* observe preceded *y,* then *x,* regarded as *present,* must have also been at some stage *past.* For *x* to have both properties is contradictory. This appears to be easy to resolve: of course, such qualities pertain to *different times.* But we were trying to discuss time in terms of pure events, and in invoking it in this way, we fall into a circularity. I shall not try to resolve this paradox, partly because it is very delicate and partly because, as you will see later in this chapter, circularity as regards time is in any case unavoidable.[5]

Despite these worries, most of us are happy to believe that time is some parameter or variable about which different people can reach agreement, and that it is as objective as the quantity of water that flows out of a well-regulated tap. We know, for example, that there are different time zones but we believe, nevertheless, (wrongly as it happens) that the time in New York is *necessarily* the same time as in London, expressed perhaps by a different number which, if we wanted, by synchronizing clocks, we could make identical for both towns. We have all seen thrillers where somebody says: "Gentlemen, let us synchronize our watches," after which they all disappear in the darkness to pursue some very macho pursuit, knowing that, if the bridge has to be blown up at 2200 hours, it will be blown up at precisely 2200 hours. Will it? Will perfect watches always stay synchronized, or is it possible that Tom the bridge blower will carry *his own time* with himself, different from that of Sam, the leader?

Contrary to intuition, the answer to this last question is *yes*, and this is not the result of mere theory but of a simple experiment, the interpretation of which is just as transparent as you can wish. Let me describe this result of experimental ontology. In October 1971, when clocks became available with an accuracy of about 3 milliseconds per year, two such clocks were perfectly synchronized; one of them was left on the ground, while the other was flown round the world on scheduled commercial flights. When back at base, the two clocks were compared, and they were out of sync; that is, the flying pilot had taken his own time with him![6] The time difference was very small, about a quarter of a microsecond (millionth of a second), but ontologically very significant. And this last expression shows how delicate scientific endeavor is: we know only too well that any measurement is only approximate, so that one could be tempted to dismiss the very small difference between the two clocks as insignificant, but this was far from being the case here. Not only was the difference larger than the experimental error of the clocks, but it also agreed with the predictions of *relativity theory* (of which more in chapter 11). The result of this experiment is that *time*, understood as some sort of an absolute time, does not exist! We are able to use it in our environment, as if it were an objective quantity of universal value, because the velocities that we normally use are fairly small, and because for the usual processes associated with pursuing our lives, we do not have to measure times with the precision of a microsecond.

This is why I said that there is a tendency to shoot ontological bullets from the hip, and thus fall into amazingly clever explications of time, ontologically as relevant as a discussion of the hardness of unicorns' horns. Which shows how wise Saint Augustine was in confessing his ignorance; and which, I hope, will gain some indulgence for the treatment that follows, based as it is on the, for some, heretical (but I believe essential) use of a circular argument.

HOW TO MEASURE SPACE AND TIME

I shall mainly be concerned with time in this chapter, but I should like to start by first discussing the measurement of space, because it will illustrate some problems that we have when measuring time, and others that we don't. One feature that the measurement of length has, which is not shared with that of time, is that in order to measure length it is easy to construct a standard; and *that this standard can be selected absolutely arbi-*

trarily. This was the way in which early standards were introduced, although some anthropomorphic relations may have been used, as in defining the foot, for instance. Also, when the meter was introduced, it was related to the circumference of Earth, but this was not at all essential: all that was necessary to know as regards the meter was that it was the distance between two marks on a platinum rod deposited at a specified temperature at the Office of Weights and Measures in France. This distance could have been chosen arbitrarily without in any way affecting our map of nature, except trivially by a change of scale.

Simplicity itself: but this is not the whole story because there is a serious catch. Imagine that I want to use this meter in order to draw a square on a piece of a very stable material, say a platinum sheet at a very constant temperature. Imagine also that space is not flat but rather that it is curved, as Einstein showed. This is difficult to visualize for three-dimensional space, but suppose that our space were two-dimensional, and that we thought that it was a perfect plane, whereas it is in fact like the surface of a rugby ball, say ellipsoidal. You will think that we would have to be jolly stupid to live on such a surface and not to know what it was like, but as you will see in a moment, this is not so easy. Because I am thinking of space in the relativistic sense, that is, as firmly tied up to matter. To say that this space is curved is to say that if you displace a rod, which might look to you as perfectly straight, along the space, the rod is constantly being deformed to hug the curvature of the space. (Think about the surface of the rugby ball. The rod has to deform because it would otherwise stick out of the surface.) So, there is no way in which we can know that it is deformed, as would be the case if it were able to protrude out of the surface in certain positions but not in others: remember that the surface, in this example, is *all* the space we have: nothing can "stick out of it," and we have no reference standard with respect to which we can observe the rod as deformed because the standard itself would be equally distorted.[7]

We now go to France, pick up the standard meter and mark the first side of our "square." Next, we rotate this rod by 90° and mark the second side: but the standard meter must have changed its length because the space is curved. (Think about the long side of the rugby ball and what happens along its much more curved perpendicular direction.) From our convention, however, the standard meter is still the standard meter, and we have no way whatever to recognize that it has changed its length. Thus, when we say that a figure is a square, we are assuming that our measurement of length is independent of the position of the object in space. This, however, is not a property of physical spaces at all, but rather of an *ideal* or imagined space, which is called a *Euclidean space*, because Euclid took it for granted that space did have this agreeable property. In actual fact, although relativity theory shows beyond possible doubt that any physical space must be curved, it also shows that our physical space *on Earth* is so nearly Euclidean that for normal purposes, we can assume that the length of any rod does not change with its position.

If what we have done as regards length appears to you as very complicated, it is in fact child's play in comparison with the measurement of time. It is very easy to see, in fact, that the definition of a unit of time cannot be based on an arbitrary choice, as it was the case for length, but that it must entail a circular argument, as I shall now show. Suppose that we want to define the minute as the duration between two consecutive positions marked on the face of an analog clock. This is possible, but for it to be possible, not any clock will do: it must be a *good clock*. But a good clock is one in which the minute hand sweeps precisely the same angle on the face in the *same time*. So, in order to define the unit of time we must be able to know what "the same time" means. More overtly circular an argument is difficult to find, and there is nothing for it except to construct a virtuous circle, the history of which is just about the history of humanity, as we shall see in what follows.

THE TIMESCALE

The first problem in trying to construct a unit of time is that, as a difference with space, time cannot be measured directly: all that we can measure is change.[8] Secondly, whereas for length any object sufficiently stable would do, whatever its length, as a standard, you cannot choose an arbitrary *change* and define a unit of time in terms of it. Suppose, for instance, that we decide to define a unit of time by putting the standard French meter vertical, and defining the unit of time as the time that it takes a specified object to fall through this length. (We know since the days of Galileo that the composition, and thus the mass of this object, is irrelevant: if you drop several different objects simultaneously from the top mark of the meter, they will all go simultaneously through the second mark, if the air resistance is sufficiently small to be disregarded.) This type of standard presents two problems. The first is the same one as for the unit of length: if we move our conventional "time unit," is it not possible that it becomes *deformed*? That is, for example, if we were to raise the meter by ten meters, would the time unit *really* be the same? Which means: would the time taken by a stone falling through the meter in its new position be the same as before? The problem is that we do not know this, and we do not have the means to know, unless somebody comes with a *better* timescale and tells us that, in fact, we were wrong (which truly enough we were: see later).

The second problem is this: if we want to measure say ten meters with the Paris standard, it is easy. But if we want to measure ten units of time with the presumptive "unit" proposed, it would be very difficult indeed. Basically, if we want to measure time, it is convenient to use a change that is *periodic*, that is that repeats itself: this is the basic feature of a so-called analog clock, where the hands repeat the same motion over and over again. The problem now is that not all periodic motions will do. The bouncing up and down of a very good elastic ball on a very smooth and rigid plane is periodic, but the period, say, between two successive hits of the plane, cannot be used as a unit of time. In order to define such a unit, we need a very special type of periodic motion, namely one in which each period is precisely of the same length of time; and here we are caught in our circular argument again: if we do not have the means to measure time, how do we know that a periodic motion is not one such that its periods are of unequal length? In the case of the bouncing ball, it is fairly evident that the motion is damped, that is, that the period between two successive hits of the plane (our presumptive time unit) diminishes steadily. But if I give you my clock, which is a bad one, you would not be able to recognize that this is so unless you had a better clock to compare it with. So how is it that we managed to obtain a unit of time, that is, how is it that humanity managed to produce a good clock?

This is a very beautiful question because it has a very beautiful answer, an answer that is an exemplar of the human endeavor to master nature. And this answer is just as historical in the *phylogenetic* sense (that is in the sense of the history of the species) as the answer to the post-Humean problem was. In other words, no amount of pure thought, or even accurate experiment, could have allowed humanity to obtain a timescale in one go: a successive-approximations, trial-and-error, process was required that entailed an argument seriously circular, a virtuous circle. I take the view that to try to explain this process on purely philosophical, rather than historical, lines is to propagate an illusion. So, let me briefly review the history.

Although *Homo sapiens sapiens, Cro-Magnon* man, emerged some 50000 years ago, they were hunters and fishermen, and it took almost 40000 years before agriculture seriously started: this was the true beginning of civilization. Without agriculture, food surpluses were not available to sustain human settlements, except very rudimentary ones. But, as a difference with hunting, agriculture depends crucially on the seasons and thus on time. I have already said in the last chapter that without repeated *consistent* stimuli, human (or even animal) minds could not have developed: likewise, agriculture could never have been sustained without some reasonable uniformity in nature. You need to know the time for planting and the time for gathering. And, as urban populations emerged, other events such as eclipses had to be foreseen, partly to avoid panic, but probably also to sustain the power of rulers. By planting a vertical rod on the ground (*gnomon*), the interval between the two successive instants when the shadow is shortest could be recognized as a suitable unit of time (the *day*): it was clearly periodic, and it was reasonable to assume that the period was uniform. It was fairly easy to verify, for instance (remember that this process took hundreds of years), that the summer solstice (corresponding to the absolutely shortest shadow) was consistently repeated after 365 such days (a calendar year). As history, and thus observations, progressed, it was realized that this was not entirely accurate: first, it was seen that the time between two consecutive summer solstices was 365 and a quarter days, which led to the first major change of the calendar. Later, further corrections were discovered, leading to further refinements of the calendar.[9]

The basic story so far is that nature provided humanity with a fairly good unit of time, the *day*, the quality of which was attested to by the constancy of such major astronomical features as solstices, equinoxes, and so on, over long periods of time. Naturally, various devices were used in order to measure shorter intervals of time, first water and sand clocks, and then, much later, mechanical clocks of varying degrees of precision. While the technology of time measurement progressed, physical science also made strides, but it must be realized that any formulation of the laws of mechanics would have been impossible without some reasonable approximation to the measurement of time.

The first important work on mechanics that concerns us was done by Galileo Galilei (1564–1642); and it could be argued that he was quite independent of the timescale constructed on the astronomical principles that I have described. Galileo, in fact, made much use of water clocks in which water is allowed to drain from a hole at the bottom of a container. He realized that as the level of the water in the container was going down, the water ran more "slowly," and I have to use quotes here because slow or fast have no meaning, unless we have a good clock by which to measure the necessary time intervals; and this is precisely what Galileo did not have. (It is possible that he was just able to *feel* the disparity, or he might have used his own pulse, which he is known to have done on occasion.)[10] Galileo cleverly used very wide vessels, which contained so much water that, in the short time intervals he had to measure, the level of the liquid was practically constant. He just weighed the water and assumed that the time lapsed was proportional to the weight.[11] Although Galileo was thus fully independent of the astronomical time unit, the question is that a good observer like he would not have used a quantity such as time in his work, unless there was sufficient evidence that this was a respectable variable to use. (Remember that he had very few physical quantities that were sufficiently well established at that time: even the concept of *force* he had to introduce, as opposite to that of *pressure* until then used.)

From the result of his experiments, it is clear that Galileo understood the law of inertia, but he never had it at the front of his mind,[12] and it was Sir Isaac Newton (1642–1727) who brought it to prominence as his first law of motion: if a body is not acted upon by any force, it will remain at rest or continue in uniform motion. And here again we get into a circular argument because in order to obtain this law, we had to be able to measure time, and the uniform motion mentioned in the law cannot be defined unless a time unit is chosen. In fact, uniform motion is one in which the particle traverses equal lengths of space in equal intervals of time. So, in stating the law of inertia, we are implicitly saying that our unit of time is a "good one," namely one that is consistent with this law, a point that was first made by the Swiss mathematician Leonhard Euler (1707–1783).[13]

We are thus closing our circle, and in so doing we have two options: one is to embrace the circularity as an essential requirement of our dealings with nature, and the other is to depreciate it by adopting a *conventionalist* approach, as first suggested in this context in the very influential treatise on *Natural Philosophy* by Thomson and Tait,[14] who claimed that the law of inertia merely expresses our *convention* for measuring time.

I shall discuss this point in the next section, but, in order to finish our historical account of the timescale, let me add that, as the laws of physics became better understood, the astronomical time unit was improved in accuracy by the construction of better and better mechanical clocks (with corresponding corrections of the calendar) until, in 1960, a much more accurate unit of time was obtained by the cesium clock. When atoms are excited, they emit *spectral lines*: an example with which most people are familiar is the sodium atom, responsible for table salt coloring flames yellow, this color corresponding to the spectral line emitted by the sodium atoms excited by the flame. It was the great insight of L. Essen at the National Physical Laboratory of the United Kingdom to realize that the cesium atom emitted a spectral line in the radio frequency region, which could thus be used to drive an electric clock. Because the periodic motion involved in this line is extremely regular, the accuracy of the clock is enormous, a second in 300 years or more. In modern practice, the unit of time is defined in terms of this cesium clock and, in principle, it can be considered to be independent of the astronomical scale.[15] The latter's accuracy, however, remained remarkable: in the modern timescale, the time of rotation of Earth differs from the day by only one second in about 100 years.

CONVENTIONALISM AND VIRTUOUS CIRCLES

Faced with the circular nature of the definition of the timescale, I have said that we have two options, of which one is to adopt a conventionalist strategy. What actually happens, of course, when we face the need to engage in a circular argument, is that we cannot appeal to logic to underwrite our discourse: thus *rationality* (a word unfortunately hijacked by the supporters of the *pure thought* approach) fails. Conventionalism is the siren song chanted to restore some credibility to the *pure thought* myth: yes, we appear to say, pure thought cannot provide the full answer to our problem, but this does not matter very much because *there is no full answer*, and all that we have are *conventions*, not *objective* properties of nature. This may not be an entirely fair description of conventionalism because there are, of course, various varieties of this doctrine, but it is probably near enough the bone.

The high priest of *conventionalism* was the great French mathematician Henri Poincaré (1854–1912), who claimed that the choice of timescale was arbitrary, dictated only by the requirement that it should be chosen so as to make the fundamental laws of physics, in particular those of mechanics, as "simple" as possible: thus the "law" of inertia. He went as far as saying: "There is not one way of measuring time more true than another; that which is generally adopted is only more *convenient*. Of two watches, we have no right to say that the one goes true, the other wrong; we can only say that it is advantageous to conform to the indications of the first."[16] Before I comment on this argument, however, I shall mention some other views along the conventionalist line.

After Poincaré, the main supporters of conventionalism were Rudolf Carnap (1891–1970) and Hans Reichenbach (1891–1953), but neither of them goes perhaps as far as the master. What these philosophers assert, although the details may vary from one to the other, is that fundamental choices in physics, such as those of the units of length and time, are conventions grounded on a desire to make the laws of physics simple. But: "This simplicity has nothing to do with truth, since it is merely *descriptive simplicity*."[17] I shall illustrate what this means by reference to two examples, the first about length. In order to define our standard of length, says Carnap, we could take a steel rod and decide that this is the standard, *irrespective of the temperature*. So, if we put a flame under the rod, because now its length *increases* (under our current nondemential knowledge of physics), the *measurement* of the Sun's diameter *decreases*. (Because fewer units of length are required to span it: marvelous action at a distance!) "We would be compelled to formulate all sorts of bizarre and complicated laws, but there would be no logical contradiction. For that reason, we can say it is a possible choice." As regards time, we could take our unit to be given by our pulse, presumably with the advantage that when we fall in love, the time that we spend in amorous activities with our partner would automatically be longer. Likewise, days will vary in length, and so on. But "there is nothing logically contradictory."[18] (The apparition of magic instances of action at a distance would apparently not worry Carnap!)

I hope that my readers will consider the praise of an argument in this context, as entailing *no logical contradiction*, more or less on the same basis that one hears Mr. Bloggs being well-spoken about for not having murdered his mother: we know very well that in discoursing about the world, logical necessities have a trivial role. The extreme weakness of Carnap's argument resides in the contention that, if we adopt crazy conventions, we shall have crazy laws and nothing worse. There is no basis whatsoever in fact or in logic for this to be the case. This is so because, however difficult it is to pinpoint what a natural law is, it seems to me that in all possible worlds (in the sense discussed in chapter 2) the expression *natural law* can be used of a proposition if and only if this proposition can systematically *predict* something about nature, and prediction requires a ruthless weeding out of the conditions under which the prediction is valid. If, in the pulse-based timescale, I had to predict the time of sunrise tomorrow by a "law" that entails knowing whether I propose to spend the night sleeping, or climbing a mountain, or enjoying my wife's caresses, that is not a law at all: it is just a bit of information that has to be extracted out piecemeal, for instance, from a huge database. There is thus no reason whatsoever to believe that different conventional choices will merely lead to different laws: it can perfectly well be the case that they leave us *with no laws at all*. That we can describe our environment by means of laws of varying degrees of accuracy is, in itself, an empirical fact, and thus clearly *contingent*. Of course, we shall also require of the laws that we use that they form an internally *consistent system*, within various degrees of accuracy, and that such a requirement can be satisfied is again an empirical, contingent, fact.

Let us consider an example of this question of *internal consistency*. As we have seen, until very recently, the timescale was defined on an astronomical basis, from the units of *day* or *year*. Such units have allowed us to obtain, in the manner discussed, the law of inertia, which is a good predictive law within a certain degree of accuracy. Take now an entirely different domain. On using the same timescale, it is an empirical fact and entirely independent of any convention (except for the choice made of the timescale), that, given a sample of n radioactive atoms, this sample will be reduced to $n/2$ atoms in a fixed time t; and that this duration will be independent of any conditions except the requirement that the sample be always made of the same radioactive atoms. There is no *logical* requirement why a timescale should afford this remarkable "coincidence," in other words, such *internal consistency.* The fact that such a consistency, which is far from trivial, can be obtained must be considered as a property of nature,[19] and not as an artifact of the conventions used in constructing our timescale.

THE CAUSAL TIMESCALE

Let us look again at the (rather absurd) timescale already considered, which I shall label the "new timescale," in which the unit of time is the time taken by a small body to fall through the standard French meter, irrespective of the height at which the latter is placed. What we want to do now is to explore Carnap's contention that a choice such as this is conventional; and that all that will happen is that we might be landed with bizarre laws of nature.[20] The first conclusion we obtain with this timescale is that Galileo wasted his time when dropping stones from the top of the tower of Pisa[21] in order to discover (in his, and our, timescale) that all objects, irrespective of their mass, fall at the same time with the same velocity. With the new timescale, we do not have to repeat the experiment: we know that, *by the definition of this timescale,* the time that the particle takes to go through the first as well as the last meter of its trajectory is one unit of time and thus constant. Therefore, the "new law" of free fall of material bodies is that their motion is *uniform*: equal lengths of space are covered in equal intervals of time. Carnap thus appears to be right: all that happens is that new laws of nature emerge. But the question, which Carnap failed to address, is this: are these new laws *internally consistent*?

Take our "old" law of inertia which allows us to predict, in principle for an indefinite length of time, the precise position of a particle in free motion, say a ball on a billiard table. (In this case the position prediction is very simple because we know that the ball must have a constant speed, from which we can calculate its position for all time.) Let us consider the "new clock" in which the particle used for the time measurement is falling through ten meters, each meter corresponding to one unit of time. In the "old" system, the time taken for the last meter is a fraction of that for the first (because the longer the particles fall, the *faster* they go). Therefore, if we compare the position of the falling particle with the positions of the billiard ball moving freely horizontally, marking along the latter's trajectory the ten points that correspond to the ten successive positions corresponding to each ("new") time unit, the last segment will be much shorter than the first one. Of course, Carnap would say that all that will happen is that we shall have to invent a force that will decelerate the billiard ball: the problem is that this force is indeterminate, since, raising the "ten-meter clock" by say, ten meters, would change the space covered in each unit of time by the free billiard ball on the horizontal trajectory. Whereas in the "old" system our physical laws are *causal*, in the sense that, given the same conditions the same effects are repeated *in the same times*, this is no longer possible to achieve in the new system.[22] It is also easy to realize that the constancy of the time by which a sample of n radioactive atoms is reduced to half that number is lost.

The need for humankind to operate on causal laws is not a matter of convention: from my discussion in chapter 4 of the post-Humean problem, it is at the core of the evolution of the mind. It must thus be considered as a major empirical fact that it is possible to construct a *causal timescale*,[23] one in which time durations can be included in the effects in such a way that these durations are constant for given causes, that is for given antecedent states of physical systems. The fact that this timescale pervades all physical phenomena in our *usual* environment (relativistic and quantum effects are not present in the environment in which our central nervous system evolved) makes it difficult to believe that this empirical fact is merely the result of convention. We have seen in the example of our "new" clock, for instance, that one can obtain conventions to fit a situation or two, but not to fit a wide range of phenomena, let alone the *whole* of the world of natural phenomena at the macroscopic nonrelativistic level.

The existence of the causal timescale allows us to enunciate an *invariance principle*; and we shall later see that such principles are fundamental in the description of nature. What *invariance* is, will become clear presently, but let us first discuss how causality works. Suppose that we do an experiment and the *causes*, that is the antecedent state, include the starting time t, whereas after an interval T, that is at the time $t+T$, a certain effect is observed. We must revise this, however, because causality entails the possibility of repeating the production of an effect, and if we were to insist that the initial time t be part of the *causes*, then the experiment which we have just described would be *unique* and not possible to repeat. Therefore, we must accept that the starting time t cannot be part of the *causes*. That is, for the conjunction studied to be causal, we must be able to repeat the experiment say at the time t', and then observe the same effect at the time $t'+T$. And if this effect is predicted by a law of nature, the initial time, t or t', is irrelevant.

This result demonstrates a very important consequence of the principle of causality, namely that *all laws of nature must be independent of the initial time at which they are applied*. We can also express this principle by saying that all natural laws are *invariant* with respect to the time at which they are applied or, more simply, that they are *time-translation invariant*.[24] More examples of invariance principles and of their significance will be seen in chapters 8 and 11.

CODA

The most important idea that has been used in this chapter is that, whereas in philosophy a circular argument is often under suspicion, the situation is different in science, because at every step of the construction of a circular argument, you have an added check, namely its adequacy in saving facticity, that is in fitting the phenomena. Because the timescale is causal, its choice entails fitting not one but *all* natural phenomena that are time-dependent. It is this that makes an otherwise vicious circle into what I have called a virtuous one. It is also important to remember that attempts made in order to limit the physical or objective significance of this work, by giving to it the status of a mere convention, are seriously misguided.

Another virtuous circle will be carefully constructed and closed in the next chapter. This will be concerned with the definition of *state* which, as with the timescale in this chapter, has also to be *causal*. In so doing we shall make our first contact with an important law of nature.

NOTES

1. A very good discussion of this point can be found in Pannenberg (1993). See, for instance, his p. 77. This question is related to what Paul Tillich called "the fight of religion against religion." (See Tillich 1965, p. 50. In Tillich 1964, p. 131, he also writes: "Without an element of 'atheism' no 'theism' can be maintained.")

2. Although circular arguments are often instantly dismissed by philosophers, as a serious breach of the basic rules of logic, Clark Glymour has come as near as using one, such as it can decently be expected, when propounding *bootstrapping*, in which one part of a theory is invoked in support of another part. (See Glymour 1980, pp. 130–31 and passim.) As far as I know, the philosopher who has more clearly grasped the unavoidability of circularity is Strawson, who avers that this is the case when discourse is supported by a network of concepts (Strawson 1992, p. 19). Strawson's network is in fact what later on in this book (pp. 148, 192, 198) I shall call the *science mesh*, *network* these days carrying information-technology connotations.

3. I use this is in a specific sense to be discussed on p. 276, but a fairly intuitive understanding of what it means is sufficient at this stage. General relativity would require further qualification of this statement.

4. Newton (1686), p. 6.

5. A detailed discussion of the paradox is given in Whitrow (1961) and in Horwich (1987). (But see also Savitt 1995.)

6. This experiment was conducted by J. C. Hafele and R. E. Keating; a detailed but simple analysis of it can be found in Will (1995), pp. 54–57. The actual experiment was a little more involved than described and required four clocks. (See Hafele et al. 1972a, b.)

7. The Dalmatian Jesuit Roger Joseph Boscovich (Ruđjer Josip Bošković 1711–1787), founder of *Naturphilosophie* and a pioneer of atomic theory, already pointed out in the second half of the eighteenth century that the assumption that an iron rod kept its length under displacement required justification. (See Alexander 1956, pp. xliv–xlv.)

8. This was a point already made by Aristotle. See van Fraassen (1985), p. 15.

9. A good account of early methods of time measurement and the calendar may be found in Feather (1959), chapter 3. For the calendar see also Duncan (1998) and Coveney et al. (1991), pp. 42–44.

10. Drake (1978), p. 20.

11. See Mach (1960), p. 160.

12. See Mach (1960), pp. 169–71.

13. Euler (1942), pp. 376–83.

14. Thomson et al. (1890), Pt. 1, p. 291.

15. Although the new second was defined to be as near as possible to the old one.

16. Poincaré (1929), p. 30.

17. Reichenbach (1957), p. 117.

18. See Carnap (1966), pp. 94 and 83, respectively, for the last two quotations.

19. Remember that by *nature* we always mean the mapping that we construct of whatever is "out there."

20. It might be objected that the "clock" that we have defined is not periodic. This is no problem: the water clock that Galileo used for his experiments was also not periodic. Despite what some people say, the requirement of periodicity for a clock is not absolute.

21. It is now generally accepted that there is no historical foundation for the reports of this experiment (see Feather 1959, p. 122; Drake 1978). Almost certainly, Galileo derived the laws of falling bodies from experiments with inclined planes in which, because the acceleration is diminished with respect to free fall, the times are longer and easier to measure accurately. I repeat the probably false story because it is such a nice one: *se non è vero è ben trovato*. In any case, as Drake suggests, it is just possible that Galileo did perform the experiment, not really to find the law, but as a public demonstration.

22. In case the reader thinks that we have actually changed the conditions when raising the "ten-meter clock" by ten meters, this is not so: the definition of the clock used was that its height above the ground was irrelevant.

23. Although Leibniz constructed a *causal theory of time*, it was Georges Lechalas (1896) who first used this expression (see van Fraassen 1985, p. 54). The reader will find in this book a very convincing proof that everything that I have done in this chapter is wrong. Bas van Fraassen (1941–), like many philosophers (Strawson excepted), utterly rejects circular arguments, and he proves very clearly that Lechalas's theory suffers of this defect, a defect that will be cheerfully propagated in this book in the next chapter. I can only offer two arguments in mitigation. (See also the reference to Strawson in note 2.) First, I have made it quite clear that I write not as a philosopher but as a natural scientist, trying to describe what the picture of nature built by scientists is, rather than what it should be. Second, it appears to me that whenever science gets entangled in a fundamental circular argument, its philosophical resolution never seems to have a degree of convincing finality. As it is clear from the main text, I have thus decided to make a virtue of a vice and admit that circular arguments had better be accepted (perhaps this is the true meaning of the fallen condition of humanity!), since this is undoubted scientific praxis, under the more respectable label of *internal consistency by successive approximations*. I am not suggesting for a moment that this implies that proper philosophical investigation of these matters is not important, as van Fraassen's own work amply demonstrates. The exposure of circular arguments, where proper and correct forms of discourse are available, is a most useful philosophical tool in expunging error.

24. See, for example Wigner (1967), p. 43. Time translation is, for example, the change of t into t' described in the text.

VIRTUOUS CIRCLES: STATES

Someone suggested that it would be ideal if, as I went along, I would slowly explain how to guess a law, and then end by creating a new law for you. I do not know whether I shall be able to do that.

Richard Feynman (1992), p. 149.

One of the problems in acquiring evidence about the development of science is that the understanding of the history of some of the most important advances largely follows from the testimony of those who have made them, and these people are often geniuses. It is in the nature of a genius that his or her mental processes are so different from ours that, when he or she tries to describe the way in which his or her discoveries are made, he or she cannot do so in a way that is always meaningful to us. Let me illustrate this with a parable. John is a highly intelligent but entirely logical person: he cannot reach any conclusion without arguing step by step. When he is faced with a lion, he reasons as follows: this is a lion, lions are dangerous animals, therefore this is a dangerous animal. And he flees away to find that Jane, who was right by him at the beginning, is already on the other side of the fence playing her bongos. Jane is as different from John as Feynman is from us. It is in the nature of geniuses that, as Jane in our example, they do not necessarily have to have recourse to verbal thinking or at least that they are not aware of any such verbal processes in their minds. No wonder that they can easily believe that what they achieve is the result of something akin to illumination.[1]

My epigraph illustrates this question. The suggestion mentioned in it came obviously from a member of Richard Feynman's audience, during the lectures from which his book was produced. The tragedy is that either members of the public ask for an explanation and don't get it (just as in the case of Feynman, who would not produce gibberish for an answer), or they do get an answer, and it is gibberish to them, and then people begin to believe that science is based on some mystical illumination. The most remarkable example of the latter is perhaps not in science but in mathematics, in the case of Srinivasa Ramanujan (1887–1920) who, as geniuses go, was perhaps the most extraordinary one in recent memory, every bit as amazing in mathematics as Mozart had been in music. More than once he discovered exceedingly complex theorems that astonished the mathematical world, and he consistently did this entirely out of the blue. When he was asked about the way in which he reached his discoveries, he would say that they were written in a dream, sometimes on his tongue, others on a scroll, by the goddess Namagiri.[2] It is not difficult to understand why many of the greatest mathematicians believe in a world of Platonic ideas which they tap into when they make their discoveries. (See chapter 14.)

The moral of all this is that geniuses in science or mathematics cannot be trusted to describe in any useful way the process of invention and discovery in their subjects, although, admittedly, even those at a much more pedestrian level in these subjects do probably share, from time to time, in those wonderful moments of inspiration that feel almost sacramental to the lucky ones who experience them.[3] But underneath all the glitter there is a lot of clear-cut work that is often telescoped in one great result, about which the rest of us can say little more than it is great.

What Richard Feynman could not achieve I shall try to do, and from what I have just said, you must realize that it does not require an overdose of chutzpah for a mere member of the public to claim that he can do what a Nobel laureate can't, as long as the task required is to go step by step in justifying what eventually *must* be a judicious guess. In order to fulfil my promise, I have selected one of the greatest episodes in science, when Newton discovered his second law of motion. From what I have said, however, it is clear that we would not learn very much by trying to reproduce exactly what Newton did or what he had in mind, so that I shall invent a *deutero-Newton*, or *Newton* for short, who stands to Newton very much in the same relation as our John above stands to Jane. We shall start our work by trying to understand the concept of *state*.

WHAT IS A STATE?

There is no doubt that the concept of *state* must be very important because there is hardly a book on physics that does not mention phrases such as *mechanical state, thermodynamical state,* and the like. I suppose, however, that I must be very unfortunate because I don't think I have ever found one such book where the word *state* was *fully* defined. Neither have I fared better looking at philosophy of science books. So, let me have a go at defining *state*.[4]

First, we need the concept of *variable*, which is any magnitude that may be assigned varying numerical values pertaining either to a system as a whole or to the individual particles that make it up. (I put things this way because, as is the case for the temperature, variables may be associated with a system, which might not have a meaning for its individual particles.) It is important to appreciate that a genuine variable is such that its values cannot be subject to arbitrary choice without other changes in the system in question. (If you change the temperature of a gas, for instance, you will produce a change of its pressure.)

Each variable may depend on other variables: in particular, variables often depend on the *time* which may (or may not) be itself a variable. For instance, the position of a particle is in general time-dependent. We mean by this that, as the time varies, the particle's position also varies in such a way that, given a value of the time, then the value of the position variable is *determined*. We say in this case that the time is an *independent variable* (because it can be freely chosen) and that the position is a *function* of it.

Some variables may change along the time but nevertheless not be functions of the time, in the sense that the time does not *determine* their value. It is not difficult to give an example. If I am pumping air into a tire of my car, obviously the pressure that I read varies with the time, but it is not determined by it: my son, who is much stronger than I am, can reach the same pressure in half the time I can. In this case, the time is not a variable but a mere *parameter*. A parameter varies like a variable, but as a difference with a variable, its values can be chosen arbitrarily, without being constrained by the values of any other magnitudes, whereas the values of variables are not entirely free but do suffer certain constraints. When the time is only a parameter, other variables, like the pressure in my example, are called *time-independent*.

Because of these constraints, if I have a set of variables for a system and I can choose at will a set of values for each of them, then I say that they are all *independent variables*. A set of variables for a system (not necessarily all independent) is called a set of *state variables* if, given *one* set of any such values, it is possible to predict, *subject to whatever constraints may exist between these variables*, the values of the same variables in any other such set. In the case of a mechanical system, one of the variables normally chosen is the *time*, on which all the other variables are dependent.

The *state* of a system is the set of all its state variables. Notice please that the word *state* is here used in a highly technical sense. If you have a gas, for instance, its state may be the set of values of the pressure, volume, and temperature of the gas, whereas any other properties in which ordinary people are interested, for instance, the color of the gas or whether it stinks, are olympically disregarded. If the variables that define the state are time-independent, the state is called a *time-independent state*. (Remember, however, that the state may depend on the time as a parameter.)

If the reader feels thoroughly confused at this stage, this is entirely to be expected. My advice is to go on reading because the examples that follow will make everything a lot clearer. It will be much better to go back to the paragraphs above, if necessary, later.

To simplify matters a little, I shall first consider mechanical-type systems and shall leave the case of parametric time and time-independent states until much later in the chapter. In a mechanical-type system, I can take the time as my independent variable, and all the other variables (such as positions, velocities, and so forth) depend on it. What I am saying is that if at the time t' the variables are, say $q, r, s, \ldots$, and t, then we can predict that at the time t' their values will be $q', r', s', \ldots$ and t'.

$$q, r, s, \ldots, t \to q', r', s', \ldots, t'.$$

The predictive equation (symbolized in the above equation by the arrow) is a *law of nature*; thus, the definition of *state* is intimately connected with the derivation of a corresponding *law of nature*. In order to understand the complex relation between these two concepts, let us have a look at figure 1, where we shall be able to see that we have the mother of all circular arguments.

Predictive equation = law of nature

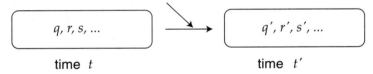

| $q, r, s, \ldots$ | $q', r', s', \ldots$ |

time t time t'

Fig. 1. States and state variables

The content of each box defines the state of the system at the corresponding time, and the variables in them are the *state variables*. Notice that when the time is a variable (*time-dependent states*), on which all the other variables depend, it can be left out of the box, as we have done, in which case it appears *implicitly* in the definition of the state. If the time is a parameter (*time-independent states*) then its omission from the boxes is significant.

That we do in fact have a circularity should be abundantly clear from figure 1, that requires us to solve two deeply interconnected problems, about both of which we are equally ignorant: we do not know what to put into the boxes, and we do not know how to relate one box to the other by a predictive equation. We cannot expect to be able to formulate the correct predictive equation until the contents of the boxes are right, and we do not know that the contents of the boxes are right (that is, we do not know how to define the *state* of the system) until we find a predictive equation that allows us to predict the content of the second box from that of the first.

Let us first consider how we might fill in the boxes. Imagine that we guess that the variable q will do. We then get its value q at the time t and, in our second step, try to guess a law that, given this value, allows us to obtain q' at the time t'. If we cannot obtain such a law, this can mean one of two things: either the variable q is not sufficient to define the state, or we have just not been sufficiently clever in finding the correct law. (The second prong of the disjunction shows how dismal the situation is; so difficult that, invariably, obtaining such a combined definition of state and law of nature is worth a Nobel Prize.) So, if we have failed, and we believe that the failure was that we did not give enough detail to the definition of the state, then we introduce a second variable r and repeat the process, until we have sufficient variables *and* we are lucky enough to guess the corresponding predictive equation.

Many philosophers[5] would consider abhorrent the circular process that I have described, as manifestly breaking all the rules of logic. I plead guilty as charged, of course, but I refer the reader to my remarks about circularity in science as opposed to philosophy, given in the coda to the last chapter: the fact remains that, whereas philosophical omelettes can be made without breaking eggs, they are, alas, nutritionally disadvantaged. I should also like to add three remarks. First, whether you like it or not, the above process is *historically* what has always happened: that is what Newton, Einstein, and Schrödinger did. (Nothing in history happens in a nice and orderly manner, and I do not claim that the process was carried out exactly as described, but rather, that the final result was the one that the recipe given would have provided.) Second, when the process of state definition is described as I have done, it appears so hopelessly chancy that it is difficult to believe that it can be made to work. If you consider the names I have given as associated with it, however, you will realize that a lot of time and an enormous amount of brain juice helped to do what otherwise appears impossible. (In any case, after you see below how *Newton* discovers Newton's second law, you will realize that the task is not so horribly impossible.) Third, notice that our approach to the definition of state is *causal*, the law of nature entailed being the instrument whereby the effect (state of the system at time t') is predicted from the cause (state of the system at time t).

NEWTON ON NEWTON'S SECOND LAW

Newton wants to define the state of a particle in motion which is at the point x at the time t. The first idea that he considers is to take just x to define entirely the state. He immediately realizes, however, that this must be nonsense. It is clear from figure 2 that if we know only x, we do not even know whether the particle is moving to the left or to the right, so that it is plain silly to expect to be able to predict its future position on such insufficient evidence.

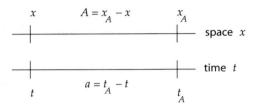

Fig. 2. Trying to guess Newton's second law of motion

The next move is pretty obvious: we must put in our "box" (see fig. 1) not only x (time t) but also the position, say x_A of the particle at a later time t_A as shown in figure 2. This way we at least know that the particle is moving to the right. *Newton*, who is not a simpleton as you might think, gets worried, in fact very worried indeed: because in order to define a state, *all* the variables which we expect to determine must correspond *precisely to the same time*. (See fig. 1.) At this stage *Newton* behaves like a good chess player who is several steps ahead when making a move.

First of all, *Newton* says: let me define an *average velocity* in the time from t to t_A, this being the space traversed in that time, A from figure 2, divided by the time interval, a from the same figure. This average velocity is thus A/a. This is not yet sufficiently good because this average velocity cannot be associated just with the time t. But what *Newton* has in mind is to reduce the time interval a gradually, thus getting nearer and nearer the starting time t, in the hope that with a bit of luck we could find a velocity that could be assigned to the time t (an *instantaneous velocity* instead of an *average* one). In order to see how *Newton* plays the next few moves, let us look at figure 3 where, in order to have a realistic example, I consider the motion of a particle falling from the top of a tower.

This figure represents the successive positions of the particle of figure 2 as a function of the time. The reader should identify in figure 3 the values A and a defined in figure 2 (40 m and 2 s, respectively). What *Newton* wants to do is to construct a sequence of average velocities

$$A/a, B/b, C/c, D/d, E/e,$$

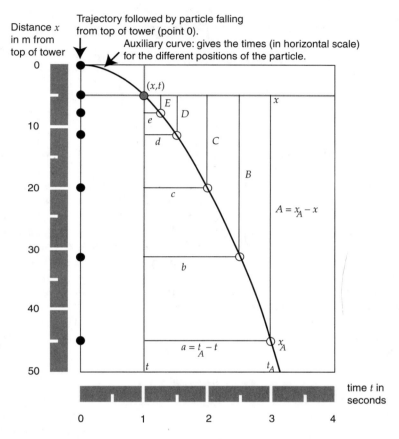

Distance x in m from top of tower

Trajectory followed by particle falling from top of tower (point 0).

Auxiliary curve: gives the times (in horizontal scale) for the different positions of the particle.

(x,t)

$A = x_A - x$

$a = t_A - t$

time t in seconds

Fig. 3. Toward the definition of instantaneous velocity

The auxiliary curve, which is supposed to be plotted after the experiment, correlates the successive positions of the falling particle (black circles) with the corresponding times read on the horizontal scale. Thus, the particle falls 20 m after 2 s. The grey circle, coordinates x (5 m on left scale) and t (1s on bottom scale) is the point at which the instantaneous velocity is sought. Coordinate differences $x_A- x$ and $t_A - t$ (see fig. 2) are written as A and a respectively, and similarly for the other positions of the particle.

for successively smaller time intervals starting from the time t, and see whether he can draw some conclusion for the value that this average velocity would take when the time interval becomes so small that we could say that we are for all practical purposes at the time t itself. The above sequence is so easy to obtain that all the reader needs in order to check it is a strip of paper on which to mark the lengths of the various segments, their values then being determined by comparison with the appropriate scale. The results, in m/s, are respectively:

$$40/2, 26.3/1.5, 15/1, 6.25/0.5, 2.8/0$$

that is,

$$20, 17.5, 15, 12.5, 11.2.$$

(Notice that all these quantities give average changes of the position of the particle in one second, that is, in unit time.)

It is easy to compute one more value, for 0.1 of a second after t, and you will get just a little over 10 m/s. One quality that *Newton* shared with his eponymous hero is courage: he expected that as the time interval is made gradually smaller, the successive differences of the average velocities might become so small that one could assume that, for sufficiently small differences from t, the average velocity will become stable to a sufficient approximation:[6] this value can now be assigned to the time t and be called the *instantaneous velocity* v at the time t. In the case above, it turns out that this instantaneous velocity is about 10 m/s.

Newton is now in a position to fill the "box" (see fig. 1) for the time t, that is, he can now expect that the following *might* be a correct definition of the state:

time t, state variables: x, v.

Before he goes any further, *Newton* recognizes that the instantaneous velocity v can be described as a particular case of a more general concept, that of *rate of change*. He thus writes:

v = rate of change of the variable x per unit time, at the time t.

In plain language, this means the number of units of x (in this case meters) by which x changes per second, calculated at the time t by the sequential process just described. The important new idea here is to realize that this definition of the rate of change can be applied to any variable u, say:

rate of change of u = rate of change of the variable u per unit time, at the time t.

The left-hand side here is no more than an abbreviation for the quantity on the right, which must be calculated by the same sequential process that gave us the instantaneous velocity before.

Having done so much, *Newton* still faces the hardest hurdle because he must now invent a predictive equation that, given the variables x and v at the time t, determines the values of the same variables at the time t'; but he is now ready to make the great intellectual jump required to try and produce a law of motion. You must realize that whatever makes men or women intelligent, it certainly depends on the database that they hold in their brains and on their ability to use it: no one makes any discovery except in relation to something that he or she already knows. There are three things that *Newton* has clear in his mind.

The first is the *law of inertia*: if a particle is not acted upon by any force, it continues in its *state of motion*, by which it is meant that, if it is at rest, it continues at rest and, if it is moving, it will continue on a straight line always with the same velocity.

The second is *Galileo's law of falling bodies*: two bodies, whatever their mass, dropped from the same height will reach the ground at the same time (provided that their air resistance be the same).

The third is *Newton's law of gravitation*: two bodies attract each other with a force that is proportional to their masses and inversely proportional to the square of the distance between them. (If the bodies are extended, rather than particles, a little more work is required to define the point from which to measure the distance: the result for a sphere of uniform composition, for instance, is its center.)

I cannot insist sufficiently on the importance of the existence of this interlaced structure or mesh of ideas and facts into which *Newton* must graft his expected discovery. Some hopeful philosophers intend to examine a law of nature on its own, and thus discuss the logic of its significance, as if a natural law could stand out naked as Aphrodite in Olympic isolation from the rest of nature. Laws of nature, as they are used in science, stand or fall not alone, but in relation to a vast mesh of ideas and results which the laws are expected to fit: the greater a law is, the greater its reach in the mesh; and we shall see that this is so to such an extent, that even when the law is falsified, it is still used! But more about this later.

We can now go back to *Newton*, trying to squeeze some good ideas out of his brain to produce a law of motion that, given x and v at the time t, will allow him to predict the values of these variables at the time t'.

To start with, he rejoices upon the fact that his guessed choice of x and v as state variables agrees with the law of inertia. In other words, if there is no force acting on the particle and I know x and v at the time t, the law of inertia allows us to predict the value of these variables at any later time. Let us verify this claim. If v is zero (particle at rest), the law of inertia says that it will continue to be zero for all time, and that the value of x at the time t will be the same forever after: thus we can predict the values of the two presumptive state variables at any time. Second case: if v does not vanish, the law of inertia says that this velocity will remain constant for all time, so that we can certainly (and trivially) predict its value. As regards x, because the velocity is constant, it is very easy to predict its value after any interval of time, which will be this interval of time multiplied by the velocity v.

(It is very important for the reader to begin to see that the apparently hopeless task that *Newton* faced, although hard, has built into it a sufficient number of procedural tests that keep the discoverer, if not on the right path, at least away from the wrong one. And this type of check is one of the major tools in scientific practice.)

Newton can now go further, because the law of inertia is very important, but he has to move forward from the case of no force to that of a force f acting on the particle. He has to relate f to his two presumptive state variables, x and the instantaneous velocity v. To make his job easier, he decides to assume that the force is constant, that is, that it be not dependent on x (and neither of course on the time, since the position x depends on the time). The simplest hypothesis one would try is:

$$\text{the force } f \text{ is proportional to the velocity } v : \quad f = k\,v,$$

where k is some constant of proportionality. Putting this assertion in a philosophical language, what we are assuming is that forces are the cause of velocities (as in fact Aristotle believed). *Newton*, however, would not have wasted a second on this possibility because it contradicts the law of inertia: from our little formula above, if f vanishes (as is the condition of applicability of the law of inertia), then the velocity must also vanish, whereas we know that in such cases the velocities can have healthy albeit constant values.

A more sensible assumption than the guess just made is to suppose that the effect of a force is not to create a velocity but *rather to change it*. So *Newton* tries:

$$\text{the force } f \text{ is proportional to the rate of change of the velocity,}$$

or, what is the same,

$$f = k \text{ times the rate of change of } v.$$

Here k is merely a proportionality constant, and the rate of change of the velocity is a useful quantity, called *acceleration*.

The next step is to try to identify the meaning of the proportionality constant k. It is easy to realize that the rate of change of the velocity must be smaller the larger the mass of the body (it is easier to change the velocity of a football than that of a cannon ball), so that if this rate of change were f/k, as it follows from the above relation, then it would make sense to take k to be the mass m of the body. (If the reader thinks that all this is very high-handed speculation, remember that we are quite free to try *any* ideas at all at this stage: by their fruits you shall know them, or, in other words, if we are wrong, nature will kindly oblige and give us a kick in the teeth.) The result of all this is that *Newton* now tries

$$f = m \text{ times the rate of change of } v,$$

as his presumptive law of nature which, if right, should allow him to predict his presumptive state variables at the time t' from the values of the selfsame variables at the time t.

As we shall soon see, the statement we have just made is the correct enunciation of Newton's second law of motion and, because I shall want to refer to it in other chapters of the book, I shall present it in a more compact notation, remembering that the rate of change of v is called the *acceleration*, to be abbreviated as a. Therefore, the more compact form of Newton's second law which we shall later use is:

$$f = m\ a.$$

We now go back to *Newton's* main work. In order to establish that his guess is correct, the first thing he has to do is to see whether he has closed his virtuous circle, which we can now symbolically write as follows, with the predictive equation over the arrow that links the two presumptive states at two different times:

$$f = m \text{ times the rate of change of } v,$$

$$x, v \text{ (time } t) \rightarrow x, v \text{ (time } t').$$

Newton has to straddle the time from t to t' in small intervals so as to be able to use the definition of the rates of change, which is valid only for such intervals. So, consider a time T very near t, and agree to represent all the variables at that time in capitals. Therefore, v at the time t is the change in position, $X - x$, divided by the time interval $T - t$:

$$\text{Time } t : v = \frac{X - x}{T - t}.$$

Here we know $t, x, v,$ and we have chosen $T,$ so that it is clear that X becomes determined. We now use the presumptive predictive equation:

$$\frac{f}{m} = \textit{rate of change of } v = \frac{\textit{change of } v}{\textit{change of } t} = \frac{V - v}{T - t}.$$

Again, we know here the left-hand side f/m. The variables v and t are part of the initial data, and we have chosen $T,$ so that it is easy to determine V. The result of all this is that, starting from x and v at the time $t,$ *Newton* has determined X and V at the time T. It is clear that we can repeat this process, now starting from $X, V,$ and T to move another small step forward, and so on, until we reach x, v at any time t'. So *Newton* has closed his virtuous circle, which means, in principle, that his definition of state is complete and that the predictive equation has a very good chance to be right.

All that *Newton* has to do is conduct some experiments to check that the values predicted by his law of motion agree with the laboratory results. This, however, is not what a good theoretician will do. She is not going to stick her neck out and ask an experimentalist to check equations that are only speculative. What *Newton* will do first of all is see whether his predictions already agree or not with the facts that he knows. To start with, he remembers Galileo's law of falling bodies. Since the bodies all start falling from zero velocity, if they hit the ground at the same time, their velocities must *change* all in step, and therefore they must all have the same acceleration, that is the same rate of change of v. *Newton* also remembers his gravitational law, and he argues that bodies fall under the gravitational force exerted on them by Earth. Remember that the gravitational force is proportional to the product of the masses m of the falling body and M of Earth, divided by the square of the distance between the body and the center of Earth. This distance can be approximated as Earth's radius $R,$ since the height from which the body falls is insignificant compared with the very large value of R.[7] So, we write, with a proportionality constant $K,$

$$f = K\frac{mM}{R^2} = m \textit{ times the rate of change of } v,$$

on identifying the gravitational force with m times the rate of change of $v,$ from Newton's second law. If we look at the last two terms in these equations, we get

$$\textit{rate of change of } v = \frac{KM}{R^2} = \textit{constant},$$

(because K, M, and R are all constants) which is precisely what Galileo required, since the acceleration or rate of change of v obtained here does not depend on the mass of the falling body.[8]

So *Newton* has done his job, and he can go home and collect his Nobel Prize. And we now know that a *mechanical* state is fully described by the positions and velocities of the particles at the initial time. There is a catch here (typical of those ambiguities that nature forces on us: you hardly finish closing a virtuous circle, when nature opens it up again). *Newton* goes to a friend's house to play billiards, and he boasts that he can now predict the position of his billiard ball at any time. He hits the ball and everything goes wrong: his friend had played a hoax on him; he had put an iron core inside the billiard ball and a magnet underneath the table! So, however good the definition of state is, it is always provisional: in the example of the magnetic billiards table, we would have to add some new variables to represent the magnetization of the iron core of the ball in order to define its state. Nature always teaches us a lesson, which is that everything is for ever open and nothing can be closed and thus be fully determined until experience allows us, for the time being, to do so.

TIME-INDEPENDENT STATES AND STATE FUNCTIONS

I shall now consider the case when the state does not depend on the time as a variable, for which purpose the example of the pumped up tire discussed before will be helpful. So, we have to discover how one defines the state of a gas. In 1662 Robert Boyle found at Oxford that at a constant temperature, the pressure P of a gas is inversely proportional to its volume V, that is, $P = k / V$, for some constant k. This is the same as to say that

$$P V = k, \text{ (at constant temperature } T).$$

In 1802 mainly through the work of J. L. Gay–Lussac (notice how long this discovery took!), the constant k was found to be proportional to the temperature T, the constant of proportionality R being the same for all gases (but see below). Therefore, we have

$$\frac{PV}{T} = R.$$

This equation is the predictive equation that confirms P, V, and T as the state variables. In fact, if I choose the initial values of P, V, and T and change T to T' and V to V', the value of P will be determined. Thus the state variables P, V, T determine the new set P', V', T', as state variables should do. (Notice that, as a difference with the state variables x and v for a mechanical system, P, V, and T are not all independent, since they are linked by the equation displayed above, so that not all three, but only two can be chosen independently.) In cases like this, when the predictive equation entails a relation between the state variables, it is normally called an *equation of state*.

Let us go back to the mechanical systems with x and v as state variables: we remember that the predictive equation was Newton's second law of motion. With a bit of ingenuity, it is possible to construct a function of these variables that is constant along the motion. The name at least of this function is widely known: it is the *energy*. This is given in terms of two functions, one that depends on the *position* x of the particle and which is called the *potential energy*, and the other that depends only on the *velocity* and which is called the *kinetic energy*. The great thing about the energy is that its prediction for an isolated system is trivial: it is constant, so that we save having to use Newton's much more complicated second law. For this reason, in order to obtain simpler predictive equations, one often forms functions of the state variables which are called *state functions*. Experience alone shows which are the functions worth forming among the infinite possibilities available: it is only when a simple predictive relation appears that the trial function is worth using.

NEWTON'S THIRD LAW: ACTION AND REACTION

I confess that this law has not got a great deal to do, at least explicitly, with the main subject of this chapter, which is the definition of state, but it is an important principle which we shall have to use later, so that a few words here are convenient. The problem that Newton faced was a very simple one. Consider, on the left of figure 4, a rigid sphere sitting on a rigid plane. (The word *rigid* here means that we are adopting a *model* for a solid, namely one where a force, applied on any point of the solid can be slid along its line of application to any other point of the body along that line. There are some solids that are almost perfectly rigid, like a perfect diamond, for instance. A modern car, instead, has crumple zones to make it nonrigid: on collision, those zones crumple rather than transmit the collision force, which would otherwise be the case.) The question is: how can you explain that this sphere is not moving, although there is a force acting on it, namely the gravitational force?

Newton's answer is illustrated in B of figure 4. Because the sphere and the plane are both rigid, the gravitational force is applied on the plane (on

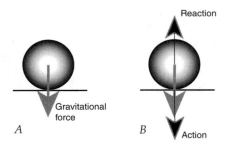

Fig. 4. The law of action and reaction

any point along the line of application of the force), and this force on the plane is called the *action*. Newton postulated that when one body acts on another with a force (action), the second acts on the first with an equal and opposite force, called the *reaction*. The reaction of the plane on the sphere is shown in the picture and, of course, must be equal and opposite to the action and thus to the gravitational force: therefore the two forces acting on the sphere cancel each other, and the sphere is in equilibrium. (You can see why Newton introduced this law, so as to have in this case a force that cancels the gravitational one thus saving the law of inertia, since we know that, after all, the sphere does not move.) This law of action and reaction is entirely general in physics; and we shall see that it must always be applied since otherwise, as in the above example, the state of equilibrium of a system would not be realized. A popular misreading of the law must be avoided: *action and reaction never cancel each other* because they apply *on different bodies*. (In the above case, the forces that cancel are the gravitational force acting on the sphere, and the reaction of the plane, also acting on the sphere.) We can now return to our main work.

ARE OLD LAWS DEAD?

I am sure that the reader will remember that the construction of the causal timescale, and thus the closure of our first great virtuous circle, was possible because humanity could start with sufficiently simple observations, under which the day appeared as an adequate unit of time, in the sense that it related to a motion that, within the approximations then acceptable, could be considered as perfectly periodic. In the same way, the definitions of *states* that we have given depend crucially on the fact that, *at the time when these definitions were arrived at*, the experimental observations were sufficiently imprecise, or the conditions applicable were sufficiently restricted, so as to be compatible with such laws as were invoked. When *Newton* compared his law of gravitation with Galileo's law of falling bodies (see p. 137), he could happily neglect the fact that the distance of the falling object to the center of Earth changes during the fall because there was no conceivable way of measuring the difference that this correction would entail. (See note 7). More fundamentally, even the very concept of *force* as used in Newton's second law has to be doctored in order to make the transition from a force such as it might be observed in an experiment to the *force* that enters Newton's equations. For a *real* force must have a certain "width" (that is, it is never applied on a single mathematical point but rather on a small area), whereas the "force" which is used in any equation must be understood as acting along a mathematical line, which is an abstraction that has *no dimension* across its length. Likewise, the initial conditions at the time *t* must be stated for a dimensionless mathematical point which, of course, does not exist in the real trajectory but it is only a part of the model used in order to describe the latter.

What I am getting at is that physicists must necessarily apply their equations not to anything *real* but to abstractions (more often called *models*), chosen so to behave as nearly as possible as the objects that scientists handle in the laboratory. We have already seen an example of a model of an object that we called a *rigid body*. Nature and history were good to us because, when humanity started this game, accuracy was very limited, and people hardly had to worry about the fact that the model of a force and the force itself might have important differences. In the case of the state of a gas, very soon, that is after a hundred years or so, people began to realize that the laws that correlated with the states defined were only roughly approximate. A *model* of a gas was thus constructed. Because it was seen that the approximate laws held good (albeit with low accuracy) for all gases, irrespective of their composition, it was assumed that all departures from these laws were due to the actual size of the molecules of the gas and to their interactions; and a *model ideal gas* was constructed in which these two effects could be neglected. In fact, the laws that I have stated were theoretically found to be precisely correct for this (unexistent) model gas. Naturally, whereas the equation of state that I have given was good enough at the time of Gay–Lussac, it had to be altered as experiments became more precise.

This little story is important: anyone who expects *truth* in science to be an absolute, valid for all time, or even at any given time valid *exactly*, expects the impossible. By definition, the laws of science can never be more accurate than the experimental errors that affect the scientific community at the time. (If, by some extraordinary chance they were, no one would know!) As accuracy increases, new laws take over. Many people like to say that the old laws are *falsified* and thus superseded by those that fit the new experiments, but you must be careful in using this language: no more can we say that Johnny, who was ten a year ago, no longer exists because he is now eleven, than to claim that old laws cease to be *true* once more accurate ones emerge. A car manufacturer would be crazy to refuse to use Newtonian mechanics in designing his cars, on the grounds that it is no longer *true* and that it has been superseded by relativistic mechanics. Just as the English language is firmly embedded in its Sanskrit and Latin roots, the whole of our science would never have been the same if we had not had the old laws *that provided the language within which the new ones were discovered.*

I must explain that when I mention *old laws* in this context, I do not mean any old law such as many which have been confined to the cobwebs of history: but any law based on a model which, however inaccurately, is still an approximation to current models, has thus contributed to the tradition and vocabulary of modern science; and even if we do not use it any more, just as we no longer speak Latin, it is still there embedded in our scientific culture. The reader will see that the description that I have given of scientific progress is firmly *evolutionary*. This is so because an important characteristic of an evolutionary system is that whatever happens at any given time, happens not in pursuit of an absolutely best possible goal but, rather, of the best goal attainable compatible with the present state of the system and with its past history. In biological evolution, this history is encoded in the genetic pool, but in intellectual and social evolution, it is encoded in the language and culture of the time.

THE IMPORTANCE OF BEING SURPRISED

All the world now knows about the meaning and the importance of information, and information is the same as surprise. When we pick up the telephone and we say that it is dead, what in fact we hear is random noise, pure disorder. If a few minutes later we hear the dial tone, already some order has been made in the chaos and we read a message: "I am ready for you to dial," the telephone says. And the more the message departs from a random sequence, the more the information you receive, and the more "surprised" you are. I place this word in quotes because, normally, you reserve it for very special events without realizing that it is only a matter of scale: you tend to use "surprise" when the information is unexpected, when it connects two events that are unlikely to be connected, but surprise and information are practically the same thing:

surprise is unexpected information, but to know that something is unexpected is already information![9]

Think now about what the world was like at the time of Galileo: there was already a set of experimental results and ideas, but this set hardly enmeshed. One knew, if the Inquisition was keeping quiet, that Earth rotated around the Sun, and even more, thanks to Kepler, that its orbit was an ellipse, and one knew the laws of falling bodies: facts and ideas were there, nevertheless, standing on their own, unrelated. Then Newton came, and his laws became perhaps the most important intellectual step until then in the history of physics because they connected several elements in the then available set of ideas and results that no one ever dreamed before were connected: nothing more audacious at that time than to claim that the same rules that govern the motion of a falling stone determine the orbit of Earth around the Sun. Galileo's discovery of his law of falling bodies provided some information, but Newton's laws carried with them the greatest element of surprise until then achieved in natural science.

The chaotic set of results and ideas that science had until then collected began to enmesh, and this was very much like what happened millions of years before when life appeared in the primeval pond: because the mesh that Newton started to knit could then grow organically, and because scientists began to realize, even if they might not put things that way, that their true job, that the pinnacle of what they could do, was not just to add knots to the mesh—a worthy task in itself—but to connect, improve, and unify it. It is interesting to speculate that this intuitive desire to bring order out of chaos was probably at its strongest within the Judeo-Christian-Islamic tradition of the great monotheistic religions, which had at their core the concept of a divine and therefore rational and well-ordered creation.[10]

One can see most clearly the crucial importance of this outlook in the development of science, when considering the enormous influence that *Naturphilosophie* much later had in motivating people to look for further connections within the scientific mesh because this sometimes bizarre school had as its major tenet the concept of the *unity of nature*. We are in the second half of the eighteenth century, in the middle of the *Enlightenment*, the rational movement, entirely antithetic to *Naturphilosophie*, immortalized by Diderot in the *Encyclopédie*. Of course, a great deal of good work came out of this rational approach, and yet the next great surprise after Newton was achieved entirely within the spirit of the *Naturphilosophie*. The founder of the school may be considered the Dalmatian Jesuit Roger Joseph Boscovich (1711–1787) who, believing that atoms (then no more than theoretical constructions) were too individual and different to be really important, tried to make them into manifestations of *forces*, the latter in their nature being more of the same kind. Even the leading British scientists Humphrey Davy and Michael Faraday, hard-headed rationalists as they appeared to be, were secret followers of Boscovich,[11] but already in 1798 Johann Wilhelm Ritter, adopting the principle of the *unity of nature*, postulated that electricity and chemical affinity had the same origin (about which he was not entirely wrong). The most dramatic result of pursuing this principle came when a friend of Ritter's, the Danish pharmacist Hans Christian Ørsted, discovered, in 1820, the *force* of interaction between magnetism and electricity.[12] Magnetism had been a mystery for almost two millennia and yet, all of a sudden, it became related to a phenomenon that even at that time was seen as one of the keys to nature. One cannot expect in this case the theological implications that Newton's discovery had, but the element of surprise that this discovery entailed was every bit as strong: the whole of the advancement of science for the next hundred years stemmed from this discovery, which linked two apparently unconnected knots in the mesh of scientific fact and theory. (The next great surprise, the identification by James Clerk Maxwell of light with electromagnetic waves, crowned this intellectual story.)

The moral of all this is that what makes a theory *entrenched* in scientific practice is that it *connects*: it links apparently unrelated facts and theories. The further these are apart, and the more of them there are, the more important the theory will be. And it will be so important that even when falsified it will continue in use!

CODA

Some very important concepts have been discussed in this chapter. We have seen that there is no logical or deductive method that can lead to the definition of state: this has to be done by a circular process of trial and error. Also, it is important to realize that the definition of state is necessarily causal. Nothing can be called a state unless it causally permits the prediction of further values of the same state variables that determine the state, and this is the most important manner in which causality enters science. This concept is so basic that we shall see that even in quantum mechanics—where indeterminism for the first time entered physics—the definition of state is still causal.

Just at the end of the chapter, we discussed how theories and laws acquire importance and status in relation to their ability to connect and harmonize the many different facts and ideas that form the mesh that sustains the body scientific. The larger and more disparate the coverage of the science mesh that a law provides, the more, I have said, it is *entrenched*. This is a word that will appear again in the next chapter when we explore once more the post-Humean problem of what is the intellectual or rational basis for prediction or forecast.

One final remark about state, which will explain why this is a concept much liked by scientists. When, given a physical system, we are able to define its state at a given time t, this is all that we need in order to describe the future evolution of the system. In other words, the *history* of the system prior to t is irrelevant. This is not so in all systems. There are some that during a given process, say heating or magnetizing, can store certain deformations or strains in such a way that, when returned to the previous temperature or magnetic value, those strains are still there, thus revealing the previous history of the sample. When one tries to define a state for these systems in terms of such variables as temperature, volume, and so on, one finds that this is not possible because those variables alone are not sufficient for predictive purposes: the history of the sample, that is its previous inputs and the length of time during which they were applied on the system, must be known. This property whereby the history of a system is important in determining its future behavior is called *hysteresis*, and is mainly found in magnetic materials and in polymers.

We are now ready to explore a new and strange world and some even stranger predicates, but all this will be in a good cause: to illuminate the concept of *entrenchment*.

NOTES

1. It has recently been found (Dienes et al. 1993; McLeod et al. 1995; see also Berry et al. 1993) that when fielders catch a cricket ball, they are in fact solving a differential equation, although of course they are not aware of this. There is here a mechanism that controls an output accessible to the conscious introspection of the subject, but whose workings are not. This might be an example of the creative situation discussed in the text. (I am grateful to Dr. McLeod for correspondence on these experiments.) That not all thought is immediately verbalized was pointed out in 1965 by the Dutch psychologist Adriaan De Groot. (See Shanker 1998, p. 136.) Snyder et al. (1999), on the evidence of the study of autistic children with exceptional gifts, concluded that they (as perhaps ordinary geniuses do) have access to brain mechanisms that we all possess but cannot access.

2. Kanigel (1992), p. 36.

3. Such well known feelings have been invoked by those who wish to claim that science shares a common fountainhead with religion. This might well be so, but a clearer and indisputable fact is that both activities are carried out by human beings, and the latter do whatever they do by drawing from the same fund of emotions and imagination. The fact that different activities share certain features in the context of discovery or creation is thus entirely irrelevant from the point of view of what they are: it is by the context of validation or confirmation that they must be judged. That William Blake's hymn *Jerusalem* contains

blatant phallic symbolism would hardly be good grounds for invoking that sexuality and religion are akin activities.

4. The earliest reasonably comprehensive, yet incomplete, statement about the concept of state was given by Georges Lechalas (1886): "Concerning the world of material bodies, the principle of mechanical determinism asserts that the state of a system of material points at a given instant is determined by its anterior states and determines its posterior states." This is quoted on p. 52 of van Fraassen (1985).

5. See van Fraassen (1985), pp. 54–55, where he neatly spots the circularity involved in the statement by Lechalas quoted in note 4. As I have already said (p. 126), Strawson has recognized the importance of circular arguments, and my definition of state is an example of his irreducible concepts which, as he acknowledges, cannot be defined without circularity. (Strawson 1992, p. 22.)

6. That such a *convergence* to a fixed value arises for the sequence of ratios, when the independent variable approaches a given value, was mathematically demonstrated by Newton (although a great deal of work had to be done, even in the present century, in order to make his proofs really rigorous). The reader must appreciate that what Newton did was almost superhuman. Already the problem at hand was difficult enough because of the circularity involved, and on top of this, in order to solve it, Newton had to invent an entirely new branch of mathematics, the *differential calculus*, a subject of such central importance that it can be considered as the birth of modern mathematics. Alas, as it often happens, his invention was independently done at about the same time by Leibniz, thus giving rise to no small amount of jealousy and bitterness. To make matters worse, Leibniz used a different notation from that of Newton which, because it was far superior, gained quick acceptance. But nothing could ever detract from the extraordinary achievement involved in Newton's work.

7. For a height of 50 m, for instance, the error thus introduced in the time of fall is one hundredth of a millisecond. See, however, p. 191 for philosophical objections to this approximation.

8. Notice that we have two ways in which we can deal with the motion of falling bodies. One is by using Galileo's law of falling bodies, and the other by using Newton's second law. The first is an ad hoc rule that explains nothing else than the fall of heavy bodies. The second, instead, is a law of much wider application, able to deal as well with all forms of motion. The use of either law entails of course an *inference* (as a deduction from the given law), but we can say that the use of Newton's law is an *inference to the best explanation* because the explanation used is much wider embracing. We shall in fact see that scientific reasoning must always be such that it covers as wide a range of facts as possible.

9. The information value of a particular event is, in fact, also called *surprisal* in information theory. See for example Attneave (1959), p. 6.

10. See Pannenberg (1993), chap. 2.

11. See Williams (1965).

12. A brief account of *Naturphilosophie* and of its influence on two great discoverers of the nineteenth century, Ørsted and Hamilton, may be found in Altmann (1992), Chaps. 1 and 2.

TIME HELD ME GRUE:
PROJECTION AND PREDICTION

Time held me green and dying
Though I sang in my chains like the sea.
Dylan Thomas (1952), p. 161, *Fern Hill.*

The reader will by now be happily resigned, I hope, to accept that when we assert a very simple proposition like *all ravens are black*, a great deal goes on behind the scenes. First of all, we look around at a good bunch of birds, and we formulate the hypothesis that ravens are black. We then make what we shall now call a *projection*. (See also chapter 3.) This entails a process of induction by which we test the hypothesis (with all the precautions discussed in chapter 3), in order to support the passage from the unquantified proposition *these ravens are black* to the quantified one *for all* ravens. The problems entailed by the induction process are, alas, painfully well known to us by now. The main question is that, however thorough we are in sampling, there are always undetermined cases of the hypothesis to be projected. As a difference with the ravens, these undetermined cases can all be in the future, as for instance in the projection from *all popes so far elected have been male* to *all future popes will be male*. Projections that entail the time in this way are called *predictions*.

The problem that we have discussed so far in this book is how to project a known predicate, like *black* or *male*, but we now want to go much further than this because we have said that the main job of science and mathematics is that of setting up a language with which to describe natural phenomena, and, obviously, creating a language must entail creating *new* predicates: how do we license their use? For that matter, how do we license the use of any predicates at all? Fundamentally, what we want to find out is when a predicate is *projectible* in the sense of *projection* described above. This is a problem that has been much illuminated by a bit of experimental work in language creation done by Harvard philosopher Nelson

Goodman (1906–1998) around the 1950s,[1] which I shall now try to imitate. I must warn the reader, however, that what we are going to do will look pretty horrible, if not entirely nonsensical, but it is only by exploring the abnormal that one can understand what the normal is. And the final results will be truly significant: contrary to what might seem to be the case in this chapter, they will also have important *practical* applications.

THE NEW PARADOX OF INDUCTION: GRUE AND BLEEN

Emeralds may not be a girl's best friend, but we all know that emeralds are green, and we all expect that emeralds will *always* be green. Of course, we base this proposition on the experience of hearsay, or of looking at a jeweler's shop windows or, perhaps, of visiting the Topkapi at Istanbul. So far, nothing to worry about, until we start fiddling with the predicate *green* whereupon our cozy bejeweled world collapses.

Try this for size. Define a new color predicate as follows: a thing is *grue* in color if it is green and the time is prior to τ (*tau*), or if it is blue and the time is not prior to τ, where τ is a specific time in the future like 12 noon, 1 January 3000.[2] Now comes the paradox. If you want to project the hypothesis that *all* emeralds are green, the evidence class for this projection (as defined on p. 57) must be all the emeralds that you have so far examined but, because all the emeralds that you have examined must have been examined before τ, this evidence class is also the evidence class for the projection of the proposition *emeralds are grue*, a bizarre but unassailable conclusion.

We have undoubtedly a paradox here, and please do not dismiss it at this stage as unmitigated rubbish because, as you will see, it is a great deal subtler than you might think. Most people faced with this paradox will try to rule the predicate *grue* out of court, on the grounds that it is time-dependent, whereas the "good" predicate *green* is not, and that we have ample evidence that things do not *suddenly* change in color. Surprisingly, however, that *green* is time-independent is not true, or at least not *necessarily* true. Because we now invent a second predicate, *bleen*, as follows: a thing is *bleen* if it is blue and the time is prior to τ or if it is green and the time is not prior to τ. Look now at figure 1, and I hope that you will realize that the first two lines in it are nothing else than the definitions of *grue* and *bleen* as just given. Once you have persuaded yourself of this, look at the now *time-dependent* definition of *green* in the third row. A thing is *green* if it is grue and the time is prior to τ (confirm this with the first patch on the first line) or if it is bleen and the time is not prior to τ (confirm this with the second patch on the second line). Likewise, blue is defined as in the bottom line of the figure. So, we discover that our trusted old friends *green* and *blue* are just as time-dependent as *grue* and *bleen*.

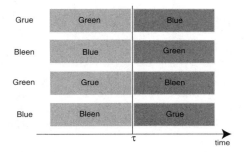

Fig. 1. Definition of color predicates

It thus appears that there is a perfect symmetry (but see later) between the grue-bleen and the green-blue languages. Why do we all use the latter and not the former? (And if you think that it is because no one bothers to invent such a stupid language, this is not so: I shall show in the coda that this type of language is unavoidable in quantum mechanics.) So, if you bear with me, I shall try to produce a scenario in which we can fruitfully investigate the features of both possible languages. Of course, the issue is whether there is some condition that makes it necessary for us to use one rather than the other language. As you probably remember from chapter 2, when we have to deal with the thorny question of necessity, it is useful to emigrate to fictitious imagined worlds in which our terrestrial prejudices can more easily be shed. So, we shall now move to a new science-fiction planet where we shall be able to discover the precise origin of the fundamental asymmetry of the two languages.

VIRBONOS

Virbonos is a planet inhabited by creatures entirely similar in their perception mechanisms and intelligence to Terrans. The same laws of nature hold in their environment except for one *changeable* natural law: whereas during the Virbonosian day (V-day) colors are identical to those in Terra, during the V-night all green objects (examined under normal light) become blue (for a Terran) and vice versa. As a result, Virbonosians do not know the predicates *green* and *blue,* and use instead the predicates *grue* and *bleen.* A Terran observer tries to understand the way in which Virbonosians use these predicates but, not being entirely politically correct, he firmly believes that his own green-blue language is, for some reason that he is sure to find out, superior to the Virbonosian one. To start with, he shows the Virbonosians colored wool threads, and he notices that when asked to point out the grue color, they pick the green thread during the V-day and the same thread (now colored blue for the Terran) during the V-night. Vice versa, when asked for the bleen color, Virbonosians point to the blue thread during the V-day, and to the same thread (now colored green for the Terran) during the V-night.[3]

As a result of these experiments, the Terran realizes that he can label a wool thread as grue and that the Virbonosians will always pick it up when asked for this color. He now thinks that he has found why his own language is the superior one, because he becomes convinced that the Virbonosians have to ascertain whether the time is V-day, in which case a green-colored thread is grue, or V-night, in which case a green-colored thread is bleen. When the Terran communicates this idea to his Virbonosian friends, they hotly deny that they have to know the time in order to predicate grue or bleen correctly. They support this statement by demonstrating a colorimeter which, when presented with a grue object, prints out the word *grue* and nevertheless does not contain any time-keeping piece in its mechanism: it merely has samples of grue and bleen colors against which the unknown color is compared. The Terran then resigns himself to accept that he himself can predicate the color *grue* without having to ascertain the time, by the simple device of carrying with him a sample of grue color, labeled with the word *grue*, and comparing always any unknown color against this sample.

Virbonosians, on their side, try to learn from the Terran how he uses the predicates *green* and *blue*, unknown to them. They notice that, when asked to point out the green color, the Terran picks the grue thread during the V-day and a different thread colored bleen during the V-night. They conclude that the Terran has to ascertain the time in order to use his predicates. In particular, they find it impossible to construct a comparison colorimeter that can determine these color values, unless a clock is built into it (because if the unknown color matches the grue sample, it will be *green* only if the time is prior to τ). The Terran vehemently denies that he requires the time in order to predicate *green* correctly: he explains to the Virbonosians that he appeals to his memory of this color. The Virbonosians are not too happy about this argument, which to them seems less objective that their use of the colorimeter, but they are polite and accept that, so far, there is no clear asymmetry in the use of the two languages, inasmuch as it is manifest that both of them can be used independently of each other and of a knowledge of the time.

It should be now abundantly clear that all the conditions are given for the occurrence of Goodman's riddle: any instance "*x* is grue and the time is V-day," which confirms the projection "all *x* are grue," is also an instance of "*x* is green and the time is V-day" which equally well confirms the (wrong) projection "all *x* are green." (Remember that *x* will be blue during the V-night.) All attempts to dispose of the paradox by alleging that the unprojectible predicate *green* is somehow defective, because its use requires a knowledge of the time, are clearly out of court in Virbonos and cannot therefore have any relevance to the logic of projection.

MORE TERRAN WORRIES

I am afraid that the Terran was one of the many philosophers from the fifties who believed that most questions could be solved by syntactic or semantic analysis. So, he was still convinced that there must be one such way of establishing an asymmetry between the two languages. Naturally, he appeals to a philosopher friend in Terra who immediately spots the problem (or thinks he does) and writes back: "Of course the green predicate is epistemologically superior to grue because, whether an object is grue, we have to find out whether it is green."[4] The Terran, however, realizes that this cannot be the answer to his worries because, for the distinction raised to have the value of a linguistic necessity, it would also have to hold in Virbonos, and he knows full well that even the word *green* does not exist in their language. He also knows that he himself can recognize *grue* by comparison with a sample, without having to remember what *green* is.

The Terran, however, finds another wonderful idea in his friend's letter, who suggests another syntactic magic wand which will reveal that *grue* is an inferior predicate: *positionality*. We already met this concept in chapter 2, where we saw that it is very dubious indeed. Since the Terran had not had the good luck of having read that chapter, he is fortunate that his friend defines it very carefully: "We shall understand a predicate to be positional if in order to find out as certainly as can be found out whether or not it applies to an object we have to find out its relations to some other *particular* thing (e.g., a particular instant of time, point of space, or physical object). We shall understand by a predicate being qualitative that it is not positional in this sense." And he goes on to claim as a "transcultural rule" "that universal nomological propositions containing positional predicates are not projectible."[5] Fortunately, the Terran remembers vaguely that a nomological proposition is a law of nature endowed with some sort of necessity,[6] and he gets excited thinking that at last he has got a good criterion to reject *grue*, as being positional. (This, although he knows full well that *grue* is manifestly projectible in Virbonos, where emeralds are grue now and for ever. He should thus have realized that positionality, even if it were significant, could not affect projectibility, except in a contingent manner, that is, a manner dependent on the particular environment in which the predicate is used.)

Thus, the Terran rejects the sensible view that positionality cannot be decided on purely semantic or syntactic grounds and confronts a Virbonosian friend. As you will see from the discussion that ensues, the distinction between positional and qualitative turns out to be untenable.

Terran. "My predicate *green* is qualitative because I do not need a sample in order to use it. I can use it, in fact, without any reference to any particular thing: I only use my memory, whereas in order to know whether something is grue you must always compare it with a sample of that color. Therefore, *grue* is positional."

Virbonosian. "You might not call your memory a *thing*, but to me it looks very much like one: surely when you use your memory of the color green, you are comparing what you perceive with some physical signal stored in your cerebral cortex. I, for my part, always compare colored objects with a small spot of grue color that we Virbonosians have in our retina. Thus, we are either both of us referring to *things*, in which case both our predicates are positional or, if you do not count brain states and sense organs as *things*, we both use our predicates as qualitative."

Even if this exchange may not have convinced the Terran, I hope better from my readers: the idea that positionality can be a semantic or syntactic property of a predicate, and thus possess linguistic necessity (which is the only type of necessity so far found!) is stone-dead. Therefore, whatever difference between *grue* and *green* that might exist, either in Virbonos or on Earth, cannot arise from any such linguistic necessity. We must look for something else because we would not like to walk the surface of the earth claiming that *grue* is just as projectible as *green*. Do not despair because the answer is now at hand and it will not come from any highfalutin linguistic nuance but from a good decent experiment.

RESOLUTION OF THE PARADOX: ENTRENCHMENT

The reader will remember the concept of a causal timescale discussed in chapter 5: if a certain phenomenon, like the burning of, say, 5 ml of pure ethanol in very carefully specified conditions takes place in a time interval T, say ten minutes, and the timescale used is such that propositions like the above one are true irrespective of the initial value of the time, then we say that the timescale is *causal*. It is also important to remember that the existence and the construction of such a timescale is manifestly contingent, being obtained by a huge circular process of successive approximations. The legitimacy of the causal timescale results entirely from a process whereby a vast internal system of successful cross checks is created, and we shall refer to this process as *entrenchment*.

The concept of entrenchment can be applied, not only to the construction of the timescale, but also to any proposition or predicate. It basically consists in producing a series of propositions and accepting or discarding them until an entirely self-consistent system emerges by a process which is akin to natural selection, as applied to the propositions used in establishing the system. Naturally, there must be a hierarchy among the mesh of propositions and predicates that are at any one time accepted as entrenched, in the sense that some propositions will at that time be entrenched beyond any possible doubt, whereas others, newcomers to the hierarchy, are, so to speak, on probation. Of course, when a proposition produces a consequence that clashes with a better-entrenched proposition, it must be seriously considered for elimination unless doubts emerge about the "better" proposition. Thus, the composition of the mesh is somewhat flexible: some parts of it will be solid as rock, others will be in a state of flux. But there is something that can never be challenged (for as long as we stick to events in our usual nonrelativistic environment): it is the causal timescale, the supremum of entrenchment, on the adequacy of which the whole of the vast edifice depends.[7]

What I shall now show is that in the Virbonosian scenario it is *only the projectible predicate that is compatible with a causal timescale*, whereas the unprojectible predicate fails to be so. It will then be clear that the same situation holds in Terra, and because of this symmetry between both places, we shall be able to conclude that the result emphasized, as applied in our environment, must be entirely independent of the contingencies of the latter. (Except, of course, for the possibility of a causal timescale existing.)

I shall define for this purpose a color-timer, designed to measure a time interval T (say ten minutes). On being set, this timer shows a patch of a certain color on its face, and T is the interval during which this color patch *retains its color*. Let us assume that the time is V-day, that the color patch chosen is grue, and that the timer will be read by the Virbonosians (V-reading) in the grue-bleen language, of course, and by the Terran (T-reading) in the green-blue language. Take as a presumptive causal proposition, for example, the following one: *in the time T and in rigorously standardized conditions, precisely 5 grams of ethyl alcohol will burn*. The experiment is done and it is repeated, and during the whole of V-day, causality is maintained: precisely 5 grams of alcohol burn, and this is so whether the V-reading or the T-reading are used. (Clearly, both observers see the color of the patch unchanged, during T, grue for the Virbonosian and green for the Terran.) But now we get near the time when V-night starts (time τ), and an experiment starts three minutes before τ. When τ comes (and thus the green patch changes color as far as the Terran is concerned), the Terran shouts '"now," and the weight reading is less than 5 grams, whereas the Virbonosian goes on until the grue patch changes and he measures precisely 5 grams. The enormously important conclusion we draw is that the grue-bleen language is compatible with causality (in Virbonos!) whereas the green-blue language is not so.

We can now propose a good criterion for recognizing when a predicate is *projectible*: it must be *entrenched*. Remember that this means that it has been tested and, not only that it has been proved projectible in the past but, most importantly, that its projection has not contradicted that of any other better-entrenched predicate. It is this second condition that makes this criterion acceptable because the first one would be somewhat circular, merely asserting that whatever has been projectible in the past will also be projectible in the future, an assertion that entails all the by now well-known difficulties of induction; plus the fact that it implicitly offers induction as its own justification. It is only by knitting a tight *mesh* of interrelated and entrenched concepts that entrenchment as a process acquires legitimacy. (That it guarantees the Truth is beyond both philosophers and scientists to say.)

Our grue timer is the best possible tool to test the entrenchment of the predicate *grue* in Virbonos. This is so because we have seen that *grue* is compatible with the causal timescale (whereas this is not so for *green*). On the other hand, as I have said, the causal timescale may be considered as the supremum of entrenchment, since nothing in human experience that clashes with it in our ordinary well-known environment can be accepted as entrenched: such a predicate or event would be akin to a miracle, and miracles, however wonderful, cannot by their very nature ever be considered entrenched. Whereas *grue* succeeds in Virbonos, it should be clear that it would clash with the causal timescale on Earth and thus that we must reject it.[8] This will not come as a surprise to any of my readers, but what is important is that we now know why such predicates are not legal tender outside philosophy books. But more about this later.[9]

Meanwhile, we should go back to the concept of empirical necessity which we introduced in chapter 4 in order to deal with causality. As you will remember, the problem was that the causal relation cannot be logically necessary, neither can it be observable. In practice, however, when a causal relation is adduced, one expects the conjunction of cause and effect to have some form of urgency, and I proposed to label this as *empirical necessity*. We also gave some empirical status to the propensity of the mind, resulting from the evolution of the human species, to apprehend such causal relations. The empirical concept of entrenchment allows us now to add strong support to the use of the principle of causality. In fact, entrenchment adds another dimension to the concept of evolution, recognizing that both the physical and the mental are subject to the same principle of natural evolution, from which mental as well as physical structures emerge in the self-organizing system that nature is. Entrenchment is the process whereby the evolution of the mental structures takes place, and it is thus the empirical way in which causal hypotheses or proposed laws are legitimized, leading to their empirical necessity. Like everything else in the entrenchment process, the latter is always in a state of flux: what may one day be conceived as an empirical necessity, may the next day be ruled out of court if a violation of entrenchment arises, that is, a clash with a better-entrenched law or predicate.

PREDICTION OR PROPHECY?

Let us spend a little more time in Virbonos. Imagine now that the V-day is very long indeed, say like one year of Terran time. Assume also that our Terran observer arrives in Virbonos at the beginning of such a day: in his conversations with the Virbonosians, neither party would be able to establish any semantic difference between *grue* and *green*. The Terran, therefore, will happily project *green* thinking that he follows the Virbonosians in their well-entrenched projection of *grue*. He will for instance write home saying "here, like in Terra, emeralds are always green, although the natives call their color *grue*." When the Virbonosian night arrives, his expectations are, of course, disappointed. Should he blame himself for having made the wrong projection?

Not at all: any causal timescale that appears entirely uniform over a certain period of time, might nevertheless contain disturbances over extremely long periods, of which we might know nothing. The task of time-projection or prediction for those who want to stick to rational evidence is to use such a period of the total timescale as it is observationally available to them. Having examined all the evidence over as long a period of time as they can, they will construct their causal timescale (perhaps, at a very primitive level, not even explicitly) and will make only such predictions as are compatible with this causal timescale. Any other form of time-dependent projection is not prediction at all: it is prophecy. No, the Terran should not blame himself: even if his result was wrong, his projection was correctly made.[10] Whether the result of a projection is right or wrong is not a fruitful problem: nature, time, and events can always play tricks. What I am concerned with, as Goodman was, is the way in which a projection is rationally supported.

CODA

One by one the bastions of logical necessity are falling down. We have learned from Hume that the relation between cause and effect is neither logically necessary nor observable; and we explicated Hume's *custom* associated with the use of causality in terms of the principle of natural evolution. This principle itself has been entrenched with reference to numerous instances where it works successfully and to the experience gained in attempting to falsify it. We have now gone further because the very vocabulary that we use also appears to be grounded on a similar process of entrenchment. This, of course, very much follows on the lines proposed by Nelson Goodman, and I am acutely aware that many philosophers profoundly disagree with these views and feel able to make strong claims for logical necessities. Simon Blackburn, for instance, utterly rejects Goodman's entrenchment criterion, and he justifies his rejection by his (correct) finding that it "is logically impossible to tell whether an explosion occurs at midnight without knowing that it is midnight."[11] This must be accepted, of course, but *you cannot know what midnight is* unless you have a timescale, the most heavily entrenched of all human constructions. So, after all, entrenchment still survives, as we should expect, since we have taken the view that it sustains the whole of language. In what remains of this book we shall see how this concept repeatedly clarifies problems in the philosophy of science and, most importantly, we shall learn how to use it properly in order to avoid erroneous consequences. Already in the next chapter, where I shall discuss laws of nature, such as they are, we shall see how the limitations of entrenchment must be understood.

The reader must grasp the relevance of our discussion to the problem of induction. If instead of *grue* we had defined *gred*, in precisely the same manner from green and red, this predicate could have perfectly safely been used to project the color of unripe tomatoes. That it is not so used is a fact of language of no importance at all. To tease the problem of induction we must deal with a situation where it is the laws of nature, and not just mere objects, that are presumed unstable, because induction in science is deeply connected with the concept of regularity of nature. It is because of this reason that we migrated to Virbonos, and thus to a situation where the change entailed in grue is not just a simple evolution of the state of an object, but the result of some deep-seated variation in a natural law.

Before we leave this chapter, it is worthwhile reflecting once more about the nature of the predicate *grue*. Our Terran habits are so engrained that it is difficult to abandon the prejudice that *grue* is somewhat defective because it can be predicated only in terms of the colors green and blue. We have seen, however, that this is not so, but I want to insist that for a Virbonosian (and as I shall prove in a moment, we are all Virbonosians to some extent) not only does he not need the colors blue and green, but he does not even know these colors, and he *perceives grue as much as a single color as we perceive green*. That this is so follows from the Virbonosian having a grue patch in his eye: because his predication of grue entails comparison with this patch, this is for him such a constant reference as for us on Earth it would be a green wool thread: we cannot see it changing color, and neither can the Virbonosian see his retina patch doing that.

The important thing is to appreciate that, because of this, grue can be construed as a perfectly simple predicate: it is certainly so in Virbonos. The reader will now say: what has Virbonos got to do with us? But Virbonos is an example of an environment where predicates get entrenched, just as they get entrenched on Earth, and the important conclusion we reach is that a predicate, entrenched in a given environment as a *simple single* predicate, may appear to those who have evolved in a different environment as a *composite* predicate. And this is relevant to us because, although Virbonos does not exist, there is in nature an environment in which *we have not evolved* (just as we have not evolved in Virbonos) and about which, however, we want to discourse: this is the environment of the microworld, of electrons and photons and the like. This environment is as alien to us as Virbonos is, and we can now understand that an intelligence similar to us, but evolved in the microworld environment, might perceive certain predicates, or even nouns, as single, which we can only describe as composite. This is precisely what happens to the so-called particles of the microworld because, as we shall see, we can only perceive them sometimes as waves and sometimes as particles, whereas in fact for the microworld-evolved intelligence they would be single objects with a single form of behavior, that might be called *wavicles*. This name, although it was coined by A. S. Eddington in 1928[12] on the same composite principles as used some twenty years later for *grue*, has hardly ever been used: humankind cannot bear very much reality, if it has to be imagined as perceived with eyes evolved in alien environments. So, in the same way that green and blue are the wrong predicates to use in Virbonos, be prepared to find that particles and waves are the wrong names to be used in the microworld, and yet that their use continues despite Eddington's wise proposal.

NOTES

1. The paper that introduced the paradox (which he called the new riddle of induction) is Goodman (1946), but to this day the best and clearest account of the subject is given in his seminal book, Goodman (1965). These works opened a veritable flood in the philosophical literature, the grue industry being a major source of employment for more than

two decades. I am afraid that I shall not be able to do justice to the literature, and that I shall only mention the works that are more directly relevant to my discussion. A very good treatment of projection and *grue* appears in Horwich (1982). A recent collection of reviews appears in Stalker (1994). See also Schulte (1999), Chart (2000), and Schulte (2000). Most of the present chapter comes from a paper which I read at the Pater Society, Brasenose College, in 1965, and which was circulated privately. I am particularly grateful to the late Michael Woods for his comments.

2. This is not exactly the definition used by Goodman, but it is the one most convenient for the purposes of this book and was introduced by Barker et al. (1960). Three major definitions of *grue* are discussed in F. Jackson (1975), but other constructions of the predicate have also been used, as in Blackburn (1973). Notice that for the definition given to make sense, you cannot come with a brush and paint the patch blue at the time τ. Neither can you act on the sample with a magnetic field, or radiation or any such thing, which would merely be a more sophisticated version of the paint brush: whatever happens at τ must not be induced by a *physical* change of the sample. This must be so because the use of projectible color predicates entails that the object to which the predicate is applied is not subject to change: when I say that a tomato is green my subject is this *unripe tomato*. If I allow a physical change, like ripening, I cannot expect that the predicate will be valid or that it can sustain a projection. What we are examining is not the stability or otherwise of objects, but rather that of a natural law.

3. The definition of *grue* and *green* that follows from the above description is not identical with the one that I had first quoted, insofar as the instant τ which I had introduced is now recurrent. If, however, the duration of the Virbonosian day is allowed to increase indefinitely, the two definitions coincide. The proposed definition includes therefore the original one as a special case and, as we shall see, has some advantages for the purpose of the discussion that follows.

4. This is a paraphrase of Swinburne (1973), p. 107.

5. Swinburne (1973), p. 106. The claim of transculturality has already been rejected on p. 44. The first philosopher to make strong claims for positionality was Carnap (1947–48). In answer to him, Goodman (1947) pointed out that positionality was very difficult to define in practice, although later (Goodman 1960), he took up a suggestion made to him by Noam Chomsky to the effect that *grue* might be held to be positional, on the grounds that instances of its application *before the time τ cannot match those after τ in color*, while this does not hold for *green*. Goodman, however, considered this property too ad hoc to be of interest. (The italicized proposition, however, is not true in Virbonos.)

6. See Nagel (1961), p. 51 for a discussion of *nomological*.

7. The concept of entrenchment, and its application for solving the new riddle of induction is due to Nelson Goodman himself, although he did not make use of the causal timescale. (Of which idea he, however, approved in a letter to me of around 1965.) The concept of a hierarchy of entrenchment is implicit in his work, as for instance in the following statement: "One principle for eliminating unprojectible projections, then, is that a projection is to be ruled out if it conflicts with the projection of a much better entrenched predicate." (Goodman 1965, p. 96, but see also Schwartz et al. 1970.)

8. The reader who wishes to explore the use of grue timers on Terra must be careful not to fall into an attractive fallacy. It might be thought that it would be perfectly possible to build such a grue timer as follows. It would have to contain a clock that could recognize the time τ (as chosen by any philosophical friend). When this time is reached, a blue patch would be substituted for the green one, and this would remain visible until the end of the interval T. Such an alleged grue timer would be perfectly compatible *at all times* with our causal timescale. What is wrong here is that the color patch is not grue at all! Remember that a color predicate can be predicated for purposes of projection as applying to one and only one object:

even if we have two red disks, each disk is *individually* red. Thus the predicate *grue* must pertain to a *single* color patch, whereas our alleged grue timer contains *two* objects (or at the very least two different states of the same object, as might be induced by an electric field), one green and one blue, which are conveniently permuted, but neither of which is grue. (When we define an emerald as grue in color, we do not consider such possible changes of color from green to blue as might be produced by an electric field or radiation: if a change in color occurs, it must not be induced by any other physical effect, as discussed in note 2.)

9. (This note is addressed to the more specialized reader and it can safely be skipped.) It might be useful to discuss a difference in projectibility between (at least apparently) time-dependent predicates as *grue* and time-independent ones. We can define *grue* as follows:

$$gx = x \text{ is green,} \quad bx = x \text{ is blue,} \quad a\tau = t < \tau,$$
$$grue \; x \Leftrightarrow (a\tau \cdot gx) \vee (\sim a\tau \cdot bx).$$

Because, in order to establish the disjunction, it is sufficient to establish the true value of any one of its prongs (see chapter 3, table 5), the evidence class of grue x *is* also the evidence class of $(a\tau \cdot gx)$. So far, this is precisely Goodman's paradox. The evidence class of $(a\tau \cdot gx)$, however, because it requires that $a\tau$ be true, is also the evidence class for the hypothesis $\sim\exists x(\sim a\tau)$, where $\exists x$ is read as "there exists an x." The definition of grue thus entails a contradiction, but this can be considered to be trivial, because the hypothesis $\sim\exists x(\sim a\tau)$ means that nothing exists after the time τ, a conclusion that can be dismissed out of hand, entailing as it does the end of the world. In the case of a time-dependent predicate such as *grue*, the logical contradiction just revealed is not of any great importance (this is why I prefer to use the concept of entrenchment to resolve the paradox), but the same is not the case when dealing with projection of time-independent predicates, a question considered by Scheffler (1964) and Swinburne (1973). The following example is inspired by Altham (1969), who reinvented Scheffler predicates. Define:

$$gx = x \text{ is green,} \quad bx = x \text{ is blue,} \quad lx = \text{ has less than } 100 \text{ facets.}$$

We now predicate of a gem that is *greecet* if *either* it has fewer than a hundred facets and it is green, *or* it has not less than one hundred facets and it is blue:

$$greecet \; x \Leftrightarrow (lx \cdot gx) \vee (\sim lx \cdot bx).$$

Here, the evidence class for $lx \cdot gx$ confirms greecet x and therefore the hypothesis that gems with no less than one hundred facets must be blue. It is easy to see, however, that this projection must be overriden because it entails a contradiction. This is so because the evidence class for $lx \cdot gx$ also confirms the hypothesis $\sim\exists x(\sim lx)$, which entails greecet $x \Leftrightarrow (lx \cdot gx)$, contradicting the disjunction. The overriding of the predicate *greecet* in this form may be considered, however, as an instance of entrenchment in the Goodman sense, the existence of gems with more than one hundred facets being (presumably) a better-entrenched predicate than *greecet*.

10. This parallels in an inverted way a statement made by Goodman himself. At the time when such inventions were fashionable, Davidson (1966) had created a compound name, *emerose*, and a compound predicate, *gred*, and had proposed the projection "emeroses are gred." In a reply to this paper, Goodman (1967) asserts that this projection is true, although unsupported. Thus, just as in the case of the Terran projection in Virbonos, it must be appreciated that there is no necessary linkage between support (or entrenchment) and the truth or adequacy of a projection.

11. Blackburn (1973), pp. 76–77.

12. Eddington (1928), p. 201.

8

THE LAWS OF NATURE

But is it not the deepest Law of Nature that she be constant?
Thomas Carlyle (1831), Book III, Chapter VIII
(*Natural Supernaturalism*), p. 247.

I must give the reader a warning: if you see a book entitled *The Loch Ness Monster*, you cannot necessarily expect that it will actually *deal* with the Loch Ness monster. It may be that the book will be quite agnostic about the monster, or that it will actually prove or disprove its existence, or that it might even deal with the characteristics that the monster should have, were it to exist. I hope that my reader will approach the subject of this chapter in the same salubrious state of mind. The question about what the laws of nature are, if they *are* at all, is a thorny one, and the less we assume about it at this stage the better it is.

To start with, some statements are used in science that are called laws and others, which appear to have similar if not identical powers, are called by other names, such as *equations*, *principles*, *rules*, *transformations*, and so on. The distinction is largely historical, and it might be partly due to the degree of credence that was put on those statements when first formulated. Thus, Newton's second *law* does exactly the same job as Schrödinger's *equation*, but there are people whose life to this day would be happier if the latter did not exist. As for the Lorentz *transformation*, when it was first formulated, it was thought to be no more than a clever bit of mathematics: to call it a *law* would probably have been somewhat offensive to the traditionalists, and the name stuck. The use of the word *principle* appears to be more consistent, as attached to a law with a very wide domain of application *and* of a qualitative or normative nature, as in the *second principle of thermodynamics*, the *principle of relativity*, or the *principle of symmetry*. Thus, although the Pauli *principle* has a domain of application not wider than that of Schrödinger's equation, it is probably called a *principle* and not a law because of its qualitative nature. But the reader must not expect these names to be used in an entirely consistent way. Do not ask me why this is so, but the fact is that most scientists would be more than happy to have a law to their name, but if it could be a *principle* then their felicity would be boundless. Also, if a rule is produced by a very famous man, it is probably more likely that people would call it a *principle*. Be it as it may, whenever I use the word *law*, the reader must realize that I include any of the variously named propositions so far mentioned, as well as many others similarly used.

So for the time being, as regards laws of nature, we do not know even their names, let alone what they are, and whether they exist at all. It is a fact, however, that most people have some expectation that laws of nature do exist, but this is a question I shall consider much later in this chapter. I shall try first to do something about the more modest problem of how to recognize that a given proposition is lawlike, for which purpose lawlikeness will have to be described (as far as it is possible, which is not very far!). There was a time a few decades ago when people went around trying to find semantic or syntactic (in plain language: grammatical) criteria that would *by themselves* permit the recognition of laws of nature. The idea that this is possible is just about as sensible as the expectation that a Martian visitor could identify the prime minister of the United Kingdom on anatomic or physiological grounds. What would you tell this poor Martian if you see him walking around with a reflectometer, trying to measure the degree of grayness or pinkness on people's skins, with a view to identifying the prime minister? I think that most people would explain to this hapless visitor that, basically, all humans have the same anatomy and physiology, and that what makes the prime minister what he or she is, is the nature of his or her *relations* to other people (use of handbags included) and that there is a whole hierarchy of such relations, between the prime minister and the Cabinet ministers, between the latter and the members of Parliament and so on. At each level in the hierarchy, what makes a person what he or she is, is not anatomy or physiology, but the particular set of relations existing within the structure. What is true for the political world is also true for propositions: they are or are not lawlike because of the way they relate to other propositions, that is, they are lawlike not *because of what they are but because of what they do.*

One of the good things one learns from philosophy is that mistakes are useful, if not important. So, having delivered my little sermon about the ways of the rightful, I shall forget my own injunctions, in order to find out how far one can go in trying to ascertain lawlikeness by purely syntactic or semantic criteria: you will see that although we shall not succeed fully, we shall learn a few useful ideas. This is so because, in the same way

that we could tell our Martian friend that, as far as physiology requirements go, the prime minister cannot (normally) be a person in a state of
brain death, there are certain minimal formalities that propositions must
satisfy for them to be candidates to be laws of nature. A little knowledge
of these formalities will be useful in dealing with practical examples.

THE FORMALITIES OF LAWLIKENESS

The first thing we notice about lawlike propositions is that they entail
some element of generality, which means that there is never a critical
instance that determines the proposition as fully confirmed. Consider a
couple of examples:

> Water (under stated conditions) boils at 100°C.
> All sodium salts burn yellow.[1]

Although the first proposition does not use the universal quantifier
all, this is implicit: what we mean is that *all* samples of water under stated
conditions (like normal atmospheric pressure, etc.) boil at 100°C; clearly,
however many samples we try, there will never be a last one that will
determine the validity of the proposition. Likewise for the second proposition. One must be careful, however, not to assume that the mere presence of the word *all* in a proposition means that the universal quantifier
is used because the set quantified by it might be finite:

> All the coins that were in Christopher Columbus's purse the day when Piero
> della Francesca died were made of silver.

It is clear here in principle that, *if we had access to Columbus's purse as
it was on 12 October 1492*, we would be able to determine this proposition:
the critical instance would be the last coin that we extract from the purse.
But we do not have access to the purse, and yet we would like to rule this
sentence out of court as lawlike. There is a criterion that helps reveal a
proposition as a candidate for lawlikeness: it must be able to support a
counterfactual. (See p. 62.) It is easy to see that the first two propositions
on this section satisfy this criterion:

> If this sample of water were to be heated in stated conditions at 100°C it would boil.
> If this sodium salt were put in a flame it would burn yellow.

These sentences make sense, of course, if one accepts that counterfactuals, despite their describing nonfacts, do say something about nature
(of which more in a moment). Let us do the same for Columbus's purse:

If this penny had been in Christopher Columbus's purse the day when Piero della Francesca died, it would have been made of silver.

This is just about the best that we can do in the way of getting a counterfactual here, and it is nonsense because we *know* that the penny in my hand is made of copper. So, we now have two characteristics to look for in ascertaining lawlikeness. One is that the proposition must not be determined by a single instance of its application, and the other is that the proposition must sustain a counterfactual.[2]

But what about *necessity*, you might now ask, because most people *feel* that laws must somehow state necessities, although we already know that to claim necessity is a very risky thing. The nearest to it that we have so far been, except for fairly trivial linguistic necessities, is in relation to causality, although you will remember that Hume's skeptical answer was that there isn't such a thing as necessity. Even he realized, however, that his alternative concept of *custom* had to have some support in nature. We recognized, in fact, that *under very strict conditions*, a degree of necessity, which we have called *empirical necessity*, can be claimed for the causal relation, this modality being grounded on the fact that the use of causal relations is legitimized (or at least it is *naturalized*) by the extended version I gave of the anthropic principle. So, in searching for some form of necessity for natural laws, it is tempting to assume that this might come from the fact that all laws are causal, but this is not the case, since we shall find examples of non-causal laws. It will turn out, however, that this concept of empirical necessity can usefully be transplanted when dealing with natural laws.

This is a question that will be settled a little later, but, in order to understand how far necessity might come into the picture, at least for laws that are in fact causal, it is useful to reword as *causal* statements the two so far successful propositions:

Heat causes water to boil in stated conditions at 100°C.
Sodium salts are the cause of the color of flames turning yellow.

It is thus clear that at least in laws that share this nature of causal laws, we can claim for them what we have called empirical necessity, as directly arising from our treatment of causality. (Remember that this means that if the cause does not exist, the effect cannot appear, a conception of causality essential in science but not otherwise accepted in other forms of discourse.) It is also useful to notice in these two cases that the original presentation of the presumptive lawlike propositions was implicitly dispositional because what we were saying at first about water and sodium salts, respectively, was really:

Water has a disposition to boil in stated conditions at 100°C.
Sodium salts have a disposition to color flames yellow.

It is clear that the statements in question should be amenable to translation into counterfactuals, since dispositions are essentially linked to such propositions, and this is precisely what we have found. We shall deal a little later with the general problem of necessity, however, because we still have to complete our treatment of the formalities of law making.

HOW TO USE LOGIC TO FORMULATE LAWS. CAUSATION

Let us use our newly acquired criteria to discover whether our old friend, the ravens, is lawlike, which I shall do by writing it followed by its counterfactual:

All ravens are black.
If this bird were a raven it would be black.

This is a plausible proposition, and it is also clear that the universal quantifier in the first line ensures that the proposition cannot be determined by a single instance. So, we appear to be alright, for the moment. Compare this, however, with the sodium salts:

All sodium salts burn yellow.
If this were a sodium salt it would burn yellow.

I hope that the reader will notice that, although formally everything seems to work well, and in the same way, in both cases, the two counterfactuals obtained have a very different status. If we look at the last sentence quoted, it is something that we might expect to hear in a laboratory: a chemist burns a sample of an unknown substance, finds that it burns green, and then mutters the proposition quoted as an explanation of why she enters in her notebook: *Sample does not contain sodium.* On the other hand, outside a play by Ionesco, would you expect an ornithologist proudly to hold a parrot up and say: *If this bird were a raven it would be black?* This asymmetry between these two propositions should alert us to the fact that, although they appear to have the same structure, the underlying logic is quite different. Let us reproduce the logical notation of the ravens, from p. 56. It was:

$p = x$ *is a raven,* $q = x$ *is black.*   *For all x ($p \Rightarrow q$).*

Remember that this means: for all x, if x is a raven, then x is black, a proposition that is true even when the antecedent p is false but the consequent q is true. (Nonravens may be black without falsifying the implication.) The situation is quite different for sodium salts. If x is not a sodium salt, then x cannot burn yellow. We should not therefore use the material implication (conditional) but rather the biconditional:

$$p = x \text{ is a sodium salt,} \quad q = x \text{ burns yellow.} \quad \text{For all } x \ (p \Leftrightarrow q).$$

This means: for all x, *if and only if* x is a sodium salt, x will burn yellow, a proposition that is false if the antecedent p is false and the consequent q is true. (See table 5, chapter 3.) Before we move any further, the reader must recollect the caveats entered in chapter 4, about the use of logic in dealing with causal relations. First, it must be remembered that the causal connection is always taken in science to entail an *if and only if* relation between cause and effect; and, second, even accepting this interpretation, the expression of this relation by the logical notation of the biconditional leaves implicit constraints (*ceteris paribus*) that are assumed in science to make this biconditional valid, but which would not be acceptable in more general contexts.

We can now safely consider two important points that arise from this analysis. As shown in chapter 3 in relation to table 6, both the conditional and the biconditional are logically equivalent to their contrapositive forms:

$$p \Rightarrow q \text{ is logically equivalent to } (\sim q) \Rightarrow (\sim p),$$
$$p \Leftrightarrow q \text{ is logically equivalent to } (\sim q) \Leftrightarrow (\sim p).$$

In the case of the ravens, we have seen that the contrapositive is of no use. In the case of sodium salts, on the other hand, the contrapositive is a most important rule in analytical chemistry. When we write: For all x, $\{(\sim q) \Leftrightarrow (\sim p)\}$, we say that for all x, *if* and only if x does not burn yellow, then x is not a sodium salt, which is precisely the way the analytical chemist reasons in the laboratory.

The second point that arises from these results is that in the case of the ravens, because we only have a material implication, the proposition in question does not sustain a causal statement (in the scientific sense), whereas the opposite is the case for sodium salts. So, although the ravens proposition satisfies the two conditions of determination and counterfactuality discussed in the previous section, it cannot sustain a causal statement; it is arguable, in fact, that the proposition is or is not a law and it is a pure matter of convention to label it as such. I certainly would prefer to call it a *property*, this perhaps being the lowest possible rung in the hierarchy of lawlike statements. The fact that it is not causal is not the reason for this demotion because we shall find in the next section a very important law, justly called a *principle*, which is not causal. It is perfectly sensible to say of the ravens that it is lawlike, but that it should be called a property, rather than a law, because it lacks both generality (it does not apply to all birds) and specificity (there are many other birds that share the blackness property with the ravens).

MORE ABOUT LOGIC AND LAWS. PRINCIPLE OF SYMMETRY

It will be very illuminating to see how a little logical analysis can be a remarkably powerful guide in understanding the intricacies of the principle of symmetry, now recognized as one of the most fundamental and general laws for the study of nature. This principle is so important that we shall come back to it later to discuss its history and origins: I only want at this stage to describe such essential ideas as are necessary to formulate the various versions of the principle that are used in practice.

If we look at the balance in figure 1, we can easily recognize that this object possesses a plane of symmetry, shown in the figure, such that each material particle at a given perpendicular distance from the plane on its right, has a precise identical counterpart at the same distance on its left. This is the case because the balance is in equilibrium and this state of equilibrium, that is this symmetry, will persist for as long as we place *identical weights* on both plates of the balance.[3]

It is very easy, when considering this example, to imagine that there is some principle of conservation of symmetry in operation, more or less on the lines that the initial symmetry of a system is maintained if nothing acts to disrupt it (like placing asymmetrical weights on the balance). As we shall see in a moment, this supposed principle is not unreservedly true, but let us try to formulate it within our logical notation. Let us call p the antecedent state of the system (that is, the state of the balance when two weights, identical or otherwise, are placed on the plates) and q the consequent state of the system (that is, the state of the balance, either in equilibrium, or with one arm down where the heavier weight is, in which case the symmetry plane has disappeared). It is traditional in this subject to label the antecedent and consequent states *cause* and *effect* respectively although we shall see that no causal relation, in the scientific sense, is entailed by the symmetry principle. I shall follow this usage, but in order to help the reader to make the necessary mental transposition, I shall place these words in quotes. (Although *cause* and *effect* are used but sparingly in science, even when their use is correct, they are freely employed in this subject, albeit illegally, by even the most hardheaded experimental physicists.) Let us call x an element of symmetry (such as the plane

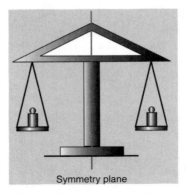

Figure 1. A symmetry plane
Only the trace of the plane is shown

shown in the figure) which may or may not persist from the "cause" p to the "effect" q, and use the following notation:

$P= x$ *belongs to p (the "cause"); $Q= x$ belongs to q (the "effect").*

If we expect symmetry to be conserved, a plausible assumption, we must expect one or the other of the following propositions to hold:

(1) *If x belongs to p **then** x belongs to q: $P \Rightarrow Q$.*
(2) *If **and only if** x belongs to p **then** x belongs to q: $P \Leftrightarrow Q$.*

You must now remember (see table 5 of chapter 3) that the material implication in (1) is true in case the antecedent is false and the consequent is true, whereas the biconditional in (2) is then false. The meaning of this is as follows: whereas (1) is true even when the symmetry element does not appear in the "cause" but only in the "effect," this is not so for (2). The latter would thus be strong principle of symmetry conservation, since it does not allow symmetry to be created in the "effect" that was not present in the "cause" (P false and Q true). It is very important for us to discriminate between these two presumptive principles because it is only the second one that has the form of a causal relation. No amount of logical or philosophical analysis, however, will allow us to know which of the two alleged principles is the true one (or at least more likely to be the true one). As always in such a case, one has to interrogate nature.

There is one fundamental fact about nature that points to the serious possibility that symmetry not present in the "cause" might be created in the "effect": this is the principle that requires that in any *spontaneous* physical process, disorder always increases. This principle (which is the mother and father of the second principle of thermodynamics) will be discussed at some depth in chapter 12, but we must understand here its implications as regards symmetry, especially because they depend on a result that most people get wrong at first: *more disorder*, surprisingly, *means more symmetry*.

You will think that I am wrong, because if you look at the symmetrical balance in figure 1 and you add a bit of disorder, for instance, by way of an extra weight on the right, symmetry is lost. But this is not a *spontaneous* process, which is one in which you are allowed to take some action on a system in order to change its state, which must then evolve isolatedly without further intervention. The relevant feature that you must understand is that in changing the state you *must not change* the system itself. (Which you do if you add weights to the balance.)

Consider a crystal, that many people might regard as an archetype of a highly ordered and highly symmetrical system. If you draw an arbitrary plane through a crystal, however, this will not in general be a symmetry plane: it is only when the plane goes through a layer of atoms that this might be so. On the other hand, *any* plane drawn *anywhere* through a liquid, which is a highly disordered structure, is a symmetry plane because liquids are regarded as homogeneous, so that a small portion of them anywhere is identical with any other portion: a liquid is thus far more symmetrical than a crystal! Which leads us to a spontaneous process whereby symmetry is created: if you melt a crystal, its symmetry will be *increased*. (Notice incidentally, again, what I meant by spontaneous: when a crystal melts, it merely changes its state, whereas in the balance, if you add a weight, you change the system.) Another very clear example in which symmetry is created in going from the "cause" to the "effect" can be understood from figure 2. On the left of this figure we represent a container which has a partition that, in principle, would be a symmetry plane of the container, were it not for the fact that the left compartment is full of a compressed gas, while the right-hand side one is under vacuum. Thus the system has no vertical symmetry plane at all. The partition, however, has a perfectly symmetrical valve which may be opened maintaining the symmetry of the partition with respect to the walls of the container. It is well known that when this system is allowed to evolve spontaneously, the ordered arrangement, whereby all the gas was on the left, breaks down and that the gas expands to occupy uniformly all the space now available to it. Disorder is increased, and *as a result*, a symmetry plane appears.

The consequence of these considerations is that the strong principle of symmetry conservation is *not true* for the following reason: the

Symmetry plane

Figure 2. Symmetry not preexisting in the "cause" may be created in the "effect."

biconditional (that is the alleged principle) is false if the antecedent is false (no symmetry in the "cause") and the consequent is "true" (symmetry in the "effect"), which is precisely what happens in figure 2. The principle that we have, instead, is the one given by proposition (1) earlier, $P \Rightarrow Q$, which requires that the symmetry of the "cause" *must* appear in the "effect" but, on the other hand, it permits a symmetry to appear in the "effect" which was not in the "cause." Because we cannot use the biconditional, the symmetry principle is not causal in the scientific sense we use for this word.

We can immediately obtain a second form of the principle of symmetry by remembering that the conditional $P \Rightarrow Q$ is logically equivalent to its contrapositive $(\sim Q) \Rightarrow (\sim P)$. (See p. 53.) We thus have two symmetry principles, of which the first is merely a transliteration of (1) above:

(A) $P \Rightarrow Q$: *If an element of symmetry belongs to the "cause," then it must also appear in the "effect."*

(B) $(\sim Q) \Rightarrow (\sim P)$: *an asymmetry in the "effect" must also be found in the "cause" or, in other words, if an element of symmetry x does not appear in the "effect," then x cannot appear in the "cause."*

Because these are important principles, it is worth paraphrasing them. (A) *means that elements of symmetry must be* preserved, but that they can also be created, that is, appear in the "effect" without having preexisted in the "cause." (B) means that asymmetries can *never* be created, that is, they can never appear in the "effect" not having preexisted in the "cause."

Although these principles were first formally enunciated by the French physicist Pierre Curie in 1894, the first of them (as we shall see later) has been in use in one form or another for almost as long as written records exist in our civilization; and both of them are in constant use to this day. The relation between the two enunciations given is a very subtle one: a reader who believes that logical equivalence of two propositions entails redundancy of one of them must disabuse himself of this misconception. The main problem is this: we must accept that symmetry, like any other property of nature, must be determined experimentally, rather than being dreamed up from spurious a priori considerations. We must be acutely aware of this requirement: it is only too easy to think of symmetry as a geometrical property; and geometry is one of the branches of human knowledge in which the idea that you can know things a priori has bitten most deeply. (I must therefore apologize for my bogus treatment of the balance in figure 1: please see p. 182).

Having this in mind, we can now examine the two forms of the principle of symmetry given above. Clearly, if we want to have an experimental view about symmetry, we must examine effects because it is they that are observable. If I use the form (A) of the principle and I observe an element of symmetry in the "effect," this does not guarantee, as we have seen, given the properties of the conditional, that it will also be present in the "cause." This is precisely the case of figure 2, where we observe an element of symmetry in the "effect" which is not present in the "cause." This form of the principle is thus not experimentally useful on its own in order to determine the symmetries of the "causes."

It is the form (B) of the principle of symmetry that lends itself to experimental determination of elements of symmetry or, rather, of their absence, because from observing an asymmetry in the "effect," we can draw a clear conclusion that that asymmetry must also be present in the "cause."

Unfortunately, it was the form (A) of the principle of symmetry that was mostly used, probably for centuries at least in an implicit form, version (B) not being understood until Pierre Curie pointed it out. I say "unfortunately" because the symmetry law (A) invites careless thinking. This is so since, in order to use it, you have to start by assuming that you know the symmetry of the "cause," and many people thought without a shadow of doubt that they knew such symmetries a priori. You do not have to be mystically minded to entertain such a belief. If you are presented with a real balance, like that of figure 1, you may think that it is obvious that it has a symmetry plane, as I have myself asserted. However: it is one thing to say that a *figure* has a certain symmetry, because this follows from geometry, but to pass on from the symmetry of a figure to that of the object that it represents is utterly erroneous[4] and it is, in fact, one of the most efficacious ways in which a person can go wrong beyond the nightmares of lunacy, as I shall illustrate in the next section with practical examples.

Before we do this, I shall consider a couple of points. On using the principle (B) experimentally, asymmetries in the antecedent can safely be determined, and from a collection of such results, it might be possible to discover the elements of symmetries of a system. It is only when these are known beyond any possible doubt that they can be fed, in future use, into the law (A) in order to predict the symmetry of the "effects." But this can only be done cautiously, when the symmetry of the antecedents has been *determined experimentally* and exhaustively.

A very important point about the principle of symmetry. If you look at the law (A), you must realize that if x is a symmetry element of the antecedent, it cannot be considered as a cause of x appearing in the consequent, because if x were not to exist in the antecedent (left of fig. 2), it can still appear in the consequent (right of fig. 2), a behavior which is *not permitted* for a presumptive cause in the strong sense of the necessary and sufficient condition that we have required for scientific use.

We are thus landed with the result that laws of nature are not necessarily causal in the strong sense. This is somewhat embarrassing, since everybody would want such laws to be endowed with an element of necessity. So far we have been able to claim some *empirical necessity* for causal laws, but we shall have to discuss eventually the question of necessity for all laws, even those that do not have a causal form, which we shall do by reference to some normative meta-physical principles. The first such principle that we shall discuss, the principle of sufficient reason, will be seen to be deeply connected to the principle of symmetry, and it will give us a clue as to how to handle the others. But before we do this, we had better look at examples of alleged symmetry paradoxes.

SYMMETRY: NOT PARADOXES BUT NOBLE BLUNDERS

Even Ernst Mach (1838–1916), one of the clearer-headed physicists of the nineteenth century, was shocked into hallucination when considering the discovery by the Dane Hans Christian Ørsted (1777–1851) of the electromagnetic interaction. I have told this wonderful story in full detail elsewhere,[5] but I shall try to present its bare bones here so as to immunize my readers against symmetry fevers. Ørsted, guided by the *Naturphilosophie* principle of the unity of nature (p. 146), assumed that the phenomenon of magnetism, until his time entirely mysterious, had to have some relation with electrical phenomena; and from about 1812 he tried to find some interaction between the two. What he did was use a wire along which he could run a current, and place on top of it a magnetic needle at rest. That is, the vertical plane in figure 3 coincides with a magnetic meridian, on which the magnetic needle is at equilibrium pointing to the North. The figure shows two configurations which Ørsted could have chosen for his experiment, but he rejected configuration *a* out of hand and for some eight years he tried instead configuration *b*. He had two reasons for doing this. As it turned out, both were wrong, but they were so persuasive that it would have been almost impossible for a man of his time not to have fallen into the trap within which he got confined for such a long time. I shall first describe what Ørsted did, and once we understand him, I shall explain why his expectations had been wrongly grounded.

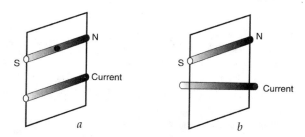

Figure 3. The Ørsted experiment

The direction of the current is toward the end of the conductor where this word appears. The black dot indicates the fixed suspension of the magnetic needle, about which the latter can freely rotate in a horizontal plane. The vertical plane is the plane of the magnetic meridian, that is, one in which the needle is in equilibrium with its north pole pointing North (N).

The reason why Ørsted (like Mach later) could not believe that any effect could be observed in configuration *a* was as follows. Look at the figure *a*: clearly, as people are wont to say when they are falling into error, the vertical plane is a plane of symmetry (anything on its right has an exact counterpart on its left). Therefore, on applying the principle of symmetry (*A*), this element of symmetry must be preserved in the effect. Therefore, the needle cannot move out of its position of equilibrium, because, if it does, it would break the symmetry with respect to the vertical plane, which the principle must conserve. (If the needle moves to the right of the plane, say, there is nothing to the left of it to match that displacement and thus preserve the symmetry.) In configuration *b*, instead, the vertical plane is not a plane of symmetry, because the current goes from left to right, and this is not symmetrical with respect to the plane. Therefore, principle (*A*) is compatible with displacements of the needle away from the vertical plane. (End of first, wrong, argument.)

There was another reason why *b* appeared intuitively to be the right thing. You must realize that such forces as were known until Ørsted's time, which were Newtonian forces, were very much like the force exerted by a billiard cue on a ball: they act in the direction of what one might assume is the cause of the force. In gravitation, this is the direction joining the two bodies and, in the billiard case, it is the direction of the cue. If the current is to be the cause of some force, then Ørsted had every right to assume that in configuration *a* this force could not possibly displace the needle because it would be in the direction of the conductor, so that all it could do was to try to move the needle along its own direction, which the suspension would impede. In configuration *b*, instead, the force would act perpendicularly to the needle, and thus push it out of the plane. (End of second, wrong, argument.)

Eight years Ørsted repeated this experiment in the perpendicular configuration *b* and nothing happened. Of course, he was perfectly entitled to believe that this bad luck was due to the fact that he had not found the right conditions, especially the right intensity of the current. Then at a lecture early in 1820, very probably by chance, Ørsted changed the setup to the parallel configuration *a*, and he observed that the needle moved out of the vertical plane toward the right. Having thus found one of the most important experimental results of his time, and one that started the science and technology of electromagnetism (and thus contemporary life), Ørsted was right out of his wits, and spent several months repeating the experiment in various ways until he could form a picture in his own mind about what was going on. But about that, later.

First of all, what was wrong about the symmetry argument? If you remember what I have said (and I have said it not because of my innate wisdom, but because I have learned this from Ørsted's mistake), you must never look at a figure or even at an apparatus and assume that you know its symmetry. This is why the principle of symmetry (*A*) must *not* be used (unless you know *beyond all possible doubt* what you are doing) because the symmetry of the causes cannot be assumed to be known a priori and must instead be determined by experiment. So, the principle of symmetry cannot tell you anything at all about how to choose one of the two configurations in figure 3. You must do the experiment in either, and if, as Ørsted, you observe that the needle moves out of the plane when the current is switched on, you can now reason properly, on using the correct form (*B*) of the symmetry principle. The displacement of the needle, that is the "effect," is asymmetrical with respect to the plane of the figure, therefore, from this principle, the system (the "cause") must have an asymmetry with respect to this plane. This, you must admit, is pretty shocking: although figure 3*a* has a symmetry plane, this *does not exist in the real system which that picture represents*.

Ørsted's second mistake, that if there was a force it had to be Newtonian-like, and thus in the direction of the current, is even more understandable, since all forces that were until then known had similar properties. If you look at the top of figure 4, the gray arrow, which indicates the displacement of the needle observed by Ørsted, means that there is a force acting on that needle that is *perpendicular* to the physical element, the current, that produces it: what he had discovered was an entirely new type of force, now called a *tensorial* force, but he could not know that because even that name was not used until the end of the century.

I said that I would try to explain what went through Ørsted's mind when he discovered the phenomenon in early 1820. I am not a mind reader, of course, and my reconstruction is based on my readings of his notebooks at Copenhagen and a textual analysis of some of his writings, which I have given elsewhere (Altmann 1992, chap. 1). So, with the limitations that any historical account entails, I hope that my evidence is reasonably well grounded, and I say this because this is a great event in the history of science, which has been seriously misrepresented more than once.

Mach was most disturbed by Ørsted's discovery because, like him, he had thought that the principle of symmetry in the (A) version could be used, thus assuming that there is a plane of symmetry in the system sketched in figure 3*a*. If Mach was so shocked sixty years after the event, anyone who might think that Ørsted observed the effect just *once* and that he then went home blissfully confident about his discovery ignores both human nature and historical evidence.[6] Ørsted was a very meticulous man, and he would not publish his result until he could "understand" what was going on, for which purpose he worked very hard, especially in July 1820, three or four months after his first experiment. Of course, like for everyone else, "understanding" for Ørsted was to arrive at a possible explanation of the phenomenon in terms of ideas that were familiar to him: you could not expect him to "understand" his effect by invoking the existence of the electron spin, which was not to be known for another hundred years. But he got a plausible picture in his mind, and it was only when this picture emerged *from his experiments* that he became sufficiently confident that they made sense and that they were not just a fluke; and he went into print on 21 July 1820. The experiments he did in order to obtain that picture were excellent, but the picture was, of course, wrong.

These experiments are described in figure 4, but they are all a part of a series in which he repeated many times the original experiment and even tried again the failed perpendicular configuration. By placing the magnetic needle in various positions around the conductor, Ørsted realized that the strange force that acts on the needle obeys a clear rule, namely that it is tangential to a circle around the conductor, as shown in the figure. Ørsted thus imagined that around the electrical conductor some perturbation was created which had a "rotational motion," and he called it the *electrical conflict*. I believe that he had in mind something almost mechanical, as the motion of some unspecified particles around the conductor. This was wrong, but opened the way to the later discovery of the proper description of the *electrical field* and, most importantly, it solved at a stroke the two great problems that exercised Ørsted's mind.

First, if you look at the circle in figure 4 and assume that although not visible, it represents some "rotational motion" around the conductor,[7] then you realize that the system no longer has a vertical symmetry plane.

Second, the whole question of the "lateral action" of the force is now solved, since it is the "electrical conflict" normal to the conductor that generates this lateral force, as it is clear from figure 4.

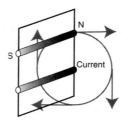

Figure 4. Further Ørsted experiments

The arrows indicate the direction of the displacement of the N pole of the magnetic needle when the latter is placed in four positions with respect to the current conductor: above (as shown), on its right, below, and on its left. Because the movement up and down of the pole when the needle is on the right and left of the current is limited by the suspension of the needle, its displacement is negligible.

Let me now briefly describe the proper way to look at Ørsted's experiment. If we go back again to figure 3, we must understand that we can say nothing, either from the figure or from looking at the apparatus, about the actual symmetry of the laboratory set up. The magnetic needle can show perfect external symmetry, but its internal structure could contain unobserved objects that rotate very much like the circle in figure 4. If this were so (which it is so: a magnet contains electrons that *spin* all in unison), the vertical plane in the figure would cease to be a plane of symmetry of the system, and there is no reason whatsoever to expect a null effect (which is an effect that is symmetrical with respect to the meridian plane) in that configuration. Once the experiment is performed and a nonnull effect appears, we obtain experimental confirmation that the needle is not symmetrical with respect to the plane that contains it. Thus, experiment and the use of the symmetry principle in its (B) form *determine experimentally* the symmetry of the needle.[8]

There is one aspect of the Ørsted saga which I must briefly mention, because it has greatly confused the understanding of what actually went on about the discovery. This was the most extraordinary scientific event of its time, and the fact that the glittering prize went to an unknown Danish pharmacist over the heads of the learned professionals of Germany or France caused untold jealousy, about which Ørsted felt very touchy. The fact that the discovery was almost certainly accidental, and that rumors about this soon became rife, greatly affected the Dane. There was no need to be nervous about that, because having been at it for years, one could hardly speak of an accident, but Ørsted did not react well to the accusations and tried to pretend that before he planned the experiment he already had the idea that a "lateral action" had to exist. In the reference mentioned, I published four different versions of the paragraph where he makes this claim, which I have found in the Copenhagen manuscripts, and this is reasonable evidence that he was covering up something by trying so hard to sound convincing. Moreover, there is no evidence whatsoever either in his writings prior to the discovery, or in his laboratory notebooks, that the idea of the lateral force, which was totally alien to the thinking of the time, guided his work. In a famous example of how dogma can affect historiography, a follower of Karl Popper,[9] however, accepted at face value Ørsted's claim, as signifying that all the way through his intention had been to *falsify* the concept of Newtonian forces. There is no scrap of evidence in any of his writings that this was so. If he had one guiding idea, that was the possible relation between electricity and magnetism, which came out of the principle of the unity of nature propounded by the followers of *Naturphilosophie*.

There are some important ideas that we can learn about scientific discovery from Ørsted's work. One is that, however poor the philosophical grounds for induction as a method, it is an unavoidable part of the work of a scientist. Having observed the effect, Ørsted repeated it dozens of times because the result appeared to him to be crazy, and he was rightly afraid of having merely observed a freak. But Ørsted was a very good scientist (a word that did not exist in his time: he called himself a "literary man"), and he realized that mere repetition was not enough. He knew that no result in science can stand alone, that you must entrench it by making it a member of a set of results that make sense as a whole; and it was only when he performed the experiment of figure 4 that he was satisfied that he "understood" what was going on. This understanding was part of the mesh of results and ideas that he constructed in his mind, and, although it was not entirely right, it gave him the confidence that he needed about the empirical necessity of his results. Later scientists added to and mended this mesh, and this is the way in which science goes. What induction cannot give us, must be obtained by internal consistencies between as many different sets of fact and ideas as it is possible. This way, a *scientific mesh* is developed, a concept central to this book, and which will be further elaborated later.

This section is getting to be too long, but there is one further point I should like to make. I have already warned you about the pitfalls of the a priori in dealing with symmetry. But when people cannot go sufficiently wrong with the a priori, they can then indulge in an activity which appears to be far more respectable but which is equally suspect, and this is to fall into a superficial form of *conventionalism*. All books state the law of the lever as saying that equal forces at equal distances are at equilibrium, as shown in figure 5a. This, however, appears to be a straightforward consequence of the principle of symmetry in its (A) version: we

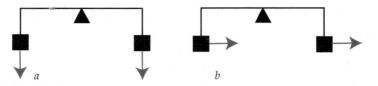

Figure 5. Equilibrium of the balance

The black rectangles represent identical weights.

have a clear symmetry plane in the picture (see figure 1), whence the effects must be symmetrical: if one arm of the balance goes down and the other up, this symmetry is broken, whence the balance *must* be in equilibrium. I hope that the reader will have spotted at this stage that this argument, good as it appears, is unsound, because we use the symmetry principle (*A*), which must *never* be used unless symmetries have previously been experimentally determined. So, the defect of the argument is that we assume a priori knowledge of the symmetry of the system which, although it appears as safe as the Bank of England, is unacceptable.

So, the conventionalists enter into action[10] to allege that the apparently a priori application of the principle of symmetry to the balance is nothing more than a *conventional* definition of equal forces: if we were to find the balance not in equilibrium, we would not invoke failure of the law of the lever: we would decide that the weights are not equal. Thus, we do not have an a priori necessity, as we thought, but neither do we have a truly contingent result: such contingency as you can claim is merely *conventional*.

This conventionalist argument ignores the real, profound contingency in our problem. Post-Ørsted, we know that forces may not be Newtonian-type but that they may be lateral (tensorial, that is at an angle to the expected Newtonian direction). Imagine that the gravitational force were perpendicular to the line joining a particle to the earth's center: there is no question that if this happens not to be the case this is a contingent and not a conventional fact. You can see in figure 5*b* that, if this were the case, the symmetry plane of the system would be lost, and the application of the symmetry principle which we had would be made utterly invalid, so that we cannot conclude that the balance is in equilibrium. (Within this scenario the balance would in fact *never* be in equilibrium!) What the conventionalist argument misses, therefore, is that the possibility of the state of equilibrium of the balance, even when the weights are the same, is a *clear contingent fact*. Given that fact, that you exploit it to define equal weights is no more conventional than the definition of the meter.[11]

After this long excursion we can now go back and look at the origins of the principle of symmetry.

Figure 6. The pendulum

The position of the bob at *b* is a state of equilibrium.

THE PRINCIPLE OF SUFFICIENT REASON

To start with, the reader must notice that the word *principle* is not used here in the same way as in the *principle of relativity*, say, a question which I shall try to clear up a little later when discussing the great meta-physical principles. Coming now to the matter of this section, I should like first to discuss briefly *states of equilibrium*, which are states that have a particular stability whereby they cannot experience spontaneous change. What I mean can be seen by considering the example of a pendulum (see figure 6): if I hold the bob of a pendulum off-center and I release it very gently (that is not acting upon it at all), then the pendulum will start to oscillate spontaneously. On the other hand, the position of the pendulum in which the string and the bob are on the same vertical is such that if I hold the bob *precisely in that position* and then gently release it, nothing happens: this is a *state of equilibrium*. If you visualize the pendulum for a moment, you will soon realize that the vertical plane through the string and bob in the just-mentioned state of equilibrium is a symmetry plane of the pendulum. We thus have an example of a general property, namely that states of equilibrium are associated with symmetries.[12] (This is so even when the equilibrium is precarious like that of a pencil balanced vertically on its tip.)

The idea that symmetry and equilibrium (or lack of motion) are deeply connected is so old that it was enunciated in the very first book on natural philosophy ever written, that of Anaximander of Miletus (ca. 610–550 B.C.E.). Unfortunately, that book is lost and all we know about it comes from a short paragraph referring to it in Aristotle's *De Caelo*.[13] The gist of it is that Anaximander, who believed that Earth is in equilibrium at the center of the universe, adduced that this must be so because, being symmetrically so disposed, there was nothing that could make it move more to one side than to any other, so that it had to stay fixed.

Almost two thousand years later, Jean Buridan (ca. 1295–1360) produced a famous example of this idea. The twice rector of the Sorbonne, who was obviously not interested in the miseries of hunger, argued that an ass, placed *exactly* between two *identical* heaps of hay, would famish to death because it would not have any reason to make a move toward the right heap rather than to the left one.[14] Bizarre as this example is, the idea behind it is both unimpeachable and most important: it led the great Gottfried Wilhelm Leibniz (1646–1716) to his formulation of the *principle of sufficient reason*, which he did in 1710 in his *Theodicy*, after discussing *Buridan's Ass* in some detail.[15] The reader might remember that Leibniz engaged in correspondence with a friend of Newton's, Dr. Clarke, and in his second letter to him, written around the last year of his life, Leibniz refers to "the principle of sufficient reason, viz. that nothing happens without a reason why it should be so, rather than otherwise."[16]

Of course, I shall have to discuss as far as possible the philosophical status of this principle, but let me first show how it naturally leads to the principle of symmetry. I shall reason as follows. First, a state in which *nothing happens* must be a state of equilibrium. Second, this state being symmetrical, as I have discussed, whatever *happens* must break this symmetry, since otherwise the system would remain in equilibrium. Thus, whatever *happens* must introduce an asymmetry in the system. Third, the principle of sufficient reason requires that this asymmetry must be considered the consequent (or "effect" in the not entirely correct language used in symmetry) of some *reason* (which is the antecedent or "cause" in that language). Fourth, this *reason* in the "cause" cannot be anything else than the same asymmetry observed in the "effect" (otherwise, it would not be a reason at all!). What we are saying, therefore, is that if there is an asymmetry in the "effect," there must also be the same asymmetry in the "cause," which is the form (B) of the principle of symmetry on p. 174.

Leibniz was a very great man, a powerful mathematician as well as an acute philosopher but, obviously, we cannot appeal to his authority to validate his principle. So, we now come to this most delicate question: why do we believe in the principle of sufficient reason, that is, why are we prepared to use it when describing natural events? I shall try to answer this question as part of the more general one about what the great meta-physical principles are (or at least what they do), and what is it that licenses their use.

THE GREAT META-PHYSICAL PRINCIPLES

One way or another we have already discussed all the major meta-physical principles, and it will be useful to collect their enunciations together in order to go more deeply into the question of their status. Because of the previous discussions, to which I shall provide cross-references, my enunciations will be somewhat compact, since I shall assume a knowledge of the terms involved. I shall list five principles although, as you will see, they are not all independent. (In fact, the only ones that are truly independent are the last two.) In order to avoid premature misreadings, please notice that what I mean by *nature* in these principles does not cover *everything that is* but, rather, whatever can be observed by macroscopic means at small velocities compared with the velocity of light, a point which will be further discussed in a moment. Also, remember that I always assume that only the strong form of causality is used in science, in which causes must be taken to be both necessary and sufficient: when we depart from this convention, I enclose the words *cause* and *effect* in quotation marks. Finally, the reader must appreciate that whenever I use the word *nature*, I mean the result of the interaction between whatever is "out there" with our rational and cognitive system. Thus, I do not claim that such regularities as I postulated as sustaining Hume's concept of causation are simply "out there," but rather that they are in the only nature that we know, as defined above. I can now enunciate the announced principles.

1. *Principle of the uniformity or regularity of nature.* There are some fundamental regularities in nature the breach of which would constitute a breach of the natural order as we know it.

2. *Principle of causation.* Every effect has its cause and the latter is such that if it is not present the effect is absent (p. 97).

3. *Principle of sufficient reason.* There is no "effect" without a "cause" (p. 184).

4. *Principle of symmetry.* There cannot be an asymmetry in an "effect" unless the same asymmetry is present in the "cause" (p. 174).

5. *Principle of natural evolution.* Nature is a self-organizing system in which the structures that emerge from the competition with other emergent structures are those best adapted to the emergent organization of the system (p. 98).

The principle of regularity of nature should be considerably sharpened because it does not mean that it shall rain everyday, but I shall not try to define it more precisely, because it is easy to fall into useless Byzantine argument over it. One important feature of the principle is that it must be clearly contingent, since it need not have held right from the creation of the universe: all we assert is that nature, as we experience it, is in a state where some major regularities exist. On the other hand, since we have argued that intelligent life could not have emerged without some regularity of inputs, we must expect this principle to have been valid for a sufficiently long time to enable the evolutionary process to take place. There is, of course, reasonable physical evidence that the length of time during which physical conditions on earth have been sufficiently stable is compatible with the timescale of evolution.

We cannot argue, however, from past regularities to future ones, so that the principle cannot be legitimized on logical lines; neither can it be the result of observation since, of course, future regularities cannot be observed. The principle of regularity of nature is nevertheless well entrenched throughout the scientific mesh, as I shall discuss in a moment, and, accordingly, we label it as meta-physical. I give the same status to the principle of evolution for the following reasons. Although it is useful to describe nature as a self-organizing system, this assumption is not subject to direct observation or experiment in it,[17] although it has the merit of not requiring the postulation of any entities that are not part of that system. (This is why I prefer not to call it straightaway *metaphysical.*) Such legitimacy as these two principles have, comes from their *entrenchment*, a process which was described on p. 156. This requires that predicates or propositions that are candidates for entrenchment never be taken in isolation, but rather as a part of a mesh of entrenched concepts,[18] the total entrenchment of the mesh at any one moment during its emergence being taken as the grounds for licensing the use of such predicates or propositions.[19] We can see, in fact, that the regularity-of-nature principle and that of natural evolution are intimately connected and support each other.

An important application of the principle of uniformity or regularity of nature appears when discussing *trajectories* of macroscopic bodies. However finely we measure the successive positions of a billiard ball, we must use this principle in order to define what we shall later call a model *trajectory*, in which the position of the billiard ball can be plotted at any *point* of it.[20]

As regards causality, its relation to the principle of sufficient reason is clear, the latter being little more than a corollary of the former. We have already recognized causality as a meta-physical principle (chapter 4, p. 97), the validity of which must be obtained from its entrenchment. We have also seen that this principle allows us to construct propositions that are *empirically necessary* precisely in the sense that their violation must entail either that the proposition is wrong, or that a violation of the natural order postulated by the principle of regularity of nature has occurred. This brings us to the meaning of such violations and thus to the domain of validity of the principles that I have enunciated.

The reader might have kept in the back of his or her mind that violations of the natural order, mentioned in the first principle stated, can be nothing else except miracles, but this is not by any means the only type of event that I had in mind. The whole thing depends on the meaning of the so far vague expression *natural order*. Remember that we are talking of nature as a self-organizing system, whose emergent organization (both physical and mental) is driven by the principle of natural evolution. Thus, such validity as my five principles have, must be derived from that emergent process, so that the *natural order*, which both emerged and created the mental processes which eventually entrenched the principles, was entirely *macroscopic* and *classical*. It is only in the last century or so that microscopic processes, atoms, elementary particles, or nonclassical velocities (large with respect to the velocity of light) were available for observation and could not have contributed, therefore, to the picture of a *natural order* compatible with the evolutionary process. We cannot, accordingly, expect our principles, or even our vocabulary, to be valid in dealing with microscopic phenomena: it is now known that the principle of sufficient reason, for instance, is violated in quantum mechanics and that symmetry breaks can occur at high energies. So, in a way, my first principle should perhaps be read the wrong way round, as stating that *natural order* is nothing more than the order that is violated when any of the five principles is broken. More appropriately, this order should have been called the *natural classical order*. Subject to the proviso that the application of our meta-physical principles must be made in the right context (that is in dealing with the same natural order in which the principles emerged) then, in the same way as we did for causality, we can use the expression *empirical necessity* to qualify their modality.

I had better comment a little on this idea of empirical necessity, lest it be objected that it is a meaningless expression because it has neither the properties of *empirical necessity*, nor those of such a modality as Leibniz's *hypothetical necessity*, which are respectable philosophical concepts. I have already said that presumptive laws of nature, propositions, and so on, do not stand or fall on their own but rather that they acquire such legitimacy as they have from the vast mesh of entrenched concepts that is at our disposal. We have also seen that not all parts of this mesh are equally solid, some parts of it being clearly on probation. It is convenient, however, to have a sort of marker in order to distinguish strongly entrenched propositions, the falsification of which would cause changes throughout the whole mesh, and this marker is the modality of *empirical necessity* which we assign to them, a modality that like the whole mesh is, however, contingent.

I should say a word or two about some of the principles listed. At one end of the scale, the principle of symmetry could be considered redundant, since it can be derived, as we have done, from that of sufficient reason. At the other end, the principle of natural evolution is a most important tool in describing how a self-organizing system can emerge. It is most important to recognize that the principle as I have stated it should not be confused with Darwinian theory. I have explained in chapter 4, p. 98, why this form of the principle of evolution is best regarded as meta-physical. Darwinian theory, of course, uses the principle, but this does not mean at all that it is not a proper scientific theory: Newtonian mechanics is certainly one, and it nevertheless entails, as we have seen, the use of the causality principle which is also meta-physical.

It is useful to notice, before I go on to discuss natural laws, that the borderline between what a law of nature is and one of the meta-physical principles described is somewhat fuzzy. Some scientists, for instance, might regard the principle of symmetry as a law of nature, but I prefer to regard it as a meta-physical principle just because, as we have seen, it is no more than a manifestation of the principle of sufficient reason which is clearly meta-physical. On the other hand, I said that the word *principle* in *principle of relativity* is not understood in the same sense as the same word in *principle of sufficient reason*, and the reason is that, although like all propositions about nature, the *principle of relativity* must stand on the shoulders of the meta-physical principles, it does a different job from them. The meta-physical principles are just like the traffic rules in Barataria, which permit us to understand how a self-organizing structure *emerges*. The principle of relativity, like all natural laws, allows us to understand how that structure *works*. But I am jumping fences because from what I have just said you might surmise that I am a firm believer in laws of nature, which I am not.

Before I go over to this question, a brief remark for those of my readers who are philosophically trained, who must have spotted sufficient faults in my arguments to drive armored tanks through them, let alone a coach and horses. I can only say in my defense that I do not believe that physical laws are amenable of an entirely rigorous philosophical treatment: many have tried and none have succeeded. So, I do not hope to have achieved more than providing a few useful notes about the subject.

Meta-physical Principles and Falsification

As I have already stated, the meta-physical principles that I have enunciated are all *normative*, and the problem of falsifying normative principles is delicate because they share their nature with *rules*. Although I know that the rule of the road in Britain is driving on the left, it would be silly to say that I *falsify* this rule by crossing the English Channel. *Norms* are defined in relation to specific circumstances, and changing the latter does not mean that the norm is falsified. The principle of sufficient reason, for example, is a normative principle that we have entrenched in the macroscopic world: it might be irrelevant when dealing with microscopic matter, as we shall discover when we discuss quantum mechanics.

What we have done might help to clear up an important point that arises in the literature about evolution. It is often said[21] that the theory of evolution cannot be falsified. This is not so: if a skeleton of *Homo sapiens* was found 500000 years older, say, than *Australopithecus* (see p. 85), then the theory of evolution as it stands today would be falsified. What cannot be falsified is the *principle of evolution* because, for instance, in order to falsify the statement that nature is a self-organizing system, we would have to produce evidence of a supranatural driving force for evolution and, although such evidence might be theologically acceptable, it would not, of course, be obtainable within the domain of *natural* science. (To hammer the nail once more: this is why I call such principles *meta-physical*, since their falsification is a question of metaphysics.)

ARE THERE LAWS OF NATURE?

There is a short story by Jorge Luis Borges about a country dominated by such obsessive cartographers that they finally constructed a map of it on a one-to-one scale, whereupon discrimination between the map and the actual terrain became utterly confused: I wish we had such a delightfully simple situation when dealing with nature. To start with, the nature that we perceive is not the nature "out there," which Kant called the *Ding-an-sich* (thing in itself), or *noumenon*. The redness of a tomato, in fact, cannot be noumenon, since if we had catlike vision such a quality would not be perceived at all. I shall call nature$_1$ this noumenal, phenomenally inaccessible, nature. What we normally call "nature" is a map of nature$_1$, call it nature$_2$ (to which I shall usually refer as plain *nature*), which is the result of our commerce with nature$_1$ through the tinted glasses of our perceptual and rational systems. We can now begin to answer the question asked in the heading to this section, because, as always, the problem is not so much to get the right answer, but rather to ask the right question; and our heading is plainly stupid. If what we meant was nature$_1$, of course there is not such a thing as laws of nature$_1$ because we do not have access to it at all. Ah, you will say, but surely we do have laws of nature$_2$. About this I can give you a quick and straight answer: we don't, and most people will be surprised on hearing this, if not worried (others might save themselves the trouble of writing books to prove that what does not exist does not work). Since I, like most people, use the word "nature" to denote nature$_2$, you can understand my agnosticism as regards the existence of true laws of nature *tout court*.

The question is that if we want to talk about laws of nature at all (and even at this level you cannot come out with clear and straight answers), the *nature* that we mean is a different one, nature$_3$, a *mapping* of nature$_1$ at *second remove*, in which most of the objects of nature$_2$ are replaced by *models*, because it is only for such models that laws of any kind can be enunciated. The art of science, of course, consists in constructing models that as far as possible retain the most significant features of their physical counterparts, but models do not exist in nature (that is, nature$_2$), except in an indirect way as belonging to our imagination.[22] Let me convince you that when we enunciate a law of nature, we actually refer to nature$_3$, by considering Newton's second law that force equals mass times acceleration. The "force" that we have to use here is not an object that you are likely to meet with when turning a corner in the street: it is an idealized object and thus it exists only in our imagination. This is so because all "physical" forces, that is forces in nature$_2$, have a cross section, which may be very small, but nevertheless a cross section. This is not so for the "force" that we must introduce in Newton's law, that must act along a *mathematical line*, which has no cross section at all (it has the same null thickness as a mathematical point). Such a model belongs to nature$_3$, and of course it does not exist in nature, (except that some mathematicians believe that it exists in a Platonic world). What about the mass? Again, this is supposed to be a *point mass* entirely deprived of dimensions. So we have already two models, but wait: I have this force and I have this mass, but this is not enough, because I have to assume that there is *no other force* acting on this system. So not only have I two models that cannot exist, but I also have to put them in a place the likes of which cannot be found in the whole of nature$_2$. (For a system to be free of any force acting on it, it would have to be the only object in the universe!)

Even the concept of a *trajectory* in Newtonian mechanics has to be carefully understood. Remember that in order to apply Newton's second law of motion you require initial conditions at an initial point. But this is necessarily a *mathematical* point: thus Newton's law is applied to a model trajectory, and the results that you obtain from it are then mapped into the corresponding trajectory in nature (nature$_2$).

I can hear on my back recalcitrant antiscientists chuckling with delight at my demonstration that science is all about rubbish, but this is entirely outside and against my intention: a piece of stone does not look like a cathedral at all, and yet if you know how, a cathedral you build, and this is precisely what makes science such an admirable activity. There is no denying that science has to start with amazingly flimsy models of reality and yet, when the time of reckoning comes, the results of working with these models *have to fit reality*, by which of course I mean nature$_2$. This is the most important feature of science: science is *accountable*, in a way in which few other human activities are. Even more, science cannot ever go permanently wrong: if it does, eventually, the wrong concept or theory will be contradicted by nature, and it will be eliminated from scientific discourse.

Going back to the use of models: Newton himself was able to produce methods whereby you could progress from the model of a point mass to that of an extensive solid. This is still a model, since the solid has to be assumed homogeneous, whereas no natural body is exactly so: but this is now an engineer's problem; if they want to use Newton's laws, they have to produce objects that fit that model as nearly as possible. (This is why engineers have produced single-crystal rotor blades for turbine jet engines!)

Which brings me to the question that, because laws must necessarily be stated for nature$_3$ and fitted as well as you can to nature$_2$, that is— because between the model and the object falls the shadow—laws, if they have any validity at all, must be valid only *within a margin of error*. To talk about laws without stating their domain of application and their accuracy (or margin of error) is just as absurd as to give international bank balances without stating the corresponding currencies. A law without a domain of application and margin of error is meaningless, and its discussion when so denuded is the sort of pernicious rubbish that generates confusion. For instance: when we discussed Newton's second law, we proved that, given it and his law of gravitation, Galileo's law of falling bodies follows, within the appropriate margins of error. No, some philosophers will say: Galileo's law of the falling bodies is not a *logical* consequence of Newton's laws.[23] To start with, if anyone wanted to make a statement that might be true for all time, there is probably no safer bet than to claim that never will man or woman or beast be able to derive *logically* any law of nature (of any of the natures I have defined!). Second, to derive any consequence of a physical law without a corresponding understanding of the error involved is completely to misrepresent what the work of science is. You can invent a medicine in one month, but it will take five years of checks before it is put on the market: it is the checks and counterchecks and the analysis of errors and of experimental conditions that makes science what it is. No one imagines that what you see on the cinema screen is just what the actors did when filming, yet some people pick up undergraduate textbooks and treat them as if what they say is the sum total of scientific activity.[24]

ARE THE LAWS OF NATURE TRUE?

Because laws must of necessity be laws not of nature but of its models, some philosophers, especially those of the Kuhnian persuasion, are apt to claim that they are fictions, and "fictions, however useful, are of course false." The problem here is that the writer assumes that "true" and "false" are universally applicable values; and that when I qualify as *false* the proposition: "The present prime minister of Great Britain is a bold woman," I use the word "false" in the same way as scientists would do if declaring "it has been proved that the Maxwell's equations are *false.*"[25] So, let me first discuss some problems when trying to assign "true" and "false" values to natural laws, although I shall then accept that these values are used in science, and I shall attempt to establish how this usage goes.

We must first remember that propositions are laws not by virtue of what they are, but by virtue of what they do. Thus, undefined use of the true and false values for a law could be as senseless as assigning them to the *actions* of a painter while sketching a landscape from nature: what matters is the adequacy or otherwise of the final product. Also, because laws are necessarily laws of *models* of nature, and because models are *maps*, laws basically work as maps in nature$_3$ of some processes in nature$_2$. But maps are as good as the use we put them to. The famous topological map of the London Underground is totally different from a geographical map of London. If you superimpose the London Underground map over the geographical one so that both Paddington Station and the north register, then Campden Town on the underground map will not register with its geographical counterpart. Would you say that the underground map is "false"? It certainly is not, if your purpose is to navigate yourself by underground between those two stations. What is fatal is to use the underground map to navigate yourself on the surface! Thus, if you misuse your maps your results will be false.

Laws of Nature and the Science Mesh

I shall now move forward from the above critical account in order to understand how the values "true" and "false" are cautiously used in science. Alex Levine, the author of my last quotation on "fiction," challenges us to answer the question: "In what sense, then, physicists continue to regard Maxwell's equations as true?" I shall try to produce some form of an answer.

A law of nature must *do* something, which is to allow us to establish a relation between two events (properly mapped onto nature from the model used). Also, whatever it *does*, it must be *empirically adequate*, that is, fit the facts within the desired margin of error. But empirical adequacy is not enough: if we took it to be so, we would be falling into some form of *instrumentalism*, in which the law or theory is nothing more than a, perhaps convenient, instrument to fit the facts. For a law or theory to begin to be a candidate for the value of "true," it is essential to place it within the *science mesh*. As we know, the elements of this mesh must all be entrenched but, naturally, the degree of entrenchment is not uniform. Some elements of the mesh are on probation, others are extremely strongly entrenched because they sustain very large and very important sections of the mesh. An example of the latter is provided by Maxwell's equations: it is they that provide the necessary *explanations* that link coherently even distant elements of the mesh. It would thus not be imprudent for a scientist to label Maxwell's equations not just as empirically adequate but also, in some sense that the scientist would understand, as "true." But before such a delicate value might be assigned to a law, more work is necessary, as I shall now show.

Objectivity

One of the features that anyone expects of science is *objectivity* which, whatever we mean by the word *true*, must be one of the attributes that people use in order to license its use: any number of people will agree on entering my room that there is a computer in it, but if I were to suggest that there is a pink elephant, I would find it very difficult to obtain a witness to corroborate my statement. Objectivity, of course, means that different observers observing the same phenomenon must be able to reconcile their findings. But objectivity in science is not, unfortunately, served to us on a platter, and it is one of the jobs of science to establish watertight rules for this purpose. Take the verification of one of our simplest laws, the law of inertia. John is sitting on a very smooth train that goes at an absolutely constant speed of 100 mph. He puts a ball on top of a table, and he perceives the ball to be at rest. Since there is no force acting on the ball (gravity being canceled by the reaction of the table top), he is glad to find that the law of inertia is verified. Jane, from outside the railtrack, observes John and notices that the ball that he has placed on the table top is moving at *uniform* speed of precisely 100 mph; and she is also glad to agree that the law of inertia is verified.

Notice, however, that John and Jane do not agree as to the speed of the ball, but, being scientists, they will try to eliminate the discrepancy by defining how they will communicate between themselves. First, they agree that each observer is using a different *reference frame* with respect to which positions and velocities are measured. For John, the reference frame is firmly fixed to the train, and that is why he perceives the ball as stationary. For Jane, it is the railtrack, with respect to which the train, and with it the ball, is moving at a constant velocity of 100 mph. Second, they agree as to a *transformation law* so that each observer can work out what the other one observes, and compare this calculated result with the report he or she receives from the other observer: if it agrees, they say that they are in *objective agreement*. The transformation law is very simple: Jane has to *add* 100 mph to the speeds measured by John (with respect to the train), and John has to *subtract* 100 mph to the speeds measured by Jane (with respect to the track). Thus, John tells Jane that the speed of the ball that he measures is 0 mph, whereas Jane measures 100 mph for it. On using her transformation law, Jane subtracts 100 mph from the speed she measures and concludes that John's result should be 0 mph, whence she knows that John is telling the truth. What these two people are verifying is what is called *Galileo's relativity principle*, which says that the laws of mechanics are *invariant* with respect to two different reference systems in relative *uniform* motion. (That is, that move at uniform speed with respect to each other. Such systems are called *inertial systems* or *inertial frames*.) *Invariance* in this context means that, when transformation laws are properly used, the results obtained by different observers in different reference frames should agree.

The reader might remember that in the last section I used the expression *domain of application* of a law, and you can now see what it means: the law of inertia, or for that matter all laws of classical mechanics, are valid only in inertial frames. But I am jumping the gun because I have done nothing so far to prove that inertia breaks down if the reference system is not *inertial*, which it is very easy to do. John's train, which was going sweetly at precisely 100 mph, now brakes suddenly to exactly 80 mph, whereupon the ball shoots forward with a speed that he measures of 20 mph, which he radios to Jane. Assume that the train has no windows and that John has no way at all to verify that it is moving or altering its speed: when the train brakes, he sees the ball, on which no force is apparently acting, shooting forward: for him, therefore, the law of inertia ceases to be valid. (This should not be surprising because, in fact, John's frame of reference is no longer *inertial*.) As regards Jane, she measures the velocity of the train with respect to her (inertial) reference frame at 80 mph and adds to it the velocity of the ball with respect to the train of 20 mph, whence its velocity with respect to the track (Jane's frame) is precisely 100 mph as before. Therefore, for Jane, the ball has not changed velocity so that the law of inertia is satisfied.

Let us go back to John and gauge his reactions about the law of inertia. He has three options. First, he can decide that the law of inertia is a con, and if he is clever, he will write a book with this or a similar provocative title and sell 100000 copies because there is nothing as good as a bit of excitement. Second, if he is a sober physicist who recognizes that the law of inertia is so well entrenched that it is unlikely to be untrue, he can radio Jane and ask for her results (whereby he would realize that his reference frame, but not Jane's, had ceased to be inertial). Third, he can reason as follows: this ball was moving at 0 mph, and I observed that in one second its speed changed to 20 mph (8.9 m/s), so that its acceleration[26] was approximately 8.9 m/s per second, that is 8.9 m/s². Therefore, if m is the mass of the ball, it must have experienced a forward force of m times 8.9 m/s², as required by Newton's second law (chapter 6, p. 139). What he is actually doing here is introducing a *fictitious force* in order to recover the use of the laws of classical mechanics in his own, noninertial, reference frame.

Very few people would do what John has just done in his own particular case, yet fictitious forces are used by everybody in a very well-known and similar case. Consider again John's very smooth train going at precisely 100 mph along a perfectly straight track and suddenly taking a bend on keeping the speed perfectly constant. Again, the stationary ball that John had on his table shoots sideways and backward; and John, like most people, will say that this is so because there acts on the ball a new, so-called *centrifugal force*. What we have is that whereas Newton's second law (*force f* equals *mass m* times *acceleration a*, p. 139) gave us $f - ma = 0$ when the train was on the straight and narrow, we should now add the centrifugal force f' to f :

$$f + f' - ma = 0 \ or: f - ma = -f'.$$

Notice that what we are saying is that Newton's law is no longer valid in the new reference system in which the train is going at constant velocity over a circle. What has happened is something just too horrible to countenance: whereas the right-hand side of Newton's equation in the way I wrote it ($f - ma = 0$) must be zero, this zero has been changed into a new nonvanishing force (the opposite of the centrifugal force, also called the *centripetal force*). Not only has the *form* of Newton's equation changed when changing the reference frame, but the monstrous sin has been committed of *changing the zero*. This cannot happen, neither in physics nor anywhere in life (as your bank manager will explain to the judge when he accuses you of having altered the balance of your account).[27]

There are several things we learn here, but before we look into them we must strengthen our definition of objectivity. We required transformation laws in order to go from one inertial frame to another; and we shall want, first, to extend this to all possible reference frames, whether or not in a state of relative uniform motion. Second, we shall require that, although the values of the various quantities may change from one frame to another, the *form* of the laws must be the same in *all systems*. This condition is necessary if we want to preserve the most precious of all our numbers, the zero, as we have seen in the case of Newton's second law. When the equations of physics are written in such a way as to satisfy this condition, they are said to be *covariant*.

Covariance, as we shall see, has important philosophical consequences which, if ignored, may lead to nonsensical claims, such as people discovering that forces do rather strange things and then complaining that physical laws fail. The question, very briefly, is this: you remember that John, when the train broke, had to introduce a fictitious force in order to save the principle of inertia (which he should have thrown away in his frame of reference). This fictitious force has, of course, no physical meaning. The question is that *all forces*, as described by me so far, and as used in elementary and not so elementary books on mechanics, *have no physical meaning*, a point which I shall now explain. When we write Newton's laws and we use our elementary notion of force, we recognize that in three dimensions the force has to be modeled by a *vector*, which has three properties. First, it is a *segment* along the line on which the force applies (this is the idealized mathematical line about which we talked). Second, it is *directed* in the same direction on which the force acts, and, third, its length indicates in some convenient scale the magnitude of the force. When we use these vector quantities, however, the laws of mechanics come out defective because their *form* changes for different reference frames (that is, they are not *covariant*). This means that the vectors that we have used in order to represent forces, and indeed the very model of the force that we have used, are not physically significant. In fact, it was found toward the end of the nineteenth century that the correct model for a force is not a vector as we have so far used, or at least implied, but rather an entirely new mathematical concept called a *tensor*. It does not matter for us to know what these tensors are, but rather what they do: it is only when tensors are introduced in the laws of mechanics that these laws acquire a covariant form, valid in all reference frames.

This is a question that is not always clearly understood, because in elementary work at schools, and even universities, people use vectors and either they deal with inertial systems, in which no trouble arises, or, if noninertial systems are used involving linear or angular accelerations, they apply hand-to-mouth procedures by introducing fictitious forces (that is, forces that are even more fictitious than those used inertially). Because of this, there are some lovely examples in the philosophical literature where reality is shown to crumble in front of our very eyes, this being the sort of reality that a well-trained physicist would handle with rubber gloves in order to avoid ontological contamination. In case of doubt: always imagine that anything normally described by a vector is at least suspect. Tensors, when properly constructed, will work well in all reference frames and may therefore be considered objective. Whether they are *real* or not it is then a matter of opinion and even taste, but at least they are as near to elements of reality as can be constructed in classical mechanics. Which means that you might need something better if you go into relativity or quantum mechanics.

HOW TO HALLUCINATE INEXPENSIVELY

As we have now seen, forces as mapped by vectors, can lead us along the garden path, but there are other difficulties, which seem to cause some philosophers a great deal of trouble. This is so because in all the laws that entail the construction of forces, one has to assume that the system in question is isolated. For instance, if we want to define Newton's gravitational force between two bodies, it is essential to assume that there is no third body nearby because this would alter the force between the former two. Likewise, if you want to define the force of attraction between two electrically charged particles, you have to assume that their gravitational force may be ignored. This is not just reckless carelessness: it is part of a very well studied mathematical procedure, which is called *perturbation theory*. I shall try to explain simply the basic idea because it will allow people to avoid philosophical error. Suppose, to simplify things a little, that the solar system contained only two planets, Mercury and Earth. If we want to compute the orbit of Earth, we need of course the force acting on our planet. This is the gravitational force between the Sun and Earth, which is large, plus that between Mercury and Earth, which is small, because the mass of Mercury is a small fraction of that of the Sun. You know very well, however, that the force acting on Earth must be the sum of these two forces, and each of these forces is called the *component* of the total force or *resultant* It is only the *total* force acting on Earth that is really physically significant (as far as forces can be so!), but you start with the largest component, that between the Sun and Earth, and then include the second component to add a correction, which is calculated by the above-mentioned mathematical method of perturbation theory.[28] So, component forces have even less physical meaning than the forces themselves have, and when you apply similar procedures in quantum mechanics, they may lose it totally. It is for this reason that in quantum mechanics, one likes to work out energies, which do have a physical meaning, knowing full well that the partial terms that appear in the expressions of these energies *have no physical meaning whatsoever*. If any philosopher reads this, I will be put immediately in stock with Cardinal Bellarmino as an *instrumentalist*, but science cannot make everything at once physically significant or objective. It would be just as silly to blame science for this as to blame nature for not being what philosophers would like it to be. Partial models have to be added up that have no physical meaning, but their resultants do, and to try to look for causal statements entailing meaningless components is nonsense. And if the reader is puzzled about Bellarmino, we shall hear more about him in chapter 10.

CODA

I have presented nature as a self-organizing system, the organization of which is, at the macroscopic level and at the present period of the development of the universe, guided by certain meta-physical principles. As a scientist or natural philosopher, the only question that can be asked about these principles is what they do, but a theologian may well make metaphysical claims as to where they come from. From our point of view, instead, the meta-physical principles describe the way in which nature emerges; and the laws of nature use the meta-physical principles in order to describe and predict how nature works. The laws of nature are strictly speaking *laws about models of nature* and, in applying such laws, models are used, such as component forces, that may have no physical meaning at all. Thus, attempts to use them in causal explanations are doomed to failure and lead to deep conceptual mistakes. The use of laws, however, is rigorous and precise, insofar as the properties of such objects as have a physical meaning must be accurately reproduced and predicted within accepted margins of error.

The use of physical law is definitely tied up with strict criteria for *objectivity*, which means that results may be exchanged and checked between observers in different frames of reference. Because of this, some physical laws that appear to be unassailable, like Newton's second law, must be reformulated in what is called a *covariant* form, so that they preserve their form in all reference frames. When this is done, it follows that vectorial forces must be replaced by better entities called *tensors*, which is an example that superficial consideration of physical law can create apparent and meaningless contradictions.

Physical laws obtain their legitimacy through entrenchment, first of the meta-physical principles on which they rest; and, second, from the entrenchment of a whole mesh of laws and of their corresponding language. Because of this mesh, which also includes all entrenched facts, there is a hierarchy among the laws, always within this *science mesh*. Those that affect the larger areas of the mesh are called *principles*. Others are *laws* and others, of a much more restrictive domain of application, are merely *properties*, but this hierarchical terminology is not always respected. Although some logical criteria can be given about the structure of laws, what laws are depends on what they do, and not on their semantic or syntactic structure. (They need not even be causal, as is the case with the principle of symmetry.) And what laws do is fit and predict physical fact. When it is discovered that they fail to do so, either they are entirely discarded, or their domain of application is redefined to exclude the regions where they fail.

The above statement must be severely qualified because in some cases models are used which are themselves aggregates of separate submodels. Within each submodel, laws may hold that have no physical meaning in themselves, but, when combined within the aggregate by means of adequate general principles, lead to results that can be factually mapped. This is a most important point, disregard of which may lead to serious philosophical misunderstandings. The application of a law to nature (nature$_2$) entails an excursion from it onto a model (nature$_3$); but this excursion has a physical meaning only insofar as the initial and final states in it are mapped back to nature. If *part* of this excursion into the model is extracted, and if it is then attempted to map that part onto nature, absolutely meaningless results may be obtained.

Because it is not their grammatical structure that makes them what they are, when an utterance is presented as a law, it is this presentation that gives it its status: thus the essence of the law is shifted somewhat from the law itself to the lawmaker (or law-user). A lawlike utterance is somewhat similar to what J. L. Austin called a *performative*, as in stating "I declare you husband and wife."[29] The essence here is that the person who utters the sentence is declaring (in this case) that a relation is established. Likewise, when a proposition is recognized as a law, it entails that it is licensed (as entrenched) for use in establishing relations between entrenched elements of the natural mesh. While the relations thus established are found to agree with well-entrenched concepts, the law is held to be valid and it is used.

This cautious approach to the meaning of natural laws does not entail that they do not carry any force of authority with them: they do so, in the sense that, while they are used, their use establishes a form of necessity that I have called *empirical necessity*. To illustrate what this entails, I shall consider an entirely artificial example. Consider the statement: "All findings of Neanderthal men have shown an average cranial capacity of 1500 cc." Although this statement carries some universality, since it refers to all previous Neanderthal findings, it does not commit us to expect that all future findings will satisfy the same result. On the other hand, if the statement were to be declared, for some (most unlikely) reason, to be a law of nature, it would entail the expectation that the following counterfactual be true: "All Neanderthal remains so far undiscovered, if they were to be found, would show an average cranial capacity of 1500 cc." It is this projection (which may or may not be correct) that entails the concept of empirical necessity. Like all humanly established relations, this one is also subject to being breached, that is to be abandoned or curtailed.[30]

Having discussed all this about laws, I am sure that some readers will feel cheated because I have not said anything at all about the *explanatory powers* of laws. But the fact that automobiles are often used for encounters of a sexual nature does not license us to describe them as assignation chambers: a car *is* a form of transport, whatever other uses it may have. In the same way, a law *is* an instrument of projection or prediction. That it can also be used in explanation may be the case, but explanation is a very subjective concept: one person's explanation is another person's muddle, whereas prediction as understood in science is entirely objective, and the main job of science is to deal with objective concepts. Explanation thus is not in the *essence* of natural law, but it forms part of the procedures whereby science muddles through in trying to go from the known to the unknown, in what sometimes is called the scientific method, which will be discussed in chapter 10. Before we do this, in fact, I want to play about a little in chapter 9 with some basic problems about space and time, so as to have a clear example of how important it is not to mix up models with their natural counterparts.

NOTES

1. Throughout this chapter, I will understand this to mean that, when the salt is burnt in a flame of the right type, the latter emits yellow light showing a spectral line of wave length 589.2 nanometers. This proviso makes the statement valid if and only if there are sodium atoms in the flame, for which this line is characteristic.

2. One criterion that was also proposed but which turned out to be false is that of *universality*. It is tempting to regard the statement: "All spheres of enriched uranium must be of diameter less than a mile" as a law because it is universal. This might appear sensible, since enriched uranium spheres of a mass larger that a few kilos cannot exist because they would blow up by fission. On the other hand, the statement "All gold spheres must be of diameter less than a mile" is equally true (because there is not enough gold on earth) and equally universal, but it would be silly to call it a law. (See van Fraassen 1989, p. 27.)

3. The argument here is dangerously oversimplified. To attempt to recognize the symmetry of a physical object by consideration of a picture of it can lead to serious mistakes, discussed in Altmann (1992), chap. 1, and sketched in the next section.

4. Even more, mere examination of the external, geometrical features of an object can be equally misleading: a man may be perfectly symmetric externally and yet his internal organs are not so. For a discussion of the principle of symmetry see Chalmers (1970).

5. Altmann (1992), chap. 1.

6. Compare with chap. 4, p. 105.

7. This idea occurs in some scratched notes in his manuscripts, which I have published in Altmann (1992), p. 33.

8. This argument is incomplete, because what I have argued about the needle could equally well be argued about the conductor. I have assumed that the latter is known from other experiments to be symmetrical with respect to the plane in question, so that the asymmetry observed can only be assigned to the magnetic needle.

9. Agassi (1959), as quoted by Lakatos (1978), p. 206. Later, this author considerably toned down his views (Agassi 1963, pp. 67–74).

10. See Nagel (1961), p. 55.

11. The reader may find a more detailed discussion of this problem in Altmann (1992), chap. 1.

12. This can be proved in general: a fundamental property of states in equilibrium is

that if you displace the system by ± m (this being some very small displacement parameter, like a small distance, angle, etc.), then the state of the system is practically unaltered. (What we mean by this is that the change in the system will be of the order of m^2 and thus negligible since m is very small.) In the case of the pendulum in position b of figure 6, you can change the angle of the string to the vertical by a very small positive or negative angle and its state will be unaltered, in the sense that the pendulum will still stick to the vertical. Because the small displacement can be taken to be positive or negative since its sign will not affect the required square, there must be a symmetry as indicated by the fact that $+m$ and $-m$ lead to the same (negligible) change of state.

13. Kirk et al. (1983), p. 134.

14. Although this parable nurtured generations of scholars, there is no evidence that associates Buridan to it. It was said to have come from an annotation on Buridan's copy of Aristotle's *De Caelo* and it is not even sure that the animal in question, despite its admirable adherence to principle, was an ass at all. A dog is also mentioned in this context, but the name *Buridan's ass* will stay for ever.

15. Leibniz (1951), p. 150.

16. See Alexander (1956), p. 16. In the *Monadology*, Leibniz (1714), Section 32, uses the concept of sufficient reason in a more general way, that is nearer a definition of *necessity*: "no fact can be real or actual, and no proposition true, without there being a sufficient reason for it being so and not otherwise." (See Leibniz 1951, p. 147, for a similar statement.) His necessities, however, are not *logical* but in a way contingent, except that they may be willed by God. Leibniz refers to such necessities as *hypothetical necessities*.

17. The reader should not confuse the principle of evolution as stated here with the Darwinian theory, of which the principle of evolution is only a normative tool. This principle is clearly circular: structures emerge because they are the best adapted, and we know that they are the best adapted because they have emerged. This does not mean that this form of the principle of evolution is empty, because its meaning follows more from what it does not say than from what it affirms. I mean by this that, as I have said before of the anthropic principle (p. 89), the principle of evolution is a conversation stopper: it stops us from asking the question: why has this structure emerged? Which would have been in order if we had assumed that the organization of nature is guided by a final cause or teleological principle. I repeat that nothing can be said in science that could rule out such meta-physical constructions but, likewise, that they have no place within proper scientific discourse.

18. We have already seen an example in this chapter in relation to Ørsted's work: he had to entrench his strange result in a wider web of results and ideas in order to acquire sufficient confidence about it.

19. Remember that the entrenchment of the whole mesh as such, rather than that of a single knot of it, is important in order to avoid arguing from past to future instances of a single proposition. Ultimately, of course, we are arguing from the past to the future: if we did not, we would not even be trying to engage in rational discourse, but, in doing so, we are using a sounder procedure than blind induction. The concept of the mesh of entrenched scientific concepts, and its significance in the validation of natural laws and causal propositions, is very similar to that of the net of causal relations given on p. 124 of van Fraassen (1980).

20. By a *point* here I mean a dimensionless mathematical point. (See pp. 205 and ff.) Also, the principle is essential in order to formulate the concept of an *object* which is not as obvious as it seems. (See further discussion on p. 575.)

21. See for instance Rose (1998), p. 47.

22. There has been recently a great deal of interest in models in the philosophy of science, but they have nothing to do with the models I am discussing here. In a way, they are *engineering models* of nature whereas the models I have in mind are *quasi-ontological*. By *engineering models*, for lack of a better name, I mean mechanistic models, in which the question

of describing the fundamental objects that make up nature is not under discussion but, rather, what is modeled is the action of such objects on each other. This is why I describe the models I deal with as quasi-ontological: they try to mimic objects, the existence of which we surmise, but which are not amenable to direct treatment by the methods available to science. Further discussion about models will be found in chapter 9.

23. Brown (1979), p. 61; Feyerabend (1993), pp. 24, 46–48. See also Feyerabend (1962) and Kuhn (1962).

24. The first philosopher to stress the importance of modeling in science was probably Whewell (1858). My approach to natural laws, models, and their mappings is similar to the one given by Duhem (1954). Duhem defines an experimental domain, a mathematical model, and a conventional interpretation. The model embodies the logic or the axiomatization of the theory. The usefulness of the theory is given by the criterion of adequacy, that is, the verifiability of predictions and the quality of the agreement. To this Duhem (pp. 138–43) adds the criterion of stability: how far the adequacy of the predictions is affected by a small change in the model. See also Thom (1975), p. 16, who largely agrees with Duhem in this respect. The "constructive realism" of Giere (1988) views physical theories as models of reality within various degrees of similarity.

25. The first quotation in this paragraph is from Levine (1999). The concept of truth is very delicate and I shall hardly skim its surface. See Horwich (1990).

26. That is, the rate of change of the velocity.

27. Because of these difficulties, the concept of *force* had been severely criticized by Ernst Mach (1838–1916) and Gustav Robert Kirchhoff (1824–1887). Heinrich Rudolf Hertz (1857–1894), following these authors, went even further, and he wrote a whole book of mechanics without using forces (Hertz 1899). See also Zylberstajn (1994) and Jammer (1957) for a discussion of fictitious forces.

28. Cartwright (1983), p. 59, understands clearly that components of forces have a metaphorical significance. Like van Fraassen (1980), chap. 3, she is concerned with saving the phenomena, that is the problem of *facticity*. This, however, is not the job of laws of nature, however absurd this claim might sound, and it is not their job because they mostly deal with models of nature, and thus not with facts themselves. Like the job of the medieval mason was to put together cathedral-unlike stones to build cathedrals, *so the practice of science is that of putting together unfactual laws in order to get factual results*. This is a point that van Fraassen (1980), p. 64, foresees: "My view is that physical theories do indeed describe more than what is observable, but that what matters is empirical adequacy, and not the truth or falsity of how they go beyond the observable phenomena."

29. Austin (1962a). See also Austin (1961), p. 220, where Austin, as I have done in the case of natural law, is concerned not with the true or false value of some utterances, but rather about what they do.

30. The counterfactual example here is inspired by the discussion in Kneale (1949), pp. 75–77.

MODELS: HOW TO HANDLE SPACE AND TIME

Wil it serue for any modell to build mischiefe on?
William Shakespeare (1600), I. iii. 385

When I say "John fires his gun," I imply that this happens "now." But what is "now"? Is it this minute, this microsecond, or what? In a *model* of time in which "this instant" stands in relation to time as "this mathematical point" does with respect to space, "now" is "this instant," but it possesses no more physical meaning than "this (mathematical) point" has. That is, just as there is no object in the world that is exactly a mathematical point, there is also nothing in the world that I can expect to be "now" or "instant":[1] mathematical points and instants are features of models and not of nature. As I have discussed in the last chapter, once you use these models for purposes, for instance, of prediction, then you have to make a correspondence or mapping between concepts such as points and instants and their counterparts in nature. And to establish this correspondence requires trajectories and durations in nature to be described within some degree of *graining*. It is enough at this stage to think of graining as the degree of detail or precision with which we want to plot the position of a point or measure the duration of a short interval in time. Later, we shall discuss this concept of graining further and the expressions *fine* and *coarse graining* will be used.

You can choose the coarseness to be as fine as you want, but even the finest grain cannot be a dimensionless "instant" in time, just as the finest gauge cannot isolate a single mathematical point. Even fine graining is a mode of our description of systems, whereas expressions like "instant" and "point" pertain only to models; thus, I shall refer to graining in the latter as *model graining*. Whereas in nature fine graining cannot be indefinitely refined, model graining, because it is done in our imagination, can be indefinitely extended. A great deal of nonsense can arise by lack of careful distinction between model and natural graining, so that I shall discuss these concepts in greater detail in the next few sections.

MODELS AND GRAINING

You might remember that, when Newton had to define the mechanical state of a particle, he had to invent the mathematical concept of instantaneous velocity (see p. 134), as a *limit* of average velocities for successively finer graining (strictly speaking *model graining*) in the space and time intervals. I must now caution the reader that this concept of a limit can be a lethal weapon unless responsibly used. Newton knew instinctively how not to fall into traps, but it took mathematicians the best part of three centuries to put the methods that he had devised on a sound basis, that is, one in which all possible cases could be treated safely and uniformly without falling into error.

Although it entails long and remarkable mathematical work, this is only half the story. What is most important for us to remember is that, in doing the work that I have mentioned, Newton was trying to formulate a law; and I have argued in chapter 8 that laws are not laws of nature but rather laws of models of nature. That this is the case with Newton's second law is evident, since in order to apply it we must know the position x and the instantaneous velocity v of the particle under study: both these variables refer to a mathematical point that does not exist in the world.

It is thus important to realize that such quantities as the instantaneous velocity, although essential to permit a simple mathematical description of physical processes in time, do not necessarily have a *direct* physical meaning because they do not refer to *actual* trajectories in space and time but rather to their *models*, and that serious confusion resulting in apparent paradoxes of motion and time can be caused by forgetting this fact. I walk at 6 km/h, which means that I take 18 s to traverse the 30 meters frontage of a house and 0.6 s to traverse a meter. I could thus "conclude," by successive reductions of the space traversed, that I must spend a null interval of time in traversing a single *dimensionless* mathematical point. This can lead us to one of the following two absurd conclusions. One is that I do not move at all, since my "velocity" at each point vanishes. (Remember that velocity is distance divided by time. In traversing the mathematical point the distance vanishes, and thus, it might be thought, the velocity also vanishes. A silly argument, since equally well the transit time through the point vanishes, and we do not know what the ratio of two vanishing quantities might be, which will be discussed on p. 208.) The other dubious argument is as follows: since a meter is a succession of an infinite number of mathematical points, and because it appears "reason-able" to conclude that the sum of however many zeroes must still be nothing, it should be "clear" that I can traverse one meter in no time at all.

There is a very simple way to identify the con underlying these ap-parent paradoxes: you must notice that I start by saying that I take a stroll, thus leading your mind into thinking about the real world, and then I cun-ningly move away from it, without saying that I do so, when I subrepti-tiously introduce in my "stroll" a mathematical point, at which stage I am substituting a model for my "real" trajectory, thus mixing too entirely dif-ferent conceptual frameworks.[2] It is this mixing that is entirely forbidden; this is nothing more than the normal precept whereby mixing of two dif-ferent levels of discourse should be ruled out of court. (Which, alas, does not mean that it is not constantly done: sometimes it seems that politicians would not be able to open their mouths if they followed this injunction!)

The reader may think at this stage that Newton fell into that trap when he introduced his instantaneous velocity. But Newton was much too clever because *everything* he did was entirely within the framework of a *model* of the trajectory; and he was entitled to give rules about how to start from a mathematical point of his model trajectory and how to then predict properties at another mathematical point of the model trajectory. As we know, in order to do this prediction, he had to use his second law, valid like all laws only *within its corresponding model*. When one is using the computations thus made to compare with experiment, one is simply verifying whether or not this model fits the actual motion in actual space *within an appropriate degree of graining*. But you must *never* mix up the model with its image in nature within the same level of discourse: this is a serious mental health warning.

It is very important to understand the relation between the model and the natural trajectory, which I illustrate in figure 1. In the model tra-jectory, we have laws that require mathematical points, which must be determined by a procedure to be discussed in a moment; and the whole work depends on the possibility of finding an *image* of a mathematical point in the natural trajectory. The mathematical point will be defined within a small segment, called a *neighborhood*, which can be *indefinitely* reduced in length because we are working within the model. The image of this neighborhood in the natural trajectory will be a small segment defined within whatever degree of fine graining one wants. This depends on the experimental precision required, and although it can be reduced in length within broad limits, this reduction cannot be indefinite.

I shall return a little later, after we understand how mathematical points are determined, to the question of why the *transit time* at a mathe-matical point (that is the time that it is alleged a particle takes in

Figure 1. Mapping from the model to the natural trajectory

The segment p that determines a mathematical point of the model trajectory may be indefinitely reduced in size. The segment a that determines a "point" of the natural trajectory must always be finite (that is, not indefinitely small).

traversing it) cannot properly be mapped from the model to the natural trajectory. It should be clear that this restriction will be fundamental in blowing up the last vestiges of credibility of the current "paradox."

We must first understand how a mathematical point is uniquely determined, and how quantities such as the velocity of a particle are defined at it. In order to clarify this question, I reproduce in figure 2, in a slightly different way, a part of figure 3 from chapter 6, used to define the instantaneous velocity at the point x as done by Newton. What we are doing in this picture is this. We have a model of the trajectory as a mathematical straight line, which is represented by the baseline of the figure. As a difference with its image in nature, the mathematical line contains mathematical points, and we are trying to *determine* one of them, x, which we do by gradually reducing the distance from a point A to x. The difference between this procedure and fine graining, which can be done within the natural trajectory, is that in the model here, because I can use the concept of a mathematical point, I can reduce the length of the segments shown well beyond that of E. I can imagine, in fact, that this process of reduction goes on *indefinitely*. This is what I have called *model graining*. Because it is an *imagined* process on an imagined model, I can do whatever I want, within the rules that mathematics might provide to keep us safe. When I consider a natural trajectory, on the other hand, however fine the graining I might take, I cannot even imagine that I can pursue it indefinitely because there will always be a natural limit to length reduction, such as the atomic dimensions, beyond which I cannot go.

Notice a very important feature of the way in which you define, in your model line, a mathematical point. In the example of figure 2 we start from the right at the point A, and we "move" toward the desired point by successively finer graining, reaching E in the figure. We imagine, nevertheless, that this fine graining continues *indefinitely*, which means that we are clearly within *model graining*. The most important feature to notice is that this model graining entails that we approach the desired point from the right (in this case), but that we *never surpass* it: in fact we never even *touch* it. We merely define the point by an *indefinite approach* to it from one

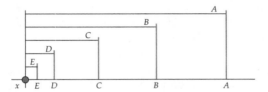

Figure 2. Model graining to determine a point in a model trajectory

The segments shown here appear vertically (in a changed scale) as in figure 3 of chapter 6, and they approach gradually the point x to be determined.

side (the right in our example). In this model graining, the mathematical point itself is a black hole (mathematically called a *point of accumulation*) that can never be traversed or even touched, but it should nevertheless be clear that the procedure described determines it uniquely.

It is most important to realize that the introduction of a model is significant only if elements of this model, and quantities computed for it, can be mapped onto natural elements. This cannot be done in an arbitrary manner, which condition is at the root of the paradox about the transit time at a point. Look again at figure 1 and think about the extremely clever way in which Newton defined the instantaneous velocity at the mathematical point represented with the gray circle. Because of the limiting process used in relation to figure 2, the instantaneous velocity is such that if I were to double the size of p (which is very small indeed) this instantaneous velocity would change but very little. (If you are in doubt about this, look back at chapter 6, p. 136.) This means that I can take it to be the instantaneous velocity defined for the natural trajectory point, this value being fairly independent of the size of the fine graining that I choose in the natural trajectory. Imagine instead that I were to define a transit time for the small segment p: if I double this segment, clearly, this transit time will *double*. Correspondingly, if I were to assign this transit time to the natural image of the point in question, and I were to double the size of the segment a, this transit time would also be approximately doubled, although the new segment is still very small. It is because of this that the transit time is not a *stable quantity* (as the instantaneous velocity was) that can be assigned to the natural "point." It should be clear therefore that the concept of the transit time at *a point has no physical meaning whatsoever*.[3]

This completely disposes of our "paradox," now revealed for what it is: a clever way of leading people into utter mental confusion. There are two ways in which such a situation may be created, each sufficient to give the appearance of a serious philosophical problem. First, models and their images in nature may be illegally mixed up, and, second, incorrect mappings may be attempted in which model quantities without a physical meaning are generated by apparent limiting processes, which do not really license such quantities to be mapped into any natural trajectory. Both confusionary techniques have been displayed in our present "paradox," but the mathematics of taking limits is itself fraught with danger, since intuition often works so well that it can lead us up the garden path by excessive reliance on it. I shall try to alert the reader about these problems in the next two sections.

CAUTION ABOUT LIMIT TAKING

There is a general problem that one must watch for when going to the limit of very small quantities, and this is that, irrespective of the danger already discussed of confusing the model with its image in nature, the apparently intuitive alleged limits are sometimes nonsensical. This is the question that was painstakingly addressed by the mathematical work that I have already mentioned, and I shall now try to highlight the problems about which one must be careful.

As you might remember from our work in chapter 6, Newton noticed that if you work with the average velocity, which is space divided by time, as the time becomes very small and so does the space, their quotient might become stable, a concept that allowed Newton to define the *instantaneous velocity*, which is nothing more than the stable value toward which average velocities tend when the space interval diminishes indefinitely (we are within model graining). The possible existence of these stable values or *limits* arises from the fact that the quotient 0/0 (toward which we make the ratio *space/time* converge) can be any number (because any number multiplied by the denominator zero will give zero and thus correctly agree with the numerator of the fraction). It is thus possible for quotients a/b, when both terms are gradually reduced to zero, to become stable, approximating some particular number, and there are sound mathematical methods for finding that number, if it exists.[4] On the other hand, great trouble can arise by trying to assign some significance to the limiting value of a quantity that is allowed to diminish toward zero *on its own*, as we already fallaciously did with transit times and as I shall now illustrate with a geometrical example.

INFINITESIMALS ARE NOT AS SMALL AS YOU THINK

Whereas the consideration of a quotient of two gradually smaller quantities protects us from danger when taking limits, arrant nonsense can arise

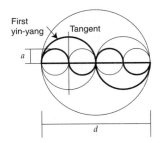

Figure 3. Forming a limiting sequence of yin-yang curves

The yin-yang curves are shown in bold lines.

from such statements as "it is evident that as we diminish the length of this segment it must become a mathematical point." Such model quantities as are allowed to be reduced indefinitely down to zero are often called *infinitesimals*, but the correct definition of the latter requires far greater care. I cannot enter here into how this is done, but, since the understanding of this question is important in order to avoid error, I shall illustrate it with an example, a very famous one, since it entails nothing less than the revered yin-yang symbol, shown in figure 3.

The way in which a sequence of yin-yang curves is formed should be clear from the figure. I have inscribed yin-yangs twice in successive semicircles but, of course, this process can be continued indefinitely. After many such steps, the segment corresponding to *a* would become so small that intuitively (but wrongly) we might identify it with a mathematical point, whereby the corresponding sequence of yin-yangs would coincide with the diameter *d* of the original circle. Therefore, their length should also be *d*. On the other hand, it is easy to prove, as I shall do in a moment, that the length of the yin-yangs, at all stages of the limiting process, equals the length *l* of the original semicircle, so that *d* = *l*. Since, from elementary geometry, the length of the original semicircle is $l = \pi d/2$, it follows that $d = \pi d/2$, whence π equals 2, a delicious piece of nonsense since we know only too well that π equals 3.1415. . . . (Ancient Chinese, who took π to be 3, used this trick to demonstrate that 2 equals 3, which permitted them to establish some required theological properties of a divine trinity.)

I shall now prove, as I have promised, that the length of the combined yin-yangs always equals that of the original semicircle $l = \pi d/2$. Consider the first yin-yang (see fig. 3), the length of which is *twice* that of the semicircle on its left, of diameter *d*/2. On using the formula just given for the length of a semicircle, that of a semicircle of diameter *d*/2 is $\pi d/4$, and twice this value will give the length of the corresponding yin-yang curve as $\pi d/2$. Since this value is equal to l, we have proved that the length of each yin-yang curve must be equal to that of the semicircle in which it is inscribed. When we apply this to the set of the two smaller yin-yangs shown, it is clear that their combined length must coincide with that of the original yin-yang and that this property must subsist at all stages of the successive constructions.

There must be something wrong in this work, of course. The mistake in the above proof is the erroneous idea that as you model-grain the sequence of yin-yangs, its length equals that of the diameter, although the two curves appear to coincide in that limit. The methods of proper mathematical analysis are required in order to pinpoint the origin of this mistake, but a fairly intuitive argument is as follows. When one curve (the yin-yang) is gradually deformed so as to coincide with another (the diameter) in the limit, they must share the same tangent for their lengths really to coincide. (This should be proved, but it is not difficult to accept that, otherwise, the two curves cannot "fit.") If you look at any of the yin-yangs where they cut the circumference diameter, their tangents are always *perpendicular* to it, and they remain so however many divisions the yin-yang undertakes. Thus, the two curves mentioned (the yin-yang and the diameter) can never share the same tangent. (Notice that in this work I refer to the straight line of the diameter as a "curve," as it is common use. Also, notice that its tangent coincides with it.) The simple rule remains, however, that merely reducing the length of a segment to "nothing" does not license you to identify it with a mathematical point without further ado; and I hope that the reader will apply caution in accepting arguments (as I have already fallaciously used) that entail such infinite reductions in length or for that matter in time. Notice from our example that even a line that can be imagined to be infinitely reduced in thickness retains a structure (in our case its tangents), and might not be identifiable with a mathematical straight line. And we are now ready to face an even greater mess.

NOT SO MUCH A PARADOX
AS EXQUISITE MISCHIEF MAKING

Proper respect for the distinction between a model and its image in nature is most important. In order to illustrate the horrible confusions that may arise when this is forgotten, I shall consider the famous paradox of *Achilles and the Tortoise*. It will be useful before we start to summarize the views that I have been propounding. A trajectory in nature is modeled by a curve (this word covers straight lines also) that contains mathematical points. In applying the laws of motion, I must place myself automatically in the model world, and I may ask such questions as are necessary to sustain those laws. Possible questions are: what is the position and what is the velocity of this particle? And such questions pertain to a stated mathematical point. The answers to those questions are then transferred (*mapped*) to the corresponding natural trajectories, within an unavoidable degree of graining. The question: what is the transit time spent by a particle at a given mathematical point, is not legitimate, since even within the model theory this transit time is not a stable quantity as the model graining is made finer. This is a requirement that all mechanical variables, such as positions, velocities, and accelerations do properly satisfy because they are assigned to mathematical points by means of well-grounded limiting processes. Neither can that question be part of the description of the trajectory in nature because mathematical points do not exist in nature. I have already shown that disregard of these distinctions may lead us to the absurd conclusion that, since the transit time at a mathematical point (alleged to belong to the real trajectory) appears to be zero, either we do not move, or the time that it takes to traverse a meter should also be zero. Such an illegitimate use of mathematical points is no more than a trap to be carefully avoided.

Traps, of course, are a favorite occupation for philosophers to lay, and as regards motion and time, Zeno of Elea (born ca. 490 B.C.E.) is the unrivaled master, having produced endless entertainment for centuries to come. Of the several paradoxes that he constructed, *Achilles and the Tortoise* is perhaps the most outrageous, because although one instantly realizes that the result proposed is absurd, it is well-nigh impossible to disprove it, despite the fact that practically every philosopher worth her salt has tried. Saltless as I am, let me try my luck.[5]

I shall state the paradox as clearly as possible. Achilles runs ten times faster than Tortoise, to whom he gives a 10 m head start. Zeno asserts that Achilles can never catch up with Tortoise, for which purpose he analyzes their trajectories in successive legs as shown in table 1. In leg 1, when Achilles reaches the point where Tortoise was at the start, 10 m, Tortoise has advanced 1 m to 11 m. In leg 2, when Achilles reaches the position where Tortoise was at the end of leg 1 by advancing 1 m, Tortoise advances 0.1 m to 11.1 m, and so on. It is "clear" that Achilles can never catch up with Tortoise. That this is a magnificent nonsense is even clearer, if we forget all about Zeno and allow Achilles to run a "new leg 1" of 20 m, as shown in the last row of the table, whereupon Tortoise moves 2 m ahead of his original coordinate of 10 m, so that he is 8 m *behind* Achilles.

If you look at the table ignoring the last line, it appears that the impossible is nevertheless undeniable: Achilles never catches up with Tortoise, a

Table 1. Achilles and the Tortoise

All positions are in meters.

	Achilles	Tortoise
Start	0	10
Leg 1	10	11
Leg 2	11	11.1
Leg 3	11.1	11.11
Leg 4	11.11	11.111
Continue		
indefinitely		
New leg 1	20	12

monstrous result that haunted me for years. One requires a lot of courage to disregard the obvious knowledge that, as shown in "new leg 1" Achilles *does* overtake Tortoise, in order to find out what is wrong about Zeno's paradox: the whole purpose of the exercise is not to find what is true, which we know only too well, but to reveal the con in Zeno's argument.

And this con will now be exposed. Zeno describes both contenders as running, and he accordingly leads us to believe that it is the natural trajectory that they pursue and *not* a model of it. That he is talking of a natural trajectory must be untrue, because the way in which he has defined the construction of its various legs is nothing else than the process used in figure 2, whereby graduated segments define a *mathematical point* by infinite approach to it, and mathematical points can never belong to natural trajectories. So, whereas Zeno creates the impression that he is talking of the real world, he is happily discussing a model of it. In a way, this is all there is to the resolution of the paradox, since there is nothing that can guarantee better instant confusion than mixing models with their corresponding natural counterparts.[6] In this case, by inserting a spurious mathematical point in the middle of a natural trajectory, Zeno is introducing a sort of black hole, out of which none of the two contenders can exit; and this is what makes the paradox so disturbing, since you know full well that they are perfectly free to move all along the trajectory. But, so as to convince you that I am on the right track, I shall pursue my argument further until we obtain some clear and solid result from it.

First of all, that Zeno is working on a model of the trajectory and not on the real trajectory, should be clear, since the various positions listed in the table, for example under *Achilles*, are themselves mathematical points. Now: a model is something imagined and, of course, one can imagine anything. The models one imagines in order to study nature must have some underlying organization because at the end of the exercise some question must be answered, and this question must permit some sensible mapping of the result of the model to what we want to find out about nature. So, what is the *real* problem that Zeno, while pretending otherwise, was trying to solve by means of his model? I shall put this problem just as precisely as an examination question:

Given that Achilles is ten times faster than Tortoise, that Achilles gives Tortoise a 10 m head start, and that they both start running simultaneously, find the point of their trajectory at which Achilles catches up with Tortoise.

If you look at table 1, you realize that Zeno had produced the perfect *algorithm*[7] to answer this question, as it follows from the fact that this algorithm defines a point at which Achilles is as near Tortoise as we want: by taking more and more steps, we can reach any desired difference, just as Newton did in the model trajectory of figure 2. The result of this algorithm is also easy to obtain; because by its definition the algorithm must be continued ad infinitum—which you can always do in a *model* trajectory. The number that comes out from the first column of the table is clearly 11.11111111111 . . . m, where the dots indicate that the decimal digits continue indefinitely. (Notice, incidentally, that this is the number that comes up also from the *Tortoise* column, thus confirming that we have obtained in our model the precise mathematical point at which the two agonists meet.) As when we use Newton, once we get a coordinate out of our model, we go from the model to the real trajectory by introducing whatever degree of graining we desire. If we want to work within the nearest millimeter, for instance, we answer that Achilles meets Tortoise at 11.111 m.

Notice the huge difference in the role of the mathematical point in Zeno's version of the paradox as compared with ours. In Zeno, the mathematical point is surreptitiously (and illegally) brought into the natural trajectory just to create mischief, and create mischief it does, paralyzing the contenders out of their motions. In the correct problem, instead, the mathematical point is used in the model so as to *identify* a point of the natural trajectory *through* which the contenders pass. And in order to identify the mathematical point, you use the same technique of gradual approach that was inaugurated by Newton without anything ever going wrong, simply because at no stage in Newton's procedure is there the pretense that it is the *actual* motion that is chopped up by the algorithm. Notice also that the model employed in the correct version of Zeno's algorithm can perfectly well be used with the concept of the natural trajectory, as followed for instance by the contenders in the last row of table 1. All this is so simple and clear that one wonders why people suffered from Achillean hallucination for almost two and a half millennia. (So did I, not alas for so long, until I started writing this book.)

You must not think, though, that all this was in vain and that Zeno was an obnoxious fool. On the contrary, it is simply amazing that twenty-five hundred years ago Zeno was able to start such a penetrating inquiry into the concept of space: the worries that he created fed the imagination of the mathematicians, especially Richard Dedekind (1831–1916) and Georg Cantor (1845–1918), who ultimately in the nineteenth century defined a mathematical point very much along the lines which, in a much simplified form, I have used. One must always remember, though, that mathematics is too beautiful a creature to be approached not previously having made spiritual exercises to achieve equanimity, about which I shall now instruct my readers.

HOW NOT TO BE MESMERIZED BY MATHEMATICS

There is a mathematical "solution" of the Zeno paradox that has much currency, having been propounded by some of the cleverest men of the last century, like Bertrand Russell and Hermann Weyl.[8] The argument, again, is very simple. If you look at Achilles' alleged trajectory in the second column of table 1, it is clear that his successive "positions" can be written as follows:

$$10 + \frac{10}{10} + \frac{10}{100} + \frac{10}{1000} + \frac{10}{10000} + \cdots$$

(Please notice the very important property that this sum also represents the successive "positions" of Tortoise.)

This is an infinite sum (called in mathematics an *infinite series*), but mathematicians easily prove that it converges, that is, that it has a *limit*, which it approaches when the number of terms is very large (*going to infinity*). This mathematical theory also allows you to compute the value of this limit as 11.111111 . . . , which happens to be the value that I proposed before on fairly intuitive grounds, as the model point at which the two agonists would meet. Having done this little bit of math, people sometimes triumphantly say: this is what good old Zeno did not know, he did not realize that this series has a limit and thus that Achilles *catches up* with Tortoise at the precise point stated.

I hope that you will find this argument suspicious, because we know only too well that Achilles does not merely catch up with Tortoise, but leaves him well behind. It is important, though, to understand what mathematics is doing here. When you look at table 1, that is, at the very clever algorithm invented by Zeno, the crux of it is that the two columns *Achilles* and *Tortoise* are *not independent* but that they are *coupled*: the entry for Achilles must always be the last entry for Tortoise. However impressive the math, the sum I have written *uncouples* the two columns: this is why we allow Achilles to catch up Tortoise in this interpretation. By adding up the two columns *independently*, we do get, of course, a single result, but we also violate Zeno's construction. Mathematicians will answer: but this does not matter because the difference between Achilles and Tortoise can be made as small as we want by taking enough terms in the series (this is what is meant by *converging to a limit*). In saying this, they have not answered Zeno's original problem: by uncoupling the two contenders, they have simply answered a *different* one. However many terms you take in Achilles' series, in fact, you should always take one more term in that for Tortoise. (This is because, in being taken by the nose by Zeno on his allegation that we are on a *natural trajectory*, we should be dealing with real and not with mathematical points, and correspondingly with *finite*, rather than infinite series. Remember that taking infinite terms is model graining.) In other words, a mathematical argument in which no distinction is made between the model and the natural trajectory, merely dismisses Zeno as a pseudoproblem: the mathematical definition of a mathematical point as *model* graining automatically disregards the infinitesimal segment by which Zeno's Tortoise is perpetually "ahead" of Achilles (in his mixed up conception).

This problem does not arise in my interpretation. To answer the true Zeno's problem, that is, at what point the contenders cross, I have to proceed entirely within the model, and all I have to do is find the required mathematical point. The sum above correctly defines, identically for both contenders, the mathematical point wanted, which is later identified in nature within any desired, but finite, degree of graining. All that the mathematical addition of the series does, once you have identified the correct problem, is to provide you with a more rigorous solution than the intuitive one that I gave you straight from table 1. After all, what we want is to determine *uniquely* the mathematical point at which the contenders cross, for which the coupling of the two trajectories is irrelevant.

In considering the status of mathematical argument in a problem like the one in hand, it is useful to remember the following. People very often say that you can never get out of a computer more than you put into it. I am not going to get involved at this stage in the rights or wrongs of this aphorism: what I want to remind the reader is that what is alleged to be the case for computers is, with even more reason, true of mathematics: to have expected mathematics to have solved the original Zeno paradox is, with great respect to my distinguished predecessors, plain silly. The role of mathematics cannot be that of telling us how to handle reality[9] and this was the challenge that clever Zeno of blessed memory bequeathed to us.

THE CONTINUUM

Many readers, I hope, will be happy at this stage to have seen the light. But others might be experiencing ontological worries, feeling that in order to bring about clarity I have thrown away some cherished views as to what is "out there." They may feel that when you consider the motion of an object, a particle say, if this particle moves from *a* to *z*, it is inconceivable that it does not traverse all the points in between, a view that appears to be contradicted if, as I have asserted, natural trajectories do not contain mathematical points at all: mathematical points are no longer horses but unicorns. It might thus appear that I have given a mortal blow to the concept of the continuum and that motion must be conceived as *quantized* in unavoidably finite steps.

The proposals I have made entail, of course, a certain amount of onto-logical purgation, but they are not as drastic as the above worries might make them appear. Let me first deal with the question of what is the status of the intermediate *mathematical points* in a trajectory from *a* to *z*. Let *b* be such a point of the mathematical line that models the trajectory. All I have to do to map it, is to form a small *neighborhood* around *b* in the math-ematical line (with whatever degree of graining I want), and then map this into a corresponding neighborhood in the natural trajectory. You can see that *each mathematical point* of the model trajectory can thus be consid-ered, and that it has a well-defined counterpart in the natural trajectory. (I shall refer to such counterparts as "points," although they are given by small segments corresponding to the graining used.) All mechanical prop-erties, like velocities, accelerations, and times, can be defined in this way for *each "point" of the natural trajectory*, as long as we recognize that *each* mathematical point of the model trajectory has to be identified within a small segment of the natural one. Because of this mapping, continuity of the natural trajectory can be ensured, in the sense that each "point" of it can be specified in a *continuous* way along the model trajectory.

So, you do not have to worry that motion as I have described it be-comes a sort of hopping along the natural trajectory. Even having agreed to some degree of graining in the natural trajectory, you can always iden-tify two "points" of it to be separated by *less* than the accepted graining, simply by so choosing the corresponding points in the model trajectory (which can always be done within model graining), and then mapping them to the corresponding elements of the natural trajectory.

Some of the concepts that I have roughly discussed in the above are now taken rigorously into account in modern mathematical analysis. Mathematical points are discussed as *points of accumulation*, small *neighborhoods* are defined around them, and a *cover* of the mathematical line is provided by means of these neighborhoods which, as the name suggests, comprises every mathematical point of the line. This is called the *Heine–Borel cover*.[10] The important point is that even the purely mathematical model has to take into account the aspects of graining which I have introduced. Just as in our discussion, the concept of continuity survives in this more rigorous mathematical analysis but it is, as in our case, not identical with that used in the naive picture of a classical natural trajectory as a sort of string of mathematical points.[11]

Notice that Zeno has forced us to purify somewhat our possibly naive ideas of continuity, even when discussing motion in classical mechanics: the price of ignoring this is to be submerged for ever in a morass of paradoxes. Of course, whatever degree of graining we use in classical mechanics can be negligible compared with macroscopic dimensions. If we were dealing with the motion of atomic particles, instead, the finest graining that can be used, which is of atomic or subatomic dimensions, is now commensurate with the size of the moving objects under study, and we must expect the classical picture of motion to break down. Break down in quantum mechanics it does, and in a glorious way, as we shall see. This break down is a confirmation of the wisdom of paying serious attention to the correct meaning of trajectories, even at the macroscopic classical level.

CODA

We have learned two very important traffic rules in this chapter. The first is that one must never take limits for quantities that allegedly become "infinitely small" without some serious thought. Newton knew what he was doing and got his results always right, but for the rest of us mortals, rigorous mathematics is required. Since I do not expect the reader to engage in such, what must remain is the need for the most cautious skepticism when such arguments are presented, especially when the results are arranged so as to appear paradoxical. The second rule is that *never* must a model be mixed up with its counterpart in nature. The model must be treated separately, and it is only when results are obtained within the model, possibly by application of the so-called laws of nature (more properly: laws of models of nature) that the mapping from the model to a result in nature can be done, within some accepted degree of graining.

Even in classical mechanics, confusion between what exists in our imagination and what exists in our perception of nature, can lead to serious mistakes and awkward but apparent paradoxes.

It is most important to recognize that in dealing with laws of nature we are establishing a connection between two events, say 1 and 2, in nature (nature$_2$). In order to do this, we must map 1 and 2 into 1' and 2', respectively, their *model* counterparts in nature$_3$, and then apply a law to go from 1' to 2.' Finally, we map back from 2' to 2. Two fatal mistakes can be committed in this work. One is to take the excursion from 1' to 2' as if it were, not a map, but in nature$_2$. An even worse one is to take a *part* of this excursion as if it had a direct physical meaning: the only responsibility of the law is to take us from 1' to 2': any intermediate steps may be devoid of physical meaning.

There are two other consequences of our treatment. First, we should acquire a healthy skepticism of *reductionism*, a method in which a problem, à la Zeno, is reduced to smaller constituent parts in the expectation that this will permit safe analysis of it. The second consequence is that it is unreasonable to expect mathematics as such to resolve problems, either of philosophy or of nature: mathematics, glorious as it is, is the servant, not the queen of the sciences.

Armed with these healthy ideas, we can now attempt to discover what people mean by the scientific method, as I shall try to do in the next chapter, but, before concluding this one a short dialogue might be useful.

ENVOY. A DIALOGUE

Mathematician. You, Zeno, were born too long ago to understand properly the concept of infinity. Because your two contenders are engaged in a never-ending race in which Tortoise is always ahead of Achilles, you think that Achilles can never catch up to Tortoise.

Zeno. Of course, if he is always behind, surely he never catches up with Tortoise.

Mathematician. That is the rub, dear Zeno. The way you decided to arrange the movements of the contenders, so that Achilles is always behind Tortoise, does not interest us. The reason for this lack of interest is that I can always make sure that, the longer the race goes, the difference between Achilles and Tortoise is smaller than any number you could challenge me to match: this is what we mathematicians mean when we say that both trajectories reach the same limit, namely 11.111111 . . .

Zeno. You may say that, but please do not believe that I do not know about other things you have invented in your century. I am very keen on your computers, and the fact is that if you run my algorithm in a computer a *finite* number of times, however long (and even computers cannot reach infinity), Tortoise is *always* ahead of Achilles. I prefer to pay attention to the computers, rather than to your imagined limits that require an infinite number of operations and appear to me to be metaphysical.

Natural scientist. Ah, dear Zeno, that is the pot calling the kettle black because you also *imagine* the Achilles-Tortoise race in your own way. But this cannot possibly be a race because, in order to define it, you require mathematical points and mathematical points do not exist in a racetrack. You are right, I believe, in your objection to the mathematician, that he does not answer the problem you have in your mind. Neither he understands for that matter that there is no race at all (because it is not a mathematician's job to know what a race is). Notice that he always accepted your description of the race at face value, and he was only interested in how to terminate it. In this you are both wrong. Achilles and Tortoise perform their *danse grotesque* only in your imaginations.

Zeno. Now you really puzzle me, because at the beginning of your book you told me that wonderful story of Chekhov: surely, if the Achilles-Tortoise race exists in my imagination and my imagination exists in nature, then the race also exists in nature.

Natural scientist. Ah, dear Zeno, nature is not as simple as that: what you have in your imagination and thus exists in nature is not the Achilles-Tortoise race, it is your *model* of the Achilles-Tortoise race. The problem with imagination is that unless you treat it just as a dream, which I don't, you have to organize it pretty carefully, so as to find what you have in it that you can map onto that part of nature which is (largely) external to your mind. You Zeno said that Achilles never catches up with Tortoise, but that is in your dream world. And your dream world is one that cannot be an image of our own, cannot be an image of a natural trajectory, because you arranged things, right from the beginning, so as to impede both contenders access to the whole of their track, having created an unsurpassable *mathematical* point in the middle of it. You know full well, of course, that they should be able to move up and down the whole of any natural trajectory or of any adequate model of it. In my world, in the world that I study, my job is to find the point at which they meet, and I find it by your algorithm, or if you want, I can use your computer program. I consider either as the definition of a mathematical point, and because I look at things that way, my friend the mathematician has defined the limits of both Achilles and Tortoise's series to be identically the same math-ematical point. He has done that because I am not a least bit inter-ested in how fast or otherwise a mathematical point is approached. All I need to know is that its definition is *unique*, so that I can identify its image in nature properly, and my mathematician friend has made sure that this be so: he is purely concerned in how to apply rigorously the algorithm that you invented, although, as you do, he does not even mention that algorithm. So dear Zeno, go back to your dream-world where Achilles is always stuck behind Tortoise. I already find my real world hard enough to understand, but thanks very much for making me open my eyes: the more I do so, the more beautiful things appear to me. It might not be the best world that can be imagined, but what is very good about it, is that I can often conceive things, and then I can make them actually work: as you can see Achilles is this very moment repairing my sandals while Tortoise is miles behind.

NOTES

1. This was a point already well understood by the Jewish philosopher Moses Mai-monides (1135–1204) who rejected the notion that time was infinitely divisible. (See Mai-monides 1904, Part I, chap. LXXIII, p. 315.) It was Leibniz's perception of this problem that led him to escape from "the labyrinth of the continuum."

2. Even in a proper treatment of the corresponding mathematical model, the concept of the transit time at a single mathematical point is avoided: the mathematical point is replaced by an "infinitesimal neighborhood" around it, which is a segment the length of which can be model-grained (see figure 1). The theory then shows how the corresponding transit times, however small, add up to a *finite value* when this addition is carried out over all the infinitesimal neighborhoods that cover the model trajectory. It is this *finite value* of the model which can now be compared with the natural trajectory. So, you can see that the description of my walk through a meter was completely wrong, even from the point of view of a mathematical model.

3. The reader should realize that the transit time at a point is a *duration*, that is, the time interval that the particle spends in a neighborhood of the point, taken to the "limit" when the length of this neighborhood is reduced to zero. This "limit" is zero, but it has no physical meaning because the value of this duration is crucially dependent on the length of the neighborhood, however small the latter might be. Instead, the time t at which the par-ticle *passes* through the point has a physical meaning (as it should be since it is an important dynamical quantity) because if the neighborhood is small enough, doubling it, for instance, will hardly change t: 0630 would change into, say, 0630 plus a millisecond (whereas the transit time might double). In this case, therefore, one can properly talk about a limit.

4. If the reader wants an example of this situation, consider the following one. Both the numerator $1 - x^2$ and the denominator $1 - x$ of the fraction $(1-x^2)/(1-x)$ vanish when x equals 1. You can check with a hand calculator, however, that if x equals 1.1 then the fraction takes the value 2.1. Take now x successively equal to 1.01, 1.001 and so on, and you will see that every time you get nearer and nearer the value 2 for the fraction, which means that this is the value of its *limit* for x equal to 1, although the fraction is of the form 0/0 at that precise point.

5. An important treatment of this paradox is Russell (1926), chap. 6. Whitrow (1961), pp. 143–52 provides an excellent review, and an ingenious treatment is given in Ryle (1962), chap. 3. Mathematicians, of course, have occupied themselves a great deal with this problem as, for example, Weyl (1949), p. 42, who nevertheless very much follows Russell here. I should repeat here on my own behalf what Gilbert Ryle, just about the most level-headed philosopher it has been my good luck to meet, says when dealing with *Achilles*: "In offering a solution of this paradox, I expect to meet the fate of so many who have tried before, namely demonstrable failure" (Ryle 1962, p. 37). Strangely, it was Gilbert Ryle (1949) who had more than anyone else exposed the dangers of *category mistakes*, a concept which I proffer in solution to the paradox. Interesting discussions of the paradox may be found in Owen (1958), Salmon (1970), Grünbaum (1967), Sainsbury (1988), McLaughlin (1994), and McLaughlin et al. (1992). The last two articles offer a very deep solution of the paradox, at the expense of abandoning standard mathematical analysis and substituting for it a new mathematical scheme, called *internal set theory*, introduced by Edward Nelson. None of these references make a clear-cut distinction between a natural trajectory and its mathematical model. The reader interested in Zeno may consult Kirk et al. (1983) or Caveing (1982).

6. This is an example of the so-called category mistake in which concepts that pertain to two different categories (in this case a model and its image in nature) are discussed in the same discourse as if they belonged to a single category. (See the note above.)

7. An *algorithm* is a systematic and well-defined mathematical procedure, usually entailing repetitive steps, devised in order to obtain some specific mathematical result.

8. See note 5. Weyl's discussion, however, appears to indicate an uneasiness about the argument.

9. It is for natural science to establish the correct maps of nature.

10. These questions are discussed in any modern textbook of mathematical analysis, as for instance Apostol (1957), pp. 49–51, which was a pioneer reference in English on this subject. This cover of the line entails the concept of *measure* which is necessary if, for instance, one wanted to examine what happens to transit times when the size of the graining is varied.

11. Even the reader who has been reasonably satisfied by my arguments might entertain a fairly serious worry at this stage. If we accept that the coordinates entered in table 1 refer to the model trajectory, we could take the points listed for Achilles, for instance, and map them as explained onto the natural trajectory; and those points would always be one step behind those similarly obtained for Tortoise. The following argument might help to dispel this worry. The Zeno set of points we start from in Achilles' model trajectory, because of the way in which they are generated, correspond to the definition of a point of accumulation (mathematical point) in the model trajectory. As we have seen, such a point of accumulation separates the mathematical line into two sets: we approach the mathematical point from the left but those on its right can never be reached. Whereas this choice of points is adequate in order to determine uniquely a mathematical point, *it cannot be used as providing an appropriate set of points from which to map a natural trajectory*. This is so because Zeno's set introduces a barrier in the model trajectory and thus in its map on the natural one, which of course cannot possess such an artificial singularity. The only use that that set of points may have is in determining the point required in what I called the real Zeno problem.

To summarize: we accept that Zeno uses an alleged model of the trajectory of the contenders, but it is not true that any choice of points on a mathematical line can be an acceptable model for a natural trajectory. Models must obviously satisfy some minimum conditions to be acceptable models of anything in nature, and Zeno's model does not satisfy this requirement because it restricts the motion to points that belong to the half of the mathematical line on the left of the point of accumulation. No natural trajectory could possibly map this feature, and this is why Zeno's model is not acceptable *except to generate a nonphysically significant algorithm*.

IS THERE A SCIENTIFIC METHOD?

Mon naturel me contraint de chercher et aimer les choses bien ordonnées, fuyant la confusion qui m'est aussi contraire et ennemie comme est la lumière des obscures ténèbres.

My nature constraints me to seek and to love well-ordered things, and to flee confusion, which is as much my contrary and my enemy as light is to gloomy darkness.

Nicolas Poussin, letter to Chantelou, 7 April 1642,
in Poussin (1964), p. 67.

I n 1978 the Tate Gallery of London exhibited a new acquisition, the sculpture *Equivalent VIII*, by the British American artist Carl André. This consisted of two identical rectangular layers of sixty loose bricks, superimposed one on the other on the floor of the gallery.[1] The mythical taxpayer was outraged at the alleged waste of public money, and the media had a field day: television cameras were brought to the gallery, and members of the public were canvassed on their views, with the expected results. It was very fortunate that among those interviewed was one of the acutest philosophers who ever inhabited these islands. When asked "Do you think that this is a work of art?" he answered without hesitation: "Of course, everything in this gallery is a work of art," and he then went on sweeping the floor, as was his job. When all else fails, and failure in such cases is more the rule than the exception, *ostensive definitions* are the resort of wise humans, and what is true about art is probably true also about scientific method.[2] If I were to operate on this principle, however, and say: scientific method is the method used by scientists, which is just about the only serious comment one can make, this chapter would be terminated right now, just as free of error as of substance. So, I shall try to review what has been said about the scientific method in the past, in order to see what the limitations are of the various proposals to formalize it, as well as to find a way forward.

I do not want to leave the example of Carl André's work without some further comment. Although the question so ably raised by that work, what is a work of art? is one that probably has no closed answer, there are certainly two moves that must carefully be avoided when facing it. One is to say "I know what I like and I do not like this," that is, one should avoid *prejudging* the issue: in the case in question, for instance, whatever art might be, it is not carved on the tables of the law that a work of art *must* be likeable. Not many philosophers would argue, on art or on anything else, on such poor grounds. Some, however, have fallen into the next trap, which is to try to define an activity under discussion in such a precise manner so as to establish beyond any possible doubt the boundaries of rational discourse in the field in question. The motivation for this is often the (vain) desire to be able to expose as fraudulent certain areas of that activity. In the case of art, some people might want to exorcise the avant-garde, or at least a good deal of it, whereas favorite areas to be condemned into utter darkness in the case of science are, among others, psycho-analysis or Marxist historiography.[3] I am not proposing that we must accept that *everything goes* but, rather, that dogmatic attempts to produce cast-iron solutions are dangerous: that one must approach the problem in a healthy spirit of tolerant analysis is obvious because, although exorcising the avant-garde might appear very desirable today, one must think what such an approach would have done (and in fact did) in the 1870s to artists like Monet and Degas, or scientists like Ludwig Boltzmann.

What I shall try to show in the next few sections is that there is not such a thing as *scientific method*, if by this we mean a closed set of rules which, unless followed, ensure beyond doubt that a piece of work, a theory, or a doctrine, must be put outside the scientific pale. To show this properly, one would have to analyze in detail all the numerous attempts at writing, so to speak, a scientific constitution that would clearly distinguish what is in and what is out of the realm of science. But this is only one aspect of the subject, because prescriptions have also been proposed as to how to conduct scientific research, which do not appear nevertheless to have been much followed in the battlefield, though some of the field marshals were scientists themselves. To review critically all this would be a gigantic task, and all that I can do is to skip quickly over some of the views that have been most influential over the years. This means, unfortunately, that I shall have to go over the rather confused history of the subject. I shall do this, however, in a very sketchy way, since all I want is to illustrate the variety of apparently contradictory ideas that have been entertained in this field.[4] Lest the reader fears at this stage that I shall fall into anarchic skepticism, this is not so: I shall provide in the last two sections a modest proposal to reestablish some order in the subject: yes, we shall see that there is a scientific method, but also that it cannot be encased in a rigid set of rules. After all, science is no more than a continuation of common sense by other means.[5]

THE PIONEERS OF THE SCIENTIFIC METHOD

I have already discussed the extraordinary—and not always healthy—influence of Euclid (active around 300 B.C.E.) in the presentation of scientific argument. He started his geometry from *axioms* that he took to be self-evident truths, from which he *deduced* all the theorems of geometry. It was an intellectual feat, and so elegant that it was adopted as a model of clear thinking for generations. Not only were British school children trained in this method until well into the nineteenth century, but even when the foundations of quantum mechanics appeared in need of a good overhaul, John von Neumann (1903–1957) used the same *axiomatic-deductive* method for the purpose.[6] Euclid's axioms, of course, had to be concerned with points of fact, such as for instance properties of parallel lines, and it should be clear to us by now that there is no such a thing as a self-evident, that is, a *necessary* truth, as regards points of fact. In the case of Euclid's postulates, it took until the nineteenth century to find that their necessity was questionable. Despite their a priori appeal to the intellect, it was discovered that perfectly consistent geometries could be constructed on axioms that violated those of Euclid: and Einstein later demonstrated that these geometries superseded that of Euclid in describing the properties of physical space. Elegant as the axiomatic-deductive method is, it cannot therefore be claimed now to be any more than a concise mode of presentation. Archimedes (287–212 B.C.E.) followed Euclid's axiomatic-deductive method in his *Mechanics*, which did not help at all in getting a grasp about how science really works.

Despite the evident problems with this type of top-down approach, it enjoyed enormous popularity among those philosophers who believed in the supreme power of unaided reason, of which the most influential was René Descartes (1596–1650), for whom science had to start from general laws and go down by deduction. This top-down tradition lasted a long time, although it was made a great deal more sensible by dispensing with the idea that the starting point must be general laws: the way to proceed was to enunciate a hypothesis, obviously a provisional one, which was then validated or otherwise by deducing from it conclusions that could be compared with experience. This method, called the *hypothetico-deductive method*, was strongly supported in the nineteenth century by the Cambridge scientist William Whewell (1794–1866), who also elaborated it further.[7] But the hypothetico-deductive method was still seen in our own time as "the essential feature of a science."[8]

I shall have to say something more about the hypothetico-deductive method in a moment, but I should like first to refer to a contemporary of Whewell's, John Herschel (1792–1871), also a Cambridge graduate, whose major contribution to scientific method was to introduce the expressions *context of discovery* and *context of justification* thus making a very clear and necessary distinction about these two aspects of scientific work.[9] It would be entirely wrong, of course, to assume that the life of a scientist is divided in two nonoverlapping periods, first of discovery and then of justification. The two go together of course and can often be hardly disentangled, but whereas the process of discovery can be chaotic and even irrational, that of justification must be rigorous and precise. Although the two contexts cannot really be held as separate, the distinction between them is useful, because it allows the elimination of the mental confusion that may arise by noticing that the context of discovery shares some of the psychological features of other activities, such as mysticism, whence such authority as science may appear to have is "borrowed" by some theologians in order to impart strength to their subject. This is a fallacy that can easily be exposed when realizing that what is truly *unique* to science is the context of justification; and it is only if comparison with this context were possible that one subject would reinforce the other. All that such misguided attempts show is nothing else than a deficit of faith. Moreover, those features of the context of discovery that are sometimes linked with mysticism and belief (such as a condition of elation or euphoria) can also be associated with other psychological states, sexual ecstasy for instance, and if the same borrowing procedure were used in such cases, the results would be bizarre if not comical.

I can now go back to the hypothetico-deductive method. Suppose that we start from a hypothesis p from which we can *correctly* draw a conclusion q, so that the implication $p \Rightarrow q$ is *true* and, moreover, the consequence q is also true. It does not follow, however, that p must be true: this is a fundamental feature of the material implication, as can be seen in the penultimate row of table 2 of chapter 3, in which p is false although both q and the implication are true. It is only when the implication $p \Rightarrow q$ is true (that is, it is correctly made), and q is found to be false, that we can draw a cast-iron conclusion, namely that the hypothesis p is false. (See the last row of the aforementioned table.)

The use of the hypothetico-deductive method, therefore, does not allow us to determine with any degree of finality the hypothesis p

advanced. All that it does is to reassure the scientist who is trying out a hypothesis, every time that the consequent is true, that his hypothesis has not been *falsified*, but I shall have to come back to this idea a little later. Naturally, when considering a hypothesis, the best thing that one can do is to consider as many consequences as possible, and it appears natural to allow one's confidence in the hypothesis to strengthen as the number of positive results increases. Although this is what happens in real life, it is not a rigorous procedure because it entails all the problems that befall the wretched concept of induction, as we have seen in chapter 3: even a million repetitions of a result does not preclude that the next one will fail. Despite this weakness, the hypothetico-deductive method is a mainstay of science, as I shall discuss later in this chapter.

Almost at the same time as Euclid started the top-down tradition in scientific method, Aristotle (388–322 B.C.E.) took the opposite line of thought, proposing the bottom-up approach, starting from observations, supported by induction, from which laws or explanatory principles can be formulated, from which in turn deductions leading to new observations arise. Despite the fact that Aristotle commended starting from observation, his approach was too passive to lead to modern science: you must realize that observation and experimentation are not the same thing, experimentation being a sort of interventionist observation. Moreover, in spite of his injunctions, Aristotle was apt to give too much weight to general principles that he considered self-evident, in order to discover first causes, and thus explain by top-down arguments why things are as they are. This is not really what science is about: the behavior, rather than the essence, of physical objects is its central concern.

It was the Oxford Franciscan Roger Bacon (ca. 1214–1292) who moved forward from Aristotle and urged for experimentation, a daring stance that cost him twenty-four years in prison. A proper understanding of the role of experimentation in science had to wait until Galileo Galilei (1564–1642) produced his spectacular results through the use of the telescope, as well as through the study of motion. Francis Bacon (1561–1626), later Baron Verulam and Viscount St. Albans, was a most accomplished communicator who helped to understand the need to move away from Aristotle's passive observation to the experimental methods of Copernicus, Kepler, and Galileo: science was to establish man's domain over nature. Not only did he support Roger Bacon's concept of systematic experimentation against the haphazard collection of data à la Aristotle, but he also had a clear insight of the dangers of premature generalization when proceeding from induction, a procedure which he nevertheless advocated strongly.

Among the opponents of Descartes's top-down approach, none is more distinguished than Sir Isaac Newton (1642–1727), who supported the use of general propositions inferred by induction. He held, quite rightly, his laws to be contingent, and he accepted that his first obligation was that of saving the phenomena, even at the cost of introducing entities, like the gravitational force acting at a distance, with a far from clear physical meaning: but he rightly refused to speculate about its elusive nature. Within this tradition, John Stuart Mill (1806–1873) occupies an important place, especially since he was in strong opposition to his contemporary Whewell. Not only did Mill propound the inductive method, but he also elaborated it into a systematic procedure for an exhaustive analysis of causal relations.

Although the bottom-up approach is even now sufficiently attractive for scientists to disregard philosophical worries and to use it when convenient, it entails some clear philosophical weaknesses. First of all, observation of what? There is certainly not such a thing as a pure fact, as I have already discussed and was long ago urged by Whewell, so that observation, to be useful, has to be placed within some theoretical context, however elementary the theoretical frame involved might be. Moreover, we know only too well the problems entailed by induction. These two aspects of the bottom-up approach were strongly rejected in our century by Popper, so that I shall come back to them when discussing this philosopher's final and complete solution to all our problems, which, as is mostly the case in the philosophy of science, is just as fragile as the doctrines that he denounced. Whatever its philosophical weaknesses are, the bottom-up approach is still used, and will always be used in science (as is also the case with some versions of the top-down approach), and I hope that the modest proposals I have already announced will justify to some extent their use.

THE VISIONARIES

Toward the end of the nineteenth century, philosophy started a period of severe restraint in the use of hypotheses, and of such a strong emphasis on experience that it was required that no unobservable quantities be introduced in theories, the dominant school being some form of the *positivism* led by Auguste Comte (1798–1857). This is to some extent understandable because just about at the same time as the French revolution swept away a vast amount of social cobwebs, some of the cosy beliefs until then entertained about matter had to be abandoned. The interpretation of combustion as the unobservable separation from its matrix of some postulated substance of *negative weight*, called *phlogiston*, had to be abandoned after Lavoisier (1743–1794) showed that the increased weight of the solid products of combustion could more simply be accounted for by combination of the combustible with the air. It is understandable that scientists, having metaphorically burnt their fingers with phlogiston, fell back into what was fundamentally an observance of the so-called Occam's razor. William of Occam (1285–1347), another Oxford Franciscan, lives forever in the famous principle: "entities are not to be multiplied beyond necessity," known as Occam's razor, a very commendable rule which was, at least implicitly, accepted during the period in question. In simple language, Occam's razor requires that, if the same facts may be interpreted by means of two different theories, the theory that involves the smallest number of assumptions be preferred.

Like all rules, alas, blind adherence to it was not always beneficial. The most remarkable example is the story of the theory of heat, for the beginnings of which we have to go back again to Lavoisier. You must remember that at that time the existence of atoms was unknown (although guesses about them existed), and though people understood that heat could be transmitted from one body to the next, what heat was was not known. Lavoisier himself invented the concept (and the name) of *caloric* as some sort of a fluid that was interchanged in such processes.[10] So, Lavoisier, having disposed of the phlogiston, forgot to sharpen again his Occam's razor, an operation that was left to Joseph Fourier (1768–1830), who was able in 1822 to give the correct *phenomenological* equations for the transmission of heat. What we mean by *phenomenological* here is that, whereas the equations provide answers that agree with experiment, they are presented without any attempt at an underlying theory or explanation, and that they do not even attempt to isolate a cause that might be held responsible for the observed effects. In the next half-century or so, a very powerful phenomenological theory of heat, still requiring no hypotheses as to its cause, was developed under the name of *thermodynamics*. At the same time, however, a parallel line of thought that completely ignored Occam's injunction (and which *in this case* turned out to be absolutely right) was developed.

This was the atomic hypothesis, about the early origin of which I shall cut the story short. The fundamental work is undeniably that of the Cumberland weaver John Dalton (1766–1844), who, in order to interpret some chemical and physical properties of materials, introduced the atomic hypothesis from which atomic theory developed, although atoms were not observable at the time. One of the most striking applications of this theory, more than half a century later, was the kinetic theory of heat. For the first time, a physical interpretation of heat was given, as due to the motion of the atoms or molecules in a material: transferring heat from a body A to another body B was simply to excite the molecules of the body B to a more energetic motion. Such ideas were already in the air early in the century, Benjamin Thompson, Count Rumford (1753–1814), doing pioneering work, as well as the French theorist Sadi Carnot (1796–1834). But it was not until the 1850s that the theory began to gain currency, first with the work of Rudolf Clausius (1822–1888), and then with the substantial contributions in the 1870s by the Austrian Ludwig Boltzmann (1844–1906) and by James Clerk Maxwell (1831–1879), then professor of physics at King's College, London. (See chapter 12.)[11]

So, toward the end of the century we have two philosophically opposed theories of heat. One is philosophically clean as a whistle, grounded on strict positivist principles, entirely phenomenological and yet extremely powerful. This is *thermodynamics*. The other, although for a long time producing no new results, neither contradicting thermodynamics in any way, involved a mechanistic model of heat as motion. This was the *kinetic theory of heat* which, if right, could become a wider-reaching theory than thermodynamics, since it covered kinetic properties of matter other than just heat transfer. This situation created one of the most striking debates in the history of science, mainly through the strong advocacy of the phenomenological approach by the Austrian physicist Ernst Mach (1838–1916). Few men had honed their Occam razors more effectively and more conscientiously than Mach, and because of this, his contributions to the understanding of mechanics and space were outstanding, but his utter rejection of atoms as unobservable and unnecessary constructions turned out to be wrong, although based on exquisitely correct arguments.[12] The brunt of the battle was left to Boltzmann, who ultimately hanged himself, probably not because of his fight with Mach and others (although this took its toll on a man prone to depression), but perhaps because of his wife having taken his suit to the cleaners, thus frustrating his desire to cut short his holiday.[13] The kinetic theory of matter remains as one of the most successful models ever proposed in science, because it could be interwoven with classical thermodynamics, and thus extended even further to produce many new results.

It was the work of a great admirer of Mach's[14] which, ironically, resulted in the general acceptance of the kinetic theory and of the *actual* existence of molecules. This was the interpretation, from 1905 to 1910, of a previously not understood phenomenon, the Brownian motion, by Albert Einstein (1879–1955).[15] Mach, however, true to his Occam's razor, refused to believe in atoms to his dying day.[16]

We have here an excellent example of how science works, sometimes through a sort of symbiosis of opposite ideas. Classical thermodynamics is phenomenological and inductive, and produced a large body of well attested results. The kinetic theory, on using instead a hypothetico-deductive method, climbed on the back of classical thermodynamics in order to borrow from it much necessary, but precarious, respectability, through agreement with its results. There is a time when things are obscure, when safety appears paramount because important theories are crumbling (as it happened in the early nineteenth century at the beginnings of thermodynamics), and in such times a phenomenological approach is prudent if not absolutely necessary. But when confidence increases and the world of science stabilizes, it becomes not only acceptable but positively desirable to go back to the leaps of imagination required by the hypothetico-deductive method. Premature rejection of hypotheses, even on the best possible grounds as adduced by Mach, can have very adverse consequences.

The failure of Mach's positivism in dealing with atomic theory did not deter those who were inclined to construct the scientific method as a closed system, and if one had to proceed from laws to their consequences, then laws had to be somehow tamed or made to fit into some pattern, although evidently they could not be a priori. Henri Poincaré (1854–1912), who was a mathematician, but whose work ranged from theoretical physics to philosophy, saved the day for the top-down approach. The problem was that if hypotheses are suspect as starting points, then laws are required, but: where do we get laws from? No worry, says Poincaré and with him the *conventionalists*: laws are conventional. In Newton's second law, for instance, the only observable is acceleration, whereas force and mass, for Poincaré, are conventional. I have already discussed conventionalism in chapter 5 (p. 122), and we saw that one of the problems with this doctrine is that it cannot explain the *internal consistency* of results from apparently independent topics. The mass in Newton's second law, for example, which is the *inertial mass*, happens to be identical to the mass that one has to use in writing down the gravitational force, the *gravitational mass*. As for the forces, I have shown in chapter 8, p. 182, that their conventionalist treatment through the equilibrium of the balance is untenable.[17]

INSTRUMENTALISM

On 24 February 1616, nine days after his fifty-second birthday, Galileo Galilei was summoned in Rome to the Sacred Congregation of the Holy Office (popularly called the Inquisition) before Cardinal Roberto Bellarmino, and admonished to relinquish the Copernican heliocentric views which he had been propagating.[18] Bellarmino was not a man to be taken lightly: in 1600 he had examined the Dominican Giordano Bruno (1548–1600), who was then condemned by the Congregation to be burned at the stake in the Campo de' Fiori at Rome.[19] The good cardinal, however,

offered Galileo a plea bargain: Galileo should state that Copernican theory
was nothing else than a mathematical model without physical meaning,
purely designed to "save the appearances"; and then his work could be
published without trouble. Thus, Bellarmino can be considered as the
founding father of *instrumentalism*, a philosophical position in which a
theory is considered purely as a mathematical device to save the facts, that
is, as a method of computation which, as tabular results might do, entail
no commitment as to the nature of any reality behind them.[20]

I have already shown that Boltzmann, for example, had used the hypothetico-deductive method, but had found support in the results of a purely
phenomenologico-inductive theory, classical thermodynamics. Instrumentalism is another approach, even more cautious than the phenomenologico-inductive one, which shares in fact its modus operandi with the
hypothetico-deductive method, but that makes no claims whatsoever as to
any physical meaning that anyone might wish to impose on the hypotheses
used, however successful they might be. In a way, all scientists are at one
time or another instrumentalists: they might prefer to assert that they save
the phenomena rather than appearances, but when the going is tough, a
minimalist, that is instrumentalist, approach is useful.

So far, I have described a number of apparently opposed philosophical positions each of which was originally presented as *the* scientific
method. I take the view, instead, that what we have is a cocktail of *strategies* which, mixed together in different proportions, and used perhaps at
different times, are all there as part of the "scientific method." A few more
drinks, however, have to be added to this potion.

POPPER AND AFTER

Mach's intellectual heirs were the *logical positivists*, who sprung largely
from the philosophical school at Vienna that came to be known as the
Vienna Circle, of which Rudolf Carnap (1891–1970) was the most influential figure within the philosophy of science. Hans Reichenbach
(1891–1953), who from Berlin largely adhered to the Vienna Circle principles, also contributed strongly to this school. The logical positivists aimed
not only at the elimination of unobservables, like Mach, but also at the
total abolition of metaphysics. We have already seen in chapter 8 that
they, and some of their successors, had tried to identify laws by their logical and syntactic structure, and also that, where necessary to avoid what
might appear to be metaphysics, Carnap, like Poincaré, was prepared to
allege that laws are no more than conventions.

Karl Popper (1902–1994), who although born in Vienna spent most of his long working life in London, is, perhaps, of all philosophers of science, the one whose name is most familiar to scientists. Although not a logical positivist, he inherited from them a desire to formalize the philosophy of science in a closed way. (See note 3.) The very title of his first book, *The Logic of Scientific Discovery*, makes a claim that is almost certainly untenable: if the way to achieve scientific discovery were to be enshrined in some sort of logic, since logic is basically an algorithm to obtain true statements from true statements and to expose false deductions, then scientific discovery would be a mere question of routine. The so-called logic that Popper proposed was based on the concept of *falsification*. We already saw in chapter 3 that if we want to confirm a proposition such as "all ravens are black," no amount of positive instances of it can determine the induction: this is the crucial weakness of the inductive procedure. Nevertheless, we also saw that a *single instance* where the proposition is falsified, such as "this bird is a raven and it is white," gives us absolute certainty that the proposition "all ravens are black" is false. So, for someone like Popper, who had inherited a belief that the mind must be able to excogitate absolutes, falsification was the answer: the whole of scientific method consisted in making *conjectures* (as far as I can see what one would call *hypotheses*) and in trying to *falsify* them. One advantage of this rule, for Popper, is that *he* can decide automatically what is pseudoscience: he rules out any theory based on hypotheses that cannot be falsified as not scientific.

I shall deal later on with the post-Popperian world, but it is useful at this moment to jump a generation and quote Paul Feyerabend (1924–1994) on falsification, especially because whole schools of scientists have grown under its umbrella: "Philosophy must be in a desperate state . . . if trivialities such as these can count as major discoveries."[21] The review I have just given from our work in chapter 3 shows that the logic of falsification is indeed trivial, and it was, in fact, so well known for such a long time that his first proponent, Robert Grosseteste (ca. 1168–1253), was perhaps the first chancellor of the University of Oxford in 1214.[22]

There was of course more to Popper's philosophy than raw falsification. To start with, the evolutionary approach to the development of science that I have adopted in this book was elaborated by him, although in a somewhat different way, based on the principle that science advances through conjectures and refutations. Popper claimed quite correctly that there is no such thing as pure observation, more or less along the lines that we have used to argue that there are no pure facts. Popper's contention that observation could not take priority unless it "disappoints some expectation or refutes some theory," however, cannot be taken seriously, as I shall prove by examples in a moment. Twice during his lectures, Popper claimed to expose the futility of observation by theatrically asking his audiences to "observe," which of course, they could not do.[23] If any of his listeners had learned any critical sense from his teaching, they should have stood up and said: "Sir, you could likewise prove the impossibility of swimming by addressing an audience at a restaurant with the injunction: 'swim.' Whenever people engage in any activity they must prepare themselves for it in some appropriate way." When scientists observe, *by the mere fact that they are scientists* (I do not mean necessarily academic ones: early Babylonian astronomical observers would do), they are engaged in a very specific act, and will use techniques that are those accepted in the particular piece of work at hand. Taxonomy or astronomy could never have gotten started without a large amount of (sometimes amazingly accurate) observation. And the word "observation" when used in science entails, like any other expression that describes a human activity, a great deal of stage-setting before its use can be understood.

Let us now consider the allegation that observation will be meaningful only when it refutes a theory or *disappoints* some expectation, which is clearly contradicted by two of the most dramatic discoveries of the twentieth century. When Alexander Fleming discovered penicillin, he observed among his petri dishes one in which the expected bacterial culture was in part dead, as shown by a large sterile patch. It would be absurd to say that the fact that the bacteria were dead *disappointed* any expectation: on the contrary, Fleming immediately recognized that there was something new and exciting happening and, accordingly, published the paper which years later led to the isolation and production of penicillin by Chaim and Florey.[24] Likewise was the discovery of pulsars by Jocelyn Bell and Antony Hewish in 1967. The only disappointed expectation might have come later when they recognized that the wonderfully regular signals observed were not the broadcast of an alien civilization, as they somewhat jokingly speculated at first. Their true nature was soon discovered, however, and there was here neither a disappointed result, nor the refutation of a theory, but the confirmation of a *new* fact.[25] If the observation of a *constructed* molecular model counts as an observation, and why not, only a lunatic could have labeled as a disappointment or a refutation Watson and Crick's discovery that their model of DNA fitted all the known facts.

Karl Popper's approach, as befitted the school of philosophy wherefrom he emerged, was top-down. Science started from *conjectures* which had to be falsifiable. The more these conjectures were tested, that is, subjected to possible falsification, the better they were theoretically, until they were refuted or *falsified*, and then replaced by new conjectures or theories, and so on. This is why he so vehemently denied the role of observation in science, and even more that of induction.[26] Whatever the philosophical weaknesses of induction, however, it is a method that a scientist cannot help using. Scientists, at least in their role as scientists, do not believe in miracles. If a scientist observes, perhaps by chance, a new result, he or she must become absolutely sure that the result in question can be *repeated* by his or her peers: the reputation of a scientist stands or falls by this. First of all, the repetition by other scientists of an experiment, and their increased confidence in the result, is nothing more than a process of induction, maybe even simple-minded induction. But, second, even before scientists announce to the world their results, they themselves will have repeated it a sufficient number of times to become "sure" that the result is not a fluke: however much philosophically minded scientists might be, and thus fearful that there is nothing much more than a psychological justification to feeling "sure," they will always repeat their experiments, as I have already illustrated in the case of Ørsted (p. 181). And in his case, once he reported his experiment, it was repeated within weeks in half a dozen major European towns, including London. Of course, Popper might claim that those that performed these experiments were trying to *falsify* Ørsted, but if this were so, why *repeat* the *same* attempt at falsification more than once.

Let us now have a look at the question of falsification. I should like to say at the start that despite my already expressed misgivings, there is a lot to commend about this concept, and that it is a *good thing* if scientists try to produce theories that can be falsified. Trouble arises, as is often the case, with plain but sensible ideas, when people try to build them up into majestic and absolute rules. To start with, although it is a good thing if a theory does not contain any elements in it that are not falsifiable, rigid application of this rule would have put some important scientific theories into great trouble. The existence of the atom was not falsifiable for the whole of the nineteenth century, and, despite Mach, atomic theory and its adjuncts, like kinetic theory, were accepted as respectable scientific theories for a considerable time while atoms and molecules were still in an unobservable limbo. In our own time, Bohm's "ontological" interpretation of quantum mechanics assumes that, inside an atom, say, electrons exist in classical states. As we shall see later, the experimental determination of such states is not possible in accordance with quantum mechanics; and this impossibility is not just technical but essential, a fact that is recognized even by the supporters of Bohm's theory. So, the situation is even worse than it was for the atoms in the nineteenth century: then no one knew how to observe them and therefore their existence could not be falsified. With Bohm's electrons, it is not that we just do not know: rather the contrary, we know that the detection of such states, and therefore also their falsification, is impossible. Although there are some people who may think that to postulate such experimentally inaccessible states may be unwise, the scientific community has learned a great deal of tolerance since the days of Mach, and Bohm's unfalsifiable conjectures are treated seriously. As regards the existence of quarks as components of certain elementary particles, which is well grounded on a hypothetico-deductive basis (because the quark hypothesis correctly explains a number of properties of these particles), it was not obvious when the quark hypothesis was proposed that it could be properly falsified: not only had free quarks not been detected, but it was possible that they might never be detected, because as part of the theory, they would be too unstable to survive free.

Another problem concerning falsification is what its meaning is in practical terms. Suppose that we would like to falsify psychoanalysis, a theory that Popper considered pseudoscientific because he regarded it as not falsifiable. The question is this: we may require to falsify the hypotheses used in this theory, or, if the theory itself is considered as a form of a medical treatment, we may want to falsify it in the sense of proving that the results of psychoanalytic treatment are negative when, for instance, proper double-blind statistics are analyzed. If by falsification we require the former, Popper may well be right, but then we would be treating psychoanalysis somewhat exceptionally, since we know quite well in medicine that false (or unfalsifiable) theories may nevertheless lead to positively useful forms of treatment. Acupuncture, because of its long history lasting millennia, contains some theories that a modern physiologist might not entertain, but which would not rule out trials to establish its usefulness as with any other treatment that is accepted (or rejected) within evidence-based medicine. Admittedly, in the case of psychoanalysis, such trials would be prohibitively expensive, partly because of the extremely long duration of the alleged treatments. Moreover, they would be prohibitively difficult to organize, given the huge problems in constructing appropriate samples, let alone in obtaining the cooperation of psychoanalysts in what is in more than one way highly confidential work. But the difficulties in this type of falsification are merely practical, and, in principle, such a trial would be possible.

So, we see that falsification is not the clear-cut test that Popper alleged, but more is to come. For, according to him, falsification is final, and a theory, once falsified, is abandoned or at least its use is restricted to areas where the found falsification is irrelevant. I shall now discuss some counterexamples. Newton's gravitational theory, as a theory of the solar system, was falsified at least three times. The first time was soon after Newton, when it was discovered that the orbits of the planets computed by his theory were not quite right and contained certain perturbations. Was the theory abandoned? Not at all; rather, better methods of calculation, which took account of the influence of other planets on the orbit of a given one, were developed that gave the correct results for all the planets, *except* for the last one, Uranus. When the theory was thus once more falsified, was it abandoned? Not so, the problem remained there on the books, so to speak, until the middle of the nineteenth century, when John Couch Adams in England and Joseph Le Verrier in France, recognizing that the theory was too well *entrenched* to be false, proposed that, rather than abandoning Newtonian theory, nature itself should be changed. This is in fact what their conjecture amounted to: that there must be one other planet beyond Uranus, which was the cause of the perturbations; and they were able to predict where that planet should be found, which was duly discovered in 1846 and named Neptune.[27]

The third falsification of Newton's theory was the most serious one. In accordance with Newton's theory, planetary orbits are ellipses of which the Sun occupies one of the two foci. The *perihelium* of a planet is the point of the orbit where the planet is nearest the Sun, and in accordance to the theory, this point should be fixed. Le Verrier discovered in 1843 that, whereas this is true for all the other planets, the perihelium of Mercury moves (this is called the *precession* of the perihelium). Le Verrier was so worried about this result that he did not publish it until 1859, well after the discovery of Neptune, when his own scientific status was unassailable. This movement of the perihelium is called a *precession* because it entails a certain rotation, which is in fact so minute that it would take thousands of centuries for the point to return to the same place. The modern value for it, slightly different from Le Verrier's, is forty-three seconds of arc per year. This may seem to you small, but for astronomers, who know their precisions well, it is a clear falsification of Newton. No one, however, raised the slightest complaint that Newton's theory had been falsified and had thus to be abandoned. Scientists went on happily as before, keeping this conundrum in the back of their minds, hoping that in time a solution would turn up. This took until Einstein, and the year was then 1915. A theory that was totally lame by Popper's standards had happily survived unchallenged for fifty-six years! Of course, the reason why no one worried about the falsification was that, like Le Verrier himself, Newton's theory was too well *entrenched* to come under serious suspicion.[28]

Newton's theory was not an exception in being used while falsified. Even worse: some theories are born falsified! When Maxwell, for the first time, could calculate heat capacities from his kinetic theory, he reckoned that they depended on the kinetic energy (that is the energy due to translation) of the atoms in the gas, so that the faster they move the more energy they can hold. He also realized that in most gases, like oxygen, say, the smallest particles are *molecules* that contain *two* atoms. They can therefore have three motions: they can translate as a whole like a bullet, they can rotate like a dumbbell, and they can vibrate like a spring, with the two atoms periodically approaching and separating. The latter motions of rotation and vibration, just like the one of translation, are also able to accumulate energy and thus heat. On the basis of these ideas, Maxwell could calculate the heat capacity of the gas. His results were so much at variance with experiment that, at the 1860 Oxford meeting of the British Association for the Advancement of Science, he announced that the whole hypothesis of the kinetic theory had to be abandoned.[29]

This dismal conclusion scared people away from the theory, but, fortunately, A. Kundt and E. Warburg found the heat capacity of the mercury gas in 1875, and its value agreed with the one predicted for a monatomic gas, for which the translational kinetic energy is the only possible energy store. Not only could they conclude that mercury was a monatomic gas (which is quite correct), but also this result gave a fillip to kinetic theory. It was left to Boltzmann in 1877 to make the bold assumption that the vibrational motion of diatomic molecules requires so much more energy than the rotational one, that at normal temperatures it would not be excited (that is the "spring" would remain fixed at equilibrium), with which hypothesis he obtained a value in perfect agreement with experiment for the cases that had defeated Maxwell.

Purists, however, were not satisfied because at the time nothing was known about the energetic properties of the various molecular motions, and Boltzmann's hypothesis appeared to be entirely ad hoc. (It was only in the 1920s that quantum mechanics provided quantitative support for Boltzmann's assumption.) Thus, many scientists insisted that it could not be denied that, as Maxwell had asserted, the experimental results instantly falsified the kinetic theory, and so they rejected Boltzmann's attempts at repairing it, as if the poor man did not have bad enough luck with Mach breathing down his neck, still negating the very existence of the atom. The followers of Boltzmann's theory were nevertheless undeterred: rather than saying that the theory had been falsified, so sure were they about its *entrenchment*—because of the vast number of good and very diverse results obtained—that they called the contested experiments *the specific heat anomaly* hoping, like Le Verrier did before in his case, that this anomaly was one day going to be cleared up. Anyone who knows a little science must be familiar with at least half a dozen so-called anomalies that were left around for all the world to see, like Victorian young ladies sitting on the sides at ballrooms waiting to be picked up by eligible young men: not failures but hopefuls.[30]

Despite the enormous influence that Popper had on the philosophy of science, the tide soon turned, and people started reacting against his dogmatic attempts at finding a scientific method which, had it been paralleled in a theory of aesthetics, would have resulted in philosophers teaching painters how to paint by numbers. Paul Feyerabend (1924–1994), who had been born in Austria, like the master—with whom he eventually became associated—but who was mainly active first in the United Kingdom and then in the United States, rebelled not only against Popper, but even against the concept that there was such a thing as a philosophy of science that had any relevance for scientists. In this I believe that he went too far, but, quite rightly, he saw the enormous danger of rigid adherence to falsification, and exposed what he alleged to be distortions introduced by Popper in trying to present Einstein as a naive falsificationist.[31]

Imre Lakatos (1922–1974), originally Hungarian but mainly active in Cambridge and London, also castigated Popper for failing to make clear the distinction between falsification and rejection; and also for his excessive belief in *crucial experiments*. These are experiments alleged to have proven a theory or hypothesis as true or false at a stroke, whereas Lakatos points out quite correctly that crucial experiments are often recognized as such only long after the event, when a new theory turns up to supersede the old one.[32] (This was the case in the most significant so-called crucial experiment of the last hundred years, the Michelson-Morley experiment, of which more on p. 286.)

A very important influence against rigid attempts to define a scientific method was Thomas S. Kuhn (1922–1996), whose book *The Structure of Scientific Revolutions* was enormously influential. Kuhn was fundamentally a historian of science, and though I would be the first to argue for such a discipline, I should like to warn the reader against sweeping generalizations that result from comparing periods and cultures separated by millennia and vast distances. To draw conclusions from the history of the Trojan War to that of the French Revolution would be considered bizarre in history, but in science historiography vastly diverse epochs are sometimes covered under the same umbrella. (And remember that science evolution in the last two hundred years has been much faster than any other cultural change.) What is even more important is that the very word "science" has almost entirely changed its meaning since the Middle Ages, let alone since the Greeks, and to label with the same word the work of Aristotle and of Galileo already entails foregone conclusions, which makes it difficult to consider such studies as an objective historiographic pursuit; but more about this in chapter 20.

A novel concept that Kuhn advanced was that scientific theories, and even the way in which phenomena are described, depend on a "constellation of beliefs, values, techniques, and so on shared by members of a given community."[33] Kuhn called this constellation a *paradigm*, an expression that became far more popular than advisable because its precise meaning was not entirely clear even to Kuhn, who only defined it (as just quoted) in a postscript to the second edition of his book. What was very important, however, was that he put behind science (and almost everything else) a backcloth with respect to which discourse had to be gauged. Whereas Popper wanted to have mental callipers with which to determine the scientificity or otherwise of discourse, this program is largely dissolved by Kuhn because different cultures will have different paradigms and because direct comparison of discourse across different paradigms is, if not impossible, at least very dangerous, since even identical words are used with different connotations.[34] There is something to be said for this when thinking of Bosnia or Northern Ireland, but the great attraction of science as we *now* understand it, is its strong transcultural value. Of course, this melts down if we face an Aristotelian with a Newtonian or, without changing periods, a Creationist with a Darwinian, but our general experience is that science evolves toward a unified, transcultural, form of discourse. As against falsification, Kuhn gives greater importance to *paradigm rejection*, which results, in accordance to him, in *scientific revolutions*.

I believe that Kuhn was, first, too impressed by the Copernican revolution and, second, as an American, a little too sheltered from the realities of revolutionary life. For anyone with direct experience of revolutions, the idea that you necessarily go through a change of "paradigm" is unrealistic. This may have been the case in the French Revolution (although even then for a limited period), but what goes on in many revolutions is not obvious to the population at large because those who engage in them may have a vested interest in obscuring their motives. That this was so during the so-called Copernican revolution is quite possible because any people who had both heliocentric ideas and a sense of self-preservation would either have kept quiet, or discussed these theories with absolutely safe friends only, or would even have induced those less worldly wise to stick their necks out for them. And although Newton, as Leibniz feared, produced grounds for removing the need for religious belief, he himself was the last person who would have accepted that this was the case.

I shall argue in chapter 20, moreover, that even the Copernican revolution was more *theological* than scientific, and if any reader entertains the idea that Einstein's relativity theory is a clear case of a revolution, it must be remembered that Einstein himself denied this, as I shall show in chapter 11. There was a revolution, of course, but it was *philosophical*: armchair thinking à la Kant had to be abandoned forever. I feel, however, that some so-called revolutions are very much like golden ages: you only know later that they happened, when happy times are remembered in present misery.[35]

THE REAL THING: THE SCIENTIST'S AIMS

The reader must carefully avoid conflating science with technology, which is only too easy to do, since it is exceedingly difficult to separate cleanly one from the other: today's science is tomorrow's technology. To try to draw an approximate line, I shall consider science (by which I always mean in this book *pure science*) as concerned with *understanding* nature, and technology with *using* it. Further than this, however, the aims of science are not easy to define, and there is even a danger of prejudging the issue in doing so. It is therefore safer for this section to be devoted to the aims of scientists, which can to some extent be judged objectively.

There is a fair amount of anecdotal evidence about those aims; no discussion of the scientific method would be complete without some reference to them. I shall produce two sublists, the first one numbered from 1 to 6 and the second from *a* to *f*. The items of the first list are given in some, very rough, order of priority, whereas those of the second list are concerned with the individual dispositions of scientists, and vary in degree of importance from person to person. This second group of aims must be considered as superimposed on the first one, floating among the items of it in a fairly arbitrary order and with varying degrees of importance.[36]

(1) Save the phenomena.
(2) Discover new phenomena.
(3) Discover new, empirically adequate, laws.
(4) Understand nature.
(5) Provide explanations, models, and/or hypotheses (theories) that lead to finding either new phenomena or new laws that unify as large a field of data as possible.
(6) Provide an ontology leading to the identification of *objective facts* or of *elements of reality*. In general, the simpler this ontology is, the better it will be considered by the scientific community.
(*a*) Obtain peer recognition and/or acceptance.
(*b*) Satisfy intellectual curiosity.
(*c*) Obtain access to any or all of the following: money, fame, power, and sex.
(*d*) Achieve the state of euphoria ("high") often experienced when creating or discovering something. (This is called in labspeak "having fun.")
(*e*) Achieve the aesthetic satisfaction often associated with obtaining an elegant proof or expressing economical, striking, or surprising ideas.
(*f*) Improve the condition of humanity and/or the environment, or expect that such an improvement will be generated by the work in hand.

It will be noticed that the items (*a*) to (*f*) are a reasonable list of the drives that motivate most normal people in most normal activities. (Although it may have happened in the course of scientific pursuits, the desire to inflict pain, physical damage, and terror, and to achieve physical domination, is not one with which we need be involved in our discussion. In most cases, when such results have arisen from scientific activities, they were probably not the purpose of the work per se, but rather the outcome of the misguided following of some of the other aims listed. The perverted goals just mentioned, however, might have to be born in mind when studying the pathology of science, since the latter, like all human activities, must sometimes lead to aberration.) In the next section, and in the coda to this chapter, I shall briefly comment on the way that the scientists' aims interweave with the methods that they choose.

METHODS, MODELS, AND EXPLANATIONS

A table of, let us say, the position of the planets along the time might be useful, for instance in navigation, but it does not attempt to allow the user to understand at all what goes on behind the data provided. (An even better example of the total negation of understanding was provided by the old artillery tables.) The next stage would be some mathematical formula that could be used as a black box to generate such a table, and this is an example of the philosophical position which I have discussed under the name of *instrumentalism*. As I have already said, there are some tight situations in science, and this does not necessarily mean that one has to face the Inquisition, when instrumentalism is prudent, and attempts at explanation are best avoided. In any case, insofar as every scientist has as a fundamental aim that of saving the phenomena, every scientist is always, at least in part, an instrumentalist.

Let us consider the second aim, that of discovering new phenomena. In principle, this could be done by a method of induction, but, disregarding for the moment the extremely delicate problems entailed by this method, pure induction is somewhat rare because a scientist would have to be extremely lucky to fish something new out of the vast sea of data in front of him or her. It happens, though: as I have already mentioned, when Fleming discovered penicillin, he did not come to the laboratory with the conjecture that antibiotics might exist. He just ran into them by a sheer chance, which his experimental experience allowed him to exploit. The actual fact is that, in order to satisfy aims number 2 and 3, some use of the hypothetico-deductive method is common; and to do this some hypothesis or conjecture is useful, partly because it considerably focuses the range of observations to be made at this stage. Even a wrong theory is better than nothing in this respect. A very good example of this is the one already often used in this book about the interaction between an electric current and a magnetic needle. It never occurred to the great electrical experts of the early nineteenth century to look for that interaction because there was no reason whatever to suppose that magnetism and electricity were linked. The prize went to Ørsted, and the reason was that he, as a follower of *Naturphilosophie*, held a belief in the *unity of nature*, strong enough to sustain him through some eight years of dogged searching. I think that it was also most important that he was a Dane, and was therefore sheltered from the need to obtain peer recognition (aim *a*) which in France, for instance, would have meant that he would not have lasted even a few years in his pursuit without being discredited.

The fact remains that in the process of discovery, conjectures, hypotheses, models, and attempts at explanation play a fundamental role. But the role must not be mistaken for the essence. (Remember that the context of discovery can be fluid if not chaotic, as a vast difference with the context of justification.) All these features are props that at a moment's notice might have to be jettisoned, whereas the new facts obtained, if any, will remain for ever on the stage. So, these props have to be treated with caution, although many great philosophers of science substituted contempt for the latter.[37] Because I have lumped models with explanations, let me remind the reader that some quasi-ontological models are unavoidable in science, as I have already discussed in reference to laws. (See p. 201.) Forces have to be replaced by one-dimensional models, straight lines; likewise, particles often have to be treated as dimensionless mathematical points, bodies as homogeneous or perfectly rigid, gases as ideal (made up of dimensionless noninteracting particles, and so on). But there are other models that do not refer to any fundamental properties of nature, but merely to properties involved in the use of nature through technology and engineering. They thus involve *mechanisms* which are nothing else than mechanical (or electrical, hydrodynamical, etc.) explanations of actions and effects. I shall not be concerned here with that type of model at all, which has been extensively treated by Rom Harré and his collaborators.[38]

The main difficulty with models and explanations is that the scientist, when using them in tackling *new* concepts, is in the same position as the generals of the First World War who based their strategy on the weapons used in the previous one, thus causing the flower of European youth to be massacred by machine guns. It is extremely difficult to find examples of explanation that do not entail the use of concepts and ideas already available in previously known theories. Thus, the kinetic theory of gases was psychologically acceptable because it started from the idea that gases are made of atoms and molecules, which was already familiar to many people. And to further postulate that heat was merely the energy stored in the motion of these objects was also understandable, because the relation between motion and work, and between work and heat, was already known.

Because of this need for explanation to graft the new onto the old, explanations and models can get science stuck with "previous generation" concepts and thus stagnate progress, sometimes for long periods, as I shall illustrate by examples. The first one, of course, comes from the work of our old friend Ørsted and I shall briefly review the account I gave in chapter 8, pp. 176–81. He thought that if the electrical current along a wire would interact with the magnetic needle, it would do so by applying a force on the needle, and he thought of this force very much as that of the billiard cue acting on a ball. He thus expected the wire to be like the billiard cue, so that, in order to "hit" the needle, it had to be placed perpendicularly to its length. (If it were parallel, all it could do was to try to move the needle support which, the latter being rigid, would resist.) He was entirely right in thinking this way because until then all forces in nature acted longitudinally. As a consequence of this model, he repeated the same experiment unsuccessfully for eight years (remember that he had to make his own batteries and that he had to change many times the operating conditions). Finally, almost certainly by accident, he saw the needle moving when both needle and current were *parallel*, which was as much of a shock for him as if you hit a billiard ball without spin and yet you see it moving *perpendicularly* to the cue. Ørsted had not only discovered the interaction between magnetic needle and current, which he had been looking for for so long, but also an entirely new type of force, which is called a *tensorial force*, that acts at an angle (in this case 90°) to its line of application.

The second example is the Bohr theory of the atom, which was almost immediately adopted because it "explained" atomic spectra. At the time of Bohr it was already generally accepted that the spectral lines of atoms originated from the motion of the electrons in it. The problem was that in such motions, as imagined until then, the energy of the electrons would have to vary continuously, whereas in order to explain the spectral lines, well separated in energy by *discrete* (that is not continuous) intervals, the movement of the electrons themselves would have to be in some way discrete. In order to achieve this, Bohr imagined the atom like a solar system, with the nucleus at its center and the electrons moving around in fixed and well-separated orbits, like the planets, each orbit corresponding to a fixed energy. This model was immediately "understandable" because of its similarity with planetary mechanics, but it turned out to be hopelessly wrong. Yet, it paved the way for my third example.[39]

In trying to obtain a better theory of atomic structure, Schrödinger realized that he had to get a *wave equation*, the reason being roughly as follows. It was clear from atomic spectra that atoms were able to emit waves of very precise frequency (this is the theory behind atomic clocks, since precise frequencies indicate periodic repetitions of the same, very small, time interval). Periodic phenomena were well known in Schrödinger's time; and it was also well known that the observed frequencies could be calculated from some mathematical equations, called in each case the *wave equation* for the corresponding phenomenon. Bohr had tried to obtain periodicity by getting the electrons to move on closed orbits, which would give equal intervals of time for each closed circuit. Schrödinger knew that this did not work properly, so he made what must have looked for him an enormous intellectual leap. He assumed that the electrons were not just fixed particles, but rather that their charge was thinly spread over the whole atom, like a cloud, with different densities at different positions. There were perfectly sound (old) methods to get the wave equation corresponding to such a variable charge density, and in this way, Schrödinger discovered his famous equation. This equation could predict his posited charge density, which he identified with what is now called the *wave function* (the role of which is exactly like that of the state function discussed on pp. 141–42). Although the Schrödinger equation is right and immensely useful because it undoubtedly saves the phenomena, Schrödinger's interpretation of the wave function as the charge density of the electron was impossibly wrong, and it was only when people moved away from models and explanations (necessarily based on outmoded ideas) that things began to fall into place. At this very difficult stage, science had to fall back into an axiomatic-deductive method, by trying mathematical axioms, the consequences of which were then compared with experience.

I shall now consider two more examples concerning unsuccessful "previous generation" models that were later exploded away by Einstein. The first concerns a famous experiment by Michelson and Morley, which attempted to measure the motion of the earth with respect to the ether (roughly speaking, its motion with respect to absolute space; see p. 286) but, most surprisingly, failed to find any discernible velocity. In trying to "explain" the result, Lorentz and FitzGerald used a model based in the then universally accepted idea of absolute space. It was only when Einstein rejected this idea that correct new results were obtained. The second example concerns the photoelectric effect. It was known from just before the end of last century that when you shine light on a metal, the light can strip out some electrons from the metal. Naturally, the more energetic the beam of light, the greater was expected to be the kick it exerted on the dislodged electrons, and thus their velocity when emerging from the metal.

The wave theory of light, universally accepted until then, had very clear ideas as to what determined the energy of the light beam, and these ideas appeared to have nothing to do with the results observed. It was only when Einstein reinvented the concept that light is made up of particles (later named *photons*), that a very simple model of mechanical collisions between the photons and the electrons led to results which Einstein predicted and which were later confirmed experimentally. For this, he was awarded the Nobel Prize in physics for 1922, the proceeds of which, alas, he could not enjoy because they had already been mortgaged to his first wife three years before as part of their divorce settlement.[40]

Just in case: all the examples I have discussed here correspond to critical points in the development of modern science, and there is a temptation to treat them as some sort of Copernican revolution à la Kuhn. Indeed, in each case, one could describe the situation as a change of "paradigm." I prefer nevertheless to regard the whole process of scientific change on *evolutionary* rather than *revolutionary* lines. As you remember, although mutations occur randomly, their results are not random, *because they are always corrections or changes on preexisting situations.* What Ørsted discovered was still a force, albeit *tensorial.* What Bohr discovered were energy states, albeit discrete rather than continuous, and so on. What is happening is that the continuously evolving scientific mesh is mended here or there, or some prominent parts of it are demoted to more restricted uses, and other low-grade parts are promoted. If one looks at science *as a whole,* however, the change in the mesh before and after these apparent upheavals is more like some form of intense darning than a ferocious tearing off of the whole structure. Even now, Newton happily cohabits with Einstein, and the old Bohr theory of the atom with Schrödinger's. Of course, Kuhn had a wider point of view than the one I am propounding here; if you look at the Copernican revolution in the mild scientific way that I suggest, the upheaval was not so extraordinary: you talked of the movement of planets before, and you still talk about their movement now, although the "details" have changed. The reason why many of my readers will raise their eyebrows at my use of the word "detail" in this dramatic context is that the drama was theological more than scientific. As science and theology learn to cohabit, such upheavals are most unlikely to be repeated in the same way.

CODA

A variety of methods—inductive, hypothetico-deductive, instrumentalist, falsificationist, and others—have historically been proposed as *the* method by which science must work. I have argued that scientists are opportunists who have certain aims which they wish to fulfil, and that in doing so they will use whatever method is better adapted to the circumstances of the time. I have tried to show that, in fact, different situations require different approaches, and that the scientific method, such as it is, should not be held separate from the *scientific activity*. The reader may feel that because of this the context of discovery in science is extremely fuzzy, and that it must allow for activities which are not to be regarded as scientific. This is a point of some importance to which I shall come back in a moment, after I review the major features of my proposals.

To start with, I have argued that the contexts of discovery and justification are not as separate in science as it was once believed: scientists must be constantly aware that they must justify themselves (if not their discoveries) before their peers. But, more importantly, I have taken the view that what scientists are doing, when obtaining new results or proposing new laws and theories, is expanding, mending, or reinforcing the scientific mesh; and that the validity of any fact, result, model, theory, or law can never be asserted in isolation but rather in relation to the scientific mesh. By definition of the latter, all its elements are *entrenched* to a greater or lesser extent; and any candidate to be accepted within the mesh must pass the test that it agrees with as many elements of the mesh as possible (the wider they go over the mesh and the more disparate those elements are, the more important this candidate is) and that it does not disagree with a better-entrenched concept.[41]

This concept of entrenchment within the scientific mesh has the important consequence that the vexing problems involved in induction are, if not resolved, at least alleviated. No amount of repetition of a given experiment can rationally give total confidence in its result, but if the latter is woven into a mesh of other facts, so that some theoretical construction can be inserted within the scientific mesh, then scientists tend to rely more confidently on the new set of facts and assumptions. We saw in chapter 8, p. 179, that it was precisely this that gave Ørsted the necessary confidence to publish his discovery. It does not matter that the interpretation he had in mind was not entirely right. Having woven his bit of the mesh, others soon came to mend and to expand it.

The next important idea that I am proposing is that the mesh is, like a biological ecosystem, always in a state of evolution: all elements of the mesh are in some form of competition, and there is a process of survival of the fittest, which is no other than the result of entrenchment.[42] At any one moment there are prospective elements of the mesh in various low-level states of entrenchment, and a new theory or a new phenomenon may either reinforce their entrenchment or entirely destroy it. As a topical example: there is at present a very important theory proposed by Peter Higgs, from Edinburgh, which, if correct, would explain the origin of the mass of elementary particles. It is a consequence of this theory that a new particle should exist, which is called the *Higgs boson*, the search of which is currently conducted at great expense because, in accordance to the theory, it should only be observable at very high energies. Should the Higgs boson be actually found, the entrenchment of the theory, at present in a sort of limbo, would shoot up strongly.[43]

We can now go back to the question whether my proposals entirely contradict Popper's puritanical desire to be able to rule out whole branches of human activity as pseudoscientific. Take this simple case: should we say that classical Chinese acupuncture is pseudoscience? Some of the hypotheses used may of course not be acceptable within the modern scientific mesh, but there are two ways in which the question could be answered, at least tentatively. One would be by means of proper statistical tests of the type used in evidence-based medicine. There appears to be already a limit body of results in this direction. These might qualify the theory for some weak form of entrenchment, but there is another possibility, namely to find whether the method agrees with some concepts already well entrenched. Some experiments have shown that excitation of the so-called acupuncture points are accompanied by the production of endorphins (natural analgesics) in the brain. Because this is now a well-known branch of physiology and a well-entrenched one, the results might be significant in entrenching acupuncture, although for the time being an element of doubt remains, since endorphins can be secreted in the brain as a result of a wide variety of stimuli. In any case, if you take an evolutionary view of the scientific mesh, there is less urgency for devising tests to eliminate false theories automatically: natural selection should weed them out ultimately. On the other hand, premature attempts to extirpate apparently nonorthodox ideas could retard scientific progress.

Whether you like them or not, models and explanations are part of the scientific method, certainly (but not always) in the context of discovery. Because science must always strive toward empirical adequacy ("saving the phenomena"), and because one knows that one's models must be suspect as necessarily entailing previous generation's knowledge, models and explanations have to be used with a healthy degree of skepticism, remembering that rigid adherence to them can often hamper progress. But they can be valuable in situations where knowledge is in a less fluid stage, like in technology.

We have now finished the first part of this book, concerned with the way in which philosophy interacts with science in order to produce some sort of a picture of what we mean by nature. The whole work is not yet finished because we must understand better what we mean by an *objec-*

tive fact or an *element of reality*: in other words, we have to understand a little how one copes with *ontology* in physics, and this can best be done by understanding some of the ideas of relativity theory and quantum mechanics. Relativity will be the subject of the next chapter, after which we shall have to return to the study of time in order to understand how certain phenomena, such as the explosion of a balloon, cannot be reversed. Two chapters, on probability and mathematics respectively, will still be necessary before we can embark on the microscopic world of quantum mechanics.

NOTES

1. See Hughes (1980), p. 392.
2. Ostension, however, is not exempt from problems, as Wittgenstein noticed. See especially Wittgenstein (1967b), 30.
3. Sir Karl Popper (1902–1994), one of the most influential philosophers of science of our time, confessed candidly that when he started his philosophical work his motivation was "to distinguish between science and pseudo-science," the latter already known to him (presumably owing to his exceptional acumen) as psychoanalysis, Marxist historiography, etc. (Popper 1969, pp. 33–34), which shows that if there might be any doubt that the Trustees of the Tate Gallery might not know how to choose works of art, they certainly knew how to select their floor-sweepers. (Just in case: the reader must not surmise that I am defending psychoanalysis or Marxism, which is as far from my intention as humanly possible. What I am defending is their right to be seriously considered, and not to be ruled out of hand by a so-called philosophy that stems from prejudice.)
4. Readers may find a helpful but simple introduction to the history of the subject in Losee (1993). Brown (1979) provides also an elementary discussion of the major problems involved in the study of the scientific method. Klee (1997) is a more advanced introduction with an emphasis on the biological sciences.
5. This view of science as the continuation of common sense was in fact held by Quine (1951), p. 42.
6. von Neumann (1932a).
7. Whewell's two major books, Whewell (1837) and (1858–1860), were among the most influential of the century on this subject.
8. Braithwaite (1953), p. 9.
9. Herschel (1830).
10. The reader may find an account of caloric theory in Hesse (1961), p. 185.
11. Dalton's book, *New System of Chemical Philosophy*, published in 1808, was one of the landmarks of nineteenth-century science. For early work on the kinetic theory of heat, in particular that of Sadi Carnot, see Cercignani (1998), p. 61.
12. Mach opposed the *realism* of Boltzmann (that is his introduction of particles that were assumed to *exist* although unobservable) in a pamphlet published in 1872 entitled *History and Root of the Principle of Conservation of Energy*. Boltzmann's kinetic theory of heat will be discussed in greater detail in chapter 12.
13. The debate had nevertheless been extremely bitter and at times not entirely fair. Thus the British Association for the Advancement of Science announced in 1891 that it had now "been proved beyond doubt" that the equipartition theorem, one of Boltzmann's major results, was wrong, a statement which had to be withdrawn in 1894 when a mistake in their

arguments was discovered. (See Nyhof 1988, p. 100.) Although it was generally accepted until recently that the circumstances of Boltzmann's suicide were not clear, the account that I have given comes from Coveney et al. (1991), p. 21, who support it with a reference to a contemporary Austrian newspaper. (See also Cercignani 1998, p. 36.) More details about Boltzmann will be found in chapter 12.

14. This was Einstein, who wrote a very warm obituary of Mach but eventually rejected his epistemological position. See Schilpp (1959), vol. 1, p. 21.

15. An excellent discussion of this work can be found in Pais (1982), pp. 93–104.

16. Pais (1982), p. 103. A very good account of the opposing philosophical positions behind the debate is given in Nyhof (1988).

17. See a more detailed account in Altmann (1992), p. 34.

18. See Seeger (1966), p. 29.

19. Truly enough, Giordano Bruno (see Singer 1950) was a propagandist for Copernicus, but this hardly needed be taken into account in his trial, since his heretical views on theology alone would have caused him to be burned, even by the protestants: he held that God is unknowable and rejected the accepted views on the Trinity. This is a fact conveniently ignored by Feyerabend who claimed that, after all, Galileo had not been badly treated by the Inquisition, in comparison with Giordano Bruno. (See Feyerabend 1993, p. 127.)

The Jesuit Cardinal Roberto Bellarmino (1542–1621) was born in Montepulciano in Tuscany, but he is almost universally called *Bellarmine* by science historians probably following Pierre Duhem, who Frenchified his name. Pius XI made him a blessed (beatus), then a saint, and then, in 1934, a Doctor of the Universal Church (on par with Thomas Aquinas!) all of which was not sufficient to ensure that he be given his correct name. I shall discuss further in chapter 20 the important ideas of Giordano Bruno and Bellarmino in relation to science and dogma.

20. Bellarmino had, of course, his predecessors. The famous theory of the motion of the planets, proposed in the second century B.C.E. by Claudius Ptolemy, was a very hand-to-mouth affair, each planet having to be treated separately in an ad hoc fashion; and in the *Almagest* he hinted that his method was nothing more than a purely mathematical device. (As often happens to successful scientists, he later claimed, however, physical reality for the motions of the planets that he had postulated.) Nearer Galileo's time, Andreas Osiander, in an unsigned preface to the posthumous work of Copernicus, *On the Revolution of Celestial Orbs*, which he published in 1543, disclaimed any physical significance for the planetary orbits posited by Copernicus.

21. Feyerabend (1993), p. 261.

22. See Crombie (1953), pp. 84 and 44. Popper does not appear to refer to the strong medieval tradition on falsification through the form of syllogism called *modus tollens*, of which the following is an example: If John is French then John is European; John is not European; then John is not French. In more general terms: given the implication $p \Rightarrow q$, and q false, then p is false. (The reader can verify this from table 2, p. 52. For q false, the only row for which $p \Rightarrow q$ is true is the last one, in which p is false.) An even more general formula for *modus tollens* can be written at once from the statement just before the parenthesis: $\{(p \Rightarrow q). \sim q\} \Rightarrow \sim p$. The more enterprising reader can easily verify by the procedure used in table 6, p. 66 that this proposition is a tautology, and thus true for all values of p and q.

23. Popper (1972), p. 259. See also Popper (1969).

24. Popperians could of course adduce that the observation of the famous petri dish refuted the theory that bacterial cultures live. Whereas this might be true, it is the type of truth that would bring a policeman in disrepute who, when finding the corpse of Mr. John Smith, were to report that his belief that Mr. Smith was alive had been disproved. Finding a new fact can always be construed as the negation of infinite alternatives: the discovery of Tutankhamen's tomb in Egypt refutes the hypothesis that his remains were buried in Westminster Abbey. Thus, falsification is not at all the specific that Popper took it to be.

25. See Will (1995), pp. 182–83.
26. I have already referred to this in chapter 3, p. 66. The reference in question is Popper (1969), p. 53.
27. See Will (1995), p. 90 for the history of the discovery. It is interesting to see what Popper has to say in this context, which flies in the face of the evidence. (He discusses a similar situation which he himself compares with Le Verrier's work.) "What compels the theorist to search for a better theory, in these cases, is almost always the experimental *falsification* of a theory" (Popper 1959, p. 108). Le Verrier had used Newton's theory in conjunction with a sophisticated mathematical method to account for the perturbations created by other planets on the orbit of a given one: the idea that he wanted to prove *himself* false is preposterous. Surely what he wanted was to show that his method worked in the hardest possible scientific circumstances, that is, when causes are unknown and have to be surmised from their possible effects. What Le Verrier did when he successfully posited the existence of a new planet, was to *entrench* in the best possible way Newton's theory of planetary orbits.
28. For the concept of entrenchment see chapters 6 and 8.
29. Brush (1976), vol. 1, p. 194.
30. A good discussion of the heat-capacities problem and its philosophical implications may be found in Nyhof (1988) and de Regt (1996).
31. Feyerabend (1993), p. 42. His denial of a philosophy of science appears in Feyerabend (1970).
32. Lakatos (1978), p. 201 and chap. 10.
33. Kuhn (1962), p. 175.
34. It is along this line of thought that Kuhn, for instance, rejects the idea that Newtonian mechanics can be derived from the relativistic one as a special case for low velocities. (Kuhn 1962, p. 101.) His view is that, whatever the quantitative aspects of the derivation, the word "mass" would be transferred from relativistic to Newtonian mechanics with a different meaning, since in the former theory mass is not conserved whereas in the latter it is invariant. This point of view, of course, agrees with Kuhn's revolutionological conception of scientific change but, as the reader will know by now, I prefer to take a more evolutionary view of science (see chapter 6). It must be accepted that in any rational argument, the meaning of words is a function of their context and, moreover, I have claimed that one of the tasks of science is that of setting up a language with which to discourse about nature. Thus, it is in the very nature of the scientific work that a scientist automatically adjusts both lexicon and semantics to the particular domain of application (and range of accuracy) implied in any given discourse. Kuhn's discovery of semantic shifts agrees as much with an evolutionary view of scientific language as with his own revolutionological interpretations. And if, as he claims, such semantic shifts affect scientific discourse, he should in the same manner realize that they seriously undermine his *own* scientific historiography: what reason do we have to believe that Aristotelian "science" and Newtonian "science," for instance refer to the same activity? (More about this in chapter 20.)
35. I am not alone, of course, in my moderate evolutionary views. Abraham Pais, one of the foremost historians of modern science, offers a very qualified view about the meaning of the so-called scientific revolutions, claiming that science does not progress by just sweeping away with the past, as well as criticizing sensationalist approaches to the subject. (Pais 1986, p. 250.)
36. Hardy (1992), p. 79, discusses these aims with much objectivity.
37. Pierre Duhem (1861–1916), professor of physics at Bordeaux, who made very important contributions to the philosophy and history of science (he discovered—others might say that he invented—the Oxford school of physics that preceded Galileo in using the experimental method in dynamics) criticized Lord Kelvin for equating understanding with a visualization of an underlying mechanism. (See Duhem 1954, pp. 71–75.) Duhem

abhorred the excessive use of mechanical models to "explain" physical phenomena as a sort of *maladie anglaise*. Although he was right in this, he went too far: for the same reason he rejected kinetic theory as a parasite. (See Nyhof 1988.)

38. Harré, however, claims a much more important status for the construction of models in science, which he regards as its central activity. He proposed this change in approach, "relegating deductively organized structures of propositions to a heuristic role only," "theory construction" becoming "essentially the building up of ideas of hypothetical mechanisms" (Harré 1970, p. 116). Harré's position is anti-Humean, of course, his mechanisms "explaining" not just causal chains, as I have argued (p. 79), but rather the one-to-one action of cause on effect.

My own position owes much to the admirable book by Bas van Fraassen, which has been most influential in the period I am discussing: "My view is that physical theories do indeed describe much more than what is observable, but that what matters is empirical adequacy, and not the truth or falsity of how they go beyond the observable phenomena" (See van Fraassen 1980, p. 64).

39. As is always the case with a theory that saves the phenomena (but not, alas, all the phenomena), even the defective Bohr theory of the atom has a range of validity. When the electron orbits are very far away from the nuclei, for which purpose they must have high energies, then the electrons behave very much like the classical particles that Bohr had in mind (that is, like little planets), and the results of Bohr's theory are empirically adequate.

40. See Pais (1982), p. 300.

41. My concept of a mesh of scientific facts, theories, laws, and meta-physical principles appeared as a germ in Pierre Duhem's criticism of Newton's reliance on induction as, for instance, in measurement. Induction, he argued, is useless without interpretation, which "presupposes the use of a whole set of theories" (Duhem 1954, p. 204). Notice as well that the idea that confirmation as necessarily relative to the mesh, is analogous to Kuhn's view that it is only within a given "paradigm" that discourse is meaningful, although of course the mesh and the paradigm are interpreted in entirely different ways. The idea that propositions cannot be considered in isolation but rather in relation to other propositions was also defended by Lakatos (1978), p. 221, who acknowledged the influence in this respect of Popper (1959), sec. 20, p. 82. Popper, in fact, always discusses the falsifiability of propositions as part of a given theoretical system.

42. The proposal that evolution exists not only at a biological but also at a social and theoretical level is not new, of course, and has been entertained by a number of philosophers, in particular Toulmin (1961), p. 113. See also Lakatos (1978), p. 51.

43. The alert reader should spot that this is logically unsound, although just as good an application of the hypothetico-deductive method as they come. The reason for this is that, of course, $p \Rightarrow q$ and q true is logically compatible with p false. (See penultimate line of table 2, p. 52.) The existence of the mesh, however, brings back sense into this situation, because the Higgs theory, even at present, acquires a respectable state of entrenchment by agreeing with a large number of other elements of the mesh. The experimental finding of the Higgs boson would finally qualify the theory as empirically adequate, and thus strongly entrenched. A Popperian could of course argue that the present experiments are merely an attempt to falsify the theory, but what they could not explain is why people are prepared to spend vast amounts of money to do so: the fact is that you are working already on some reasonable level of entrenchment, and you know that the possible gain *if the theory is not falsified* is huge, since all of a sudden the scientific mesh would grow wider and more solid, which would not be the case if one were to chase the Bloggs meson instead. This is another example of Popperian falsification being only a small part of the story, and not a very significant one for all that. In other words, a crude falsification approach would not provide any guidance whatever why one should try to falsify one conjecture rather than another.

NATURE'S TOOL-KIT: SPACE, TIME, FIELDS

The only justification for our concepts and system of concepts is that they serve to represent the complex of our experiences; beyond this they have no legitimacy. I am convinced that the philosophers have had a harmful effect upon the progress of scientific thinking in removing certain fundamental concepts from the domain of empiricism, where they are under our control, to the intangible heights of the a priori. For even if it should appear that the universe of ideas cannot be deduced from experience by logical means, but is, in a sense, a creation of the human mind, without which no science is possible, nevertheless this universe of ideas is just as little independent of the nature of our experiences as clothes are of the form of the human body. This is particularly true of our concepts of time and space . . .
Albert Einstein (1950), p. 2.

One of the fundamental concepts in the development of science, I have claimed, is that of the entrenchment of the scientific mesh. I have alerted the reader, however, that we have so far discovered entrenchment only in relation to our interaction with the normal, macroscopic, human environment. Once we move away from the latter, that is, once we start exploring the microworld, or considering objects moving at very high velocities, say of tens of thousands of kilometers per second, then there is no reason why the principles that we so lovingly nurtured for thousands of years should still be valid. What we perceive and even the fundamental meta-physical principles that we use mirror, not the whole of nature, but only such part of it as is normally available to our senses. Thus, when we try to go beyond that cozy domain of the world environment, we must expect surprises. Almost every remaining chapter of this book, starting with this one, will produce examples of this situation.

I have so far eschewed ontology by taking a fairly naive attitude to what is "out there," but the time has come to talk of serious things. I shall therefore try to look in some detail at the tools that are used to describe nature, in the hope that the language that we thus deploy, if properly tested, might reveal such things as *elements of reality*, an expression which I shall later try to clothe with a precise meaning. In doing this work, I shall describe a few of the results and concepts of relativity theory, which will be a splendid test for some of the ideas that we have discussed about scientific method. I hope that, after readers peruse this chapter, they will appreciate the need to be prudent in accepting some of the claims made for conventionalism, for the significance of scientific revolutions, for the paramount importance of falsification, and for the role of crucial experiments as the spearhead of scientific progress.

Looking back from the comfort of our armchairs, it is only too easy to believe that Albert Einstein (1879–1955) was a revolutionist, if not a philosophical terrorist, that he blatantly engaged in the use of conventional definitions (thus with a very limited physical meaning), and that he fired his scientific rockets from Bern, sheltering in the safe enclave provided to him by what we now consider the archetypal crucial experiment of the century, the *Michelson–Morley result*. I shall discuss all these points in detail, but let me say straightaway that what Einstein did was, in fact, conservation work of the highest order: the apparently outrageous proposals that he made about space and time were calculated to keep intact and well as much as possible of the scientific edifice until then known. The man in the street, however, is told only of what is new and shocking, whereas the enormous amount of conservation work entailed by the new theory is often left undiscussed, as fit only for the specialists. Most people know that, in his mature years, Einstein appeared as a scientific conservative, attacking the ideas of the new quantum mechanics. What is not appreciated is that Einstein was *all his life* as much of a conservator as a good surgeon, who knows that in order to preserve the living tissues he has to excise, or at least isolate, the dead ones: he himself entirely rejected the view that relativity theory was revolutionary.[1] Einstein's genius, as we shall see, lay in his extraordinary ability to recognize what was really well entrenched in the scientific mesh as he found it, and, on building from this conception, to extend solid entrenchment into entirely new and, until his time, uncharted territory.

One important result of Einstein's work, many philosophers would argue, is that he falsified Newton's mechanics, but even this view is one that a scientist would not necessarily entertain: what he did was give precise conditions, until then missing, under which Newton's mechanics could accurately be used: thus it might be argued that rather than falsifying Newton's theory he complemented it.[2] The only branch of human knowledge that Einstein utterly falsified was a great deal of the armchair philosophy until his time revered as the apex of human intellectuality.

There is one other point about which Newton's and Einstein's theories are not as contradictory as one might think. Newton thought about time in two ways. One was the absolute time characterized by its flow "without relation to anything external."[3] This is time as an *absolute dynamical variable*. The other is what he called *duration*, measured by parameterizing motion, which is the time used throughout classical mechanics, the absolute time remaining behind the scenes as inaccessible as the *Ding an sich*. Admittedly, Einstein eliminated, as we shall see, the absolute time, but he instead reincorporated time as a dynamical variable with the same standing as the geometrical coordinates of a system.

All this will be treated in more detail in the rest of this chapter, but before I can safely discuss Einstein's ideas, I shall have to revise some of the conceptions that preceded him about space and about some of the objects, especially waves and fields, which are used to fill it up. I shall not attempt, however, an exhaustive treatment of these matters. As regards space, in particular, I shall briefly review only such concepts as were familiar to Einstein, mainly through Mach's work.

SPACE: NEWTON, LEIBNIZ, MACH

Leibniz believed that what characterizes space is that it entails *relations* between objects; thus, it is their distances rather than their absolute positions that are made significant by space. One of the good things about the time when Leibniz lived was that people had no inhibition in invoking God in scientific discourse: not that I advocate this, but such an attitude is much preferable to proposing muddled concepts that are meaningless unless a deity is posited—which is shunned, lest a supposedly scientific argument were revealed as a sham. Be it as it may, Leibniz argued that the creator would have no sufficient reason to place the universe here rather than, say, ten miles away, whence the absolute position of a point in space could not be relevant: only its relations to other points could be meaningful. As you can see, this is a pretty good argument, even if you were not prepared to accept its deistic slant, and as it turns out, Leibniz was right and Newton wrong; for Newton did believe in absolute space or, what was the same for him, in absolute rather than relative motion. If all that matters is the *relation* between two objects, as Leibniz held, it is the same to say that *A* moves relative to *B* as to say that *B* moves relative to *A*, and, likewise, it does not make sense to describe an object as absolutely stationary, that is stationary not with respect to some other object but with respect to absolute space.

As always, Newton grounded his opinions on experimental facts.

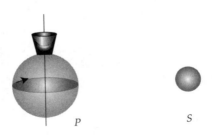

Fig. 1. *Newton's* bucket experiment

The better to understand what Newton meant, I will not, however, discuss his actual experiment, but an imagined one in an imaginary universe in which only two celestial bodies exist, a planet P and a distant star S, as illustrated in figure 1. I shall thus use again our old friend, the deutero-*Newton* introduced in chapter 6.

Suppose that it had been asserted through measurement that the planet rotates about its axis at the rate of sixty revolutions per minute, and that *Newton* had been challenged to demonstrate that this is the "true" rotation of P with respect to space and *not* the orbiting of S with the same angular velocity around the stationary P. In order to answer this challenge, *Newton* brings in a bucket full of water (grossly exaggerated in the figure) and places it on the planet's pole, as shown. The centrifugal force throws the water toward the sides of the bucket, but, being incompressible, all that the excess water can do is pile up on top of the existing water, whose surface thus becomes concave, as shown in the figure. *Newton* would say that this is incontrovertible proof that the bucket, and thus the planet, is rotating *absolutely* with respect to fixed space because the surface of the water would otherwise have been flat. This, however, cannot be the final word, since it is perfectly possible to conceive that P is stationary, and the orbiting of S around it causes the water in the bucket to be raised, very much like the Moon causes tides on Earth. It was thus that Mach demolished Newton's argument in favor of absolute motion; and Einstein was aware of Mach's views, which he admired. So, at the end of round one, absolute space is out. We must now have a look at the things that we put in such space as we are left with.[4]

WAVES VERSUS PARTICLES

Events or disturbances happen in space, and if they are such that along a straight line the disturbance is *periodic*, that is, if it repeats itself at regular intervals, then we say that we have a *wave*. The first point, of course, is:

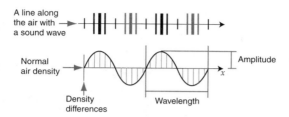

Fig. 2. A sound wave propagating in the direction x (A longitudinal wave)

At the top of the figure, the air densities are represented as follows. The short lines indicate normal density at the corresponding points. The thicker the black lines, the greater the excess density over the normal. The gray lines indicate densities below the normal, the thicker the line, the greater the departure from the normal. At the bottom of the figure, the density differences are plotted along the vertical axis, the normal air density corresponding to the horizontal line, and the densities above and below the normal being represented as positive and negative (above and below the horizontal line respectively).

disturbance of what? And this, as we shall soon see, is not a question with an easy answer. I shall first consider a very simple case: a *sound wave*, in which the disturbance is just the change in the density of the air along the direction of the wave, as I illustrate in figure 2.

In the lower part of the figure the wave is represented in the usual way, the maximum positive or negative value of the disturbance being called the *amplitude* of the wave, and the periodicity of the wave being given by its *wavelength*. It should be noted that figure 2 represents the perturbation at one instant of time and that this perturbation changes along the time, moving forward with some velocity *v*.

Sound waves are very easy to understand, because the fact that they travel on air is undeniable: if a sound source is placed in a vacuum, no sound pervades. This type of wave is called a *longitudinal wave* because the perturbation is created on the same line along which the wave propagates. You can also have waves that look exactly like the picture at the bottom of figure 2, but in which the perturbation is *perpendicular* to the direction of propagation at each point of it: imagine a very viscous medium and a "generator," which is simply a bar with a rough surface that goes up and down, so that by friction the medium begins also to move up and down, whence this perturbation (which is the velocity of the particles of the medium) propagates perpendicularly to the direction of motion of the "generator." Such waves are called *transversal waves*, and an important example will come in a moment, but I also show in figure 3 a very simple and familiar case.

Newton thought that light was a phenomenon caused by particles

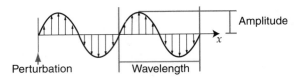

Fig. 3. A transversal wave propagating in the direction x

You can imagine this as a wave on the surface of a pond, in which the horizontal line corresponds to the level of the undisturbed water, and where the perturbation is the height of the water, positive up or negative down, with respect to that level. This figure is formally identical with the one at the bottom of fig. 2, but notice the following difference. In the present figure, the vertical lines leading to the wavy curve give not only the numerical value of the perturbation but also its direction, perpendicular to the x axis. Thus, those lines may be considered as depicting the perturbation itself, and this is why, as a difference with fig. 2, they are represented with vectors (arrows). In fig. 2, instead, the apparently identical lines give only the *magnitude* of the perturbation, this magnitude corresponding to, but not depicting, the perturbation at the pertinent point of the x axis.

that propagate along the light beam. This view was soon abandoned, largely because light shows the phenomenon of *interference,* which until the 1920s was thought to be exclusively characteristic of waves. (We shall criticize this view later, but for the time being, let us accept it.)

When light was recognized as a wave, it was also found that it had to be *transversal* rather than longitudinal, but I shall discuss later the nature of the perturbation involved. The phenomenon of interference is illustrated in figure 4 for two light waves that are out of step, one with respect to the other, by precisely half a wavelength. It can be seen from the picture that the total effect of the two waves shown cancels at the intersection of the planes shown. Imagine, on the other hand, a wave belonging to the same beam as the one on the left, and thus in step with it, which is translated parallel to itself until the whole left-hand plane reaches the first peak on the right-hand plane. That new wave, when it combines with the one on the right-hand side, would double the magnitude of the perturbation because the two interfering waves are in step at that point. When you consider the screen itself, it is not difficult to imagine that there will be vertical lines on it that correspond respectively to maxima and minima of the combined two beams. It is also fairly clear that the lines of this set (usually called bands, since they have some thickness) will be alternately bright and dark, corresponding to reinforcement and cancellation of the waves, respectively, and that they will be disposed periodically on the screen.

The resulting pattern is thus very distinctive and characteristic of the phenomenon of interference of waves.[5] So much was it believed that inter-

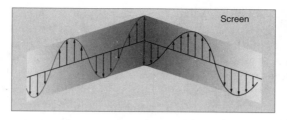

Fig. 4. Interference of waves

As in fig. 3, the curves shown here are such that their ordinates (vertical lines) represent the periodic perturbation itself. This is so because the waves are now transversal. The two waves lie on the respective planes shown in gray, and they propagate toward the screen in the back. When the waves hit the screen, the perturbation corresponding to the left-hand-side wave, represented with the vertical arrow going up, is canceled by the equal and opposite one from the other wave.

ference could only be produced by waves and never by particles that, when X rays were discovered and their nature was still somewhat doubtful, the result by Max Laue in 1912 that they could experience interference was taken to be incontrovertible evidence that they were waves rather than particles.[6]

Despite this, it was early recognized that the distinction between waves and particles could not be quite as clear-cut as it appeared to be. If two waves of the same wavelength are totally in step and propagate along the same line, perturbations add up all through, as discussed in figure 4, and produce a wave which, it is fairly easy to imagine, will have the same wavelength but double the amplitude. The situation changes nevertheless when one adds up a great many waves with *very slightly* different wavelengths, and it can be proved (but I shall not try) that there is indeed a reinforcement in the region of the original wavelength,[7] but because the waves are not precisely in step for some wavelength that may be very little different from the original one, the sum of all the perturbations practically cancels out and becomes negligible. One is then left with a sort of blob, which is a very much distorted wave, large in some very small region of space, but which effectively cancels outside it. This blob, which is called a *wave packet*, will superficially look like a particle, in the sense that it is largely localized in a certain region of space. Thus, the possibility cannot be discarded that what at first appears to be a particle is in fact a very tight wave packet, a question which will be important when we deal with quantum mechanics.

Corpuscular radiation differs from waves in an important way. If I have a loudspeaker and I move it forward, the sound generated will still travel with the same speed: if we look at the top of figure 2, it is clear that for the speed to increase, the whole cylinder of air in front of the loudspeaker would have to acquire the velocity of the latter. (Wind, in fact, adds its own velocity to that of sound, whereas what we are assuming here is that the generator moves *without creating any wind*.) The same is the case for transversal waves. Thus, for all waves, their velocity is independent from that of the source. For corpuscular radiation, instead, that is, for the model of light favored by Newton, the velocity of the source *adds* to the velocity of the radiation: if you fire a machine gun that is moving forward rapidly, its velocity is added to that of the bullets, a feature which is appropriately called the *ballistic effect*. This distinction between corpuscular and wave radiation is one that we shall have to take into account later.

One final point about the propagation of light. Even in Newton's own lifetime it was realized that light does not propagate instantaneously, an illusion which is easy to entertain on doing experiments on Earth, since the velocity of light is huge. The Dane Olaf Roemer was the first to realize from astronomical observations in 1675 that the velocity of light is finite, and he worked out that it took some eleven minutes to reach Earth from the Sun (the correct time is in fact a little over eight minutes). In 1849 Armand-Hippolite-Louis Fizeau actually determined the velocity of light to be 315000 km/s. The modern value is nearer 300000 km/s, but Fizeau's inaccurate result is historically important, so that you should bear it in mind.

Well before Fizeau's time the nature of light had begun to be much better understood through the work of Augustin Fresnel (1788–1827), a French civil engineer, and Thomas Young (1773–1829) of Somerset, a medic by training. In Newton's own time those who supported, against him, the wave theory of light thought of it as a longitudinal wave, very much like sound is, an idea that Newton utterly rejected. Although his corpuscular theory of light had to be abandoned when the phenomena of interference were studied (the prevailing doctrine being that such phenomena could only occur with waves), his view that light could not be a longitudinal wave was confirmed by Young, who proposed in 1817 that it was a transversal wave,[8] a view that was quickly adopted because it explained a number of phenomena associated with light propagation.

And we now have to answer the thorny question: propagation in what medium? Because it was clear that, as a difference with sound, light could traverse any amount of void space (as between the Sun and Earth), it was thought, no physical perturbation could propagate in just vacuum. At this stage, scientists engaged in imaginative ontology: they postulated that space was pervaded by an undetectable medium, which was called *ether*, along which light propagated, and it was this medium that could sustain at each point the perturbation (as yet unknown) that produced the light wave. This medium had to be subtle, so that no one could observe it, but, horror of horrors, it turned out after Young that this subtle medium had to be nevertheless highly viscous, since it was impossible to understand how transversal waves could otherwise be sustained. (Remember the generator moving up and down in a viscous medium, which I described before.) This is an example that premature attempts at *understanding* lead to nonsense, and that you do better in such cases by sticking to *prudent ignorance*.

Once the ether was invented,[9] it was realized that it could behave in two different ways. When a body moves, it could drag with it the ether, very much as a moving train carries with it the air inside and around it. This model, however, had to be abandoned early because dragging the ether would have entailed dragging the light rays propagating through it, and there were certain astronomical observations that did not agree with the concept of telescopes dragging in their motion (that of Earth) light rays from stars. It was thus finally accepted that ether pervaded the whole of space (considered at that time absolute), that as the latter it was in absolute repose, and that it was utterly undisturbed by the motion of any objects going through it.

As I have said, the velocity of light c (understood *with respect to the ether* or, what is the same, absolute space) is independent of the motion of the source in the wave picture. If we have, say, a mirror moving away from the source at a velocity v, in the same direction as the light signal, the effective velocity of the light *with respect to this mirror* would in this classical picture be equal to $c - v$, the minus sign reflecting the fact that the mirror is moving *away* from the light. We shall soon see that all this theory of the ether is rubbish, but wait while we look at the parallel story of *fields*. Even then, although the invention of the ether appears to be an ontological travesty, we shall see that a certain amount of ontological fantasy is not necessarily frowned upon, even in the most modern and sober physical theory.

FIELDS

Besides sound and light, other perturbations were known to propagate through space. Gravity was one, and it created serious intellectual problems because it appeared to entail instantaneous action at a distance. Remember that when we discussed causality we required that distant objects be related by a *contiguous* causal chain, whereas there is no hope of constructing such a one to bridge the distance between Earth and the Moon. Moreover, consider an object on Earth, say a car, at a distance r from a given rock on the Moon, so that the gravitational force between these two objects is proportional to the inverse square of that distance. If we now move the car ever so little, its distance to the rock will change instantaneously, of course, and therefore the gravitational force between the car and the Moonrock would also change instantaneously, although the intervening distance is huge. This is most disturbing because we argued that it is the existence of well-established contiguous causal chains that makes the difference between science and voodoo. An answer will come, but we shall have to wait on the borderline of sanity while we consider other actions at a distance, and how the corresponding problems entailed were tackled.

The most important examples were the forces of repulsion or attraction between equal or opposite electrical charges, respectively. Although the forces entailed could be measured and a law was easily obtained to determine them, these electrical forces still acted at a distance, very much like gravitation does in our example above. When in 1820 Ørsted (see p. 179) found a similar force between a magnet and an electrical current carried by a conductor, he expected the current to be surrounded by what he called the "electrical conflict," a region of space at every point of which there was some perturbation generated by the current, this perturbation thus acting *contiguously* on the magnet. A step forward in order to begin moving away from the awkward need of postulating actions at a distance was taken by Michael Faraday, who improved very much on the idea of the "electrical conflict." His method was essentially *phenomenological*, postulating a new concept, but without trying to go further with it than fitting the facts. Faraday's idea was the concept of a *field*, for which I shall give a later and simpler definition: at every point outside an electric charge a force is posited to exist, called the *electrical field*, the intensity of which is so calculated that, if you place a charge at a certain point of space, the field at that point precisely reproduces the experimental value of the force acting on that charge. The advantage of this construction is that now the force acting on any given body is contiguous with it, so that no action at a distance is involved, since the electrical field covers the whole of space. Also, this concept of a field can be applied to all the awkward cases of gravitational, electric, and magnetic forces. There are problems, though, with this idea of a field of forces: first, the field forces must hang around on nothing, so to speak, although the ether supporters could "understand" them as ether perturbations. Second, if you move ever so little an electric charge, say, the electric field changes *instantaneously* all through space, which is not an easily acceptable possibility.

The language of fields, nevertheless, has come to stay, and, like all languages, it has been honed along the years so as to save the facts and get rid of all the awkward conceptual problems discussed. Whether this means that the *fields* described by this language are *real*, or whether the field language happened to be easier to use and to adapt to the facts (that is, to *entrench*) than other possible languages, I shall leave it to the reader to decide. The question might be purely rhetorical, scientific ontology perhaps being nothing else than the ontology compatible with the language in use as currently successful. Those who are ontologically hungry, however, might reckon that, after all, entrenchment of an ever-wider scientific mesh increases confidence in its corresponding ontology.[10] (This is despite the fact that, in the present case, the fields acquire a life of their own, "existing" at every point of space without the need of any substratum, as the ether was wont to provide.) How this wider entrenchment came into being at the close of the nineteenth century is largely the result of the work of one man, Maxwell.

MAXWELL AND MAXWELL'S EQUATIONS

If one had to select the names of the four or five men or women from Britain who contributed most to nineteenth-century science, James Clerk Maxwell (1831–1879) would certainly be on that list. He was educated first at Edinburgh and then at Trinity College, Cambridge, where he started to develop a theory of electromagnetic fields already at the age of twenty-four, when he was elected a fellow. Electromagnetic fields are the fields created by electrical charges or currents or magnets, and the problem that Maxwell wanted to solve was, basically, as follows. A given medium may have some known charge and current densities, which may change with the time, and one would like to be able to calculate the values of the electric and magnetic fields at every point of space and at every required time. The final answer to this problem did not come to Maxwell until he moved to a chair at King's College, London, from where his major results were communicated in 1861 and 1862. It is unnecessary for us to get into the details of Maxwell's equations, but a very brief account of what they entail will show how important they are.

The first piece of news is that Maxwell's work shows that electromagnetic perturbations (that is, changes in the electromagnetic fields) do not propagate instantaneously through space, but rather that they do this with a fixed velocity c, which is a *constant* that appears in Maxwell's equations. What this means is that, if we modify the value of an electric charge at a given point, then the value of the electric field at a distant point in space does not change instantaneously, but only after a definite time, determined by c and the distance from the source to the field point. This is very important, indeed, because it removes the awkward difficulty that we discussed in relation to the gravitational field, where propagation of a perturbation appeared to be instantaneous and, also, because it makes the fields vastly more physical than the ad hoc array of forces that we previously defined. But a great surprise was to come: very soon after Maxwell published his first papers, in 1856, Weber and Kohlrausch measured c and found a value of 310000 km/s, almost identical with the value of the velocity of light measured by Fizeau at 315000 km/s. Nobody could accept this as a coincidence, and it was concluded that the electromagnetic perturbation traveled through the field at the velocity of light. (The reason why these two results for the velocities were both wrong, and wrong in the same manner, was understood later as due to a different definition of the velocity of the wave used by all three authors, as compared with the now current definition, which gives c equal to about 300000 km/s.) As the ether theory prevailed at the time, it was accepted that c was the velocity of light with respect to the ether or, what was the same, with respect to absolute space.

An even greater surprise was in store because it was discovered by Maxwell soon after that his equations admitted of a solution which was clearly a *wave* propagating through the field: that is, it was possible to find specific perturbations traveling through the field that were periodic and which, like all such perturbations, traveled at the speed of light. The corresponding wave solution was very curious: at each point of space, the perturbation consists of two field vectors (or forces), one electric and the other magnetic, both at right angles to each other and to the direction of propagation of the wave. The wave was therefore transversal, as Young (p. 260) had expected the light wave to be, but a little more complex than originally thought, since the disturbance propagated by the wave entailed two perpendicular vectors rather than only one. It was entirely natural for Maxwell to identify this *electromagnetic* wave with light, since it propagates with the velocity of the latter, and this was the first time that a satisfactory theory of the light wave emerged. The crowning triumph of Maxwellian theory came in 1888 when, by designing a wave generator leading to the correct wavelength, Heinrich Hertz produced for the first time radio waves, which are nothing else than light waves (that is, electromagnetic waves) with a wavelength in a specific range.

So, by the turn of the century, Maxwell's theory was just as well entrenched as any theory could be, since not only it agreed with all the known facts, but it had also produced new ones of extraordinary importance. Yet, the theory entailed what appeared to be a very serious contradiction: all light sources must generate signals with velocity c with respect to the ether, and this velocity must *be the same in all directions*, since the theory contains *one and only one constant* that may be identified with the velocity of light. So far so good. Suppose now that we view the signals from a rocket, which moves at a fairly large velocity v in the same direction as the signals. From what we have said before about a mirror moving away from a light source (p. 261), we would expect the signals now to move with velocity $c - v$, with respect to the rocket, but this is not possible, since, in accordance with Maxwell's equations, the wave solution can have one and only one velocity, c. (The minus sign in the above example would become a plus if the rocket moved in the opposite direction to the light signal.)

This is the moment when Albert Einstein comes to rescue Maxwell's theory from this contradiction. But in order to understand what he had in mind (and this is most important in order to comprehend how scientific method really works), we must briefly go back to Newton's equations.

NEWTON'S EQUATIONS AND
GALILEO'S RELATIVITY PRINCIPLE

When discussing the question of *objectivity* in chapter 8, we remarked that certain laws might change when the reference frame is changed,[11] but if the law keeps the *same form* in different reference frames, it is often said that the law in question is *covariant* with respect to those frame transformations. Galileo had studied frames of references in uniform motion one with respect to the other, frames that are called *inertial frames* because the law of inertia holds good (that is, it is covariant), in all those frames. (See p. 194.) But it is not only the law of inertia that is covariant for all inertial frames: all the laws of classical mechanics, that is, all three of Newton's laws, are covariant for all inertial frames, and this is called *Galileo's relativity principle*.

Because this is important, I shall explicitly prove that this statement holds true for Newtons's second law. We know (see p. 139) that if u is the velocity of a body, variable in time, and m its mass, then at a given time the force f acting on the body is given by Newton's second law as:

$f = m$ times the rate of change of u. (Train frame.)

Suppose now that we measure the velocity with respect to a different frame of reference. That is, the previous statement could have been made with respect to a train moving with *constant* velocity v with respect to the track. (The velocity of the train must be constant for it to be an inertial frame.) Obviously, the velocity of the object of mass m moving on the train will now be $u+v$ with respect to the track, so that the force acting must be

$f = m$ times the rate of change of (u+v). (Track frame.)

It is very easy to realize that the rate of change of $u+v$ must be the rate of change of u plus the rate of change of v; and since the rate of change of v must be zero, because this velocity is constant, we have in the *track frame*

$$f = m \text{ times the rate of change of } u,$$

exactly as it was before for the *train frame*. Thus Newton's equation is covariant with respect to all inertial frames, as we had announced. We can now go back to the Maxwell equations and to the Bern patent office, where an unknown clerk by the name of Albert Einstein is defrauding the federal government by using his office time in scientific speculation. The year is 1905.

EINSTEIN THE CONSERVATOR

Most people who have heard something about Einstein tend to believe that what he did was conduct a frontal attack on the concepts of space and time as held until the turn of the century. That space and time were never the same after Einstein's work may be true, but so was theology after Darwin's theory, and no one would ever imagine that Darwin's work was motivated as a frontal attack on theology. Darwin and Einstein did what they did because they had a problem to solve, and they both were courageous enough to look for a solution irrespective of the philosophical or theological problems that they created.

That Einstein's problem was the resolution of the difficulties inherent in the Maxwell equations is evident from the unpretentious title of the paper in which he first introduced relativity theory: *On the Electrodynamics of Moving Bodies*.[12] He starts his paper by remarking that the problem with electrodynamics, when applied to moving bodies, is that it establishes false "asymmetries," by which he means that it makes a distinction between objects, such as magnets or conductors at "rest" and others in "motion." In rejecting such distinctions, he is, of course, rejecting the notion of absolute space and postulating that motion has always to be considered as relative. In classical mechanics, Galileo's relativity principle already entails that there is no distinction between a frame of reference at "rest" and another one moving with uniform velocity with respect to it: as we have just seen, the laws of mechanics are exactly the same in both frames (covariant). So, all that Einstein is positing is a *new relativity principle*, that is one in which the laws of nature (whether relating to mechanics or to electromagnetics) must be the same in all inertial frames. This principle is the basis of what is called the *special theory of relativity*, to distinguish it from a more general theory, also due to Einstein, to be discussed later.

In proposing this, he is asserting the validity of Maxwell's equations in all inertial frames, and since Maxwell's equations entail *essentially* the velocity of light c as a constant of the theory, a corollary of Einstein's relativity principle is *that the velocity of light must be constant in all directions and in all inertial frames*. Einstein, however, produces as an independent postulate "that light is always propagated in empty space with a definite velocity c which is independent of the state of motion of the emitting body."[13] The remarkable feature of Einstein's work is that he does not try to do anything ad hoc in order to produce a result compatible with the covariance of Maxwell's equations. Rather, he recognizes that the confused views about those equations are due to an erroneous postulation of absolute space, which lies behind Newton's laws, so that the only way forward is by reexamining the fundamentals of space and time and by obtaining a new form of the laws of motion.

As I have said, in formulating his postulate of the constancy of the velocity of light in all inertial frames, Einstein was guided by the need to do so in order to preserve the covariance of Maxwell's equations. But he knew that he was along a good line of thought because experimental results existed to support that assumption, in particular the aberration of the light of the stars (which he later discusses in his paper, showing that his theoretical results agree with the experiments), but also some observations due to Fizeau. Although he probably knew of the Michelson-Morley experiment (of which more later), he does not even mention it in the paper; and in later life, he was unsure of having known it at the time of writing up his work.[14] I shall discuss these points later as an example of the problems entailed in placing excessive reliance on the existence of crucial experiments.

I shall consider in a moment what Einstein had to do about space and time, once he accepted the constancy-of-light postulate, but first of all, I want to show the result of this work from the point of view of the preservation of Maxwell's equations. The crucial question is that of the composition of two velocities, that of a particle, u, with respect to the track, and that of the train, v, also with respect to the track. We illustrate this in figure 5, where the train is the frame of reference F' moving with velocity v with respect to the frame of reference F (the track). Notice please that in this example and throughout this chapter all velocities are assumed to be along parallel lines, in order to make the work simpler.

The classical result is easy to obtain in the example of the figure. If we take all velocities to be positive to the right of the figure, we have, writing "particle/train" to represent the velocity of the particle with respect to the train, and so on:

particle/train	u',
particle/track	u,
train/track	$-v$.

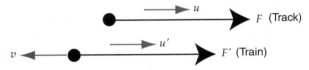

Fig. 5. Classical composition of velocities

The symbol *v* is the velocity of the reference frame *F'* with respect to the reference frame *F*. The vector *u* is the velocity of a particle (or wave) in *F*, and *u'* is the velocity of the *same object* as measured in *F'*. Because *F'* is separating from *F* with velocity *v*, this velocity must be *added* to *u* to give the velocity *u'* of the particle with respect to *F'*. (See also the proof of this result in the text.)

Since, clearly,

$$particle/track = particle/train + train/track,$$

we get

$$u = u' - v, \text{ that is, } u' = u + v.$$

When *u* equals *c*, it appears to be utterly impossible for this velocity to remain invariant in *F'*, that is, to be equal to u', since the last equation requires that *c+v* be equal to *c*. This impossibility, amazingly, dissolves when Einstein's new mechanics is used because his result is:

$$u' = (u + v)/\rho,$$

where

$$\rho = 1 + (vu/c^2).$$

If the relative velocity *v* of the two frames is very small with respect to the very large velocity of light *c*, it is clear that ρ (rho) takes the value of unity, and Einstein's formula coincides with the classical one. On the other hand, if *u* is the velocity of light *c*, it is very easy to prove[15] that ρ equals (*c+v*)/*c*, which, with u equal to c on the right-hand side of *u'* above, gives the latter equal to *c*. Thus Einstein has done two good things: first of all he has shown that, starting with the constancy of the velocity of light (a postulate which we have not yet used, since we have skipped great chunks of the theory), an *addition-of-velocities formula* is obtained that is *compatible* with that postulate. (This proves that his assumptions were *consistent*.) Second, because his addition formula shows c to be constant in all inertial frames, the Maxwell equations are now valid in all such frames, as, in fact, it was Einstein's original intention to establish. But, instead of having pulled this assumption out of the blue, he has shown that it is compatible with a new and general treatment of mechanics, which includes his addition-of-velocities formula.

There is one point about the velocity of light that I should like to clear up, although it is not essential for our philosophy, because even Einstein got it wrong.

THE VELOCITY OF LIGHT AS A LIMITING CASE

"Velocities greater than that of light have no possibility of existence,"[16] said Einstein, a statement that is still often repeated although it is not correct. Substitute u equal to kc (for some constant k, the significance of which will be later explained) in the equations for u' and ρ above. Assume also that v is sufficiently smaller than c so that it can be neglected. It is then easy to prove[17] that

$$u' = kc.$$

If k is smaller than 1, whence u (which equals kc) is smaller than c (which means that the object is moving in frame F with velocity smaller than c), then its velocity u' in any other frame cannot surpass that of light. In other words, a *subluminal* particle in one frame will be subluminal (travel below the speed of light) in any other frame. On the other hand, if k is larger than unity, so that the particle is *superluminal*, that is, it is traveling with speed greater than light in one frame, there is no other inertial frame in which the particle will *not* be superluminal. However, superluminal particles (also called *tachyons*) have never been observed: whether they share or not ontological properties with unicorns has not yet been determined. So, Einstein's wrong statement appears to satisfy facticity for the time being.

The fact that for subluminal particles—that is, all the particles that are at present known—the velocity of light is an upper limit, has an important consequence about Newton's second law, for, in accordance to it, if the force f acting on a particle is constant, then the acceleration is constant and equal to f/m, where m is the mass of the particle. This cannot be valid in relativity since, if the acceleration is constant, the particle would increase its velocity indefinitely, even surpassing c. In order to have a limiting velocity c, it is clear that as the speed of the particle increases under a constant force, its acceleration must gradually diminish, reaching the value zero when the speed reaches c, so that no further acceleration takes place. This requires that the mass of the particle increases with its velocity, up to an infinite value at c, so that its acceleration then vanishes, thus preventing further augmentation of the velocity. This increase of the mass of particles with their velocity is an everyday occurrence in particle accelerators.

RELATIVISTIC TIME: SYNCHRONIZATION OF CLOCKS

We can now go back to discussing the way in which Einstein obtained his law of addition of velocities, which required his restatement of the concepts of space and time. As regards the latter, the first question one has to treat in dealing with clocks is their synchronization. In classical work, this is a pretty trivial operation: you bring two clocks together, set them to precisely the same time, and—if they are good clocks—you know that whatever the way in which you separate them, they will always keep the same time. This assumes that time is absolute, that is, valid for all frames of reference whatever their motion, whereas Einstein realized that if space could not be taken to be absolute, the same might be the case for time. So, we must find a way to synchronize clocks that does not assume any such thing as the constancy of the timescale for different frames. The required procedure, however, is a bit of a technical matter which I would gladly skip, except that it has been ferociously exploited by conventionalist philosophers to "revise" the status of relativity theory. Thus, since I shall have to comment on their efforts, I have no option but to describe how to synchronize clocks, in which I shall follow Einstein with mere changes of notation in order to make the procedure as simple as possible. The procedure is described in figure 6.

We have two good clocks at the points A and B and, because we cannot assume that their time readings are identical, I shall denote them with t_A and t_B respectively. We set the clock A to the time t_A equal to zero and *simultaneously* send a light signal to the clock B. This light signal arrives at B at the time t_A equal to a certain value t_1, *which we do not yet know*, and *simultaneously* the time t_B, registered as the value T by clock B, is recorded. It follows therefore that when the time at A was zero, the time at B was $T - t_1$. In order to synchronize the readings of the two clocks both should show zero when one of them shows zero. To achieve this for the time when t_A was zero we must subtract $T - t_1$ from t_B for any value of the latter. We are almost there: all that remains for this procedure to work is to determine the as yet unknown time t_1, which is done as follows: the clock B is provided with a mirror which reflects instantaneously the light beam that hits it. On reflection, the reflected beam arrives at A at the time t_A equal to t_2. The latter is therefore the time that the light beam takes in going from A to B (t_1) and then from B to A. I shall now use the following:

Isotropy postulate: The velocity of light in going from A to B equals that from B to A.

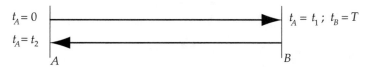

Fig. 6. Synchronization of clocks

The top and bottom parts of this picture should be read as superimposed on each other on the same line. They have been separated for clarity.

I shall discuss this postulate a little later, but it should be clear from it that t_2 must be twice the value of t_1, whence

$$t_1 = (1/2)\, t_2.$$

Clearly, the problem of synchronizing clocks is now solved. Before discussing the status of the postulate used, I shall show that the concept of simultaneity of events has to change substantially in relativity theory.

MORE ABOUT TIME: SIMULTANEITY OF DISTANT EVENTS

John and Jane stand at either end of the corridor of a fast-moving train. Their watches are perfectly synchronized and, as agreed, at precisely twelve noon, they blow their brains out. Jack, who is standing precisely at the center of the train, witnesses that the two events took place at exactly the same time, so that there is no possibility that the action of either John or Jane could have been the cause of the suicide of the other partner. Jacqueline, however, who is standing by the track so that both she and Jack have exactly the same coordinate with respect to the track when Jack witnesses the shots, sees very clearly that John shot himself *before* Jane.

The consequences of this disparity are of course legally significant, but the above story is not possibly true, *if the train is moving at normal speeds*: simultaneity for Jack must entail simultaneity for anybody else, Jacqueline included, as we have known all our lives. At relativistic speeds, that is speeds that are a large fraction of that of light, Jacqueline, however, can well be right, which shows that our whole classical idea of time must be substantially revised with obvious consequential problems for the concept of causality, as in the above example. We shall understand how simultaneity of distant events is not an absolute concept (that is valid for all observers in all frames) by means of figure 7.

We consider in the figure two frames of reference, of which F can be regarded as fixed, whereas F' moves with respect to it. In the frame F

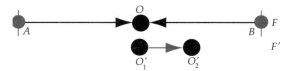

Fig. 7. Relativity of simultaneity for distant events

The black arrows represent light paths, whereas the gray arrow indicates the movement of the reference frame F' during the time when the two light signals from A and B reach O.

there are two clocks, A and B, beautifully synchronized. The clocks A and B also contain a device to fire a light beam at any specified time. Precisely at the midpoint of the distance AB there is an observer O, so that if the two synchronized clocks fire at the same time, the light beams will arrive at O, both at exactly the same time, and O will observe the two events as simultaneous: this will be our relativistic definition of simultaneity for distant events, namely that the signals from two events, A and B, must arrive simultaneously at the midpoint between these two events (remember that we accept the isotropy postulate of light propagation).[18]

Because the velocity of the frame F' is known to the observer at O, she can compute the exact time at which her horizontal coordinate will coincide with that of the observer at O_1' in F' and arrange for the two clocks at A and B in F to fire at precisely that time. Clearly, as I have said, the observer at O will perceive these two events as simultaneous, but this cannot be so for the observer in F' since, if the two light beams arrive at O at the same time, they cannot do so at O_2' (see the figure).[19]

The consequence of this result for causality is serious because simultaneous events in one frame, which could not be causally related as in the example of the suicidal activities of our late friends John and Jane, could appear not to be simultaneous in another, thus raising the possibility that the earlier event be the cause of the other. Causality thus begins to crumble, but I have repeatedly said that we are the children of our environment, and with us so are also the meta-physical principles which had served us so well until now. I had never claimed, however, that they would endure transplantation to new environments; and here we have our first example. We shall soon see, however, that even within relativity, causality may be used as long as some new necessary conditions are satisfied.

We must now consider another important consequence of the results of this section: if simultaneity is not valid for distant events observed in different frames, the timescale must change for moving frames, as we shall now show.

TIMESCALES FOR MOVING FRAMES

It is easy to see that the time interval between two events will be measured by different values in moving frames, which means that there must be a corresponding change of the timescale. I illustrate this in figure 8 where a spaceship (frame F') moves with velocity v with respect to another frame F. The observer at O' (whom I shall conventionally call the moving observer) causes a light beam to traverse his ship, as shown by the vertical arrow, in the time $t,'$ as measured aboard the ship. During the time that the light takes to traverse the ship, the latter has moved with respect to O by, say, the distance indicated with the gray arrow, so that the stationary observer at O sees the path of the light beam as shown by the thick arrow. Because the latter is clearly longer than the vertical one, *and because the velocity of light is constant*, the duration t of the event for the stationary observer must be *longer* than that for the moving observer, which was t'. (If we were using classical physics, the velocity of light in the frame F *would* be that velocity in the space ship plus the velocity of the space ship itself: this larger value would be found to compensate exactly for the longer trajectory.)

Although the relativistic result is pretty clear, its consequences are hard to swallow because, what we are saying is that, even if the two observers were to start with perfectly synchronized clocks, when the spaceship moves, they both will measure the same event by their clocks, but they will read different results: their clocks tick faster or slower depending on the movement of their frames! Surprising as this result is, we have already seen incontrovertible evidence that this is the experimental situation.[20]

CONTRACTION OF LENGTHS

Let me remind the reader of what is going on. Einstein wanted to save Maxwell's equations, for which he had to postulate the constancy of the velocity of light, but if he had said that this was all that was necessary, no one would have paid any attention to him. What he realized was that, if the velocity of light had to be the same independently of the velocity of the frame of reference, an entirely new law of composition of velocities was required. Velocities, however, are space (length) divided by time: the daring plan of Einstein was, by analyzing what happens to time and to lengths when frames move, to try to obtain a law of composition of velocities *consistent* with his postulate, at which point he would happily close a virtuous circle. We know that he succeeded in this enterprise, and we have already seen that time, surprisingly, changes from one frame to

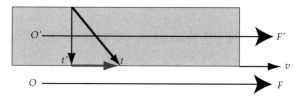

Fig. 8. Change of timescale for a moving frame

The symbol *v* is the velocity of the frame *F′* with respect to the frame *F*.

another, rather than being the absolute entity previously conceived. That time plays games with us is not too dolorous, however, because we are all a bit like Saint Augustine, that is, people who in principle would like not to touch it with the end of a ten-foot pole. Give us a good solid iron rod and we shall be comfortable: this is another illusion that Einstein had to destroy in order to get a good law of composition of velocities.

That lengths, even of solid iron rods, change with their motion, is very simple to verify. In figure 9 I show what in modern practice is the best way to measure lengths, by measuring the time it takes for light to traverse them. (This is so because such times can now be measured with extraordinary accuracy, far better than similar measurements of lengths.) For this purpose, we put a laser source at the end 0 of the rod and a mirror at the other end *x*. If *T* is the time it takes for the light to return to 0, then

$$t = (1/2) \, T$$

is the time the light takes to traverse the length *x*, whence

$$x = ct$$

is the *length* of the rod, *c* being the velocity of light. Since the time *t*, as we have seen, changes for different inertial frames (frames that are moving with constant velocity one with respect to the other), lengths also change when measured in such frames. This phenomenon is known as the *Lorentz-FitzGerald contraction*.[21]

We have now broken the back of the relativistic ideas that we need, but before we go any further, we must exorcise a possible philosophical error. For what we have been saying appears to be founded on arguments that can easily be misinterpreted: you do not need to be excessively naive to reason that after all, you know (in your bones, you may add if you like a certain amount of drama), that length and time are "out there" or at least that this is the case for rods and clocks; and that all that we have been doing is discussing how *we* observe and measure them. Yes, it is per-

Fig. 9. Measurement of length

fectly possible to concede that length and time change as we have discussed, but that, nevertheless, this is "all in our mind," that is, it is all in the numbers that we use to describe what is "out there"; and that we have not said anything at all about what "is really out there." In a more elegant philosophical language, it could be argued that all I have done is discuss the *epistemology* of space and time (that is our knowledge of them), but that what I have said is irrelevant to their *ontology*. (And as regards the latter, some people might even claim that they know *in their bones* what *really* exists or not, and that there is no need of any Einstein to tell them what to think in this respect.)

Ontological hallucination is one of the most common ailments of humanity: if just one thing could be said in defense of philosophy it is that *good* philosophical training prepares your mind to recognize and avoid such a disease. There is something even simpler: to look at nature, although this sometimes does not work because there may not be experiments available at the time to guide us. This is not, fortunately, the situation with the questions I have raised in the above paragraph. As regards time, it is not the case that our *readings* of clocks merely vary in different frames of reference: the clocks themselves *tick at different rates*, as the simple experiment discussed on p. 116 clearly demonstrates. Similarly, there is no question that the Lorentz contraction might be a mere epistemological artifact: that iron rods *shorten* their lengths in motion is as true as to say that they shorten when I file them. All this is very good, but even if it were possible to argue that ontology is not the primary concern of science, there is at least the minimum requirement that any ontology as may be, explicitly or implicitly, supported by a given theory should at least be consistent with its principles, a point which I shall illustrate in the next section.

SPACETIME AND APPLIED ONTOLOGY

In order to understand the job in hand, I shall start by considering the transformation rules used in Newtonian mechanics to change from one inertial frame to another, for which purpose we have to look at figure 10.

It is quite clear from the figure that the required rules of transformation are:

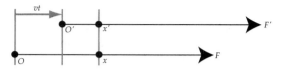

Fig. 10. Classical transformation of coordinates

The frame *F'* moves with constant velocity *v* with respect to the frame *F* and its position at the time *t* is shown on the upper part of the figure.

$$x' = x - vt,$$
$$t' = t.$$

The second equation here, which says that time never changes from one frame to another, is not usually written in classical mechanics, but I include it so as to stress the difference between the classical and the relativistic transformation equations. Obviously, the form of the above equations will change in relativistic theory; the correct equations, derived by Einstein but previously found by Lorentz, are called the *Lorentz transformation*:

$$x' = function\ of\ x,\ t,\ and\ v.$$
$$t' = another\ function\ of\ x,\ t,\ and\ v.$$

I do not give here the actual form of the functions entailed because it is not necessary for our work. The important thing is this: in classical theory, time plays a very different role from that of the space coordinate because it is automatically an invariant of the transformation. In the Lorentz equations, instead, time has lost its privileged status, and it transforms in a similar way to that of the space coordinate. Thus, the distinction between time and space, which was so clear-cut in classical mechanics, loses meaning in relativity. (Which agrees with what we have seen on p. 274, that a measurement of length is in fact conducted as one of *time*.)

I shall discuss now the question of ontology or rather ontologies, since classical mechanics and relativistic mechanics differ fundamentally in this respect. In doing this, one problem is that the word *ontology* means different things for different people. In order to have a concept of existence adequate for science, I shall accept that spotting *elements of reality* is a part of the activity of any physical theory. The words in italics here were used by Einstein[22] with a very specific meaning which I shall adapt to our present needs. Different observers in different frames of reference measuring the same quantity may or may not come up consistently with the same value. If the former is the case, we shall say that the quantity in question is an *element of reality*.[23] Within this definition, lengths are elements of reality in classical mechanics for all observers in all inertial frames, for the following reason. It is clear that if we use the first of the two classical transformation equations, that for x', for two values x'_1 and x'_2, their difference in the primed frame, $x'_1 - x'_2$, equals that in the unprimed frame, $x_1 - x_2$, since the two vt terms cancel out. It is thus part of the ontology of classical mechanics that lengths, and in the same manner time intervals, are elements of reality, that is, that they have some objective meaning.

As scientific theories are refined and modified, the same happens to their ontologies. It should be clear from our discussion so far that neither lengths nor time intervals have an objective significance in relativity, and it is important to try and find out what the new ontology is. This entails, of course, finding an *invariant* under the Lorentz transformation; and although I have shunned writing the latter explicitly, I shall sketch an argument that will lead us to the necessary invariant. When we discussed how to measure the length x of a rod on p. 274, we wrote for it x equal to ct, where c is the (constant) velocity of light and t is the time that light takes in traversing the length x. Because c is constant this relation must be valid in all frames:

$$x = ct \text{ entails } x' = ct' \text{ in all frames.}$$

(Notice how neatly this result can be written without having to know the details of the Lorentz transformation.) We have here the idea of x as the length of the rod, but if we look at figure 9, from which we got the relation under study, x is really the coordinate of one end of the rod, which could be negative (if, for instance we exchanged 0 and x, keeping coordinates positive to the right). In order to avoid problems with negative values, it is useful to square the relations above, since the squares will always be positive. I shall also bring the terms coming from x to the right-hand side, for reasons that you will see in a moment:

$$c^2t^2 - x^2 = 0 \text{ entails } c^2t'^2 - x'^2 = 0 \text{ in all frames.}$$

You might remember that I said on p. 195, that the conservation of the zero of a physical quantity with respect to the transformation of frame was essential[24] for that quantity to have a physical meaning, which in the new terminology means that we hope that the quantity in question, when nonvanishing, will be an element of reality. You must realize that in the above relations t has a very specific meaning, since it is not just any value of t as can be associated with x in the Lorentz equations of p. 276, but rather the time that light takes in going through x. For values of t not so restricted, therefore, it is not to be expected that $c^2t^2 - x^2$ will vanish, but it is not unreasonable to hope (because of the conservation of its null value) that it will be an invariant of the Lorentz transformation:

$$c^2t^2 - x^2 = c^2t'^2 - x'^2 \text{ in all frames.}$$

a fact that can readily be verified if one uses the detailed form of that transformation.

We now have a very important ontological result: within relativity theory neither space nor time are elements of reality. The only elements of reality are the combinations $c^2t^2 - x^2$ of space and time, which are called *spacetime intervals*, although more precisely the *spacetime interval S* is the square root[25] of this quantity, or what is the same,

$$S^2 = c^2t^2 - x^2.$$

The discovery of the spacetime intervals as Lorentz invariant, and of their profound significance in relegating space and time to the mere status of ghosts without ontological significance, was made not by Einstein but a few years later, in 1907–1908, by his former mathematics teacher, Hermann Minkowski (1864–1909). The new "space" made up of the old space coordinates plus the time one, which is the only space with ontological significance, is often called *spacetime* or *Minkowski space*.

I shall develop these ideas a little in the next section, but I do not want to leave the reader with the impression that the rather restricted definition of ontology that I have given is the only type of ontology used in science. Science uses many models which postulate the existence of certain objects that, as a difference with those discussed here, are not open to observation. As regards the meaning of their existence, you can take two views. One is that their mode of existence is the same as that of unicorns or of Prince Hamlet. The other is that their existence is linked to how far they save facticity, which is not a very logical position but nevertheless scientifically almost unavoidable. It is not very logical because, as I have said repeatedly, the fact that the consequences of a proposition are true does not entail that the proposition be true. It is almost unavoidable because it is often difficult to make progress in physical theory without inventing unobservable objects, such as quarks or the vacuum. We thus see that scientific ontology in a way has two subsets, one of quantities that through their invariance for all observers may be considered as elements of reality, and the other containing far more tenuous objects about which one has modestly to say "through their fruits thou shall know them." When later in this chapter we deal with the vacuum, we shall indulge in such an ontology. We must always remember that as science advances, so its ontology also changes: if you are very optimistic, you may think that science will soon advance so much that its ontology will be the *real* one, and that thus we shall know the mind of God. You may well believe such a thing, but about this I cannot possibly comment:

all I can say is that we are laboriously building a picture of nature that is each time empirically more adequate, and that, correspondingly, it is not inappropriate to place increasing reliance on its emerging ontology.[26]

If we accept that the ontology that science constructs is, like all physical theory, subject to refinement, we might also accept an even more disagreeable thought, namely that ontology, like physical laws themselves, might also be *approximate*. You could say, as I have done, that the ontology of Newtonian mechanics is falsified by relativity, but there is no denying that, at low velocities, lengths may be considered for most uses and purposes to have an objective meaning as elements of reality; and thus that, *within the domain of validity of Newtonian mechanics*, its ontology is valid, albeit approximately. Of course, the reader who is in search of absolute truths will find this suggestion repugnant, almost as if I had proposed that a unicorn could approximately exist! There is no denying, though, that to think one impossible thought every day before breakfast is a useful preparation for life.[27]

SPACETIME AND CAUSALITY

In order to make things easier, I have so far considered a single space coordinate x. If we consider all three, we must appreciate that x, y, z, and t form a four-dimensional spacetime, and that it is only this spacetime, rather than the three-dimensional space of classical mechanics, that has a physical meaning (that is, the one that generates *elements of reality*). For simplicity, however, I shall continue to consider spacetime in two dimensions, x and t only, and shall choose convenient units so that the velocity of light c be unity. (That is, equal to 1.)

I represent spacetime in figure 11 with a view to understanding how events, which are described by their x and t coordinates, are related. We must first discuss the construction on the left of the figure. Consider events the representational point of which move along the gray line on the top quadrant, which is its diagonal, at 45° to the x axis. The velocity x/t corresponding to point 1 is a/b, which is clearly unity, that is, the velocity of light. A line that links the origin to point 2, instead would have a velocity e/b, which is smaller than unity (the speed of light), e being smaller than a. It is thus possible to send a material signal (subluminal particle) from O to the point 2. This means that these two events can be *causally related*, as I shall now explain. First, O precedes 2, since the time at 2 is later than at O. Second, we know that the relation between cause and effect must entail some causal chain, that is the propagation of some material objects or signals, which is only possible in relativity if this

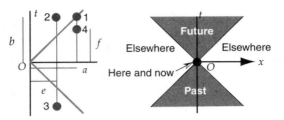

Fig. 11. Spacetime

For simplicity, only one space coordinate is shown, and the velocity of light c is taken to be unity (the number *one*). The figure shown is understood to correspond to one reference frame with an observer at the origin (*here*) and at time zero (*now*). In order to extend the figure on the right-hand side to three dimensions it would have to be rotated about its vertical axis whereupon the areas in gray would become cones. The diagonal lines form an angle of 45° to the x axis.

propagation can be effected at velocities below that of light. This is pre-cisely what we have seen is possible between the two points O and 2. Since the latter is at a later time than O (*here and now*), the possible effect at 2 is in the *future* with respect to the observer. (Remember that, because simultaneity is gone, the classical time-relation of events breaks down.) Exactly in the same manner, the velocity from O to point 3 is also e/b (on taking the absolute, that is the positive value of the time coordinate which is $-b$), and thus subliminal. Therefore 3, because it antecedes O, could be causally related to it, the passage of a material signal from 3 to O being possible. Clearly, 3 is in the *past* of O.

It is very important at this stage to appreciate that there are events that, although pertaining to a later time than O, cannot be causally related to it because any signal that connects them to O would have to travel at superluminal speeds. This is the case for point 4, for which the speed from O is a/f, clearly larger than unity, since f is smaller than a (remember that a and b are equal).

I collect all these results on the right of the figure. All the events in the areas in gray can be causally connected to the origin (*here and now*), but the events in the white areas cannot cause or be caused by O, and they are in relativity called *elsewhere*. We have seen (remember the story of Jack and Jacqueline), that one of the possible problems with causality in rela-tivity is that the time ordering may not be invariant for different ob-servers in different frames: if event A precedes B in one frame and can thus be its cause, B can precede A in other frames thus making nonsense of the causal relation. The situation in relativity is far more orderly, how-ever: because all events in the gray areas of the figure may be reached from the origin by a subliminal signal, the time-ordering of the events with respect to *here and now* must be the same for all observers in all frames[28] and thus, as we have already said, causality works normally in such areas. It is only in *elsewhere* that causal relations lose meaning.[29]

GENERAL RELATIVITY AND GEOMETRY

Some ten years after his special theory of relativity, Einstein generalized it very profoundly. Although special relativity produces new dynamical laws of motion valid at all velocities, it does not concern itself with some of the deeper problems of nature, especially the origin of gravity and of the gravitational mass. Einstein's first observation was basically this. Suppose that in deep space, where there is no gravitational attraction, we have a spaceship moving with *constant* velocity. A ball left in the center of the spaceship will float at rest as required by inertia. So far, we are within special relativity in an inertial frame (constant velocity). But we now note that, if the spaceship suddenly *accelerates* forward, the ball "falls" toward the back end of the ship. Einstein concluded from this remark that acceleration and gravity are the same thing; and that the "gravitational mass" introduced in Newton's gravitational law is nothing more than the inertial mass used to describe the motion of the ball on acceleration. (Curiously, no one had explained until then why the mass of a body used in Newton's second law, or *inertial mass*, is equal to its *gravitational mass* as used in the law of gravitation.) Einstein's program was therefore to extend his earlier relativity principle by asserting that the laws of physics are valid in *all* frames, not just inertial frames. (More properly, that they are *covariant* with respect to *all* transformations between different frames of reference, that is, that they always have the same form after such transformations.)

This program required some very heavy mathematical work to develop a method for writing down dynamical laws covariant with respect to all changes of frames of reference, and in order to obtain such equations, Einstein found that he could not use the ordinary geometry of three-dimensional space until then employed, that is, Euclidean geometry. (Naturally, Einstein was dealing at this stage with four-dimensional spacetime, but even in four dimensions geometry can be Euclidean, by a fairly straightforward extension of three-dimensional Euclidean geometry.)

I do not wish to get here into the complexities of non-Euclidean geometry, but it will be sufficient to produce just one example. The reader will surely know that if you add up the three internal angles of a triangle you get 180°. Suppose now that you do not realize that Earth's surface is a sphere, but you think it is a plane, and you form a huge triangle with a base on the equator and a side along the Greenwich meridian from the equator until the North Pole. Clearly, the angle these two lines (which you will think are straight lines) form is 90°. Take the third side of the triangle from the North Pole along the 90° East meridian until it cuts the equator. It is evident that the three angles of this triangle are all of 90°, whence their sum is 270° in contradiction with Euclid. Although the discrepancy here is due to our wrong assumption that the surface of the earth is a plane, it was discovered, since the nineteenth century, that consistent geometries could be constructed in which the defined "planes" could sustain triangles the sum of whose angles differs from the Euclidean value, for which reason these geometries are called *non-Euclidean*. Such a geometry had to be used by Einstein, and he found that if you take a small region of space and you work out the sum of the internal angles of a triangle sustained there (or some such similar quantity), then this number varies from region to region, which indicated that the space is *curved*. (Just as Earth's surface was in our example.) A flat space (an expression used even in three dimensions) is, instead, Euclidean, and its main property is that wherever you form a triangle in this space, the sum of its internal angles will be 180°. Einstein's result, instead, was that the curvature of space changes from region to region and, moreover, that it depends on the masses placed on it. Thus, space is not an empty container in which objects, that is masses, are placed: matter and space are irreducibly linked, space without matter being largely a meaningless concept.[30]

The very beautiful result that emerged from Einstein's theory is that gravity appeared now as an inherent property of space, and you no longer had to introduce unsatisfactory gravitational forces acting at a distance. When you say that an object falls because of the gravitational attraction of the earth, for instance, what really happens is that because of its large mass, the earth distorts space over a large region, and the object moves along the space so distorted following its curvature so as to hug the shortest possible route (called a *geodesic*) in curved space. If space is curved, even light must follow these curved geodesics, and since they must be very curved near a large object like the sun, light should follow a clearly curved path near it. This was actually observed by Eddington during a total eclipse of the Sun on 29 May 1919, the deflection of the light rays being precisely that predicted by general relativity, an event which was one of the most exciting points of twentieth-century science, and which shot Einstein into unprecedented scientific glory.[31]

Einstein's general relativity put an end to a long-standing philosophical tradition whereby the basic structure of space was not an empirical question, but rather the subject of some sort of a priori knowledge: the structure of space (that is, its *curvature*) is now no longer given by any preconceived geometry, however rational, but it is determined by the distribution of matter in the universe and thus entirely empirical. Moreover, Einstein's radically new view of space, not as an inert frame of coordinates, but as something that can bend, and strain, and ruck, provided an immediate and deeper understanding of the concept of a field. Remember that, so far, we described such fields as being defined by certain perturbations acting on each point of space. This is an extremely vague concept since, if space is just inert, the perturbations appear to be perturbations of nothing (unless an ether is invented). These perturbations can now, instead, be understood as strains in space, affecting its curvature.

Einstein's general relativity, however, was not able to provide such a picture for any other field except that due to gravitation. In other words, it was only the gravitational masses that both created and distorted space. It was the German physicist Theodor Kaluza (1885–1954) who, in 1921, took the next step forward by introducing a fifth space dimension in spacetime, with the remarkable result that Einstein's gravitational field equations written in this space produced now not only the gravitational, but also the electromagnetic field as precisely described by the Maxwell equations. No wonder that Einstein was so impressed by this work that he communicated it himself to the Prussian Academy. The problem remained of course that spacetime, observationally, appeared to have only four dimensions. It was the Swedish theoretician Oskar Klein (whose room in the Department of Theoretical Physics at Stockholm I occupied in the winter of 1972, when he was still only seventy-eight years old) who proposed a solution to this conundrum in 1926. Imagine that what we call a point in space is really a tiny circle: this means that the two dimensions of the circle will appear to us as a single dimension, especially since Klein found that the diameter of such circles would be millions of times smaller than the atomic nucleus. As it was the case with Kaluza's space, Klein's five-dimensional space with one of its dimensions thus *compactified* still sustains the electromagnetic as well as the gravitational field.

This is not the end of the story because other fields are necessary in nuclear physics in order to account for the forces that bind nuclear particles together, and extensions of the Kaluza-Klein theory have been very much developed in the last ten or twenty years, mainly through what is called *string theory*, strings being the tiny circuits around a compactified dimension that Klein had postulated. The reader may ask: how it is that this compactification has arisen? One has to accept that, among the many wonders of the big bang, one of its results was that not all dimensions of space grew in the same manner, but rather that some (because current theories postulate more than one compactified dimension), were, so to speak, stunted in their development and thus effectively unobservable.

THE ILLUSIONS OF CONVENTIONALISM

There is something in the human mind or in human cultural history, account for it as you will, that makes some people desire that the universe be known to us by pure reason: the empirical, for them, is a sort of original sin to be kept at a distance. You cannot go very far along that road, but you can nevertheless, if you are so minded, throw a good many wrenches into the wheels of empiricism. Empirical results are of course accepted by such people as required, but they claim that the theories or laws by which we describe experience are conventional: thus any number of theories are all equally satisfactory and will easily explain all the known facts, as long as one be prepared to put up with different and perhaps more involved laws of nature than those currently discussed. I have already shown on p. 123 that it is most unlikely that, by changing entrenched conventions, one might be able to obtain any laws at all; and although I have already said enough about conventionalism, I should like to stress a few points about it, since this doctrine, which was a great love of my misspent youth, is dangerously attractive intellectually.

Conventionalism, like not a few other philosophies of science, ignores one of the most profound aspects of the latter, namely that it is the result of evolution. (I mean of course cultural rather than biological evolution.) A most important feature of evolution is that no structure emerges at any given time, except by changes of the structure as it is found *at that time*. Thus the whole of the previous history of the evolving structure is telescoped at each evolutionary step: to allege that at any given step things could have been otherwise, although it is right in the sense that the structure could have evolved along different lines right from the start, can be utterly wrong if changes are predicated that would contradict the *previous history* of the system. You can modify a language at a given evolutionary step by changing some, probably small, features of it, but it would be irrational to assume that you could at any one moment of change wipe the slate clean and start from the beginning, however elegant and rational that beginning might be. Thus whereas evolutionary steps in science might entail conventions, it is absurd to predicate that science is free to choose such conventions at will: it is like assuming in biological evolution that an ear could in one giant step evolve into an eye. Yes, there might be free will involved in a convention or a postulate, but it is a free will that must move along predetermined tracks and not on a skid pan. The very conventions that can be chosen are most often in practice an *unavoidable* consequence of the empirical structure of the evolving system, the scientific mesh which I described in chapter 10.

We have already met in chapter 5 the leading figure of conventionalism, Jules Henri Poincaré (1854–1912), who was one of the heroes of my teens. In trying to understand space, and thus geometry, Poincaré admitted that, in applying geometry to describe experience, there was an empirical element as well as the abstract (a priori) one associated with the geometry used. He also claimed, however, that when faced with an empirical fitting, one was free either to modify the geometry or to modify the physical theory (keeping the geometry intact) so as to preserve facticity. As one would expect, he claimed that, faced with this situation, physicists would always modify physical theory rather than abandon Euclidean geometry.[32] The year was 1902, and the fact that Einstein proved Poincaré's approach utterly wrong in the next decade did not, of course, deter some philosophers from following Poincaré's lead whenever a chance arose.

Special relativity was fertile ground for this sport, the postulate of the constancy of the velocity of light providing, it was alleged, a demonstration of its conventional structure. I have shown that the definition of the synchronization of distant clocks is central to the theory and accepted that, in so doing, we needed the *isotropy postulate* (p. 270), which allowed us to write the time t_1 of the light passage between the two clocks as $(1/2)t_2$, where t_2 is the time the light takes in traversing that distance *twice*, once in each direction. Hans Reichenbach[33] claimed that the factor 1/2 here "is not epistemologically" necessary and suggested that a value ε (epsilon) between 0 and 1 would do just as well. We have seen, however, that the value 1/2 was absolutely essential for Einstein: any other value would have contradicted the isotropy postulate, and if Einstein had thrown away the latter, he would have destroyed the very piece of the scientific mesh, Maxwell's equations, which he considered well entrenched and in need of further protection. Of course, if Maxwell had never lived, and yet a deutero-Einstein had invented relativity, he might perhaps have followed Reichenbach's proposal, but such a speculation is nonsense because we have to live with science as it *has* evolved and not as it *might* have evolved: *in this case the nature of the evolving structure is as much a part of our empirical facts as the results of experiment.*[34] In one way or another, Reichenbach's critique, however, was reiterated more recently by Adolph Grünbaum and by Bas van Fraassen.[35] Einstein himself would have rejected the idea that his choice of 1/2 was a mere convention. On discussing his work much later, when he was still at the peak of his powers, he claimed that the statement that light has the same speed in two opposite directions is "neither a *supposition nor a hypothesis about the physical nature of light,* but a *stipulation,*" the latter word carefully chosen to indicate a necessity.[36]

THE ETHER IS DEAD: LONG LIVE THE VACUUM

Einstein's relativity theory sounded the death knell of the ether, but many physicists would at present consider the experiment conducted in Cleveland in 1887 by Albert Abraham Michelson and Edward Williams Morley as the most conclusive proof that the velocity of light is constant, independently of the frame of reference, and thus that no ether can exist to define absolute rest. If you consider any point on the surface of Earth, it is subject to two motions. One is that of the diurnal rotation of the planet with a linear velocity of about 0.5 km/s. The other is the motion of Earth on its orbit around the Sun, with a much larger velocity of about 30 km/s. We can thus ignore, to a first approximation, the rotational motion: if we measure the velocity of light parallel to the orbital motion of Earth, this velocity, with respect to the fixed ether, should have added to it (or subtracted, depending on the direction) that of Earth's orbital motion, 30 km/h. The same measurement along a perpendicular direction would not, on the other hand, be affected by such orbital motion, whence the difference between the two results should give Earth's orbital velocity, that is about 30 km/s. Michelson and Morley found instead a null result: the time taken by light to traverse the two perpendicular arms, of equal length, of their instrument was identical.

Those who believe both in crucial experiments and in falsification would expect that the Michelson and Morley null result should have been the sufficient cause for abandoning the ether theory. Before I go any further in discussing what really happened, I should like to remind the reader that I do not deny that crucial experiments and falsification do play some part in scientific method; but that in my view regarding any one of the weapons in the scientific armory to be all-determining is to fly in the face of evidence. First, as I have already mentioned, Einstein himself was entirely unimpressed by Michelson and Morley[37], and it is only now, when we consider Einstein's theories sufficiently entrenched, that we believe that Michelson and Morley should be regarded as the crucial experiment to disprove the existence of the ether. In saying this, however, we are not claiming that it *was* the crucial experiment that disproved the ether. So, effectively, we are changing the meaning of the word *crucial*, because a crucial experiment should indicate a bifurcation in the road of scientific progress that proves or disproves a fact or theory *at, or at least very near, the moment* the experiment was performed. That is, a genuine crucial experiment should not just be the best way in retrospect to teach one's pupils what the result is. The battle of Saxa Rubra *was* the crucial battle that established Constantine as the ruler of the Roman Empire, and it is crucial now because it was crucial *then* in 312 C.E., not because we regard it *now* as the reason why we hear no more about Maxentius as the ruler of the empire after that time.[38]

Scientific battles, alas, seldom lead to instant routs, and, like the Michelson-Morley experiment, most so-called crucial experiments (unless this expression is purely used for didactic purposes) are no more than very important experiments. And we now come to our second point: was Michelson and Morley's result actually taken to falsify the theory of the ether? Not at all. Five years after the Michelson-Morley result, Sir Oliver Joseph Lodge (1851–1940) mentioned in a note in *Nature* an ingenious idea of George Francis FitzGerald (1851–1901) to resolve the apparent paradox. In the direction of the orbital motion of Earth, within the ether theory, the velocity of the light must be *reduced* by that of the orbital velocity of Earth, to get the right velocity with respect to the ether.[39] Therefore, if it is postulated that the *length* of the path in that direction is *contracted* by an appropriate amount, then the time taken by light to traverse that shorter path, at the reduced velocity, could be the same as that taken in the perpendicular direction. It is of course quite easy to calculate the precise value of the contraction required to achieve this result, which FitzGerald did. Five months later this proposal was adopted by the Dutch physicist Hendrik Anton Lorentz (1853–1928), who produced a more complete, although still heuristic, theory including the famous Lorentz transformation; the contraction thus postulated is now called the Lorentz-FitzGerald contraction.

Neither FitzGerald, nor Lorentz, nor Lodge, were fools unable to see beyond their noses to detect the possibility of a good bit of falsification when it came their way. Obviously, they judged the ether theory so well entrenched (because otherwise they could not conceive of the possibility of the propagation of electromagnetic waves) that they were prepared to accept an entirely ad hoc and quite implausible effect in order to preserve the ether. As I have said before, it was only because of Einstein's correct intuition of what should safely be entrenched, that he arrived at a far more satisfactory resolution of the problem: he put his money on the entrenchment of Maxwell's equations instead of that of the ether. It must be appreciated that Lorentz and FitzGerald were as prepared as Einstein to falsify hallowed concepts: to suggest that lengths of real objects vary with their speed is no mean denial of Newtonian mechanics. What really matters is that they did not consolidate, in any serious way, the total entrenchment of the scientific mesh in making their ad hoc proposal. The reader should notice again one of the major weaknesses of Popper's falsification theory, namely that whatever is to be falsified is not *unique*.

Let me now have a look at what happened to space after the ether went. You must realize that, in the wake of Einstein's general relativity theory, space as an independent entity, and I mean in this context the only space that is left, that is, spacetime, was also gone: spacetime is merely a manifestation of matter. So, if matter is not there, if we have a vacuum, we are left with nothing at all. Because spacetime cannot be absolutely empty, vacuum is not empty space, vacuum has no place and no time, vacuum is not here nor there, not now, nor ever. Yet, as you will see in a moment, the vacuum plays a remarkable role in modern theoretical physics.

VACUUM AND FIELDS

We have seen that the new views about spacetime allow us a model or a picture of what fields are, as perturbations in spacetime revealed by curvatures in it. When quantum mechanics came into operation, a new view of the fields emerged, and it was Einstein who again led the way, through his study of light emission. I shall briefly sketch how light is produced: if you take an atom such as sodium at normal temperatures, the electrons in it, which occupy energy states around the nucleus, keep to the lowest states available to them, and it is then said that the atom is in the *ground* state. This does not mean that all electrons in such a case are in the lowest possible energy state, since there are certain rules that restrict the number of electrons that can occupy any one state. Consequentially, in an atom in the ground state, electrons occupy states of successively higher energies, until all the available electrons in the atom are used up. (See figure 12.)

If you put the atom in a flame, the atom gets excited, which means that one of its electrons, the one that was at the top of the energy range in the ground state (*a*) jumps onto a state of higher energy which would normally be empty, (*b*). (In saying all this I assume that you accept that, as a difference with classical physics where changes of energy can vary continuously, this is not so in an atom, and that the energy levels that are possible are separate and discrete, *quantized* as one says.) Again, many readers will be aware that light is emitted when the excited electron falls back onto the state that it occupied in the ground state, releasing its extra energy as light, which is in any case an easy idea to grasp. As we shall see later, Einstein revived Newton's corpuscular theory of light, and he invented a particle, eventually called the *photon*, beams of which constitute what we call light. So, we agree that we have an excited atom, and that after some time it goes down in energy emitting a photon. Many books, some of them even expensive, tell you this story but miss the interesting bit: why does the excited electron ever go down to a lower energy state? You may think that this is obvious, but it was not obvious to Einstein, and his curiosity started a line of thought that gave rise to some of the most exciting ideas of the century. And we are now not talking about relativity but quantum mechanics.

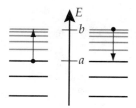

Fig. 12. Emission of light by an atom

The energy levels normally occupied in the ground state are shown in black and those unoccupied in gray. On the left, an electron is excited to an unoccupied level. On the right, this electron falls back to the ground state releasing the energy *a* as light.

Let me first explain why we have to worry about what it is that makes the electron fall down to a lower energy state. Imagine first of all a cone. If you put it on its base on a *perfectly* firm and rigid plane, it will stay like that forever: we say that it is in a state of *stable* equilibrium. Now comes the juicy bit. I put the cone on the same plane but carefully standing it on its tip in a totally still atmosphere (see fig. 13) and then release it very carefully without perturbing it at all: you might think that it will soon fall down, but *this is not true*; it will remain *forever* in equilibrium. And please do not jump up and down telling me that I tell lies, because you will upset the poor cone. (Because the equilibrium is in this case *unstable*, so that a very small departure from it will produce a large and irreversible displacement: the cone will fall down, contrary to what would happen to a cone firmly sitting on its base. So, I assume that drafts, vibrations, sneezes, and the like are entirely eliminated.) No, I have not told a lie: remember that physics cannot address nature except through the mediation of models, and I did precisely that. Look a few lines above and you will see my plane described as *perfectly rigid*. That is, it is a plane that is not subject to any form of vibration or any other perturbation.

Obviously, a perfectly rigid plane does not exist in nature, although physicists can now construct planes that are amazingly free of vibrations. So, you have to imagine a model plane, a perfectly rigid plane, in order to be able to state an important counterfactual (see p. 62): *if the plane that supports this cone were perfectly rigid, the cone would never fall down.* And this counterfactual is very important because it is only by formulating it that the problem of how it is that the cone falls down can be addressed at all. The answer is very simple: if the plane were perfectly rigid, the cone would stay up for ever, but it falls down from its state of equilibrium because in *real* planes there is always some perturbation, as provided by thermal vibrations, that jiggles the cone away from equilibrium.[40] (I am

Fig. 13. Unstable equilibrium of a cone

assuming in all this, of course, that drafts and other such perturbations are totally eliminated.)

So, back to our excited atom: what is it that makes it fall down from its state of equilibrium? The analogy with the falling cone requires that we look for some perturbation that jogs the excited atom out of its high-energy state. Einstein recognized that the electric charges (the electrons) in the atom create an electromagnetic field. That is, the charges act on the field and by the principle of action and reaction (p. 143), the field *reacts* on the charges. It is this reaction of the field on the excited atom that induces the transition to the ground state shown on the right-hand side of figure 12. This, nevertheless, was only the beginning of this story. Once the idea became firmly established that light had a corpuscular nature owing to the existence of the photon, the problem of the so-called *ballistic effect* had to be solved, and a more detailed picture of photon emission was required. Remember that the picture of light as a wave was a good thing because its velocity, like that of sound (wind excluded, of course), would not depend on the velocity of the source, a fact that was accepted even within the ether theory. If a light ray is instead made up of particles, when the source moves its velocity must be added to them, as that of a moving gun with respect to the shots emerging from it. This was called the *ballistic effect*, which of course bedeviled the corpuscular theory of light by making its velocity dependent on that of the source, contrary to relativity.

This is the beginning of a new view about fields that started mainly through the work of the English theoretical physicist Paul Adrien Maurice Dirac (1902–1984). This work gave a new importance to the vacuum, although keeping its essential properties that I have described. What is new is that, although the vacuum is not just simply space without matter or fields, it is what it is not by virtue of merely not being anything at all, but rather because of being the equilibrium of opposites that cancel each other. This equilibrium, though, is rather active in the sense that objects, such as photons, are constantly created and constantly annihilated in the vacuum. Such photons, because they are created and annihilated within an exceedingly short time—thus the appearance of nothingness—are called *virtual photons*. It is this albeit virtual activity of the vacuum that interacts with the excited state of an atom and jiggles it out of its state of equilibrium: the atom goes down in energy to the ground state and a photon, which otherwise would have been annihilated in the vacuum as a virtual photon, becomes free and it is thus created. The important concept is that the photon is created in the vacuum and not in the atom: thus the state of motion of the atom is irrelevant, and its velocity is not added to that of the emerging photon. This solves the ballistic problem.[41]

The concept of virtual photons provides also a graphical "explanation" of why two electrons repel each other. Electron *A*, say, is interacting all the time with the vacuum and creates a virtual photon which is sent toward electron *B*, which "absorbs" this photon, that is, it interacts with it and causes its annihilation, as it is necessary for the balance of the vacuum. The whole thing is very much as if two ice-skaters go on parallel tracks throwing balls at each other. On sending the ball toward the other partner, each skater recoils very much like a cannon on firing, and thus the two skaters appear to repel each other.[42] Although this model requires the postulation of unobservable virtual photons, this is a price worth paying, since it allowed physics for the first time to provide a picture for the interaction between two charged particles.

The reader may be asking in his or her mind the question: why have I gone into this story about the vacuum, obviously a little premature since some ideas of quantum physics are involved, but also, interesting as it might be, apparently without obvious philosophical implications. If you think about it, however, we have here a model, and accordingly an explanation, that looks forward rather than backward. Connection with the past, however, is maintained, for instance in requiring a perturbation to get something out of an equilibrium state, or in the picture of the electron recoil when emitting a virtual photon. (I have some time ago said that one of the problems of employing models and explanations is that we tend to use pictures coming from the past, just because we understand them, in order to explain new facts.) The models entailed in the description of the vacuum and fields, instead, were developed as the very complex mathematical structure that is called *quantum electrodynamics* emerged. This was actually able to explain new facts, and it did so with an accuracy unprecedented in the history of science. But because the mathematics was so heavy and dry, it was important to develop a parallel somewhat intuitive picture of it, for which Richard Feynman (1918–1988), one of the pillars of the theory, had an outstanding gift. And what is interesting for us is that this picture entails an elaborate new ontology, about which I shall comment in the next and last section of this, alas, too long a chapter.

But I cannot close this section without a health warning to the reader. You have here an example of an extremely complex mathematical theory that, because of its complexity, is embellished with graphical models and with a rich ontology. What reaches the general public, as the last page or two exemplifies, are the models and the ontology of the theory, which thus acquire a perhaps unmerited weight, whereas what has made the theory acceptable to the scientific community is its facticity, that is, its ability both to entrench large sections of the already known scientific mesh, and to extend this entrenchment to entirely new areas. The parables and the ontology are useful and perhaps fascinating, but what you must judge the theory by is its facticity. "Understanding," if that means understanding the parables, the models, and the ontology, is only significant insofar as it is an instrument to generate new ideas that will reinforce and extend the entrenchment of the scientific mesh. Remember that brilliant people certainly "understood" Ptolemaic theory, and nothing remains of their painstaking understanding of it or of its accompanying ontology: at the day of reckoning, it is the facts that sing their lasting song.

ENTRENCHMENT AND COMMON SENSE

In discussing relativity, we found our first problem with entrenchment because we had to abandon apparently perfectly well entrenched concepts such as space and time. This is, however, very much to be expected, since these very primary concepts, used by humankind for countless generations in order to organize experience, were entrenched by evolution (for instance in the perception of dimensions and in the biological clock) in an environment in which velocities never reached anywhere near the relativistic range, that is, the limits at which classical mechanics ceases to be valid. Thus, we could never have expected to have entrenched *space-time* as part of our perceptual apparatus. This does not mean that entrenchment is wrong but, rather, that as science progresses and we have to deal with environments that are different from the one in which the species evolved, we must expect that some of the old and intuitively satisfying concepts have to be revised, sometimes drastically.

You might remember that I suggested in the introduction to chapter 10, echoing Quine, that science is the continuation of common sense by other means, and you may think that the disenfranchisement of space and time by relativity which we have witnessed shows me wrong. Some of my readers, however, may have detected that in expanding a little Quine's phrase I brought back other and less gentle resonances, those of the strategist and military historian Carl von Clausewitz. I was thus keeping in my mind that "other means" might entail such a ferocious analysis of commonsense concepts as conducted by Einstein, and as necessitated when moving away from the cosy environment wherefrom common sense arose, to that prevailing in far more stringent new situations.

Of course, common sense is a good thing, and you will find distinguished scientists arguing "by the seat of their pants" or by feelings "in their bones," but you must not pay too much attention to this. Creativity is a very delicate thing, and if a poet needs the smell of rotten apples or a scientist a belief in rubbish in order to move away from a creative block, so be it: what in the last resort matters is the purity and strength of their analyses. That such analyses might bring further crises of entrenchment is to be expected, especially when we try to understand the microworld, which is of course well beyond our sensorial experience.

CODA

We expected in this chapter to witness a great scientific revolution, and what we have found is as if the National Convention had appointed Louis XVI as mayor of Bordeaux, instead of sending him to the guillotine. For Einstein, rather than ruling Newtonian theory out as defective and thus unusable, gave us the precise boundaries of error to be expected when classical mechanics is used, and by making this range of accuracy dependent on the velocity of the system, defined precisely the domain of application of classical theory. Thus, in the same way that the great Copernican revolution was most important as a theological rather than a scientific revolution, Einstein's magnificent relativistic "revolution" is, first and foremost a philosophical one, insofar as it exposed beyond reasonable doubt some previous philosophies as erroneous and unacceptable. Whereas Newton, in fact, is still in business after Einstein, Kant's philosophy of space can no longer be entertained and must be relegated to the pigeonholes of history.

As for scientific revolutions, so also for crucial experiments: their cruciality can be more in our contemporary minds than in their historical significance at the time when they were performed, as we have seen with the Michelson-Morley experiment.

What so far has remained on fairly solid ground, I hope, is the significance of entrenchment. Both Lorentz and Einstein did entrench such areas of the scientific mesh as each thought important. Lorentz failed because he forced the known facts to conform with the ether theory, whereas Einstein recognized the much wider import of Maxwell's equations. Einstein's entrenchment succeeded because he did not just preserve those equations piecemeal, as Lorentz had done with the ether, but because he did so by claiming a vast and so far uncharted area of the scientific mesh as his own. Not only did he finish his first paper on relativity with some equations of the motion of the electron which were new and testable, but he also, less than three months after this paper, published another one with the famous relation between energy and mass, two quantities until then invariant, which he instead showed could one be transformed into the other—a result of which Hiroshima is the most dramatic and unfortunate confirmation.

The success of Einstein can be contrasted with the failure of Lorentz's attempt to preserve the concept of the ether. That this was a poor scientific exercise, is clear from its ad hoc character, the contraction of the arm of Michelson and Morley's interferometer being a device calculated to avoid the falsification of the ether hypothesis. Popper is correct, in this respect, in denouncing as just about the ultimate sin in scientific practice the sheltering of any proposition from falsification, thus castigating Lorentz.[43] Remember, however, that Einstein deliberately chose not to falsify the Maxwell equations. Exposure to falsification is, on the other hand, automatic in any proposal that widens the scientific mesh and thus predicts results of proposed new tests, whereas the mere impossibility of falsification is not always regarded in scientific practice as an overwhelming defect. By its definition, the vacuum is a concept even less susceptible of falsification than Freud's unconscious, yet not only is it widely accepted, but is it a very useful model that has considerably extended our understanding of many natural processes.

Which brings me to the question of ontology. As a theory progresses, its ontology becomes more sophisticated: fields, for instance, started as a device to get round the question of action at a distance, in the form of some perturbation hanging about in some unspecified way at each point of space. Later, it was discovered that the perturbations in the field propagate with a finite speed, which makes them more realistic physically. In time, the nature of the until then unspecified perturbation became understood as some strain in spacetime, as the gravitational field is, creating the curvature of the latter. We then realized that fields interact with matter, and that in so doing are subject to the principle of action and reaction, which again adds strongly to their physical significance. Finally, the very forces that the fields were meant to manifest were understood as the exchange of virtual particles, like photons and gravitons. This is typical of how many entities are created in scientific theories, with hardly any claims as to their "real" existence at first, whereas they are later endowed with so many physical attributes that one begins to believe in their "true" existence. Of course, to believe that as a theory becomes more and more entrenched its ontology can be given more credence, entails all the difficulties associated with induction. That this is nevertheless extremely tempting cannot be better demonstrated than by the fact that Popper, the archenemy of induction, which he exorcised not only as illogical, but also as not ever used as such in science, went himself a notable distance along that way.[44] I take the line, however, that although the ontology of a theory must take second place to its facticity, as the latter is further entrenched, it also licences its ontology to be used—in the construction of models, for instance—with increased confidence.

Before we bid goodbye to Einstein's special relativity theory, notice that his work is another example of a virtuous circle. He started with the postulate of the constancy of the velocity of light, and, when he obtained his new mechanics, he derived a law of composition of velocities which indeed gives the velocity of light as a constant in all inertial systems, irrespective of their velocities. As you can see, the logic of physics is not very logical: at the moments that really count, what you must look for is some internal *consistency* of your assumptions, keeping at the same time an eye on their supreme test, *facticity*, a test which is entirely beyond the bounds of logic to provide. This unavoidable circularity can feed of course the dreams of conventionalism, but for this doctrine to have any value at all it would have to demonstrate, in this case, for example, not just that the same circular path followed by Einstein could be successfully completed with different "conventions," but also that the *explicit* derivation of the entirely new natural laws which would then be required to satisfy facticity can actually be produced. Anyone can dream up any number of alternative "conventions," but the fact that the corresponding alternative natural laws have never been formulated is sufficient to relegate the whole enterprise to the role of an idle pipe dream.

There is one aspect of time upon which we have not touched in this chapter, and it is the obvious asymmetry which makes me now half the man I was, with no hope of waking up tomorrow with the other half amazingly restored. This (not entirely unfortunate) state of affairs is what people call the *arrow of time*, which we shall discuss in the next chapter.

NOTES

1. He asserted this explicitly in a lecture he gave in 1921, when he claimed that relativity theory was the continuation of the work of Maxwell and Lorentz. (Einstein 1921.)

2. Many philosophers would of course reject this view, arguing along the lines which I have already illustrated in considering the relation between Newton's theory and Galileo's law of falling bodies (p. 191): because Newton's mechanics is not a logical consequence of relativistic mechanics, it is claimed to be falsified by the latter. This is an argument that I have already rejected, on the grounds that every theory has its domain of application and

range of accuracy, and that if a more accurate theory can lead to the results of a less accurate one in a given domain of application with a given range of accuracy, then it must be said to *confirm* the latter theory in that domain of application within the stated accuracy. This is precisely what happens: for a given desired range of accuracy, relativity theory agrees with Newtonian mechanics for a particular domain of velocities that can be specified. What relativity theory falsifies is *not* Newtonian theory but rather its *ontology*. (See p. 279.) Facticity, however, is what any scientific theory or natural law must primarily be judged by, and although the ontology behind a physical theory is of course important, it is more the concern of its philosophy than of its science. For Kuhn's views about the problem discussed here see chapter 10, note 34, p. 251.

3. Newton (1686), vol. 1, p. 7.

4. The reader interested in Newton's view about space and time should consult Thayer (1953). Zylbersztajn (1994) provides a readable account of the ideas of Newton, Leibniz, and Mach. See also Mach (1960), pp. 277, 284. A good introduction to Leibniz's ideas may be found in Mates (1986), but the most illuminating treatment remains his own in Alexander (1956).

5. The experimental situation has to be qualified. Ordinary light sources emit light in short outbursts (lasting fractions of microseconds), which generate not continuous waves but rather, so to speak, wave segments that are called *wave trains*. Thus, wave trains originating from two different sources would produce in fig. 4 waves that are randomly in or out of step, so that no interference would be observed. For this to be the case the two beams must be *coherent*, that is, they must keep a constant relation all the time. The simplest way to produce this is to have a single source and split the beam in two by some means, so that at any one time the waves reaching the screen come from the same wave train. Lasers emit always coherent light because the wave trains that they produce can be considered infinite in length for practical purposes.

6. Work by Charles Glover Barkla in 1904 also pointed out in that direction. Soon after the discovery, Laue became styled as Max von Laue. (See Whittaker 1953, p. 19.)

7. By the "original wavelength" I mean the wavelength that corresponds to the center of the band of continuously varying wavelengths that are added up.

8. See Whittaker (1951), p. 114. This work contains a very complete account of the history of the theory of light.

9. The hypothesis of an ether was first proposed by Robert Hooke (1635–1703) in his *Micrographia* (1667).

10. Just as you could not support induction on any logical argument, such a position cannot be logically sustained. As I have repeatedly said, to expect logic to tell us what reality is like is absurd: all it can do is control a little the way we think about it. I hope very much, however, that the ontological tolerance that I here suggest is not taken to be a licence for ontological debauchery. Ontological constructs that provide a model for a well-entrenched physical theory are one thing: the other are ontological inventions to force interpretations of reality to save, not just the facts, but preconceptions arising from psychological attachment to defunct theories.

11. A frame of reference must not be mistaken with a system of coordinates. I can have many different systems of coordinates to mark the position of a point on or near the surface of Earth but Earth is the single frame of reference. That is, a frame of reference must be a material body: even a single particle would do.

12. Einstein (1905*a*). References to this paper will always be from Sommerfeld (1923).

13. Einstein (1905*b*), p. 38.

14. See Pais (1982), p. 172. (But also p. 133.)

15. Since $\rho = 1 + (vu/c^2)$, for u equals c you get $\rho = 1 + (v/c) = (c+v)/c$, as stated.

16. Einstein (1905*b*), pp. 63–64.

17. On substituting kc for u in the equations for u' and ρ, we get

$$u' = (kc+v)/\rho,$$

with

$$\rho = 1 + (kv/c) = (kv+c)/c.$$

Therefore,

$$u' = c\,(kc+v)/(kv+c),$$

which for v negligible in comparison with c gives the value kc for u' as stated.

18. This definition is not circular, because the second appearance of "simultaneous" in it (as "simultaneously") pertains to events that *coincide in space* as well as time, for which the definition of simultaneity merely requires the equality of the readings of a local clock.

19. A diligent reader who might wish to treat the above experiment classically must be warned that the definition of simultaneity is far more complicated in this case. Because of the assumption in classical physics of a universal timescale, the observer on F', which will receive the right and left time signals at different times, must work back from the times he measures to the times at which the signals were produced: if you do this, which is simple but not trivial, you will find that the two signals were created simultaneously. In doing this you must take the velocity of the left signal as $c - v$ and that of the right one as $c + v$, where v is the velocity of F' with respect to F.

20. In the experiment referred to on p. 116 the time differences were very small, of no practical interest for us sublunar people. Relativity, however, is rapidly becoming part of everyday life as, for instance, in the computations behind the use of the Global Positioning System (GPS).

21. That fig. 9 entails a contraction if the rod is moving to the right is clear because the origin 0 of the rod will also move to the right, and thus catch the return signal earlier: t will be smaller and thus x shorter. The names of Lorentz and FitzGerald are associated with this phenomenon because, as we shall see on p. 286, it was they who introduced it heuristically before Einstein, in order to save the existence of the ether in the Michelson and Morley experiment.

22. See Einstein et al. (1935).

23. This definition will have to be somewhat modified when we discuss quantum mechanics.

24. By this I mean that it is necessary but not sufficient.

25. The square of a negative number is always positive: thus, the square root of 4 can be either +2 or –2. Since, as lengths are in classical physics, the spacetime interval must be unique, the convention is used that the square root be always taken as positive.

26. This does not mean that one necessarily must *believe* in it, but rather that one is increasingly more confident in using such emerging ontology in order, for instance, to construct new models that expand the facticity of the scientific mesh.

27. The possibility of an ontology to be considered as approximate agrees with the point of view expressed in the last note, to the effect that what really matters is the use to which you are prepared to put any given ontology.

28. This is a point that requires elaboration. If O is *here and now* and A is in the *future*, a subluminal signal can reach A from O whence, with even more reason, a light signal could arrive at A *ahead* of the time pertaining to A, so that the latter event is *subsequent* in time to O. Because the time that the light has taken in the abovementioned path is invariant for all observers in all frames, A must be subsequent to O in all frames.

29. The whole of our discussion on causality depends on the condition that the effect can never precede the cause which, surprisingly is not accepted by all philosophers, the pio-

neering paper to this effect being Dummett (1954). (See also a discussion in Horwich 1987.) This, however, is not a possibility that need be entertained in physics, unless some fancy time-travel effects, which are at present highly speculative, are contemplated (Deutsch et al. 1994).

Many readers might find it useful to reflect upon some interesting properties of the spacetime interval S. Remember that $S^2 = c^2t^2 - x^2$ so that S must be obtained by forming the *positive* square root of the right-hand side here. (See note 25.) On the other hand, whereas for $c^2t^2 - x^2$ positive it is possible to find a real number such that its square be equal to $c^2t^2 - x^2$, if this quantity is negative, call it $-a$, such a real number does not exist, since the square of all real numbers, whether they are positive or negative, must always give a positive number. The way out of the difficulty is to invent an *imaginary unit*, called i, so that $\sqrt{(-a)}$ is defined as i$\sqrt{a}$. This must be known to many readers, but even those with some familiarity with imaginary numbers tend to worry about their meaning, for which the name *imaginary* does not help at all. The significance of imaginary numbers here is nevertheless very clear: they act as a *marker* that, when you have to deal with quantities that are as different as chalk and cheese, will allow you to keep them apart and safe, as well as to instantly recognize which is which. A splendid example is afforded in the spacetime diagram. Looking back at fig. 11, it is very easy to see that $S^2 = t^2 - x^2$ (remember that we are taking c as unity) is positive in the gray areas and negative *elsewhere*. Thus, the spacetime interval is *real* in the gray areas (*future* and *past*) and *imaginary elsewhere*. This alerts us immediately to the fact that in the gray areas (S real) time ordering is significant, whereas this is not so *elsewhere* (S imaginary). Thus, when using the spacetime interval, there is never a danger of mixing these two fundamentally different cases thanks to the presence of the imaginary marker i. Another case where the same role is played by the imaginary unit is in distinguishing the mass of subliminal particles (which is real) from that of superluminal particles (which can be shown to be imaginary). More about imaginaries in chapter 14.

30. The equations of general relativity have nevertheless solutions corresponding to empty spacetime, the *vacuum* equations, but it is not at all clear that they correspond to anything that could be described as pure spacetime with nothing added to it. In particular, it does not appear to be possible to recognize as distinct two "points" of such a space, because the very concept of a *metric*, which is a measure of the "distance" between two nearby "points," fails. (See d'Inverno 1992, p. 178.)

31. The very long preparation for these experiments had to be done when World War I was still raging, and Eddington was discharged from military duties for this purpose, although his proposal was that of confirming the work of an enemy alien.

32. Poincaré (1905), p. 50.

33. Reichenbach (1957), p. 127.

34. Even in the hypothetical Maxwell-free world history suggested, someone at some time would have had to establish some laws of electromagnetism, and there is no reason whatsoever to assume that this task could have been possible without a change of the deutero-Einstein's equations. Thus, it cannot even be argued convincingly that such conventions as are used in science are merely a result of its history.

35. See Grübaum (1973), p. 682, and van Fraassen (1985), p. 155.

36. Einstein (1952), p. 23. This question is fully treated in Angel (1980), chap. 5.

37. See note 14 and also Holton (1969), who adduces evidence against the alleged crucial character of the Michelson-Morley experiment. A preliminary discussion of this question was given in chapter 10.

38. A vigorous and well-documented attack on the alleged role played by crucial experiments in scientific progress is given in chapter 10 of Lakatos (1978).

39. That the orbital motion of Earth with respect to the ether should *reduce* the velocity of light can be seen as follows. Light is moving along the ether with velocity c, but the mirror in the Michelson-Morley experiment toward which it is directed is moving *away*

from the light with a velocity of 30 km/h. Thus the effective velocity of the light with respect to the mirror is c *minus* 30 km/h. (Reflect that if the mirror were moving away with velocity c, then light would never catch up with it, that is its velocity would be zero. Remember that this argument is not valid in relativity theory.)

40. This is merely an application of the principle of sufficient reason or, what is effectively the same, the principle of symmetry as enunciated in chapter 8. An asymmetry in the effect, in this case the cone falling from its perfectly symmetrical configuration, must pre-exist in the cause. So, although the antecedent configuration of the cone appears to be perfectly symmetrical, there must be somewhere a hidden asymmetry that must be entailed by the postulated perturbation.

41. Think of a man on a train throwing tennis balls forward: the velocity of the balls with respect to the track is that of the train plus the velocity of the ball with respect to its source (the thrower). If instead, as the train moves, the man releases a mechanism on the side of the track that activates a tennis-ball-serving machine, then the velocity at which the balls are emitted from the machine is obviously independent from the velocity of the train.

42. A very simple account of modern field theory may be found in Ridley (1995). If you want to extend the parable of the skaters to two attracting particles, imagine that the skaters throw out the balls boomerang-like in the opposite direction to that of the other partner. This concept of a particle exchange being responsible for the appearance of forces in a field is quite general. In the gravitational field, for instance, it is believed that masses exchange particles called *gravitons*, and that it is this exchange that manifests itself as the gravitational force.

43. Popper (1959), p. 83.

44. In Popper (1974), vol. 2, p. 119 he accepted as a "whiff of inductivism" that "reality, though unknown, is in some respects similar to what science tells us or, in other words, with the assumption that science can progress toward greater verisimilitude."

TIME'S ARROW

Quisiera
que mi envejecer no fuera
envejecer de tiempo.

Would
that in aging I could
age without time.

Abel Martín (1940), p. 27.

'I saw this TV thing—a second law
Of thermo something.' A duet: 'Dynamics.'
A solo: 'Entropy.' The girl's dropped jaw
Went 'Jesus.' Brian went on: 'Smashed ceramics,
Spilt milk, bombed cities all add one bit more
To universal breakdown's jar of jam. Ex-
Cept in films you can't wind back the action.
Addition's never fashioned from subtraction.'

Anthony Burgess (1995), p. 68.

The beautifully graphical expression *time's arrow* was coined by the famous British astronomer and mathematical physicist Sir Arthur Stanley Eddington (1882–1944): "I shall use the phrase 'time's arrow' to express this one-way property of time which has no analogue in space."[1] This chapter, however, should more properly be entitled *time's arrows*, because several arrows of time are usually discussed, such as thermodynamical, cosmological, and psychological. This list can even be extended, but I am not aware of anyone discussing the *linguistic arrow of time,* and I should like to spend a little time doing that.[2] To the natural philosopher, language provides a record of our phylogenetic development every bit as important as archaeological evidence is for the historian. This is so because we know that language has evolved through the interaction between humankind's thoughts and observations; and that the very facts that we observe acquire their full human significance by virtue of the specific language used in attempting to describe them.

LANGUAGE AND TIME

Consider a few interesting sentences: "The stone falls from the top of the tower of Pisa," "Wood burns in the stove," "John beholds Christine," and "John loves Christine." Without getting into too much detail, the nouns in these sentences (stone, wood, John, Christine) all entail some concept of permanence or invariance, whereas it is the verb (fall, burn, behold, love) that carries what we recognize as a flow of time.[3] But we could not recognize that flow if it were not for the linguistic convention that the noun expresses some sort of invariance. It could have been possible to have a language without verbs but in which nouns are inflected with suffixes as, for instance, in *stonevone, stonevtwo, stonevthree*, and so on, used to describe the *different objects* perceived in different positions along the Pisa tower, the *v* in the suffix indicating the "set" of objects as vertically aligned.[4] But this is not what we do: rather than "perceiving" a multiplicity of objects in different states, we imagine the object to be invariant (although this is sometimes manifestly untrue, as I shall show) and allow it to sail through a series of states, denoted in more or less detail by the verb.

Once we realize that the flow of time both originates in, and is reflected by, our language structure, we must understand that for this to be the basis of an understandable language, we must use a considerable amount of *coarse graining*. This concept, with its companion one of *fine graining*, has already been discussed in chapter 9; as we consider more examples, these concepts will be endowed with further technical requirements in specific cases. For the time being, it is enough to accept that coarse graining entails ignoring a certain amount of detail in our cognitive field. Thus, when I say "wood burns in the stove," the noun "wood" is assumed to have a single referent, whereas the object that is actually burning in the stove changes drastically along the process, and it can hardly be said to be the same object throughout. But even the stone falling in Pisa must have shed not a few atoms during its fall, although coarse graining here reflects pretty closely the underlying fine-grained structure. You must not think, however, that coarse and fine graining apply only to nouns: the verbs also have their form of these modes, and a great deal of misunderstanding about time arises from ignoring this fact.

When I say "John beholds Christine,"[5] I imply that this happens "now," or "this instant," but I have already argued in chapter 9 that these expressions stand in relation to time as "this mathematical point" does with respect to space. That is, just as there is nothing in the world that corresponds exactly to a mathematical point, there is also nothing in the world that I can expect to correspond to "now" or to "instant": unless you want to make mischief, these expressions must automatically be understood within some degree of *graining*. You can choose the graining to be as fine as you want, but even the finest graining cannot correspond to the dimensionless "instant" in time, just as the finest gauge cannot pinpoint a single mathematical point. At the other end of the scale, it is useful to remember that some forms of verb usage necessarily entail very coarse time-graining. As a difference with "beholds," "loves" in "John loves Christine," if it does not imply the sugary eternity of the poetaster, it certainly conjures up a decent interval of time.

Interesting as this concept of love might be, it must be displaced in our study by that of the symmetries of the timescale.

TIME'S SYMMETRIES IN MECHANICS

There is a symmetry property of the timescale that we have already discussed in chapter 5, p. 125, and it is that of the invariance of all causal statements (and thus of laws of nature) whenever the time t which appears in them is replaced by $t+T$, an operation that is called *time translation*. Let me discuss this property for Newton's second law, and in doing so you must remember that whereas t is a *variable*, which can assume a variety of values, T is a constant (say 10 minutes). Consider now Newton's equation:

$$f = mass\ times\ the\ rate\ of\ change\ of\ v,$$

where the velocity v is

$$v = rate\ of\ change\ of\ position\ x.$$

Remember that, as we have done in chapter 6, the rate of change of a quantity is the amount it changes during a given interval in time, divided by the time interval, and extrapolated for a very small (infinitesimal) time interval. It should not be too difficult therefore to accept that, whereas the rates of change might have different numerical values if I change t by $t+T$, they must still be rates of change with respect to t (since T does not change at all.) It thus follows that even if f and v have different numerical values in the first equation (Newton's equation), the *form* of the equation is unaltered. (Causality would otherwise break down, and because of

this, this invariance property under time translation is valid for all physical laws, as it can be proved case by case.)

All this should be, if not crystal clear at least reasonable, I hope, but
proceeding in very much the same way, we shall now find a very strange
fact. Whatever the origin of this belief be, most people accept that the past
and the future cannot be exchanged. This is basically what is called the
psychological arrow of time, which requires that $+t$ (future) cannot be
exchanged with $-t$ (past). Newton's law, however, does not discriminate
between future and past in this way. In order to understand why this is
so, let me remind you that when you change the sign of the time, rates of
change must also change sign. (If a water tank *increases* its weight at the
constant rate of one ton per hour, then its change of weight in 1 h from 11
A.M. to 10 A.M. is *minus* 1 ton.) Having agreed to this, let us return to the
two displayed equations. When you change t into $-t$, v in the second
equation changes sign,[6] which introduces one change of sign in the first
(Newton's) equation. There is a second change of sign in the latter equation, however, due to the sign inversion of the rate of change in it: the
whole equation is therefore *invariant* when the sign of the time is
changed, that is it *is symmetrical with respect to time inversion*.

It can be proved in the same way that all the equations of mechanics
and for that matter all the fundamental laws of physics have the same
property (although this might break down for some elementary particles,
like neutral kaons). This is an amazingly counterintuitive result, but it is
one that most people have experienced, and if you have not, try the following simple experiment. Get a pendulum with a *very smooth suspension*
and place it where the air is extremely still. Even better (if you do not
mind a little suffocation), remove all the air around. Now take a camcorder, start the pendulum oscillating, and record the motion for a *few*
seconds. Once you have done this, run the tape the wrong way around,
that is, inverting the time: you will not notice any difference at all: past
and future are exchangeable! Which, after all, is precisely what the symmetry of Newton's equations with respect to time inversion entails. (I
shall come back to this in a moment, and you will understand why I have
italicized some words in the above account.)

THE TRUTH ABOUT LIFE: FRICTION

Many people, philosophers included, tend to be somewhat overawed by
the power and beauty of mechanics. And when you think that scientists are
able to aim one satellite at another hundreds of thousands of miles away,
and even dock them when moving at 50000 km/h, you might believe that

Newton's equations, a little bit extended, should be able to tackle every-thing in heaven and on earth. If you believe this, I urge you to look at Abraham et al. (1978). This is a superb treatment of mechanics, where the full panoply of the mathematical methods developed in the second half of the twentieth century is beautifully explained and applied. It is eight hun-dred and six pages long, and its last chapter treats the most complex system for which classical mechanics can be used in an exact manner. This system is one where *only three bodies* are involved. Even then, the motion considered must be restricted: it must be assumed that the plane deter-mined by these three bodies is constant in time; and that they are isolated from other bodies. This is not to depreciate either this splendid book, or the science of mechanics, but to remind you that the latter has profound limi-tations, and that computations of orbits and the like, which can be amaz-ingly accurate, entail numerical rather than exact solutions of the problem, for which apparently very rough approximations must often be used.

Look again at the pendulum example of reversibility that I gave you, and you will realize that I was lying through my teeth. Remember that I mentioned a *smooth suspension*. However smooth the suspension, if you look at the pendulum long enough (not just for a *few* seconds as I had suggested), you will notice that the amplitude of the motion (that is, the maximum displacement of the bob from its equilibrium position), which appeared at first to be constant, begins to *decrease*: here you have your time's arrow, because you would be able to recognize the inverted tape in your camcorder, since the amplitudes in it will appear to *increase*.

We can now face the facts head-on: of course, the equations of mechanics do not apply to the real world but only to a model, and one must understand how to use this model if one wants to be able to com-pare with experience. The first restriction that the model entails is that the bodies to which it is applied (no more than three as explained above) must be *isolated*, which can most adequately be achieved by eliminating all other objects in the universe. If you are not so destructively inclined, you must then modify your Newton's equation a little, to make allowance for the disobliging nature of reality; and what you do is to introduce *friction*. Now, for a physicist to ignore friction is like for a the-ologian to deny sin (as in fact Pelagius, not too successfully, did). So, we now accept that the suspension of our pendulum, however well oiled, entails friction. Let us see how we include friction in Newton's equation, and what it does to it. Let me write again that equation:

$$f = mass \text{ } times \text{ } the \text{ } rate \text{ } of \text{ } change \text{ } of \text{ } v.$$

If the force f were constant, the right-hand side would also be con-stant, whereas we know that the pendulum's bob must eventually come to rest, thus causing the right-hand side to vanish. So, the existence of friction must affect the left-hand side here, which means that I must add to it a *frictional force*. I shall first quote what the physicists do, and you will then see that it is sensible. We change the above equation as follows:

$$f + kv = \text{mass times the rate of change of } v,$$

where k is some constant (which can be left to be determined later from experiment) and kv is the frictional force. That this force must depend on the velocity is sensible, since we know that if the pendulum is still ($v = 0$), it vanishes. Naturally, what physicists do is to verify that, with a suitable choice of k, the above equation agrees reasonably well with experiment, which is just about all we can say to justify its use.

The important property is that the new equation is no longer invariant under time inversion: the right-hand side, as in the last section, does not change sign under time inversion; but this is not so for the left-hand side. We have seen in that section that all rates of change change sign when the time is inverted. The velocity on the left of our last equation, therefore, being a rate of change, introduces a negative sign on inversion, as stated. The "true" equation that we must use, that is the one that includes friction, is *not invariant* under time inversion, which is what gives us the *time asymmetry* that we observe and thus the arrow of time.

All I have done in this section is this. I have noticed that the model entailed by the original form of classical mechanics was too crude, and I improved it by creating a new model, including friction. This model allows us to reach better agreement with experiment, and it is asymmetrical under time inversion. So, if you want, you may think that the origin of time's arrow is the unavoidable existence of friction. You may well think that, but there is a lot more to it, as I shall now try to begin to unfold.

ORDER AND THE ARROW OF TIME

We have seen that the motion of a real pendulum is irreversible, and, in order to understand what happens when you have an irreversible phenomenon, let us have a look at the lottery type of device that I illustrate in figure 1.

The working of this lottery machine is clear. You start with the nine balls in the correct numerical order, you turn the handle several times, the balls bounce about in the spherical chamber, and when you stop an entirely new order emerges. Why should this happen? If you believe in the ultimate applicability of classical mechanics, which is symmetrical with respect to time inversion, you can argue as follows. Assume that you had an incredible machine in which you can store in a computer every

Fig.1. An irreversible process: a lottery machine

position and every velocity of all the minutest particles of the machine for every infinitesimal instant of time.[7] Once you have operated the machine and got a given new order of the balls in your tube, then you run the computer the wrong way around and, since everything in the machine is mechanical, and since mechanics is reversible, you will then go back to the correct order 1 to 9 that you started from. And pigs will fly.

Let us look at a few numbers. The number of different ways in which you can order the digits from 1 to 9 is 362880. If it takes you 10 seconds to reset the machine in each of these permutations, it will take you about 42 days to run through all of them but, most importantly, this number grows enormously with the number of balls, 42 days for 9 balls; shall we say 10 years for 14? You may think that that is a plausible guess, but the correct figure is 27600 years! This will give you some idea of the unfathomably large number of permutations you can have, even with a fairly small number of objects. Thus, the chances of hitting precisely one out of that huge number of permutations (for *all* the particles of the machine!) is unbelievably small.

There are two ways in which you can persuade yourself that the idea is well-nigh absurd that classical mechanics could, in principle, even in philosophical principle, lead you in this case to reversibility. The first is a question of errors in your data. However good your computer, and however good the machines that you use in order to measure all the positions and all the velocities of all the particles in order to invert them, even the most minute error in each entry, when propagated through billions of events, is bound to introduce unexpected changes in the sequence of balls resulting. We have seen that there are billions upon billions of permutations possible: the probability that the errors involved be so perfectly compensated as to go back to the original order is nil.[8]

The second way to look at the problem is by considering friction. For the previous work with my miraculous computer to make any sense, I had to assume that the whole machine was completely isolated, that I could measure every minute particle of it, and that I could then apply classical mechanics ab initio. Assume now that I humanize my model a little, ignoring (coarse-graining) all the particles in the glass tube and sphere, thus concentrating only on the motion of the numbered balls. But, because the latter *are no longer isolated*, I will have to include friction and friction is not time-symmetrical.

One way or the other, there is a very important conclusion that the above work leads to, which is that as time goes by *order decays*. You must not be confused by this slogan. After all, give it a few decades and even Gaudi's magnificent *Sagrada Familia* will be finished, with a considerable increase in order from that of its raw materials: but, meanwhile, the well-ordered food that the workers ate will have gone down the sewers, many of the workers themselves into their tombs, and the fuel used will have decayed into ashes and innumerable pollutants. Thus local order in a small area of Barcelona will have increased, but *global order will have decayed*. (This latter statement is the result of the so-called entropy principle which I shall discuss a little later.)

I must make it clear that I do not claim to have proved in the above that order *must* decay, a slogan which of course would give us the arrow of time. What I hope to have shown is that, even taking into account that classical mechanics is time-reversible, it is most implausible that time-symmetry would hold when dealing with macroscopic systems.

MORE ABOUT FRICTION

Even I, when I reread the above pages, am prone in my weaker moments to think that I have provided a jolly good explanation of time asymmetry. But the reader must have realized by now that I have a healthy skepticism about explanations: my treatment is in fact a calculated example of how easy it is to deceive yourself into thinking that you have "explained" something. For many purposes, and didactically they are splendid purposes, explanation is no more than excavating into our minds until we find some layer of knowledge about which we are all comfortable, and upon which we can build foundations that reach out to the explanandum. This may be good didactics, but it can be bad philosophy because it can prevent us from seeing that all that we have done is to reach a point where what we wanted to explain is not explained at all, but merely bypassed. This is precisely what I have done: I have invoked friction as responsible for the arrow of time, but it is quite possible that friction itself is no more than a manifestation of that very arrow.

That this might well be the case should be clear, because my so-called explanation has so far been *phenomenological*: I have introduced some terms in my equations to regain facticity which, however intuitive they might be, are not given a precise physical meaning. What the inclusion of frictional terms in our problem entails, in fact, is nothing more than coarse graining of the model: a great many interactions between particles have been subsumed in a single and simple frictional term which, with the use of the frictional constant k, can be empirically adjusted in order to fit the facts.

I have to confess sadly, therefore, that I have not explained time asymmetry at all because the phenomenological correction introduced in classical mechanics, being calculated to recover facticity, had necessarily to fit the fact that there exists an undoubted asymmetry in the time. So, we have to look around to obtain a wider point of view that can accommodate both time's arrow and friction within the same framework.

HEAT, THERMODYNAMICS, AND DISORDER

We can get a hint as to where to start by going back to our pendulum. We agreed that there was friction at the suspension, and if we were to measure the temperature at its axle, we would find that after some time it increases, which means that the pendulum has generated some heat. This is very reasonable, because the pendulum started with a given quantity of energy, *kinetic* from the motion of the bob, but also *potential*, which is the energy that bodies, for instance, can transform into work when going down the gravitational field of the earth, like water falling down from a high level through a turbine. Clearly, part of the energy that the pendulum had at the start is lost because the pendulum slows down (less kinetic energy) and also because it reaches gradually smaller heights (less potential energy). It is reasonable to assume that this lost energy must correspond to the heat that through friction passes to the suspension. This idea is the famous *first principle of thermodynamics*, which merely says that, when mechanical energy is transformed into heat, the total energy is conserved. Such a proposition is, of course, nothing else than a *principle of conservation of energy*.

How is it that mechanical energy can be exchanged into heat? This is a question that enraged purists in the nineteenth century because they felt that energy was a very good concept to use in physics, since you could measure it and accordingly write down perfect energy balances. This, they argued, was all that could be observed and all that we should talk about. Anything else was like the uncivilized behavior of a bank customer who, having obtained his balance at the counter, insists that he wants to go down to the vault in order to count the actual bank notes. Ernest Mach, to whose work we have often referred, fought a marvelous battle for purity;

and although he was right, he did not win. He was right because a phenomenological theory, as he supported, is a theory calculated purely to save facticity, and after all saving facticity is the first obligation of a physical theory. It is general experience, however, that unless science goes beyond bare phenomenology, vast spheres of knowledge can in the end remain ignored, which was the great failure of Mach's logic.

The non-Machians, of course, wanted to formulate some theory that would go deeper into the nature of heat, and the only sensible way they found to do this was by using the idea that matter is made up of atoms, which were still entirely unobservable in the nineteenth century. Mach insisted, therefore, that they were metaphysical concepts not to be used. History, alas, proved that if science had followed Mach's diktat, it would have remained stuck in a blind alley.

Once the atomic structure of matter was postulated, it was easy to understand what heat meant. Consider a copper crystal; when it is very very cold, the atoms in it are all beautifully ordered in a cubic array, but as the crystal heats up, the structure becomes a bit disordered, since the atoms vibrate around the positions they occupy at very low temperatures: the more you heat the crystal, the more they vibrate. And we can now be daring and postulate that the temperature of a crystal, or for that matter that of a gas or any material body, depends on the average of the mechanical energy of all its moving particles.

You can now understand how friction works. As the pendulum moves, its suspension rubs against the axle, and the motion of the suspension is gradually transmitted to that of the atoms of the axle. Every time that the suspension excites an atom of the axle into motion by hitting it, energy conservation requires that another atom in the suspension must lose kinetic energy, and so must the pendulum itself. This will be a very small amount, since the kinetic energy that one atom can exchange is very small. But, as time goes by, more and more atoms of the axle get excited, and the quantity of kinetic energy lost by the pendulum reaches macroscopic levels.

You now realize that the focus of our "explanation" of time asymmetry must change. If we agree that classical mechanics, which is time-symmetrical, cannot lead to the arrow of time, it is not sufficient to invoke friction as the reason for time-asymmetry, since friction is itself a phenomenon in which *mechanical* energy is transferred from one body to the other; so that we are roughly where we were right at the beginning. (Keep in mind what I have said here, because I shall return to this later with some rather surprising ideas. This is my *first clue*, but all will be revealed in due course à la Hercule Poirot.)

The next step, therefore, is to try to understand the rules that govern such energy transfers as are entailed in friction through the propagation of heat from one body to another. The science of thermodynamics obtained such rules in the last century in a purely phenomenological way based on empirical considerations. The final rule, which was discovered in 1850 by Rudolf Julius Emmanuel Clausius (1822–1888, born as Rudolf Gottlieb), was enshrined under the name of the *second principle of thermodynamics*. This is one of the most influential results of the nineteenth century and, as we shall see, it can be enunciated in three different ways, depending on the quantities that one is focusing on. The first enunciation, based on the consideration of the behavior of heat, is:

Heat can never spontaneously flow from a cold to a hot body.

This is an extremely sad constraint on humanity: if we could reduce the temperature of the Atlantic ocean by a fraction of a degree and transfer that heat to the high-temperature boilers of power stations, then the whole world would get practically free heat and light.[9] There is of course no shortage of human benefactors who conjure up schemes to perform such feats, and not all of them are in mental homes. I hardly know a scientist who has not received such proposals from people in the community.

Some of my readers will feel that I am a little unfair to those who try to falsify the second principle; after all, I myself have said that laws come and go, and if the second principle would go, it would be all to the good. The reason why scientists have little patience for such dreams is that the second principle is one of the best-entrenched laws in physics. Should it fail, the whole of the scientific mesh would be torn asunder, such is the range of applications it has, and such the number of well-tested predictions that it has provided. So, at least within the present state of knowledge, it is one of the laws most unlikely to be breached.

Let us see what the second principle has to say as regards the transmission of energy from one crystal to another: it says that if I put a copper crystal at low temperature—in which therefore the atoms are all well ordered—against a hot copper crystal, in which the atoms are disordered, what will happen is that heat will flow from the hot to the cold crystal. The atomistic view of this situation is very simple: the excited atoms of the hot surface hit those of the cold surface, which then get excited and in their turn excite the internal layers of the cold crystal. Eventually, the atoms of the hot crystal will lose some of their energy and thus move less, the crystal itself becoming more ordered. The cold crystal, instead, will become more disordered. Thus, what the second principle is telling us is that, in a natural process of heat flow, the flow of heat is accompanied by a flow of disorder. We shall see, in fact, that the relation between heat and disorder is even deeper, and that the second principle of thermodynamics can be understood as legislating about disorder. So, we shall come back to the observation we made about the lottery machine of figure 1, namely that in a natural process order decays.

ENTROPY AND THE MAXWELL GAS MODEL

Although I have so far talked of the conduction of heat in solids, the first important advances in thermodynamics were made in the study of gases, much simpler to understand than crystals because the particles of which gases are made up—mostly molecules, although in some cases like the inert gases they are just atoms—are sufficiently separated so that their interactions do not play a crucial part in their properties. We have already seen in chapter 6 how the state of a gas can be defined in terms of its pressure P, its volume V, and its temperature T. As you remember, once you know how to describe a state, it is useful to define *state functions* such that they are fully determined whenever a state is given. Naturally, people were interested in trying to get a state function associated with the quantity of heat a system receives in order to reach a certain state. This, alas, was not possible: the same values of P, V, and T could be reached either by heating the gas, that is, by giving it some quantity of heat, or by subjecting it to *friction*. This can be done, for instance, by moving a piston up and down, generating heat (as you notice when you pump a tire), in which case the quantity of heat received is nil, because, although the temperature is raised, the only energy received is mechanical, not thermal.[10] Yet, quantities of heat are technologically and economically very important: we pay for them.

One of the great discoveries of the nineteenth century, indeed, was how to construct a state function related to the quantity of heat transacted. This is a story I cannot go into, since it requires some delicate mathematics, but it was found that if a system received a certain quantity of heat *at a constant temperature* in certain conditions, then that quantity of heat, divided by the temperature, was a state function, which Clausius in 1865 called the *entropy*. Before I go much further, it is useful to understand what state functions do for us. If I know the state of a gas, I do not know the quantity of heat it has received: I need also to know its *history*, that is, whether the state was reached by the use of friction or not, and how this was done. This is the case because the quantity of heat is not a state function. A state function, instead, allows us to dispense with history altogether: if I know the state of a gas, I know its entropy, full stop.

The definition of entropy allowed Clausius to propose the second enunciation of the second principle:

In a spontaneous physical process entropy can never diminish.

This enunciation, of course, can be proved to result from the first one that I have given, but I am not going to discuss it very much, first, because the definition of entropy which I have so far used is one that is not so easy to grasp and, second, because it will soon be replaced by another one that has a more intuitive physical meaning, and which will lead to the third, and thank goodness last, form of the second principle.

The next stage in our understanding of gases and thermodynamics came through the *kinetic theory of gases*, the pioneering work for which was done already in the eighteenth century by Daniel Bernouilli (1700– 1782), who already described gases as elastic molecules colliding with each other. This work remained little known until revived by Rudolf Clausius in 1857, at a time when atoms and molecules were far from respectable.[11] It was James Clerk Maxwell (1831–1879) who established in 1860 the basis of modern kinetic theory by producing a model of an *ideal gas*: this is one in which the molecules are sufficiently separated, so that their interactions may be neglected. (At a later stage, of course, they are considered with increasing degrees of sophistication.) Because of this lack of interaction, the molecules of Maxwell's gas move freely, colliding *at random* one with the other. The result of these random collisions is to create a distribution of the velocity of the particles: this is the important feature, which means that, at a given temperature, not all the molecules have the same velocity, but that the gas is a *mixture* of molecules with different velocities. The proportion of the molecules of the mixture that have a given value of the velocity may be considered as the *probability* of this velocity value.

The reader should look again at the last half dozen lines or so because something exceedingly important has happened. For the first time in physics, a model was proposed that invokes *random motion*, which thus led Maxwell to use the concept of *probability*, until then the preserve of gamblers. We thus begin a story that will pervade the whole of our modern understanding of nature. I shall discuss probability in more detail in the next chapter, but for our present purposes I shall ignore all worries about it and assume that everybody has some idea of what words like *random* and *probability* mean. If you visualize a *random process* as one that is haphazard or rule free, it is sufficient for the time being; and we can leave headaches until the next chapter.

As I have said, Maxwell's model describes the gas as a mixture of molecules moving with different velocities. This mixture may be described by a set of velocities, each of them with its own probability, the so-called probability in effect being nothing else than the ratio of the number of molecules with a given velocity to the total number of molecules. As the temperature of the gas is raised, the probabilities of the higher velocities increase, which means that the number of molecules in the mixture with these higher velocities is also increased. The result is that the *average* (or *mean*) velocity of the gas increases with the temperature. The beauty of Maxwell's model was that such quantities as the pressure of the gas were given a transparent physical expression, and that the elementary laws of gases were well reproduced by the theory. The ill-fated Austrian Ludwig Boltzmann (1844–1906), whom we have already met, refined Maxwell's work in the 1870s, and the velocity distribution for molecules in this gas model is called the Maxwell–Boltzmann distribution. Boltzmann's fundamental contribution, however, as we shall show, was his discovery that entropy and disorder are both manifestations of the same thing, which will bring us back to the arrow of time. (His important work on heat capacities, which was significant in getting his ideas accepted, was already reviewed in chapter 10.)

A classical example of irreversibility in a natural process, that is, of the effect of the arrow of time, is the milky cup of tea: no amount of stirring and waiting will return the cup to the two original separate layers of milk and tea; and it was Boltzmann who first came with a view about this problem. His theory depends on the relation between the states of the individual molecules of a system and the state of the system as a whole. This was, of course, daring at the time, when the very existence of molecules and atoms was in a scientific limbo, but it also entails a deep philosophical problem. For, in order to understand the relation between the states of the individual molecules and those of the system as a whole, we must be able to identify or distinguish those molecules, for which purposes we must put *names* on them or, what is the same, *number* them. And we cannot do this by stamping numbers on them as we did with the lottery balls in figure 1.

So, the reader will have to bear with me through what might appear to be a pedantic detour before we embark on our main work, but if one wants to have *clear* explanations, the worst thing one can do is to try to cover up *essential* details under the lazy pretext of "simplicity." And you will see later that the work we shall start here will have amazing consequences when we deal with the physics of elementary particles, where counting particles as we are now going to do cannot be taken for granted. Thus, a little worry at this stage will help in the future.[12]

IDENTITY AND NAMING

The practical problem that we face is how to *name* the particles of a gas or liquid, which can be held to be *identical* (about which, more later). The first question is what we mean by a *name* in this context, and I shall simplify things greatly because for our purposes *naming* and *counting* may be considered as the same activity. If I have any aggregate, and I can map the members of the aggregate to the integers 1, 2, 3, . . . , I shall take those integers as the *names* of the elements of the aggregate. The fact that, having shuffled the aggregate, I could then obtain a second mapping and thus a second group of names, is unimportant: what is essential is that, at least during the period in which I must deal with the elements of the aggregate, I must always be able to identify uniquely each element by its corresponding number.[13]

The substantive question that must now be answered is how it might be possible to establish the mapping that I have first described, which is not trivial. This work is not made easier by slight ambiguities that arise in the literature when comparing particles variously described as *identical* or *indiscernible* or *indistinguishable*, a problem that goes back to the great days of the medieval schoolmen. In order not to get into a scholastic bog, I shall distinguish the properties of particles as *intrinsic* or *extrinsic*, intrinsic properties being those that the particle possesses irrespective of the physical state in which it may be found, and extrinsic, those that are dependent on its state.[14] The so-called rest mass of a particle, which is the mass it has in a frame of reference with respect to which it is at rest, is an intrinsic property, as it is the type and geometrical dispositions of the atoms within a molecule, or the charge of an electron. Examples of extrinsic properties, instead, are the kinetic energy of a particle, and also, any *relational* properties, such as *topographical properties* whereby the relative positions of particles are specified.[15]

I shall call two particles *identical* when they have the same intrinsic properties. Two particles will be *discernible* or *distinguishable* (both expressions being synonymous for my purposes) either by virtue of not being identical, in which case they differ in intrinsic properties, or by virtue of having different extrinsic properties, despite being otherwise identical.

The important question to understand Boltzmann's ideas is how it can come about that identical particles may be distinguishable, which is necessary in order to assign *names* to them; an operation which for our purposes is the same as counting them in the sense previously described. The answer is very simple in classical mechanics, which after all is the foundation of Boltzmann's *statistical mechanics*.

If you look at the two identical particles considered in figure 2, they have distinct initial conditions (black circles) and these initial conditions uniquely determine (from Newton's second law) the classical trajectories of these particles. Since these trajectories are continuous,[16] each particle may be uniquely identified, and thus named, by following it continuously during the trajectory. The particles are therefore named by means of their initial positions, and it is they and the particle trajectories which

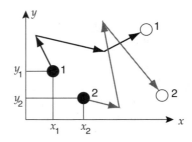

Fig. 2. Naming identical particles

Given the initial topography, the trajectories determine the particles uniquely.

are the extrinsic properties that make the identical particles distinguishable and thus denumerable. Because the names depend on the initial condition, they may be regarded as *topographical names*. You must remember that once you stick labels on the particles in this way it is no good to change your mind: particles 1 and 2 could have their labels exchanged in figure 2, but this has no more physical significance than, say, translating their numbering into French.[17] We can now safely go back to Boltzmann.

BOLTZMANN'S CUP OF TEA

One important problem that Boltzmann had to tackle was the cup-of-tea question, that is why, once disorder emerges in a system, the latter never appears spontaneously to revert to its original ordered state. Boltzmann's argument is illustrated in figure 3, where the example considered is that of a gas initially compressed which, like the cup of tea, once expanded never goes back to its original state. Consider first the part of the picture on the left of the thick vertical line. In order to understand the ideas involved, I shall consider an unrealistically small number of molecules, nine, and I shall assume that the container is divided into nine cells, through whose walls, however, the molecules can move absolutely freely. Imagine also that at some initial time all the nine molecules of the gas are confined to a single cell, as in A, and that they are somehow released. After a very short time, one will observe a configuration like, for instance, the one illustrated in B, with one molecule in each cell: the gas that was very highly ordered has now become entirely disordered.[18] I am not theorizing here, but merely describing the facts. The question that we must address in order to understand the "cup-of-tea problem" is this: why is it not the case that in the course of time the gas will spontaneously go back to the ordered configuration A?

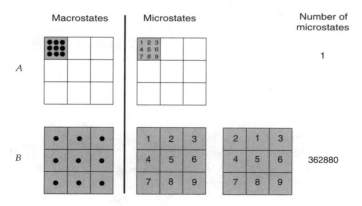

Fig.3. Macrostates (coarse graining) and microstates (fine graining) in Boltzmann's statistical mechanics

Only two of the 362880 possible microstates, obtained by permutation of the named molecules, are shown. The numbering given for the microstates in *A* indicates the initial condition that labels the trajectories entailed in all other microstates; this numbering should be kept fixed throughout the counting of microstates.

I shall describe Boltzmann's theory by formulating three hypotheses. The first is concerned with the relation between *macro* and *microstates* or, what is the same, between coarse and fine graining, all of which I shall now explain. The states *A* and *B* shown on the left of the figure are *macrostates*, which are nothing else than what so far I had called the *states* of the gas. It should be clear, for instance, that the volume occupied by the gas in *A* (which is one of the state variables) is different from that in *B*. It is important to realize that for simplicity of the picture I have considered only two macrostates but that very many more are possible. For instance, in one macrostate I could have two molecules in one cell and one empty cell, or three in one cell and two empty cells, and so on. Clearly, an immense number of such macrostates is possible.

What Boltzmann realized was that in so describing the states of a gas one is *coarse graining* them, that is, ignoring details that might be physically significant. Each coarse-grained state or macrostate can be fine-grained by recognizing that, although the molecules of the gas are all identical, they are distinguishable, and that indeed they can be *named*, or what is the same *numbered*, just as described in the last section.

Once we accept that the molecules of the gas can be numbered, we realize that each macrostate may be *fine-grained* because exactly the same state will be obtainable by permuting the molecules without in any way altering their geometrical disposition. Such fine graining is shown on the right of the thick vertical line of the figure; the fine-grained states are also called *microstates*, the meaning of which will become clearer in a moment. The number of microstates that corresponds to a single macrostate is in general enormous. Even with nine molecules and nine cells, there are in *B* 362880 ways in which the molecules can be arranged in the cells. And of course there are many more macrostates in which one or more cells are empty and others are multiply occupied. Thus, the total number of possible microstates is huge, even in this unrealistically small example. We have already seen that as the number of objects involved increases by even small amounts, the number of possible arrangements shoots up astronomically.

So far, the examples I have given entail unrealistically small numbers, but when we are dealing with the molecules in, say, one liter of gas, it is even difficult to imagine the numbers involved. (If you could count them at ten per second you would take 100000 times the age of the universe!) So, you realize that there are billions upon billions of microstates and of these, one and only one in which all the molecules of the gas are tidily collected in one corner of the container, as in *A* of figure 3. Remember that permutation of the labels here is not physically significant, because it is merely a "linguistic translation" of the set of names of the particles, which would rename the whole of this single microstate, and would lead to precisely the same counting of other microstates. In other words, the only changes of names that matter are those which, from an arbitrary initial set of names in the initial microstate, are effected by changes in the trajectories therefrom, as we shall now discuss.

Before we go any further, in fact, we must understand the meaning of the microstates, because the reader might so far think that they arise by some exchange of the particles which we do "in our heads." This is not so, and the physical meaning of the microstates depends on the second hypothesis of Boltzmann's theory, namely that the trajectories of the particles are *random*. Let me try to explain what we mean by this. If you look at the trajectory of particle 1 in figure 2, if this particle had been entirely isolated; then given its initial conditions the trajectory would have been entirely determined by Newton's equations. But the particle is far from isolated; it is subject to billions of interactions which deflect the particle from the previously imagined trajectory, a fact that I have already represented very schematically by the kinks in the trajectory, which correspond to collisions with other molecules. (If the particle had been really isolated the trajectory, by inertia, would have been a straight line.) When we say that the trajectory is *random*, we mean that instead of the final position of the particle being predetermined by the initial conditions, the particle can in fact land anywhere. That is, looking back at figure 2, particle 1 from *A* could end at the final position of particle 2 in *B*, and vice versa. This is the physical meaning of the particle *exchanges* that we were talking about in describing microstates. They are not "bookkeeping" operations but real physical processes. ("Real," of course, understood in our somewhat skeptical connotation.) That is, each microstate represents the result of a *physically possible* set of trajectories of the particles.

It must be understood that the various macrostates that we have discussed will in principle correspond to different energies since, for instance, the repulsive interactions between the molecules in *A* must be much higher than in *B*, owing to their closer distances. We remember, however, that in the ideal gas model, such interactions are ignored, and it is acceptable for our present purposes to do the same. This allows us to enunciate the third hypothesis we need in order to understand Boltzmann's theory, which is one of symmetry as between all possible microstates (that is, the microstates that correspond to all possible macrostates): in our simplified model *we accept that no physical distinction can be made between all the possible microstates*.

This hypothesis, which for convenience I shall call the *ergodic hypothesis*, has a very clear physical consequence.[19] Consider all the microstates of the gas, of which only a very small sample is represented on the right of figure 3 (since in any case many more macrostates should appear on its left). As time passes, the gas changes from one microstate to another, and we shall call the *sojourn time* of a microstate the average time that, over a reasonably large period of time (say a few seconds), the gas spends in that microstate. Our ergodic hypothesis entails that the sojourn times are equal for all microstates.[20] And this immediately answers the "cup-of-tea question": because there are 362880 microstates corresponding to *B*, and only 1 corresponding to *A*; if the average sojourn time were 1 second, then for each second the system spends in *A* it spends 4 days in *B*. Since as I have shown these numbers grow enormously for anything like realistic models, it means that for the many billions of years that the gas system spends in the states that cover the whole gas container, it will spend only minuscule fractions of a millisecond in the compressed state. Thus, having started in the compressed state *A*, we would have to wait billions of years for the random arrangement of the molecules to bring the gas back to it.

This is all very clear, I hope, but I am making here a huge assumption because all I can assert is that, on the average, the system will spend, say, a fraction of a millisecond in *A* over a period of billions of years: but this does not rule out that this millisecond will happen today rather than billions of years hence. We shall see in the next chapter, however, that the laws of probability make such an eventuality highly unlikely. This, alas, does not mean that I have solved the cup-of-tea problem: all I have done is to pass the buck to the laws of probability which, as you will see, are exceedingly difficult to justify philosophically. Even more, when I introduced the second assumption, that of randomness, what I did was provide an alternative to the deterministic trajectories that hold in classical theories for *isolated* systems. We have already seen that one could account *phenomenologically* for this lack of isolation by introducing friction. The hypothesis of random motion operates at a deeper level than that of friction; that is, it is less overtly phenomenological, in the sense that it does not require the immediate use of an experimentally fitted parameter. It does nevertheless a similar job, and it has the status of an empirical model to be superimposed on that of classical mechanics.

ENTROPY, DISORDER, AND THE SECOND PRINCIPLE

Boltzmann's great achievement was the discovery that entropy is related to disorder. As regards entropy, we already know that it is a function of state related to the quantity of heat that a system receives in some stated conditions. Since the energy is also a function of state, each state will have a certain energy as well as a certain entropy. This means that different macrostates, which, as I have said, if considered precisely would have different energies, will also have different entropies.

So much about entropy: let us now look at the disorder to be associated with a given macrostate. If we consider the difference between the macrostates *A* and *B*, it is, I hope, clear that the latter is far more disordered than the former.[21] We can see of course that this more disordered macrostate is also associated with a much larger number of microstates, and we can tentatively assume that the number of microstates corresponding to each macrostate is a suitable measure of the degree of disorder of the latter. That this is plausible, follows also from the way in which the microstates are generated by random exchanges of trajectories: the higher the number of such exchanges, the larger the number of microstates is. Also, the higher the number of trajectory exchanges, the larger the effect of randomness is, and thus the degree of disorder.

The important point is that Boltzmann postulated that the degree of disorder was given in terms of the number of microstates corresponding to one given macrostate; and he was able to establish in this way a quantitative relation between the entropy of the macrostate and its degree of disorder, a relation that not only agreed with all known thermodynamic facts, but that could also be further extended to predict some new results. This relation, which is not necessary for me to quote, provides the second principle of thermodynamics in its third and last form:

The degree of disorder of an isolated system never decreases.

This is one of the great normative physical principles required for the description of nature. As with the rest of them (chapter 8, p. 185) although sound arguments can be given for the second principle, as I have tried to provide, these arguments cannot be considered as a rigorous proof. The use of the principle, however much it can be grounded on experience, cannnot be directly justified by such experimental work as can be done, for instance, for Newton's law.[22] Rather than claiming a putative status for my imperfect proof, I take the line that one should accept it as a normative principle of natural phenomena, just as, for instance, I have done with the principle of causality.

BOLTZMANN AND AFTER

It was Johann Joseph Loschmidt (1821–1895) who first put his finger on a weakness in Boltzmann's argument.[23] Although the microstates in figure 3 are depicted in terms of the last point in the trajectory of each molecule, their respective velocities should also be considered. As we have agreed that all microstates are physically equivalent, this must also be the case for those in which the final velocity of the molecule is inverted, under the time inversion symmetry of classical mechanics. If we do this with the trajectories in figure 2, what happens is that the final state becomes now the initial state and, by reversibility of classical mechanics, precisely the same trajectory must be followed in reverse, thus leading all molecules back to the initial cell as in A. (Here, Hercule Poirot twirls his mustache and writes in his notebook: *second clue.*)

Boltzmann's answer to Loschmidt is the one that we have already largely discussed, namely that the laws of classical mechanics apply only to isolated systems, that the molecules of a gas are not isolated, and that the effect of all the other molecules of the gas on the given molecule, whose path it is attempted to reverse, cannot be described except by some sort of random shuffling that would interfere with the integrity of such reversal for the totality of the gas molecules. Although this argument is probably as valid as any empirical argument can be, it cannot dislodge the force of Loschmidt's criticism.

I shall soon refer to another important argument that Boltzmann used in order to counteract the criticisms raised against his proposals. Boltzmann, however, was plagued not only by those who took Loschmidt's argument seriously, but also by the fundamentalists who refused to accept any arguments at all that entailed the existence of "unobservable" molecules. In the light of this, it must be remembered that poor Boltzmann had to clutch at straws just to keep his enemies from utterly destroying his theories. Even twenty years after Loschmidt's work, at the 1895 meeting of the *Deutsche Gesellschaft für Naturforscher und Ärtze*, Friedrich Wilhelm Ostwald (1853–1932) repeated the *reversibility-of-mechanics* objection in order to conclude that entropy could not possibly reflect an underlying mechanical structure (which would be entailed if molecules were postulated). Boltzmann's eloquence, and the fact that physicists tend to be practical people—who would not expect reversibility to be kept after billions of events that would each require astronomical precision—won the day at that meeting, but not without serious damage to his mental outlook.[24]

Two years after the sad end in 1906 of Boltzmann's life, at Duino, near Trieste (then under Austrian rule), triumph came for him when even his scientific enemy Ostwald accepted the fundamental value of his theory. (Mach, instead, died still unconvinced in 1916.) This change of mind came through evidence supplied by Einstein on the so-called Brownian motion. In 1828 the botanist Robert Brown (1773–1858) had observed suspensions of pollen under the microscope and noticed that the particles moved about in an apparently random manner. Einstein argued in 1905 that such particles were free from interactions, like the molecules of an ideal gas, and were in fact displaying the very random motions that the kinetic theory of Maxwell and Boltzmann had postulated: but now these motions were clearly observable under any microscope. What was important enough to convince the scientific community of the significance of this proposal was that Einstein was able to use kinetic theory to show that it did agree numerically with the experimental observations. The crowning result of this theory was that Einstein was now able to propose a method for determining experimentally the number of molecules in a liter of gas, which, as we have already mentioned, is related to the fundamental constant that is called the *Avogadro number*. (The only difference between the two is that the volume specified in the Avogadro number is larger and chosen so as to have its own physical significance.)

In the next few years experimental work was done, largely by Jean-Baptiste Perrin (1870–1942)—for which he received the Nobel prize in 1926—to determine this important constant. The wonderful result was that there was complete agreement in dozens of observations that came, not only from Brownian motion, but also from radioactivity, from studies of the blue color of the sky, and various other methods. Perrin was able to say in 1909: "I think it is impossible that a mind free from all preconception can reflect upon *the extreme diversity of the phenomena which thus converge to the same result without experiencing a strong impression*, and I think that it will henceforth be difficult to defend by rational arguments a hostile attitude to molecular hypotheses." [My italics.][25]

I have quoted Perrin's statement at length because it illustrates very clearly the meaning of the scientific mesh: it was not the fact that the molecular kinetic theory had been exposed to falsification[26] a dozen times and survived that matters. It was the fact that there was now a theory linking together parts of the scientific mesh until then entirely unrelated that won the day (and the century) for the theory.

The triumph of the molecular kinetic theory of matter does not mean, of course, that Loschmidt's worries about Boltzmann's theories were surmounted. Neither were they forgotten: Reichenbach (1956) revived them and attempted a solution, as also did Grünbaum (1973), but their proposals are far from providing a complete resolution of the Loschmidt argument.[27] One possible way out of the dilemma stems from a proposal that Boltzmann himself had made, and which I shall discuss in the next section, although as we shall see, it does not provide a final answer to the problem of irreversibility.

Before we go into this, I must mention another archetypal example of the arrow of time that one should be able to understand. This, again, is about a cup of tea, but now broken, not stirred. When a porcelain teacup falls on the floor and breaks, it is unfortunately more sensible to have a jolly good nervous breakdown, than to wait for the cup to jump back from the ground to the table and reconstitute itself to its pristine glory. Since everything that happens here is mechanical, the latter behavior is what the reversibility of mechanics would lead us to expect, as with a perfect rubber ball that falls on the floor and then bounces back to its original height. So this is another instance where the Loschmidt argument appears to be unassailable, and requires an explanation of how the obvious time asymmetry displayed arises. (The reader would do well in examining this sentence with a magnifying lens: it contains my *third clue*.)

THE COSMOLOGICAL ARROW OF TIME

The reader who has bravely followed me so far in this chapter may have some basic worries about why we look for the arrow of time in something as esoteric as increase in entropy or disorder: if we pluck a bow and release an arrow in free space (no friction, etc.), the arrow will move for ever in the same direction neatly separating past from future at any instant of time. This, however, is not playing the game we wanted: we were considering *spontaneous* processes (that is processes in isolated systems) and there is nothing spontaneous in sending an arrow away.

The universe as a whole, however, does behave a little bit like our arrow, having started from a very small beginning (the big bang) some ten to fifteen billion years ago, and continued ever since to expand, as Edwin Hubble first observed in 1929. In trying to answer Loschmidt's criticism, Boltzmann (who was not of course aware of such results) had already suggested that the universe was, in our own time, trapped in a low-entropy configuration, which might explain why systems start from low-entropy states which then have to evolve by increasing their entropy.

Of course, just as with the arrow with which I started this section, an expanding universe would itself entail its own arrow of time, and, when this expansion became known, it was thought to be also a possible explanation for the thermodynamic time's arrow. (The expansion of the universe in relation to the big bang would entail a sustained increase in disorder, which one might assume as the origin of the second principle.) For this type of explanation to be valid, however, it would be necessary that a (presumed possible) reversal of the Hubble expansion entailed also a reversal of the arrow of time (that is, a reversal of the second principle); but although Stephen Hawking thought so for a time, he later recanted.[28] Thus, at least for the time being, and until this matter is cleared up, the cosmological arrow of time entailed by the Hubble expansion cannot be taken as an obvious explanation of the thermodynamic arrow. So, within present knowledge, the only prudent thing that can be said is to state the empirical fact that we live in a universe, or in a state of the universe, in which the second principle holds.[29] We thus appear to have left poor Boltzmann stuck, with nothing he can do to get Loschmidt off his back. We shall now come to his aid, and in so doing the clues that I have scattered in the text will all be gloriously "explained."

CONTRA LOSCHMIDT: ALL IS REVEALED

I shall start my work by describing the scene of the crime when we found our second clue, which was the moment when Loschmidt knifed Boltzmann in the back with his appeal to the reversibility of classical mechanics. So, he starts with a molecule at the end of its trajectory, and he argues that he can reverse its velocity; and that the molecule will wind back its trajectory step by step until arriving at the origin. And since this must be valid for all molecules, they should all get back obediently to the initial configuration, a possibility which Boltzmann on his probability considerations had shown to be so unlikely as to be just as good as impossible. Let us see what this entails, for which purpose it might be helpful to have a quick look back at figure 2. What Loschmidt says is that the motion of the molecule 1 at the end of its trajectory can be reversed, and that the same is valid for all the other molecules of the gas. I am prepared at this stage to disregard the problem of errors and the consequent impossibility of controlling the data with enough accuracy so as to make the whole reversal possible. In a spirit of tolerance, we concede Loschmidt's point, and we assume that his trajectory reversal can be done with total accuracy. Therefore, when the molecule 1 arrives at the point of the first kink (which in the original trajectory was the last), it will find there the molecule with which it had collided in precisely the same classical-mechanical state as it then was. The two molecules now collide again and because they are in the identical states as they were before (except reversed), the outcome will be the identical outcome we had before, but reversed. This is so because classical mechanics is causal, and given the same states we must have the same effects.

You may well believe in this but please, wait a moment and take a deep breath. We believe in causality because we had entrenched it, but we had entrenched it in the macroscopic world, since our mothers and nannies never allowed us to play with high-energy particle accelerators and such other desirable toys. Thus, we never acquired sensorial experience of the microworld (neither would we have acquired it, if the aforementioned toys had been available to us!). Thus we have no right whatsoever to expect the principle of causality, or any of our other fundamental meta-physical principles, to be valid in the microworld. So I put it to you that it is here that the villain is lurking, precisely at the point where the two molecules collide because, if causality were to break down at that collision, the whole Loschmidt argument would not be worth the minuscule discharge of the computer battery with which I write his name. We must seriously consider that possibility, and in order to do this we have to jump a bit forward in our studies, because if we have to look at the possibility of different principles that rule the microworld, then a little quantum mechanics is necessary. Fortunately, all that we need to know can be said in a few lines.

As we have seen, the laws of classical mechanics are valid for *rigid* bodies. These are bodies which, on collision, cannot deform and thus they only exchange kinetic energy, that is the energy which they have by virtue of their velocities. (I am ignoring such complications as might be introduced by the presence of a gravitational field.) This is not so for automobiles, which never bounce intact but deform on collision; and this deformation is due to the fact that part of their kinetic energy is spent in moving around the atoms that make up the metal plate. Molecules, like automobiles, are not rigid, and therefore they can borrow kinetic energy from other molecules with which they collide, not to gain the same amount of kinetic energy, but rather to deform. I use the word *deformation* here with a bit of latitude, because if, for simplicity, I think of a diatomic molecule, this can absorb energy in two ways, such molecules having two forms of internal mechanical motion. One is a rotational motion, which although it does not really deform the molecules, permits them to store energy very much as a flywheel does. The other internal motion is one in which the two atoms separate and approximate alternatively in a regular *vibration*.

The energy states corresponding to these motions, like all states in the microworld, are *quantized*, which means, for instance, that if you think of the rotational motion as that of a flywheel, the latter cannot absorb or deliver energy in a continuous form, that is, in arbitrarily small quantities, but only in *jumps*, which take the molecule from one to another of the *discrete* (that is, not continuous) energy states that it can assume. The same is the case for the vibrational motion and its corresponding quantized vibrational states.

Of course, the fact that the energy is now quantized is quite remarkable and new, but, as far as I can see, not many people lose their sleep over it: the real shocker is to come. If one molecule collides with another one twice, say, in *precisely identical* conditions, the energy jump which the molecules will experience may be *different* in both occasions, which means that *now the principle of causality is not valid*. This also means that the quantum jumps, which are random, are not time-reversible. This is so because *there is no causal link* between antecedents and effects, so that the same effects, when reversed, cannot be expected to lead to the same antecedents.

This is all pretty strong stuff, but we seem to have struck gold, because we wanted to help Boltzmann to throw back at Loschmidt some time-asymmetry, and that is what we have done. Obviously, if individual molecular collisions cannot be assumed to be reversible, then the whole of the Loschmidt argument collapses, and it collapses because it ignores the fact that when we move from the macro to the microstates, classical mechanics is not the only ruling theory. So clue number two has given back a victory to Boltzmann and quantum mechanics is responsible for this.

We can now go back to our first clue. I had then said that friction is a phenomenon in which *mechanical* energy is transferred from one body to the other: appeal to friction, therefore, could not exonerate us from the exigencies of classical mechanics as regards time-symmetry. Let us think again, however, how the transfer of energy entailed in friction works. As you remember, the moving suspension of the pendulum excites the motion of the atoms of the crystal material (such as copper) of which the axle is made. But this is also motion of atoms, not of macroscopic bodies, and it is well known that the vibrational states of crystals, like those of molecules, are quantized and subject to the same quantum jumps. Therefore, there is no reason, since the process entails quantum-mechanical jumps, for the transfer of energy involved in friction to be reversible. So, again, we find that we were much too impressed by the macroscopic world to understand what was going on here.

Finally, our third clue, the teacup that falls on the floor and breaks. Surely, you will say, this is as macroscopic a phenomenon as they come. But wait a moment: what does it mean that the porcelain breaks? To answer this we must know why porcelain holds together: surely, whatever its structure, there must be atoms and molecules in the material which are linked by strong bonds, just like the atoms in molecules are bonded. For the porcelain to break, interatomic bonds must break. This means that quantum jumps must take place in the material, thus absorbing the energy that was provided to it by the impact with the floor. Again, it is the irreversibility of the quantum jumps that we must invoke to explain the macroscopic irreversibility observed.

It is possible to believe at this stage, it is possible even for me to believe, that I have completely solved the problem of irreversibility; and that therefore we now "understand" where the time arrow comes from. I shall be much more modest in my claims. All that I can honestly claim to have done is to have circumscribed the problem: appeal to classical mechanics to show that irreversibility should not exist is futile, I have argued, because all the problems involved in irreversible processes entail the microworld and thus quantum mechanics, which admits of irreversible processes. But to say that we now "understand" the problem would require us to know why there are certain processes in quantum mechanics that appear not to obey causality and which entail random changes. This is not in any way a clear-cut situation, because we shall see later that the main equations of quantum mechanics, that is, the equations that take the place of Newton's equations as applied to microsystems, are like those equations *time-reversible*: so, we shall be back to square one. As some of my readers will know, there is a principle of conservation of human misery operating here. But take heart: we had three different problems as revealed by our three clues, and at least we now know where to look if we hope to find a solution. I shall try my best to sort this out in due course (chapter 19), although this is one of the hardest problems of our times and no fully satisfactory final solution is yet available.

In the meantime, we can do with some light entertainment.

THE PSYCHOLOGICAL ARROW OF TIME

We have seen that, although one might expect some relation between the cosmological and the thermodynamic arrows of time, the origin of such a relation, if it exists, is not clear. Is there a relation, instead, between the thermodynamic and psychological arrows of time? I take the view that this is a much more straightforward problem. We have to consider how we experience the passage of time, for which purpose it is enough to be able to distinguish the present from the past.

I shall explore this by means of a counterexample, based on a short story by Jorge Luis Borges, *Funes the memorious*.[30] Funes, as a result of an accident that crippled him, acquired an indelible memory, which allowed him to "reconstruct" (rather than "remember") any day of his past, the "reconstruction" taking a whole day, since he remembers or relives every single instant in total detail. (Funes's memories were so good as to be even sharper and more vivid than our own perceptions.) Imagine him in one of those reconstructions: at three fifteen of our time a fly lands on his nose, and at the same time of his reconstructed day another fly lands on his cheek. Since his perception of the first fly is no more vivid than that of his "reconstruction," his memory and his perception are not distinguishable for him: so, Funes cannot distinguish between present and past.

For our good fortune to be able to do so, we must thank our defective memory. Even my recollection of my window as seen a second ago is not the same event as its perception was as I averted my eyes from the keyboard. Consciousness, the mind, and the soul, are all ferocious battlefields of philosophical controversy, but I am probably right in saying that no one denies that memory, whatever it is, is *physically* based somewhere in our brains (although not necessarily associated with a single site in them). Even computer memories deteriorate, driven by the inflexible arrow of time that dictates that well-ordered information must eventually become disordered; for the same reason our memories fade, and this fading provides us with our perception of the passage of time.[31]

Given that the psychological and thermodynamic time arrows must be parallel, the anthropic principle might be invoked to claim that, were the thermodynamic time arrow inverted, we would not be here to tell this story—as I shall show in a moment—so this "explains" why that time arrow is what it is. (As I have already said, the anthropic principle never really explains anything, but rather helps us to accept that things are as they are.) The anthropic argument required is basically as follows. Imagine the arrow of time inverted: then our memories would be more ordered and therefore more vivid than our present. But if we live, we must live in the present. Since we would be perceiving a faded-out present, we would not be able to cope with it, as it happens to us in real life, where we cannot manipulate our faded memories.[32]

CODA

I hope that the reader has acquired a picture about the meaning of the second principle of thermodynamics, both with respect to the propagation of heat and from the point of view of the decay of order. You noticed that there is a tension in the whole work arising from the fact that classical mechanics is time-symmetrical, whereas the principles just mentioned undeniably entail time asymmetry.

It is very important for us in this respect to reflect about Boltzmann and his treatment both by friends and enemies. To get some perspective on this, compare him with Newton. Of course, Newton was surrounded by controversy, but you will not find a single book on the philosophy of science that chastizes him because his work leading to the statement of the second law of motion is not *logically* justifiable. Yet, Boltzmann proposed a number of hypotheses that led him to his enunciation of the second principle and to his discovery of the relation between entropy and disorder: and *all* the philosophical references that I have given in this chapter (quite correctly) find logical faults with his arguments. Imagine that this were not so, that Boltzmann from the comfort of his armchair had hit on a series of unassailable a priori arguments that led him to what is, after all, a point of fact about nature. Would that not be amazing? Would that not be trying to achieve facticity without facts?

And saving the facts is after all what science is all about. We tend to think of phenomenology only when we are operating at a crude level, as when friction is introduced. Boltzmann's model was far more elaborate, but he had to formulate some phenomenological hypotheses to save the facts, and the crucial hypothesis in his work is the introduction of *random interactions*. We know that classical mechanics can only be applied to isolated systems, and we know that a molecule of a gas is not isolated, but engages with billions of other molecules. To replace the effect of the latter by a random interaction is the crucial concept, and although this idea, as Boltzmann did and I have tried to illustrate, can be somewhat justified, its validity has to be taken as much as an objective "fact" about nature as Newton's second equation is. In any case, we have seen that whenever irreversibility appears, some quantum-mechanical process lies behind, and that there is no reason why causality should be totally transferred from our entrenchment in the macroworld to the unexperienced realm of the microworld. Because of this, some random processes appear in quantum mechanics that I have made responsible for macroscopic irreversibility, but this is not the end of the story, as we shall see in chapter 19.

Having said this, I hope that my readers will not protest too much that I have pulled wool over their eyes because I have never said precisely what *random* really means. And I am sorry if I have abused their good nature, hoping to make do with a degree of wooliness here, which I shall try to remedy in the next chapter. If one were to invert the understanding of the second principle as one does when pulling an old sock inside out, the most important idea that it has brought to us, via Boltzmann, is that random interactions do play a crucial role in nature, a point which was magnificently demonstrated by Einstein in his study of Brownian motion.

NOTES

1. Eddington (1928), p. 69.
2. The question of time and language, however, is considered by van Fraassen (1985), p. 82.
3. This is not always so, as in the sentence "Everest *towers* over the Nepal plains." It could be argued even here that a flow of time is entailed, but that it is over such a long period, a geological era, say, that it appears to be static.
4. The Hopi Indians of North America did have, in fact, a language without verbs, nouns being declined with suffixes. Whorf (1956) claimed thereby that their perception of time was entirely different from ours, but such a claim is not one that is seriously entertained today. One of Borges's characters, Funes, whom we shall meet again later in this chapter, "resented that the dog of the three and fourteen (seen in profile) bore the same name as the dog of quarter past three (seen full face)" (Borges 1956, p. 125).
5. I have to use the rather stilted "behold," because in current English usage "to see" might imply variegated activities such as "going out with" or even "having sex with."
6. If you are worried about this, remember that velocity is the rate of change of position; and that rates of change invert their sign when the time is inverted.
7. I am assuming that accurate numerical methods are used in the amazing computer, so that the motion of billions of particles (remember for instance all the atoms in the glass container) can be followed with perfect accuracy, a very tall order, of course.
8. See Morrison (1966).
9. This energy transfer could be done, of course, by spending energy (whereby nothing would be gained): what the principle forbids is the transfer to be *spontaneous*, that is, not driven by any energy source.
10. When you compress the gas, as when pumping a tire, you force its molecules to collide with each other and increase their velocities. This is a phenomenon of friction: before, the molecules were well separated and did not interact with each other to any large degree. On compression they get nearer and "rub" each other.
11. See Clausius (1965); de Regt (1996).
12. Further discussions on the problem of counting elements of an aggregate will appear in chapter 14. Here and elsewhere I use the word "aggregate" very much as what most people call a *set*, but I avoid this word since it will later have a very specific mathematical meaning.
13. Notice that an aggregate of *named* particles, from the point of view described, must be *denumerable*, that is one must be able to *count* them. The reader must realize that this is not an obvious property of aggregates, since it entails that one must be able to distinguish in some way between their elements.
14. I follow here a proposal made by Jauch (1968), p. 275.
15. Intrinsic and extrinsic properties correspond roughly to the philosophical notions of *essential* and *accidental* properties, respectively. My definition of identical but distinguish-

able particles is in the spirit of St. Thomas Aquinas's doctrine that identical objects share the same essence, but can be distinguished by their accidents. After Aquinas's death in 1274, his views were attacked by John Duns Scotus (ca. 1266–1308), (see Wolter 1990), who held that distinct things must always differ in their essences, a view which was later defended by Leibniz (see Leibniz 1696). Reviews on this question may be found in Schwartz (1977). Further discussion about identity and distinguishability of particles will be given in chapter 16.

16. This question is discussed in chapter 9, p. 216.

17. Whether or not particles may be labeled by the principles stated cannot be understood independently from experience. Going back to figure 2, the two particles there could collide, and the collision could be so ferocious that it might not be possible to decide after it which particle is which. In the case of molecules in a gas, two circumstances allow us to disregard this possibility. First, the molecules travel at relatively low speeds, so collisions are more in the nature of one molecule bouncing off the other. Second, the collisions are not strictly speaking by contact of the colliding molecules, because there is always some repulsion between the external parts of the molecules, so that they never approach one very near the other. Thus, molecular collisions are more like deflections in their trajectories. It is because of these circumstances that the integrity or continuity of the trajectory of a single molecule is maintained even during collision.

18. The reader should not be overimpressed by the tidy elegance of *B*: imagine that the dots were marbles: is it not nursery-like disorder to have the marbles lying all over the floor rather than being neatly placed in one box as in *A*?

19. Although the concept of ergodicity was introduced by Boltzmann in 1871 (see Earman et al. 1996), the word *ergodic* was coined by him in 1887. The importance of the ergodic principle in the study of probability will be discussed in chapter 13.

20. More precisely, the ergodic hypothesis should be applied only to those microstates that correspond to a single macrostate. This is so because, different macrostates having in fact different energies, those macrostates with lower energy would be favored, that is, would be visited for longer periods, and thus their corresponding microstates, which should then be characterized by longer sojourn times.

21. Remember note 18.

22. Because the second principle entails that something never happens, every time that you test it corroborating the principle, the question remains whether the lack of performance is due to defective or at least not sufficiently imaginative planning. Thus, we have all the philosophical problems inherent to induction, plus the continuous nagging doubt about the quality of each experiment. Although the latter doubt may be eliminated by careful scientific protocols, it is philosophically troublesome. The possibility in each individual case of better but unknown experimental procedures cannot, of course, be subject to direct evidence. The validity of the second principle, as I have said, rests instead on its profound implications throughout vast sections of the scientific mesh. Popper, of course, would say that this validity rests on the huge falsifiability of the principle. These two points of view are *almost* synonymous: falsifiability is all or nothing because you either falsify or you corroborate. When I talk about entrenchment over the scientific mesh, I take into account not only the large number of times when an aye or nay answer could have been obtained, but also the *quality* of the question tested, that is, its significance for the integrity of the mesh as a whole. If you like to talk about "understanding," the quality of a question tested is indicated by how far it contributes to the understanding of large and important areas of the mesh.

23. Loschmidt (1876).

24. An account of the dispute around Boltzmann's work has already been given in chapter 10. See for more details Broda (1955), Klein (1973), Rosenfeld (1955), and Cercignani (1998). That Boltzmann was a brilliant lecturer, we know from the glowing testimony of his pupil Lise Meitner, who later codiscovered fission (Broda 1955, pp. 9–10). As for his ability

in debate, the then-president of the Accademia dei Lincei referred to him as a "dreaded polemicist" (Broda 1955, p. 12). Except during his bouts of depression he was lively, humorous, and loyal: throughout their long scientific disputes, he kept his friendship with Loschmidt, for whom he wrote a warm obituary, and with Ostwald.

25. Quoted by Pais (1982), p. 95.

26. As far as I have been able to find out, there appears to be no evidence that any of these experiments were conducted with a view to disprove the molecular theory. Even if it were to be adduced that the intention of the observers plays no part in the philosophy of science, this example is probably a good illustration that, if falsification ever takes part in the drive to scientific discovery, it is certainly not the only mechanism to this effect. In fairness to Popper, however, it must be mentioned that, by joining together different theories in order to determine in different ways the Avogadro number, statistical mechanics is thus increasing its *degree of falsifiability* as defined by him (Popper 1959, p. 119). Despite its name, this is not a concept that can easily be put in a quantitative scale, and its application in a practical case like the one in hand is no more precise than the simpler and more graphical description that I give in the text.

27. A very thorough account of these ideas may be found in Horwich (1992), pp. 68–71. See also Reichenbach (1956) and Grünbaum (1973), pp. 240, 248.

28. Hawking (1988), p. 150. This is a point that was already made by Penrose (1986), pp. 47–48.

29. Various developments which try to connect the two arrows of time can be seen in Horwich (1992), p. 71. Useful general references are: Flood et al. (1986), Zeh (1992), Coveney et al. (1991), Ridley (1995), Le Poidevin et al. (1993), Hinckfuss (1975), Davies (1977), Earman (1974), Gold (1962), Morrison (1966), Newton-Smith (1980), and Savitt (1995).

30. Borges (1956).

31. The reader must realize that this argument, attractive as it might be, requires further elaboration. The second principle must be applied to an isolated system, which the brain is not, since blood supplies it with nutrients precisely to avoid the otherwise unavoidable decay. Thus, whereas the brain evolved to be self-healing to some extent, and "fuel" was provided to keep such memory mechanisms as exist at an acceptable level, evolution did not favor the possibility (if it existed) of keeping memory always at the top of vividness. Funes's example can perhaps explain this. Substitute two ferocious lions for the harmless flies: it would have been fatal for Funes to be unable to discriminate between them. It is probably because of this that the psychological arrow of time had to evolve as to be parallel to the thermodynamic one.

32. The use of the anthropic principle here is discussed by Hawking (1988), pp. 144–48, although he appears not to have considered the point raised in the last note.

NATURE'S LOTTERY: PROBABILITY

> *It is incorrect, then, to say that any phenomenon is produced by chance;*
> *but we may say that two or more phenomena are conjoined by chance, that*
> *they coexist or succeed one another only by chance; meaning that they are*
> *in no way related through causation; . . .*
>
> John Stuart Mill (1843) Bk III, Ch. xvii. §2, p. 526.

Readers may skip the more mathematical arguments of this chapter, and some of the tables, without detriment to the further reading of the book.

Although John Stuart Mill (1806–1873) is here defining the word *chance* very much as it is used colloquially, this is not the sense in which I use it in this book. We all accept, for instance, that the arrival of the newspaper at my house and my brushing my teeth, are two events not causally connected; and we could perfectly well say with Stuart Mill that they are conjoined by chance. But I shall not be concerned with such *accidents* or *coincidences* that, although anecdotally interesting, tell us nothing about nature. What I have in mind are instances in which two events are so conjoined that, in principle, one could try to construct a causal chain to relate them; and yet the effect obtained is not causally determined by the antecedent state. Hitting a billiard ball with a cue, in fact, could appear to be somewhat similar to tossing a coin, but whereas in the first case the result is fully determined (give or take some minor perturbations), in the second case it is not.

Even then, the contraposition "causation-chance" is not as clear-cut as just explained. Remember, to start with, that although philosophers may consider single instances of cause-and-effect relations, causation in science is always conflated with some form of induction. We accept that the force exerted by the cue is the cause of the acceleration of the billiard ball if we are sure that, however many times we repeat exactly the same antecedent actions, identically the same consequents will arise. This, we all know, is not the case when we toss a coin a large number of times. If chance were exactly the opposite of causation, one would expect that nothing, absolutely nothing, could be predicted about the sequence of tosses. And yet everybody knows that in most cases (that is, unless we were dealing with very pathological coins), we may expect roughly the same number of heads as of tails to appear in that sequence.

And this is one of the most mystifying features of nature: "Experimental scientists imagine that it is a mathematical theorem and the mathematicians imagine that it is something experimental." Thus Poincaré; this is not, alas, an obsolete saw: it is quoted by the Polish American mathematician Mark Kac (1983), who appropriates the dolorous sentiments expressed by Poincaré, although few people in our time understood probability and statistics better than Kac.[1] Just in case: many readers may know, or at least, may have heard, that vast advances in the mathematical theory of probability have been made, and this is entirely true. But it must be understood that these advances are based on an axiomatic formulation of the subject. If you believe that those axioms are true full stop, that is, that they are true by virtue of belonging to a world of mathematical verities that somehow underlies nature, then you will believe that the problem that worried Poincaré and Kac has been licked beyond a shadow of doubt. But if you take the more skeptical view that the axioms are accepted as valid, merely because they give results that are empirically adequate, then the nature of what is empirically the case is just about as obscure now as it was before.

I wish I could proudly announce to the reader that all these preoccupations will be dissolved by the treatment on which we are now going to embark, but fantasy is a poor substitute for reality. I hope, however, that we are not going to waste our time. First, even the simplest ideas about probability give rise to phenomenal misconceptions that feed the pockets of lotteries, and allow people to break traffic rules with the same carefree innocence as a burglar who believes that he is doing his victim no harm because the insurance will pay. Second, I hope that I shall be able to explain what the main problems are as regards the meaning of probability. Third, and most important, we shall see how probability is used in classical physics so as to be prepared to understand the changes that arise in the microworld.

There are various ways in which probability is treated,[2] but I shall take up as far as I can Poincaré's challenge, and concentrate in looking at probability from the point of view of natural science. Thus, the natural objects that lie behind probability will be for us *random sequences* and their *generators*; the latter being physical, not mathematical objects.

AN APPROXIMATION TO HISTORY

One of the problems with probability is that it means different things to different people (as my remarks above on the use of the word *chance* show). In order to understand this, one has to look back at the history of the subject, which is bedeviled by the fact that it started within a context that is not the best from the point of view of natural science, namely gambling. Gambling, alas, can be irrational, in the sense that expectations of gain, however much the gambler may invoke them, are largely discounted: the gambler is essentially paying a fee to have some form of excitement, or to be able to entertain some pleasurable daydreams. At the other extreme, gambling can be wonderfully rational; there are people who gamble day after day at Monte Carlo and *always win*: they are called the *bankers*. And make no mistake, every time you place your bet at the roulette, the banker is gambling, but obviously he has found a way to do so while consistently making money (at least in the long run).

Imagine you want to bet at the roulette making *sure* not to lose: your strategy might be to place one dollar on *each* of the 37 numbers 0 to 36. At least, you are sure to "win" since one of them must *certainly* come up. But whatever number does come up, the banker will pay you only thirty-six dollars; this is the rule of the game, and thus the banker will be one dollar in pocket, and despite your "win," you will have lost one dollar! The problem here is that the banker had a good adviser, and that his bet of thirty-six dollars for each of your single dollars was wise.

It was to achieve some such degree of rationality that Antoine Gombaud, Chevalier de Méré, consulted Blaise Pascal in 1654 about a game of dice. This led to an exchange of letters between Pascal and Pierre de Fermat, from which probability theory was born. Shortly afterward, in 1657, Leibniz's teacher, the Dutch scientist Christian Huygens, published the first book on probability: *De ratiotiniis in ludo aleae* (*Rationality in Aleatory Games*). The next two important books came half a century later. In 1713, *Ars Conjectandi* (*The Art of Guessing*), by the Swiss mathematician Jakob (also called James or Jacques) Bernouilli (1654–1705), was published. Just at about the same time, the French Huguenot mathematician Abraham de Moivre (1667–1754) was working on probability in England, where he lived most of his life, and in 1718 he published *The Doctrine of Chances: or, a Method of Calculating the Probability of Events in Play*. The reader will notice from the title of these works that they are all concerned with gambling, which is why I quote them.

Such activities are, however, concerned with problems well outside those where probability is important in science. The gambler may be thinking of a single event, and he may want to make the best of it, that is he might want some sort of a rationale behind a single bet. This situation has informed a number of philosophical problems of the same nature as, for instance, the question: "How likely is it that the Prime Minister will still hold office in the autumn?" The basic feature of such situations is that one has to take a decision or make a forecast from a position which is largely of ignorance, however much additional information might be available. I shall not be concerned, nevertheless, with such approaches to probability theory. Neither will I be concerned with how to justify decisions, or with the strategies that people follow in order to defend them in such situations. My purpose, instead, is to examine the ways in which chance and probability manifest themselves in the study of natural phenomena. But let us go on with our historical skeleton.

One of the great intellectual events in the second half of the eighteenth century was the publication of the French encyclopaedia, the *Encyclopédie ou Dictionaire Raisonné des Sciences, des Arts, et des Métiers*, in twenty-eight volumes, a veritable monument to the age of the Enlightenment and to a belief in human rationality. Its editors were the philosopher Denis Diderot (1713–1784) and the scientist-mathematician Jean Le Rond d'Alembert (1717–1783).[3] The final edition of the *Encyclopédie* was published in 1772, but a volume including D'Alembert's article on probability, entitled *Croix ou pile*, already appeared in 1754, containing one of the juiciest mistakes ever made in the subject, to which I shall devote one of the next sections. And we must not forget that, as a forerunner of the French Revolution and as perpetual secretary of the Academy of Sciences of Paris, d'Alembert was perhaps the most influential scientist in his country during his lifetime.

Only one generation after d'Alembert, we get at last into the real thing with Laplace, who not only developed the mathematics of the subject well beyond the work of his predecessors, but who also contributed profoundly to its philosophy. Pierre-Simon, marquis de Laplace (1749–1827) was, as you can see from the dates, an aristocrat who managed to survive the French Revolution; and he even held an office for a time as an ineffectual minister of the interior under Napoleon. Laplace's two major works on probability are *Théorie Analytique des Probabilités*, published in 1812, followed in 1814 by *Essai Philosophique des Probabilités*. Discussion of the ideas that he presented will allow us to begin digging our teeth into these problems.

TWO VIEWS OF PROBABILITY

It was Laplace who, in studying celestial mechanics, completed the great chapter in science started by Newton. Remember that the crux of Newton's method was his causal definition of a mechanical state; and

Laplace, who was not a believer like Newton, constructed around that concept a materialistic and deterministic view of nature. In the introduction to the *Théorie Analytique des Probabilités*, which appeared in volume seven of his *Oeuvres* (1812–20), he wrote one of the most influential statements of the nineteenth century for natural science, which despite its so frequent quotation it is useful to repeat here:

> Given for one instant an intelligence which could comprehend all the forces by which nature is animated and the respective positions of the beings which compose it, if moreover, this intelligence were vast enough to submit these data to analysis, it would embrace in the same formula both the movements of the largest bodies in the universe and those of the lightest atom; to it nothing would be uncertain, and the future as the past would be present to its eyes.

This sentence was a guiding light for many thinkers until well into the twentieth century; and because it was so very important, I must break chronological order and emphatically state that it is entirely wrong. The first breakdown of determinism came not from macroscopic phenomena (of which more in a moment), but from the microworld. When atomic phenomena were studied in the 1920s, it was soon realized that Newtonian mechanics was not applicable to particles such as electrons; and that it was not possible to state initial conditions as Newton had done, which would always determine, for instance, the later position of the particles. An archetypal example of what is called a *random* process was the decay of radioactive atoms, which was found to obey no rule whatsoever in time: given one radioactive atom, it is impossible to predict when it will decay. (For the time being, I shall use the word *random* to indicate any process that is not deterministic, not ruled by any law, and thus unpredictable. Obviously, probability is closely associated with randomness since, if a process is deterministic, its outcome is precisely known, so that probability statements are irrelevant.)

But even in the classical world, Laplace's determinism is a chimera. If you hit a billiard ball, Newton's equations will work, although in practice the initial conditions will be subject to some, in fact, small errors. This is so because one knows that those equations are *linear*, which mathematically entails that small initial errors do not grow very much along the time. The equations that rule the development of other more complex macroscopic systems can be *nonlinear*, which means that a small error in the initial conditions multiplies so many times in the evolution of the system, that deterministic prediction is impossible. Most readers will be familiar with the famous example of weather prediction, in which the initial conditions, in order to predict let us say a tornado in Florida, would accurately have to represent the beating of the wings of one single butterfly in the Australian bush. Systems that display this property are called *chaotic*, and the study of *chaos theory* has been one of the major triumphs of mathematics in the last few decades. In chaotic systems, it is impossible to give initial conditions that accurately determine the full evolution of the system. And it is not sufficient for a follower of Laplace to dismiss this on the grounds that it is a mere technical impossibility. Since initial conditions are required in the minutest detail, the random events of the microworld are sufficient to make their precise determination impossible. Laplace's "vast intelligence" requires information about microscopic particles which, as far as current knowledge goes, is *necessarily*, and not accidentally, unavailable.

I shall not be concerned with chaos, although we shall return in full force to the microworld in later chapters: but we must understand that the negation of Laplace's determinism is one of the major features of modern science. Naturally, Laplace's determinism did not allow for the existence of random events in nature, which strongly colored his view of probability: chance could not be a feature of the natural world, so that the only way out for Laplace was to regard probability as a *subjective* phenomenon. Another important element in his entertaining this view must have been the original beginnings of the science of probability as a study of gambling, in which probabilistic considerations are used in subjective decision making. Laplace's subjective probability deals with events whose premises are incompletely known and, thus it is "relative to our knowledge and ignorance."

Laplace's subjective view of probability is still much supported among philosophers and other theorists, although natural scientists are probably less enthusiastic about this approach. The most important supporter of subjective probability this century was the Italian mathematician Bruno de Finetti (1906–1985), who nevertheless elaborated his theory so as to admit an element of objectivity. (See de Finetti 1990.)

At the opposite end of the philosophical scale from subjectivism, it is sufficient to quote Ernest Nagel: "Probability statements are on a par with statements which specify the density of a substance; they are not formulations of the degree of our ignorance or uncertainty."[4] This is the purely *objective* view of probability but, of course, there are numerous positions in between these two extremes, which I shall discuss a little, later. I am afraid, however, that the reader may be getting impatient at this stage because I have managed to use the word *probability* dozens of times never having defined it. This I will now begin to do, despite the title of the section that follows.

IS PROBABILITY A PRIORI?

Suppose I want to define or work out (for the time being I cannot make a difference between these two alternatives) the probability of getting heads when tossing a coin. There are two possible outcomes when doing so, heads (H) or tails (T), and the number of cases favorable to the desired outcome (H) is only one. We could define the probability of heads as the number of favorable outcomes, 1, divided by the total number of outcomes, 2, which would give this probability as 1/2. Laplace, however, realized that this definition is unsatisfactory. If I have a loaded coin one of the sides would appear more often, and the above definition would not work: in the extreme case, the probability of heads might be nil! Thus, Laplace proposed the following definition: "Probability is the quotient of the number of favorable cases by the number of *equally possible* cases" [my italics]. And, because *equally possible* is the same as equally probable, we have fallen into a circularity.

Although I have accepted in the past that some circular reasoning is unavoidable, if not positively desirable, this does not license arrant carelessness, and we shall have to investigate this question. To start with, if you read carefully what I have written, it is a mess, and it is a mess because I am talking about two entirely different things at the same time. First, I think about the coin in the abstract and I work out a number which I call *probability*, call it $p(H)$, the probability of a heads throw, and this is entirely, apparently, a priori. But then I compare this number, or at least I imagine that I might compare this number, with an entirely different thing, the *frequency* of the desired event, say $f(H)$, which is the proportion of the number of heads that appear in a sequence of throws. Thus if in 100 throws 45 heads appear, then $f(H)$ is 0.45. Notice that this concept was implicit when I said, in order to explicate Laplace's definition, that for a loaded coin "one of the sides would appear more often." Herein lies one of the deepest problems about probability, namely that one starts from allegedly purely a priori considerations and then, lo and behold, one appears to be able to apply them to empirical results.

The reader might remember that we already met one case, in the study of symmetry, when a similar temptation arose: we appeared to be able to think in the abstract, basically on purely geometrical lines, about a balance, and then "deduce" the experimental result that it would be in equilibrium. (See chapter 8.) But we soon realized that this is nonsense and that symmetry must be experimentally determined.[5]

This will give us a clue as to where lies the weakness in Laplace's definition of probability, because this definition circularly entails the concept of equiprobability, which is essentially a symmetry argument: the coin, we might say, is symmetrical with respect to heads or tails. It is tempting to sort this out by an *indifference principle*: if I cannot find any difference between the two possible outcomes, heads or tails, I could claim to be entitled to take them to be equiprobable. This, of course, goes well with the subjective view of probability because one is working on the basis of one's knowledge or lack of it, and Laplace, in fact, called two events equally possible if we are equally undecided as to their occurrence.

We should be able to do better than this, however. First of all, the *indifference principle* that we have mentioned, which is sometimes called a *principle of insufficient reason*, is nothing else than Leibniz's principle of sufficient reason (p. 183) applied in a case when such a sufficient reason does not appear to exist. And we know that however a prioristic the principle of sufficient reason is, it can correctly give us the principle of symmetry, when dealing with experimentally determined, rather than a prioristically known, symmetries. So, there is some hope that here again we shall be able to avoid a situation where unaided thought appears to lead to empirical results.

We must first of all understand that the principle of indifference might be a very blunt weapon. To start with a crude example, suppose I ask this question: "What is the probability of life existing on Mars?" On the basis that there are two possible outcomes on Mars, life or lack of it, and that I know nothing, and therefore believe that they are equiprobable, then my question would have 0.5 for an answer. But this is just too silly, since we know that (perhaps) there is no water or oxygen on Mars, and therefore that the two possible outcomes, life or no life, should be far from equiprobable. There are far better examples to show that counting the allegedly equiprobable cases is far from trivial, as we shall now see.

John lives in a cottage halfway up a hill, and three footpaths come straight out of the front door of his house, one going uphill and two going down. What is the probability that John will go uphill when coming out of his cottage?[6] If you consider only two possible outcomes, uphill and downhill, and take them to be equiprobable, then the probability is 1/2. But if you judge the three footpaths to be equiprobable, the probability is then 1/3, because only one goes uphill. It is evident that in order to discriminate between these two different results one needs a great deal more than just a priori reasoning.

Let me now think aloud for a moment, going back to the, possibly biased, coin. I might argue that, in the absence of prior knowledge, I could assume heads and tails to be equiprobable, whence the probability of heads $p(H)$ equals 0.5. I then toss the coin one hundred times, say, and find $f(H)$ equal to 0.3, whereupon I reject my equiprobability hypothesis and declare the coin to be biased. Such an approach would commend itself to us, as demoting the equiprobability postulate to a working hypothesis, to be accepted or rejected by experiment. Also, because it shifts the emphasis from a prioristic to empirical considerations, which appears plausible, but it entails lots of problems. First of all, why compare probabilities with frequencies at all? Second, why toss the coin one hundred times and not ten, say. Third, what happens if I toss the coin one hundred times and the frequency is 0.45; am I entitled to say that the coin is biased because I did not get the dreamed up value of 0.5? All these are very serious questions which I shall address in this chapter, but, rather than dealing with them now, I shall first consider a couple of practical problems, which will help us build up our understanding of probabilities.

D'ALEMBERT'S BLUNDER AND INSIGHT

I shall now treat the problem that led to d'Alembert's famous mistake in his 1754 article of the *Encyclopédie*. In the game of *croix ou pile*, a coin is thrown twice and a punter can bet, for instance, on obtaining at least one head. The two successive throws can be two heads, which I shall denote by HH, or one head followed by a tail, HT, or TH, or TT. Since there are three cases favorable to the bet mentioned (at least one H) out of four cases, its probability is 3/4. This, however, is not d'Alembert's analysis. He argued that, under the rules of the game, once the first toss produces H the throw stops, whence both the HH and HT outcomes he considered as a single one, H, which with TH and TT gives two favorable cases out of three, and thus a probability of 2/3 for the bet in question.

D'Alembert's mistake is as subtle as it is instructive, and to reveal it I shall produce a sequence of coin tosses *experimentally* obtained (see row 2 of table 3). A few will suffice:

TTHHTHTTHHHH.

Notice that I am not yet playing the game, but the first principle that we must learn is this: that the coin tosses, or whatever other random phenomenon we are observing, do not care a bit for any game we wish to play. Therefore, they must be treated as they come *without any alteration*. (I am trying to introduce you to the important concept of a *random sequence*, of which more later.) We now play *croix ou pile*, in this sequence, each game being separated by a hyphen. (Remember that each game is two throws, except when your bet, which I take to be H, comes first.) The results are:

TT – H – H – TH – TT – H – H – H – H.

D'Alembert claims that because the game stops when you get an H, the first throw that comes after it must be ignored. He thus looks at our original sequence and divides it neatly in successive pairs, ignoring the second element of a pair (which I shall write in brackets) if the first is H:

$$TT - H(H) - TH - TT - H(H) - H(H)$$

A serious sin is committed here, which is interfering with the *natural* sequence of events that one is handling. If we go back to our original sequence, it had twelve elements, of which seven were H, giving their frequencies as 0.6. In d'Alembert's doctored sequence, there are only nine elements left, of which four are H, with a frequency of 0.4, quite different. Obviously, if you start tampering with natural sequences of probabilistic events, you can get anything; at worst, if you ignore all the heads their frequency vanishes!

What we have learned is this: that dealing with probabilities is not something subjective, something that we do entirely in our minds. There is something "out there" that has to be respected and treated properly like any other natural event, and this something is a *random sequence*, such as the first displayed line in this section. But it will take us some time to get to analyze what a random sequence is.

Now we come to a good idea of d'Alembert's. Challenged by a Geneva mathematician that the correct probability was 3/4, and not 2/3 as he had stated, d'Alembert's answer was that in any case equiprobabilities required *experimental* analysis. In claiming this, he was rejecting the prevailing view that they were determinable a priori.

This is a very important point, which I shall illustrate by considering a slightly different game from the one that d'Alembert was treating. Instead of having a single coin tossed twice, I shall consider two objects, each of which can have one of two states, which I shall label as H or T (to allow readers to keep coins in the back of their minds), and that are considered *simultaneously*, as if two coins were tossed at the same time. I show the possibilities arising in table 1.

Table 1. States of two objects

Object 1	Object 2
H	H
T	T
H	T
T	H

It is clear that the results in table 1 depend crucially on being able to *label* the two objects in question. Notice that in order to make the distinction between the outcomes HT and TH in the last two rows of table 1, we have to agree that the first-named outcome, H or T, must pertain to object 1 and the second to object 2, and this requires that the objects be capable of being named or counted. We have seen, however, that labeling objects cannot be taken for granted. If the two objects are coins, this is very easy to do: coin 1 is tossed by John and coin 2 by Jane, and there cannot be any dispute as to which is which. The probability of anyone of the possible four outcomes, say HT, is clearly 1/4 in this case.

Imagine now that the two objects in question are photons. We shall see in chapter 16 that photons can be found in either of two states (called *polarization states*) which we may identify with the symbols H and T used in the table. The question of labeling photons is now delicate. We have seen in chapter 12 that molecules are labeled by means of their trajectories, but the trajectories of photons are far more elusive, since they travel at the speed of light, and it might in certain cases be impossible to tell one photon from another to label them. Whether this is or is not case is unimportant at the moment; but it is evident, nevertheless, that it is a question that could only be answered by experimental analysis. (See discussion in chapter 16.) The implication of this on the probabilities arising from table 1 is most important. If the two objects cannot be discriminated, the last two rows of the table must be collapsed into a single outcome, written as HT, although it could equally well be represented by TH. In this case, table 1 entails only three equiprobable states, HH, TT, and HT, and the probability of anyone of them is 1/3, not 1/4 as before.

I hope that the *enormous importance* of what we have now discovered will be understood, because, much as I admire d'Alembert, I am just as likely to raise a monument to Hecuba as to try to vindicate his memory. No: what this example shows is how much we take for granted *about nature*, even when we consider the simplest possible probabilistic example. Equiprobabilities, of course, depend essentially on being able to count the allegedly equiprobable states, and we have seen that even this cannot be assumed without reference not to our thoughts, nor to the *Encyclopédie*, but to that other and only reliable book, nature. This dismisses at once any notion that probability entails some arcane sort of a priori knowledge. And in honor to d'Alembert, he understood that perfectly well. It is not enough to invoke a principle of insufficient reason to determine equiprobabilities. They are at most a working assumption that has to be experimentally tested. Apparent probabilistic paradoxes might result from the inability to distinguish between two otherwise identical objects, and, for convenience, I shall later refer to such problems as *the d'Alembert paradox*, although their relation to d'Alembert is tenuous.[7]

Before we leave the example of the two coins, it is useful to warn the reader that exactly the same sequence of trials may be regarded in probability in two different ways; and that failure to understand this difference can lead to fallacies, as we shall see in the next section. When, on looking at table 1, I ask what the probability is of one head and one tail, I can mean one of two different questions. I can assume that it does not matter which coin is which, thus ignoring some detail and using what is called a *coarse-grained* approach. Or I can mean, say, that coin 1 must be heads and coin 2 must be tails, or vice versa, in which case I use a *fine-grained* treatment. This makes, of course, a substantial difference: in the coarse-grained approach, the probability of one head and one tail (that is, either HT or TH out of four possible cases) is 1/2, and in each of the two fine-grained instances of the same result, HT and TH, the probability is 1/4. A great mistake that people often make is to use the coarse-grained approach when the problem in hand demands fine graining, as we shall now see.

THE MONTE CARLO FALLACY

People who toss coins for professional reasons, like cricket captains, and other gamblers, are apt to jump to the wrong conclusions when they face long sequences in which the same result is repeated. Suppose that in three successive test matches, out of a sequence of five, the English team has got tails. The captain will perhaps think: "I should assume that I shall get a head at the next match." Likewise, when *noir* turns up in the roulette at Monte Carlo ten times running, most of the punters put their money on the *rouge*. In both cases this is utter nonsense: coins or roulettes *do not have a memory and every throw is like the first one*. However many times tails will be repeated, the probability of tails turning up again is *exactly one half*, and the same for the roulette as regards the colors. Although what I have just said is undeniably true, as I shall discuss further later, it is an uncomfortable fact, since it appears to be rational to act as the punters do. The problem is that the punters have the right idea, albeit about the wrong facts, as we shall now see.

Let us consider the cricket captain who after three tails decides that another tail is unlikely. What he has in his confused mind is shown in table 2, where all the possible sequences of four throws are displayed. The captain looks at the table and complains to me about my nagging about his faulty probability theory: "I have told you, he says, there are altogether sixteen configurations of which only one with four tails, whereas there are four with only three tails, so when I had three successive tails I was damn right to bet against another tail."

Table 2. Tossing sequences of four coin throws

H = Heads, T = Tails.

The probability of each sequence, in which the order of the results is significant (fine graining), is 1/16, the denominator being the total number of outcomes. The probability of a given number of tails, disregarding the order of the results (coarse graining), is given in the last row.

Number of tails:	0	1	2	3	4	Total no. of outcomes
	HHHH	THHH	TTHH	HTTT	TTTT	
		HTHH	THTH	THTT		
		HHTH	THHT	TTHT		
		HHHT	HTTH	TTTH		
			HTHT			
			HHTT			
Number of outcomes, *r*	1	4	6	4	1	16
Probability, *r*/16	0.06	0.25	0.38	0.25	0.06	

The mental confusion involved in this argument is that the cricket captain is using coarse graining when fine graining is essential. This is so, because, if having reached TTT he puts his money on TTTH, he knows only too well that this outcome is different from HTTT, since he did not get an H on the first day of the sequence: the *order* of the results is vital. Thus the fact that there are four outcomes with three tails is entirely irrelevant because only one of them contains the three tails that the poor captain has so far endured. This shows that fine graining must be used, and in fine graining, of course, all the sixteen outcomes are equiprobable. TTTT is just as likely as HTHT, say, as indicated at the head of the table.

There is another way in which you can convince yourself that our cricketeer was wrong, as you can see from the table. When he had three running tails, he was in the way of generating either the last sequence TTTH displayed under "3," or the single sequence TTTT displayed under "4." These are *two* possibilities and both are equiprobable, so the chances of getting another tail are exactly identical with that of getting another head. As coins have no memories: however long a sequence of tails might be, the probability of heads in the next throw is always one-half.

One almost final remark. The way in which people use probabilities is very often more an exercise in psychology than mathematics. As I have just said, the sequence TTTT is identically as probable as THTH, but, because people tend to remember such results as "4 tails" or "2 tails" respectively, they automatically use, wrongly, coarse graining (ignoring the order) when fine graining is required. Moreover, because an all-tails sequence is so conspicuous, they notice it and react to it, without realizing that, in the fine graining that they should be using, there is absolutely nothing out of the ordinary with such a sequence: *all* sequences are equiprobable, as we accepted in table 2. Another strong psychological problem is that, however much people are told that such devices as tossed coins and roulettes have no memories, they nevertheless act as if such memories existed, partly on the implicit grounds that, after all, when we toss a coin a hundred times, we do approach one frequency, 1/2, and not any other, so that the probabilistic system must have some sort of inner memory that keeps a running count of how things go. This, I repeat, is a serious fallacy, which I hope to clear up soon. Meanwhile: keep your eyes open the next time you visit Monte Carlo.

And now for the really last remark. The concept of equiprobability appears to be a priori, which seems to throw the whole concept of probability into an unnatural limbo. We have already seen in the d'Alembert example that determining equiprobabilities is a serious task, and I shall now provide a further example. If you count the total numbers of H's and T's in the body of table 2, you will see that there are thirty-two of each (the total of sixty-four is of course the number of fine-grained outcomes, sixteen times four). Suppose that I take thirty-two balls labeled H and thirty-two labeled T and put them in a bag, and that I take four at a time from the bag. I will observe of course, lots of four-ball combinations each with the same labels as in table 1. Will they be equiprobable? Not at all: if my first draw is THTT, before the second draw there will be only twenty-nine T and thirty-one H left in the bag, and the second draw will be taken from a *different sample* than the first. If you want to reproduce table 1, in which each coin toss is done in entirely the same conditions (the events are thus said to be *independent*), it is necessary to *return each ball to the bag* after each single draw.

It follows from the two examples discussed that equiprobability has to be decided, like any rational empirical decision, on a carefully analyzed empirical basis. Once such analysis is exhausted, equiprobability may be asserted as an assumption to be tested empirically, by comparison with observed frequencies. This, however, is not as straightforward as it sounds, and although it is certainly a desirable program, it entails some problems of mathematical rigor, which I shall discuss later (note 19). But before we do anything, we must have a look at the concept of frequency.

FREQUENCIES

In order to discuss frequencies, I have done a bit of coin tossing in table 3 and, so as to compare with our previous work, I present the results as sequences of four consecutive throws, repeated altogether fifty times.[8]

To start with, and in order to dispel the fallacious Monte Carlo notion that a sequence such as TTTT is rare, it is easy to analyze the table in the fine-grained sense to find the frequency of that sequence (five events, frequency 0.10) and compare it with, for example, THTT (four events, frequency 0.08). As you can see, there is nothing much to choose between the two, and the equiprobability assumption for all sixteen sequences, which gives to each a probability of 1/16, that is, 0.06, is reasonably compatible with these results.

Table 3. Fifty sequences of four coin throws

This table should be read from left to right and top to bottom, and it is the same as two hundred consecutive tosses.

TTTT	TTHT	HTTT	HHTH	TTHT	HHHH	TTHH	THHT	THTT	HHHH
TTHH	THTT	HHHH	HHTT	HHHH	THTT	HTHH	THHH	TTTT	HTHH
HHTH	HTHH	HHHH	HHTT	THTH	THTH	HTTT	THHH	THHT	HHHH
TTHH	HTTH	HHTH	TTTT	HHTH	TTHH	HHTH	TTTT	HHTT	HTHH
HHHT	THTH	TTTT	TTHH	HTHT	THTT	TTHH	HHTT	THHT	HHTT

It is useful to analyze table 3 in coarse graining, so as to be able to compare the corresponding frequencies with the probabilities listed in the last row of table 2. The *frequency* of a given event is the number of times this event happens (*r* in the table) divided by the total number of trials *n*. We calculate these frequencies in table 4 for two values of *n*, 25 and 50 respectively (always counting consecutively), and it is clear from the table that, as the number of trials increases, the frequencies approach the probabilities, as is best seen from the graphs in figure 1.

We should also notice in figure 1 that the black line that gives the probabilities has a very characteristic bell-like shape. This curve is called a *probability distribution*, since it gives the probabilities pertaining to a number of different outcomes. These outcomes are all *independent*, which means that the realization of any one of them does not affect in any way the occurrence of any other. (This is clear when we are dealing with sequences of coin tosses, but remember that in the equivalent example of balls drawn out of a bag, they had to be replaced in the bag at the end of each draw so as not to affect the next outcome.) The bell-like probability distribution curve is characteristic of such independent events, and it is called a *normal probability distribution*.

Table 4. Frequencies and probabilities
for the sequences of table 3

The frequencies f are counted for the first 25 ($n = 25$) and for the total of 50 ($n = 50$) of the sequences in table 3. The variable r is the number of sequences with the number of tails stated in the first column. Coarse graining is used, since the order of heads or tails in each sequence is disregarded, only the total number of tails being considered. The probabilities stated are obtained from the last row of table 2.

No. of tails	n = 25		n = 50		Probability
	r	f	r	f	
0	5	0.20	6	0.12	0.06
1	6	0.24	12	0.24	0.25
2	6	0.24	19	0.38	0.38
3	6	0.24	8	0.16	0.25
4	2	0.08	5	0.10	0.06

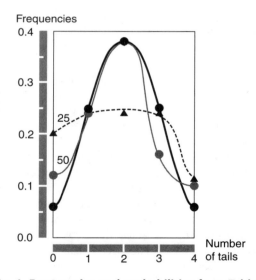

Fig. 1. Frequencies and probabilities from Table 4

Frequencies and probabilities must be read from the scale on the left. The gray and broken lines are frequencies for twenty-five or fifty sequences as shown. The black line gives the probabilities.

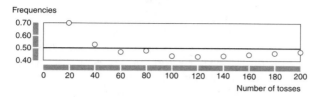

Fig. 2. Frequencies of tails as a function of the total number of tosses for the two hundred single-coin throws of table 3

The question of the relation between probabilities—which are working assumptions based on as much understanding as possible of the nature of the possible outcomes of an experiment—and the frequencies, which are entirely experimental, is of course philosophically very important, so that we shall look at one further example. The events recorded in table 3 can of course be treated as two hundred single tosses of one coin and, when this is done, it is easy to count the number of tails in the first 20, 40, 60 . . . 200 throws and thus calculate the corresponding frequencies. These are given in figure 2, where we see that, indeed, the frequencies appear to approximate the assumed probability one-half, as the number of events increases. We appear to have once more a situation similar to the one that Newton discovered when he defined instantaneous velocities, namely that, as a given point is approached, a certain quantity becomes stable, that is, it *converges*. In the present case, we would hope that, as the number of tosses increases well past the value of two hundred given on the right of the horizontal axis of the figure, the frequency "converges" toward the assumed probability of 0.5.

As a fairly intuitive picture this is not too bad, but the sad mathematical reality is that the question of convergence for the frequencies as the number of events increases is far more delicate than in Newton's case. It entails, in fact, some very thorny problems that must be discussed, if we do not want to fall into philosophical error as to the nature of the probabilistic objects that we are trying to grasp. But everything (alas, a *bit* of everything) in its own good time.

It is high time, however, that we grapple with perhaps the most important of the problems that we have so far sketched. I have said that it is not quite right to contrapose probability to determinism, because the opposite of a deterministic action on a coin would appear to be a situation when absolutely nothing at all could be said of the outcomes of sequences of coin tosses. We know, instead, that chance has its rules, and that it is an empirical fact that frequencies do not vary *sans façon* but, on the contrary, *appear* to have some sort of convergence as the number of events increases, as figure 2 illustrates. Since this is perhaps the *pons asinorum* of probability theory it is entirely appropriate that we bring in an ass to help.

THE ERGODIC PRINCIPLE: BURIDAN'S ASS GOES PLACES

One of the problems behind the difficulty in understanding what probabilistic processes are, such as the results of coin or dice tossing, is that one tends to put too much emphasis on the coins or dice or such similar objects, rather than on the whole system, hand-coin, hand-dice, roulette-ball, and so on, from which a probabilistic result arises. I would like now to begin to investigate systems that are capable of producing probabilistic events—let us call them *probabilistic generators*—in order to understand their nature. What I want to do is to construct a general model of such systems. As with all model construction, although its characteristics and functioning rules are taken as plausible, the validity of the model depends crucially on how far it can be used to map empirical facts. In order to construct such a model, I shall bring back the hapless ass, discussed in chapter 8, who was metaphorically placed by his creator, Buridan, between two identical heaps of hay and died of hunger, because he could not find a sufficient reason to feed from one heap rather than the other.

Seven hundred years later, the resuscitated ass comes up with a bright idea: he will feed from any one of the two heaps for just one minute at a time, then rush to the other and feed again for one minute, and repeat this delectable activity until the two heaps are empty and his belly replete. In doing this, he reasons that all that the principle of sufficient reason required from him was to treat both heaps equally, and thus spend exactly the same amount of time on each. Buridan, from his empyrean, reproaches him: "My ass, my ass, have you not forgotten that you had no valid reason to start on the right, as I saw you doing, rather than on the left?" Centuries of deep thinking have not failed the ass: "Master," he says, "I have spent well over twelve hours munching hay, and of these 720 minutes I spent only one acting without reason. Had my legs not been stiff after so much time reading in public libraries, I would have spent only one second at a time on each heap, and my philosophical aberration would have lasted only one second out of 43200. Not a bad score for an ass, Master, and one that could be improved almost indefinitely."

Imagine now Buridan's Ass Mark II at the precise center of a perfect circle of identical heaps of hay separated by identical arcs of $10°$ each, on which he performs his equitable mastication: here we have an example of an *ergodic system*, which has a large number of states, all perfectly equivalent, and such that the total sojourn time of the system in each state[9] is always the same. Obviously, if such systems existed, they would be straightforward manifestations of the principle of sufficient reason, in the sense that all the states are identical, and all are treated in an identical fashion.[10] In physical systems, the number of states is huge, and individual sojourn times of visitation of each state can be extremely short, so that the asymmetry introduced by the starting step becomes negligible.

Notice, however, that in order to diminish this asymmetry it is important to have a very large number of visitations of each individual state: in other words, we require a *very large* number of events for an ergodic system to be an adequate probabilistic generator.

I must discuss further our good ass's activities. First, in order to satisfy Buridan, he must make his visitations of the various states (hay heaps) without any recognizable order, because if some order existed in his perambulations, then this order would require a sufficient reason, which of course cannot exist in such a perfectly symmetrical system. This means that the ass must move in an entirely spontaneous manner, never keeping track of what he has done, because such a track or record would break symmetry. To put this in a slightly different way: Buridan's ass must keep no memory of his movements, because if he does, he could then repeat exactly the same total trajectory many times, and this repeated trajectory would break the condition that there is no sufficient reason for one trajectory to be preferred over any other. Clever as our useful ass is, he must act as a complete but efficient idiot: the implacable Buridan would immediately castigate any deliberation in whatever he does.

If we consider as an outcome the event which corresponds to one state of an ergodic system being visited (which is no more than a metaphor for the state producing a corresponding event, such as a coin falling on heads) then, because of the complete equivalence of the states, which licenses the application of the principle of sufficient reason, the outcomes are all presented without any recognizable order. Nevertheless, the events that correspond to each individual state must appear in the same numbers, that is, with the same frequencies, in order to satisfy that principle. I must stress that I understand the states here in the fine-grained sense, so that it is possible that many such states might correspond to exactly the same outcome in the coarse-grained sense.[11]

I am not suggesting that I have proved anything at all: what I have done is produce a *model* of a probabilistic generator. Thus, if the system of my hand and my coin were an ergodic system of the type described, because this system has only two states, heads or tails, then heads and tails should appear with the same frequencies over reasonable long periods of time, that is over reasonably large numbers of outcomes (because it is only then that the model is valid). Like with any other physical model, its validity must be checked in any individual case, and, naturally, we must expect the model to fit the facts within some margin of error.

The introduction of the ergodic model helps to relieve the pressure on the circularity of Laplace's definition of probability. In fact, if we invoke that model, we may define the probability of an event as the ratio between the number of outcomes favorable to the event, divided by the total number of outcomes. We do not have to qualify the outcomes as being equiprobable because that is built into the model. I hope that the reader will appreciate that all that I have done is, not to eliminate the thorny question of circularity, but rather to push it back into the underlying model of probabilistic systems. Like with all such models in nature, we require this model to stand or fall by its ability or otherwise to save the facts. The most important result is that, by using the ergodic model, the concept of equiprobability ceases to have an a priori meaning or to reflect a degree of ignorance about the system. Although this model is of course an ideal one, based on the perfect symmetry between fine-grained states, the features of the model have as much objective significance as those of any of the major models used in science, in the sense that it is used because it can be mapped onto some natural systems while saving facticity.[12]

There is one other feature which I have assumed for ergodic systems that begs one of the fundamental questions of probability. The way I have described sojourn time in the model, or what is the same, repetition of the production of events, is important, because it is only if we take a long period of time that we can expect the ergodic system to equalize in treatment the various states in it. There are indications toward this situation in my description of the system. Equalization is required by the principle of sufficient reason, at least for a model that is theoretically perfect (as models are), with perfectly equivalent states. But we have introduced a completely new feature, namely that we allow that principle to break down over a short period of time, as long as it works on average for long times. This is the way, in fact, that Buridan's Ass Mark II was able to dispose of its initial breaking of the principle, and it is clear that, by the same token, he can be permitted to break the sufficient reason condition from time to time, for short periods of time, as long as *on average* the condition holds.

Why can we afford to take this freer view of the principle of sufficient reason? I shall try to answer this point: I have repeatedly said that the validity of such a principle stems from its entrenchment during the history of our species; and that this entrenchment arises necessarily from macroscopic phenomena only. We are now beginning to recognize that macroscopic phenomena are like rafts floating on a sea of underlying microscopic events.[13] Thus, all that we can require of the principle in order to save its facticity as macroscopically entrenched, is that, although it might break down at the microscopic level, these breakdowns should *preserve the average* behavior of the system, which is what determines its macroscopic properties. The rafts must appear to behave classically as saving the principle of sufficient reason, although this is only because microscopic events, for which we are not really entitled to use the principle (since it is not entrenched in this context), average out any break-downs of it, so as to make them unobservable at the macroscopic level. As you can see, all that I am doing here is appealing to a multilayered approach to our understanding of nature, in which we save entrench-ment, at the outer or macroscopic level, by postulating some form of behavior for the underlying microscopic level, for which the rigid appli-cation of the principle of sufficient reason could not be justified.

In summary: we map probability generators by means of ergodic models, which automatically define probabilities for the states of the gen-erator, that is, for the possible outcomes. On making the generator work, frequencies will be experimentally obtained. These can be expected to map adequately the probabilities, as long as the total number of outcomes is suf-ficiently large to average out any underlying sufficient-reason breakdowns. (This trend was illustrated in figs. 1 and 2.) It is significant that the mathe-matical theory of systems for which equiprobabilities may be postulated produces laws that predict this behavior. These are called the laws of large numbers and will be briefly discussed on p. 361. The reader must beware, however, that even in the limit of large numbers, frequencies cannot be rig-orously identified with probabilities. (See p. 362 and note 19.)[14]

One final remark: I hope that the reader has got a picture of the per-ambulations of our good ass through the different states of his ergodic system: they are characterized by having no recognizable order, and by being entirely deprived of memory. The sequence of steps by which the states of the system are visited is called a *random sequence*; and we are now at long last in a position to try to define the word *random*.

RANDOM SEQUENCES

I had claimed on no less authority than that of the Rand Corporation that the sequence of H's and T's in table 3 is random, whatever random means. I shall now begin to illustrate what that concept entails by pro-ducing the following sequence:

TTTTHTTHTHHTTHHHTHHTTHTHHHHHHTTHHHTHHTH.

If you look superficially at this sequence of forty entries, it is very much like the first row of table 3, but on closer examination you will realize that I have introduced one extra H every four entries in that table. As a result, we can say that the new sequence is *reducible*, meaning by this that we can give a rule to write the first *n* entries in this sequence which is shorter than merely repeating that sequence.

Let me put this statement in a more precise way. Call *sequence* 1 the one in table 3 and *sequence* 2 the one just printed, and imagine that I want to have a computer program capable of printing the first forty entries of either sequence. (Of course, the forty entries of sequence 2 entail only the first thirty-two entries of sequence 1, the other eight entries being the interspersed Hs.) We assume in each case that the necessary entries of either sequence are stored in the computer, each entry occupying one address, by which I mean one store location. The two very simple programs are displayed in table 5.

It should be clear that we could also use program 1 to write the sequence 2, simply by storing the forty entries given at the beginning of this section. On the other hand, we could also use program 2 as shown in table 5 to print the same sequence. The first approach would require the use of forty-five computer addresses and the second one, instead, would need only forty-one, as shown in the table for the corresponding programs. We can now put the concept of reducibility on a firmer basis: a sequence is *reducible* if (and only if) it is possible to write a computer program shorter than the one that would require the storage of the full sequence. Within this definition, sequence 2 is reducible, since it can be printed using forty-one addresses rather than the full forty-five. We can now define a *random sequence* by the condition that it must be *irreducible*. What this means intuitively is that there is no way whatsoever to predict what the number in a given position of the sequence should be because, if you could predict it, then you could make use of this information to write a program that would reduce the sequence.

Going back to our ergodic model with m symmetrical equivalent states, a random sequence of these states entails that no information is given at any stage of the sequence that discriminates one state with respect to any other. This is the concept, of course, that entails what we otherwise call *equiprobabilities*.

The programs we have here will help to understand a useful concept, namely that of *complexity*, which is very deeply related to randomness. The programs we have here will help to understand a useful concept, namely that of *complexity*, which is very deeply related to randomness. It should not be too difficult to accept that programs such as those sketched above could be standardized, by stating a minimum set of instructions necessary to carry out any computation whatsoever. Once you have produced this standardization, any computation or ordering operation, such as described in the table, will have a standard length corresponding, in the case of output of sequences, to the specific sequence that it is desired to output. The *complexity* of a sequence can now be defined as the length of the corresponding computer program, so that a random sequence will

Table 5. Two computer programs

Programs 1 and 2 output the first forty entries of sequences 1 and 2, respectively. The variable n is defined to be the contents of some computer address (that is one entry in the computer store). For simplicity, I have not given the sequences required to terminate both programs which, being of the same length, would not affect their comparison.

Instruction no.	Program 1	Program 2
	Store entries 1–40	Store entries 1-32
(1)	Set $n = 1$	Set $n = 1$
(2)	Print entry number n	Print entry number n
(3)	Set $n = n + 1$	Print entry number $n + 1$
(4)	Return to (1)	Print entry number $n + 2$
(5)		Print entry number $n + 3$
(6)		Print H
(7)		Set $n = n + 4$
(8)		Return to (1)
Number of computer addresses used	$40 + 4 + 1$(for n) = 45	$32 + 8 + 1$(for n) = 41

be one of maximum complexity. I mean by this that this complexity cannot be reduced, as is the case with sequence 2, for which we have seen that program 2 is of lower complexity than program 1.[15]

THE BARE BONES OF THE MATHEMATICS OF PROBABILITY

As the reader may have realized, it is not my purpose to cover the practical aspects of probability, neither for that matter the fundamentals of the subject. What I am interested in is understanding the status of probability vis-à-vis nature, that is whether it is purely epistemological, thus revealing some inherent ignorance in our knowledge, or whether it is an objective feature of nature, as manifested by the fact that the ergodic model appears to be, like any other model used in the description of nature, both necessary and satisfactory from the point of view of facticity. As the reader must have found not too difficult to guess, I am inclined to the second view. There are some difficult problems that remain, in particular how frequencies may do their job in mapping the probabilities assumed within any given ergodic model. For this to be so, frequencies must first of all *converge* to some stable values, and these stable values must be comparable to the assumed probabilities of the model. These two questions are *very difficult* to discuss, and it is impossible to do so with a modicum of honesty without understanding, a little, some mathematical problems that arise when these questions are posed. So, I shall try to discuss in the simplest possible way what the whole worry is about. Even then, we shall see that the program I have sketched cannot be fully realized, at least at present.

The first question we must tackle is how the probabilities of compound events are obtained, for which purpose we have all the information that we need in the case of the two-coin tossing discussed in table 1. I shall consider the various cases that arise, and although all this will look trite, you will soon see that we shall learn of a very important condition that probabilities must satisfy, which is not at all understandable unless we build the background step by step, as I shall now do.

Independent Events

Two events A and B are said to be *independent* when the occurrence of one does not affect in any way the occurrence of the other. In the example of table 1, if I write H(1) for head in coin 1, and so on, then if A and B are H(1) and T(2), respectively, they are clearly independent. I shall write the probability of the compound event in which A and B occur simultaneously as Prob(A **and** B), the **and** here having the same role as the logical **and** in table 3 of chapter 3. The important result is that

$$\text{Prob}(A \textbf{ and } B) = (\text{Prob}A)(\text{Prob}B),$$

where ProbA and ProbB are the separate probabilities of A and B respectively. I shall not prove this result, but it will be useful to check with the example of the two-coin tossing. The probability of H(1) **and** T(2), we have seen in discussing table 1, is 1/4. The probability of each of the separate events H(1) and T(2) is in each case the famous 1/2 characteristic of coin tossing, and their product, as required on the right-hand side of the formula, is correctly 1/4.

Exclusive Events

Two events A and B are *exclusive* if the occurrence of one is incompatible with that of the other. If, for simplicity, I write H(1) and T(2) as H(1)T(2) and similarly for H(2)T(1), both of them, clearly, cannot simultaneously occur. In a fairly obvious notation, Prob (A **or** B) will denote the compound event in which either A or B is the outcome. The rule governing this case, which is called the *additivity condition* is:

$$\text{Prob}(A \text{ or } B) = \text{Prob}A + \text{Prob}B.$$

Let us check this with our example from the two-coin toss. We have proved in the previous paragraph that $\text{Prob}\{H(1)T(2)\}$ equals $1/4$, and the same should be the case for $\text{Prob}\{H(2)T(1)\}$. Therefore:

$$\text{Prob}\{H(1)T(2) \text{ or } H(2)T(1)\} = \text{Prob}\{H(1)T(2)\} + \text{Prob}\{H(2)T(1)\} = \frac{1}{4} + \frac{1}{4} = \frac{1}{2},$$

which is the correct result that we have used in relation to table 1.

Additivity and Normalization

Suppose that A, B, C, and D are not only *exclusive* events but also that they are *complete*, in the sense that they exhaust the range of all such events that can be considered. This is the case for instance for the four possible outcomes for the two-coin toss, HH, TT, HT, TH, now listed in the simplified notation we used in relation to table 1. These four events can be taken to be A, B, C, and D; and they certainly exhaust all the possible outcomes of the experiment: if we toss two coins, we are absolutely sure that either A or B or C or D will come out. This result is entirely certain, and thus must have probability equal to one. (Zero corresponding instead to an entirely impossible outcome.) Thus

$$\text{Prob}\{A \text{ or } B \text{ or } C \text{ or } D\} = \text{Prob}A + \text{Prob}B + \text{Prob}C + \text{Prob}D = 1.$$

This condition, whereby the sum of a mutually exclusive and complete set of events must be unity (the number one) is the *normalization condition*. This is a condition of extraordinary importance: it is a warning that we must be very careful with our bookkeeping. Probabilities must be carefully measured so that they add up to unity when required by the normalization condition. Even 1.0001 will not do! In the cases that we have so far considered, it is quite easy to make sure that things go well, and that the so-called additive property required by normalization is satisfied. We do not have to reach very far in navigating through probability theory, alas, before finding in our map the frightful warning *here be dragons*, as I shall now illustrate.

Infinite (Countable) Additivity

Our complete set of mutually exclusive events contained only a total of four in the example above, so that it is not too hard to ensure that their probabilities add up to unity correctly. But even in simple problems the number of such events can be *infinity*, and I hope that the mere thought of this will not make my readers recoil in horror. Consider the following example.

Toss a coin repeatedly until either two heads or two tails appear in succession, and find the probability, call it P(n), that you stop at n tosses, where n can take the values 2, 3, ... up to infinity. The following are examples of a sequence of such multiple tosses:

HH, TT, HTT, THH, HTHH, HTHTT, THTHH, ...

I shall work out P(n) by inspection rather than by using any sophisticated mathematics. And I shall do this simply by systematically constructing sequences of tosses in order to find the successful ones, which I do in table 6. I must first of all explain how the table is constructed. The first column gives the four possible terms entailing two tosses. To obtain the second column all you need is to copy each term in the first column twice, and add either H or T. Clearly, the first column has 2^2 terms (that is 4) and the second twice this number, that is 2^3. Further columns would be obtained by the same principle and have 2^4, 2^5, and so on terms.

Table 6. Generating the sequences of tosses

The successful sequences are given in bold.

2 tosses	3 tosses	Number of row
HH	HHH	1
	HHT	2
TT	TTH	3
	TTT	4
HT	HTH	5
	HTT	6
TH	THH	7
	THT	8

The next question is the selection of the successful terms at which a sequence of tosses would stop. The first two terms in the first column are obvious, and it is also clear that the remaining two in that column, are not good stopping terms. In order to obtain successful terms in the second column, we must disregard those that have been generated by successful terms of two tosses. For example: the term TTH in row 3 would never have been reached, since the sequence of tosses would have stopped at the successful term TT. Clearly, only the unsuccessful terms of two tosses generate successful ones, at the rate of only one each.

Let us calculate probabilities. There are 2^2 entries in the first column, whence the probability of each entry is $1/2^2$ (one quarter). Since there are two successful sequences, the probability that you stop at two tosses, P(2), is $2/2^2$. In the same way the probability P(3) is $2/2^3$, which is two out of eight, as you can verify from column 2 of the table; and we can happily assume that this goes on so that P(n) is $2/2^n$ (an expression that can be proved correct). And we now come to the moment of truth: all the events we are discussing are mutually exclusive and normalization requires that

$$P(2) + P(3) + P(4) + P(5) + \ldots + P(\text{infinity}) = 1.$$

This is a tall order, but we can first of all write:

$$P(2) + P(3) + P(4) + P(5) + \ldots = 2\left(\frac{1}{2^2} + \frac{1}{2^3} + \frac{1}{2^4} + \frac{1}{2^5} + \ldots\right).$$

The bracket here is an *infinite sum*, because it must go on until infinity, and this object is called a *series* in mathematics. This particular series is known to *converge* in the sense that the more terms you take, the less you change the total sum, which thus approaches a fixed value called the *limit*. Mathematics shows that this limit is $1/2$, which means that as you take more and more terms in the sum you approach $1/2$ with any degree of precision that you wish. (Already after nine terms, the next one is less than 0.001.) It is now clear, therefore, that the right-hand side of the sum above equals $2/2$, that is 1, satisfying the normalization condition, although we are dealing with an *infinite* number of possible outcomes.

In sum: probabilities of mutually exclusive events are additive, and this additivity must be valid even when the total number of events goes up to infinity. There is one proviso. In the example that I have given, the events are labeled by a number n that takes the values 2, 3, 4, ... Therefore those events, although infinite in number, can be *counted*, because counting is nothing else than putting whatever you have in correspondence with the integers 1, 2, 3, 4, ... The condition that is in fact imposed on the additivity of probability is that this must be valid as long as the events are *countable* (or what is the same) *denumerable*, whether finite or infinite in number. (Although it might appear at this stage that counting must always be possible, this is not so, as we shall see on p. 435.) Because of this requirement that the individual probabilities be countable, the condition stated is called the *countable* (or *denumerable*) *additivity condition*. Without it, normalization might fail and thus the individual probabilities would entail numbers without physical significance because the essential bookkeeping would be violated.

Kolmogorov's Theory

The countable additivity condition was stated for the first time by the Russian mathematician Andrei Nicolayevich Kolmogorov (1903–1987) as part of a comprehensive program to set up the mathematical basis of probability on sound principles. In particular, he was the first who, by using the methods of a branch of mathematics called *measure theory*, was able to produce principles that ensure that what I have called the book-keeping of probabilities can correctly be done. His book, Komogorov (1933), was the most influential mathematical work on probability of the century. Important as this work is, its limits have to be understood because, as I have more than once said, mathematics alone cannot say anything about nature. Kolmogorov himself stated in that book (pp. 8–9) that it was an important problem for the philosophy of natural science "to make precise the conditions under which any given real phenomena can be held to be mutually independent" which, as I have already said is an important hypothesis in dealing with probability.[16] Even more, Kolmogorov wrote in 1956 that in his 1933 book "he did not state how probability is applied because he did not know it."[17]

The most important consequence of Kolmogorov's work is that he was able to state clear and precise mathematical conditions that a function has to satisfy in order to be treatable as a probability function. Some are obvious, like that it must not be smaller than 0 or larger than 1. The denumerable additivity condition we have already seen, but Kolmogorov also recognized that the normalization condition, which requires that the probabilities of mutually exclusive events add up to unity, means that those probabilities must be so given that they preserve certain carefully defined *measures* in order to satisfy bookkeeping. The result of all this is that now it is no longer necessary to think of probabilities in terms of tosses, frequencies, or the like. As long as you are prepared to work formally, it is sufficient to pick up any function that satisfies Kolmogorov's conditions and that function may be identified as a *probability*. The important thing is that you now have a rigorous and powerful *mathematical* tool that allows you to assign probabilities to a *single* system (since frequencies are not required), just as you can assign a volume to it. This is most important in quantum mechanics, as we shall see. Of course, what one is doing in this formal approach is bypassing the awkward problems of equiprobabilities and of the physical meaning of the formal probability functions defined, which as I have shown, baffled even Kolmogorov himself.

Although I cannot pretend to solve that problem, we must try to obtain an empirical view about the question of frequencies. This is necessary, of course, because even if you manage to define probabilities formally by using Kolmogorov rules, when you want to apply those formal probabilities, there is nothing you can do unless you deal with experimentally determined frequencies. I shall try to go back to ergodic systems and see how far we can make progress in order to give a physical meaning to probabilities assumed for them. The comparison of such probabilities with observed frequencies is not transparent, although the laws of large numbers that we shall now discuss help a little in doing this.

THE LAWS OF LARGE NUMBERS

We have already seen in table 4 and figures 1 and 2 that, as the number of trials increases, frequencies approach the assumed probabilities. This is so, of course, as long as the underlying ergodic model is valid since, for instance, if a coin is biased or a roulette crooked, the results will be different. The two mathematical results about frequencies that I shall review in this section may be considered, in fact, as treatments for the frequencies of ergodic models. Of course, each ergodic model will have built in a probability for the corresponding events, and the laws of large numbers predict, within certain constraints, the frequencies observed for these models when the number of events considered is very large. There appears to be for the frequencies a phenomenon of convergence somewhat similar to the one that we have encountered in Newton's definition of instantaneous velocity in chapter 6, but this similarity is deceptive.

I shall illustrate this by reference to the first law of large numbers that was discovered, which was given by Bernouilli. He considered the case of mutually exclusive events of the same type as the examples just discussed, and what he found is that, if the number of trials N increases, then the *probability* that the frequency differs from the model probability by less than a small number, say ε (epsilon), can be calculated as a function of N, and goes to unity as N goes to infinity (that is as N is given arbitrarily large values). The relevant feature here is that it is not just the frequency that converges to the probability, but rather that the difference between the two for a given N has a *specific probability* of becoming small. Admittedly, this probability goes to 1 for very large values of N, but for normal sequences of trials, say one hundred, it will still have a nonvanishing difference from unity. Suppose that for a given choice of ε this probability were 0.9. This means, roughly, that if we do ten sequences of one hundred trials each, in nine of them the difference between the frequency and the probability will be less than ε, but that in the remaining one it can be any value larger than this number, which means that the frequency could be quite different from the assumed model probability. The consequence of this is that the type of convergence of the frequencies toward the probabilities is nothing like the well-regulated and orderly process that we had in Newton's case.

It is also most important to realize that, in trying to compare frequencies with probabilities, a *new probability* appears which relates to the difference between the two terms that are compared.

Of course, much progress has been made this century on this subject, and this has resulted in what is called the strong *law of large numbers*. This law asserts that the probability of the difference between frequency and probability being smaller than a chosen ε can be made to go to unity as a function of the number of trials m, whenever m is larger than some number n, and n is allowed to increase indefinitely. This appears to be complicated, but it has a simple and important significance. As a difference with Bernouilli, we can now make sure that as we increase the number of trials, the moment will come (for a given n) when frequencies and probabilities will differ by less than ε with probability equal to 1. This statement of unit probability might appear to be beautifully final, but it is still a probability statement, and, as such, it makes frequencies somewhat fragile, because however near to unity a probability might be, if we wait long enough, anything however improbable might happen. As Kolmogorov has to say in dealing with the laws of large numbers: "But here one has to stop and remain content with the imprecision of the notion of probability."[18]

There is here a most important point that must be understood. We have seen that the relation between the frequency and the probability, as made by the law of large numbers, depends on a statement that is itself given within a stated probability. As a result, it turns out that the limits of frequencies, although they do generally behave like probabilities, *do not satisfy the condition of denumerable additivity* (except when dealing with a *finite* number of events) and cannot therefore be rigorously identified with probabilities. It is thus important to realize that *the law of large numbers does not license frequencies to replace probabilities on their own right*. All that the theorems of large numbers do is provide a foundation for the empirical use of frequencies, but the philosophical problems inherent in the definition of probability are still there.[19]

THE FREQUENCY DEFINITION OF PROBABILITY

You will remember the circularity entailed in Laplace's definition of probability, insofar as it required this definition to be applied to equiprobable states. I patched this up a little by requiring that model probabilities be defined for model ergodic systems, the states of which are not susceptible of discrimination. Although this was postulated on the basis of some symmetry property, I have to admit that the whole thing is no more than dressing up the concept of equiprobability in more acceptable clothes; and people can still claim that the circularity remains behind a new façade of respectability. Others might be quite happy to accept the ergodic model as a good background from which to construct assumed model probabilities. I shall not take sides on this matter, but, even then, I have to recognize that the presence of assumed probabilities, even if we are prepared to keep them in a limbo until they are checked with frequencies, may be considered by some purists to entail an undesirable degree of a priorism.

This was of course a view held in the heyday of positivism when everything had to start from hard fact (whatever that might be). A program was thus started in 1919 by the Austrian mathematician Richard von Mises (1883–1953), which culminated in his famous book, von Mises (1936), the purpose of which was to deal entirely with frequencies, without having to make use at all of any assumed probabilities. Von Mises based his work on the concept that you repeat an experiment indefinitely, thus generating a sequence of numbers that he called a *collective*. This constitutes what we have already defined as a random sequence, for which he asserted two postulates. (1) Limits of frequencies exist in a collective. (2) These limits are stable with respect to subsequences of the collective.

This second postulate is open to misconstruction. If we look at the random sequence of H's and T's in table 3, it is easy to obtain a subsequence in which the frequency of heads is unity by just picking up the heads alone in that table. To stop this from happening, it is necessary to assume that one knows how to generate a random sequence (if not, you must ask the Rand Corporation); and then the positions of the letters that are extracted in order to compose a subsequence must be carefully chosen at random.[20]

Despite the fact that Kolmogorov was only too well aware of the fact that frequencies are not countably additive, and therefore that they cannot be properly used, except when dealing with finitary situations, he himself attempted to revive in the 1960s the frequentist approach to probability, a not very well known event which Jan von Plato reviews in a section not inappropriately called "The Curious Reappraisal of von Mises Theory."[21]

We have now finished the main work of this chapter and all that remains is to discuss a few problems that might create difficulties of interpretation.

STATISTICAL PROBABILITY

We are all familiar with statements like "Men in the United Kingdom have 50 percent probability of reaching seventy-eight years of age," which roughly means that if we follow over a number of years a sample of, say, one thousand men now alive, 50 percent of them will die before they reach seventy-eight. There are two problems entailed here. First, the alleged follow up would be a waste of time and money if a large number of members of the sample were, for instance, HIV positive, cancer patients, or stunt men. Second, what is the relevance of such a statement for Mr. John Jones? Both these problems are better understood in the following example, where we shall see precisely who is the beneficiary, so to speak, of the probabilistic statement.

John Jones, aged sixty-two, applies to a company for life insurance. The insurers are not at all interested in Mr. Jones, just as a Monte Carlo banker is not interested in Sheikh Abul al Sadir, because they must operate on a very large sample of punters (so that the law of large numbers can be applied reasonably well). Thus, the insurance actuaries, on using the data on their files, construct a sample of, say, one thousand "identical" copies of Mr. Jones. Of course, this is not possible, but the actuaries do their best: they pick up men who are sixty-two years old, 1.75 m in height, 63 kg in weight, who live like Mr. Jones in the East Midlands, who smoke less than ten cigarettes a day and drink six units of alcohol a week, are married and not divorced, and who engage in no more dangerous sport than alpine skiing, like Mr. Jones does. From their computers, the actuaries pick up from that sample a subset of recently dead men, and they find that 50 percent of them died within sixteen years of being sixty-two. They know now that within 16 years they would have to pay the sum insured to 50 percent of the men in the "Jones" population. They can now calculate a suitable premium for all the one thousand men in the sample, so that they will be able to pay the sum insured to five hundred of them, and still make a whacking big profit. This is what they mean by saying that Mr. Jones has a life expectation of sixteen years (although the actuarial details might be much more complex). So, Mr. Jones pays his premium; and drops dead the next day flattened by an elephant.

The conclusion of what I have said is that individual statistical probabilities only make sense when the individual is made part of a sample of "identical" individuals. "Identical" in this context is taken fairly literally if one is dealing with car carburetors, or with a much greater latitude if one is dealing with living beings, for which a sample as homogeneous as possible must nevertheless be selected. In many cases, even in physics, creating the sample might mean a great deal of coarse graining, since many individual characteristics might have to be disregarded. The frequency within a *sample* is usually taken as an approximation to a probability, but a better fit may be obtained by theoretical consideration of the so-called population, of which the sample is a small part. Thus in the case of Mr. Jones's sample of one thousand men, the population might be the aggregate of all men in the United Kingdom with the same characteristics as Mr. Jones or, even better, a theoretically constructed sample of infinite size. It is sometimes possible to extrapolate from the sample to the population to improve the probability estimates from frequencies. In physics, the theoretically constructed population is usually called an *ensemble*.

PROBABILITY IN CLASSICAL PHYSICS

We have already seen that probability entered physics mainly through the Maxwell-Boltzmann statistical theory of gases. Consider a gas, say hydrogen, at a given temperature. We can assume that all its molecules are identical (which they are not, because some, albeit a very small number, will be made up of the heavier deuterium isotope), and thus, that they produce a perfectly good example of an *ensemble*, as defined in the last section, because the number of molecules in the gas is huge. We know that the temperature of the gas is a reflection of the kinetic energy of the molecules, that is of their velocities, but this does not mean that all the molecules of the gas have an identical velocity. Rather, there will be an average velocity of the molecules, with most of them at or near it. Accordingly, the molecules of the gas will have a distribution of velocities, which, like their average, will depend on the temperature. The Maxwell-Boltzmann gas is therefore a *mixture* of molecules at different velocities. The ratio of the molecules with a given velocity v to the total number of molecules can then be considered as the *probability* p_v of finding molecules with this velocity in the gas. If we represent with ψ_v the state of a molecule with velocity v, then the state Ψ of the gas as a mixture can be written as

$$\Psi = p_{v1}\,\psi_{v1} + p_{v2}\,\psi_{v2} + p_{v3}\,\psi_{v3} + \ldots .$$

Although most molecules will be found in a fairly narrow range around the average velocity, there will be molecules, albeit with small probabilities, over a range of velocities. This range may be considered to be infinite because, in any case, we know that the probabilities are countably additive. (Which means that even an infinite number of them add up correctly to unity.) Thus,

$$p_{v1} + p_{v2} + p_{v3} + \ldots ,$$

if the range here is taken to be infinite, must give the probability of finding a molecule with *some* (any) velocity in the gas; and this probability must be unity:

$$p_{v1} + p_{v2} + p_{v3} + \ldots = 1.$$

This condition is called the *normalization condition*. (I hope you now realize how important the countable additivity condition of probabilities is in practice.) Remember that what we have here is a *mixture*, and that the probabilities give the composition of the mixture. We shall see similar expressions in quantum mechanics but with an entirely different meaning.

CODA

Few subjects in science suffer the existential ambiguity that affects probability. Is it a branch of mathematics or of natural science? Is it an expression of our ignorance or is it a hard property of nature? How is it that, from apparently a priori grounds, we obtain hardheaded frequencies? Do frequencies converge in the same way as average velocities converge to instantaneous velocities? How is it that some people seem to think that the laws of large numbers make frequencies subservient to probabilities, and others believe that a frequentist approach is important? These are some of the questions that I have tried to tackle in this chapter, and I hope that the reader will be able to think back about such answers as I was able to provide.

There are many ways in which probability may be treated and used, as readers may find in the references cited. I have tried, however, to present probability as a topic in natural science, which is what we need for the rest of the book. The *physical data* required in the subject is then *random sequences*, which are generated by *random generators*, and which must never be interfered with. As always in natural science, we have mapped the random generators by a *model*, the *ergodic system*. How far this model is empirically adequate will depend on experience.

Finally, the reader must remember that Kolmogorov allowed us to define probability functions in an entirely formal way, bypassing the problems of equiprobabilities and frequencies. Because of this, such probability functions make sense for a *single* system, although of course the problem remains how to compare them with experiment, which must be done somehow in terms of observed frequencies.

NOTES

1. Both Poincaré and Kac (pronounced Kax) are in fact writing not exactly about the example that I have given but of a mathematical law of probability which parallels the result in question. Poincaré's quotation is from Poincaré (1912), p. 171, but I give it in Kac's own version.

2. For a review of other views on probability see Cohen (1989), chap. 2, or Horwich (1982).
3. See Morley (1886).
4. Nagel (1955), p. 365. Notice, however, that his approach to probability (first published in 1939) suffers seriously from the defect pointed out in note 19.
5. Symmetry, in fact, has a lot in common with equiprobability. We could say that equiprobable states are in some sort of symmetry relation. Thus, if we consider the symmetry plane parallel to the face of a coin and exactly halfway through its edge, reflection through this plane takes heads into tails and vice versa. (We have to ignore the details of the figures for this purpose, but this we have to do in any case if we want to regard the coin as symmetrical.) If the coin is biased, the symmetry plane mentioned fails to exist.
6. This example is based on one quoted by Ayer (1972), p. 35.
7. For the *Encyclopédie* article see d'Alembert (1754), p. 512. This problem is correctly discussed in Boyer (1968), p. 497, but Lines (1994), p. 181, misrepresents the game treated by d'Alembert as the simultaneous toss of two coins.
8. I confess that I have cheated: I have used some tables of random numbers in which the digits from 0 to 9 appear in a random order, which is an order that mimics that in which results appear when drawing these numbers out of a lottery machine. (See p. 313 for randomness.) In order to do this, I have taken all odd digits in the table to represent T and all even digits to represent H. I shall discuss randomness a little later, but the usual tables are extracts from tables generated by the Rand Corporation using the output of vacuum radio tubes, which is generally accepted to be random. Such tables are reproduced in many books on statistics. I have started at the top left of p. 337 in Lindgren et al. (1978).
9. This is the sum of all the times corresponding to each individual visitation of the state. Such individual visitation times need not be all the same, as Buridan's ass had at first thought, but their total sums must be identical in order to satisfy the principle of sufficient reason. The reader may refer back to chapter 12 for further discussion of the ergodic hypothesis.
10. In this context, one sometimes speaks of insufficient reason, this concept guiding the lack of difference in sojourn time between the various states. I had already mentioned (p. 340) that insufficient reason entails philosophical problems, which I have avoided in my ergodic model by requiring all the states therein to be *distinguishable* but *identical*. Notice that, although I did not mention the condition of distinguishability in the text, this is necessary bearing in mind the d'Alembert paradox.
11. Think of a die with a single 1 and five blank faces. The five fine-grained states corresponding to the blank faces all entail a single coarse-grained outcome, that is a blank as opposite to a 1.
12. This is a point well discussed by van Fraassen (1980), p. 167.
13. This somewhat poetic metaphor takes an almost literal sense in the case of the Brownian motion discussed on p. 322.
14. No better or more complete critical history of the relations between ergodic and probability theories can be found than von Plato (1994). Although the concept of an ergodic system was already given by Boltzmann for statistical mechanics in Boltzmamm (1868), he first used the words *Ergode* and *ergodisch* in Boltzmamm (1884). The first important mathematical treatment of ergodic systems in statistical mechanics was given by Poincaré (1894). It was Weyl (1916) who first equated the relative sojourn time with the probability. (See also Weyl 1914.) After this, the two fundamental papers are Birkhoff (1931) and von Neumann (1932b). An important philosophical insight was given by the Russian mathematician Alexander Khintchine (1933), who showed that probabilities as defined in terms of frequencies were a special case of the ergodic theorem. Much of the approach that I have taken is influenced by Eberhard Hopf (1934), whose work, as von Plato (1994), p. 174, remarks, later fell into oblivion. Yet, Hopf had the merit of highlighting the problems that the more mathematical approaches to probability tended to ignore. In the work mentioned (p. 51), he

wants to find out "the true origin of the fundamental laws of probability" and to move toward a "true" law of large numbers, that is, one that deals with "what is actually going to happen when an experiment is repeated a large number of times" (p. 94). And in Hopf (1937), p. iii, he wanted to give "an understanding of the stable frequency phenomena in nature," a problem that falls very largely within the Poincaré statement quoted on p. 334. Rosenfeld (1955), pp. 10–15, has a useful discussion of Boltzmann's ergodic hypothesis and of Hopf's work. See also Amaldi (1979).

I must warn the reader that the ergodic model that I have provided must be taken, as is sometimes the case in physics, as a purely heuristic model. To make it rigorous would be a major undertaking, both mathematically and philosophically. An extremely good and comprehensive discussion of these difficult aspect of the model is given in chapter 5 of Sklar (1993). See also his very clear nontechnical discussion of probability in chapter 3 of Sklar (1992).

15. The proper study of randomness in relation to algorithmic (that is programming) complexity was started by the American logician Alonzo Church (1940). This work was considerably extended by Kolmogorov (1963a) and further in Kolmogorov (1965) and (1968), where he discussed the relation between entropy and randomness. The idea of complexity in relation to randomness was very much developed by the American mathematician Gregory Chaitin (1988) and (1990). See also Kac (1983) and (1984). Good and simple but authoritative discussions on these questions may be found in Casti (1993) and Gell-Mann (1995). See also Davies (1993), p. 130.

16. Even in simple coin tossing the idea of the independence of the successive events is merely approximate. Every time you toss a coin you detach a few atoms from its surface. Although this is of course insignificant in, say, one hundred tosses, repeating tosses one million times would certainly affect the coin, and thus raise suspicions about the (even approximate) independence of successive tosses.

17. I am quoting from von Plato (1994), p. 23.

18. See for instance Horwich (1982), pp. 46–47.

19. Because of this, the link between frequencies and probabilities established by the Bernouilli theorem is far from being as strong as Lucas (1970), p. 100 suggests, thus rendering frequentist theories "otiose" in his opinion, which, as proved in the text, is far from being the case. Lucas, alas, is not the only philosopher to have misunderstood the significance of the laws of large numbers: Nagel (1955) falls squarely into the same trap. The delicate relation between frequencies and probabilities is a point very clearly made by van Fraassen (1980), p. 186.

Even if the sequence f_n converges to p for n large with probability 1, there is a noncountable number of sequences with a frequency not converging to p. In fact, the number of sequences for which this happens equals that of those that converge, since both sets are uncountable. See Cassinello et al. (1996), p. 1360.

20. As you can see, this approach leans very heavily on the concept of randomness, which is itself so closely linked to probability that one feels whiffs of the old circularity. Remember that, after all, an ergodic system is essentially a random generator, so that although it is not explicitly used by von Mises, its main feature is still lurking there.

21. Von Plato (1994), pp. 233–37. The frequentist theory of probability, nevertheless, should not be lightly dismissed. See a discussion in Howson (1995).

14

MATHEMATICAL HEAVENS AND OTHER LANDSCAPES

As far as the laws of mathematics refer to reality, they are not certain; and as far as they are certain, they do not refer to reality.
Albert Einstein (1922), p. 28.

A physicist goes off to a conference. After a week his suit's gotten soiled and crumpled, so he goes out to look for a dry cleaner. Walking down the main street of town, he comes upon a store with a lot of signs out front. One of them says "Dry Cleaning." So he goes in with his dirty suit and asks when he can come back to pick it up. The mathematician who owns the shop replies, "I'm terribly sorry, but we don't do dry cleaning." "What?" exclaims the puzzled physicist. "The sign outside says "Dry Cleaning'!" "We do not clean anything here," replies the mathematician. "We only sell signs!"
Alain Connes in Changeux et al. (1995), p. 7.

Natural science, we hope, deals with objects like tables, bacteria, and quarks, but in ordinary discourse people also refer to qualities, such as redness or goodness. Whereas *this* tomato is red and singular, some people will say that it has the property or quality of *redness*, which is a *universal*. Belief in objects is roughly grounded on causation, in the sense that my belief that there is a table in my study is caused by there being a table in my study. My belief in universals, if I were to have such a belief, is not so easily grounded because I cannot so readily claim that my belief in whiteness follows from the fact that I perceive whiteness: I might perceive a white object, of course, and this will ground my belief that I perceive a white snowflake. And some people might claim that this is the end of the story.

Universals, however, have greater generality than the qualities to which they are attached. There are many shades of red, but there is only one *redness*. Even a blind woman who does not understand what *red* means can learn to use redness correctly: all she needs to know is that she can apply it to tomatoes and rubies but not to bananas. (In fact, Jeffrey Elman's neural networks have learned to use universals so understood.)[1]

Because of their great generality and abstract nature, people sometimes believe that, whereas ordinary properties are merely *contingent*, universals transcend our contingent reality, existing so to speak in a world of their own, where time, and space, and the ordinary constraints of nature do not apply. There are of course nonmaterial objects about which people believe, such as angels, the tooth fairy, and mathematical points; and if we have to try to understand the nature of such beliefs, we are very likely to have to move away from philosophy to theology, a move which I would warmly welcome, since one would at least be able to define fairly clearly the level of discourse that is engaged. This is not, however, a move likely to be approved of by the many mathematicians who believe in the "reality" of abstract mathematical concepts.

Readers will remember that there are two views of nature, a *monist* one which holds it to be made up of a single, material, substance and a *dualist* view, which superimposes on the latter a second, nonmaterial sphere, to which I shall roughly refer as the *spiritual sphere*.[2] What is important is that dualists claim for that sphere the same ontological status as for the material one: it exists just as my desk exists, and its existence is, and always was, and always will be, fully independent of the human mind. It would be oversimplifying, on the other hand, to believe that those who stick to the single-substance approach, let us call them roughly materialists, reject the existence of spiritual objects. This is not so: the main difference is that (at least some) monists hold the view that spiritual objects, as life itself, are so to speak by-products (or more elegantly *epiphenomena*) of matter, whereas the dualists claim that they have an entirely independent form of existence, one that is sometimes held to be of a superior nature, being necessary as opposite to contingent and eternal as opposite to temporal. It is for this reason that mathematicians who hold that abstract mathematical objects have this type of "real" existence are usually called *realists*, an expression that might mislead the inexperienced, since their form of realism entails asserting that what most people hold as real is less real (because it is less durable) than the world of mathematical objects.[3]

I hope that this brief introduction will not drive too many people into despair or contempt for its blatant disregard for detail. If I have created paradoxical confusion, this is perhaps not too inappropriate, since it probably maps the state of affairs in the subject. It is perfectly possible, for instance, for a mathematician to hold the view that mathematics does not engage with reality but only deals with signs; and to be at the same time a convinced "realist," as the author of my second epigraph, the French mathematician Alain Connes, a Field's medalist (the nearest thing to a Nobel Prize in mathematics), is.

Be it as it may: curiously, it is in dealing with mathematics that we shall experience our first serious encounter with the problems entailed by the nature of nature; and we have a lot of hard work in front of us. Contrary to my general inclination, it will be necessary to discuss in some detail some of the opposing views so briefly mentioned here, before I can begin to propose my own form of order into the subject, which of course will be nothing but disorder for others. This, alas, is all to the good: if you want to reach unanimity in problems philosophical, you could just as well wait for rigor mortis.

My first task will be to try to describe what grounds people use in order to justify their beliefs: this will be all the more necessary if we want to understand how people can claim an ontological status for abstract objects.

EMPIRICISM AND BELIEF

If challenged, I will defend my statement that "there is a table in my room" on the grounds that there is actually a table in my room, and that it is possible to find a causal chain between that event and my belief, in the Humean sense, that whenever there is a table in my room I believe that there is a table in my room. The justification of belief is nevertheless delicate. Suppose that a is a certain belief, such as "Jane owns a blue Volkswagen." This may be true and yet my statement: "I believe that a" may be ill grounded as a belief claim, in the sense that there is no causal connection between a and my belief.[4] In fact, I can assert "I believe that a" on the grounds that I have seen Jane driving a blue Volkswagen, and it may be actually true that Jane owns a blue Volkswagen: but it was in the garage being serviced; and the car that I had seen Jane driving was a rented one. Therefore, there was no correct causal chain between my perception and my belief, although my belief was correct in point of fact. Naturally, this worry about the grounding of belief should really affect only those who entertain an empiricist position, but even theologians most often take the view that the mere truth of a statement should not be invoked as sufficient grounds for believing in it. (Historicity, revelation, and so on are required steps to be very carefully analyzed.)

UNIVERSALS AND PLATONISM

Let me first explain why I am interested in universals at this stage. It can be argued that when I observe three lines that form a triangle, and I associate with that perception the idea of a *mathematical triangle*, the latter is a universal that embodies whatever all observed triangles have in common (having three sides which are straight lines that cut at three points). As already adverted, however, not everybody believes in universals: *nominalists* are those who deny that anything exists over and above individuals, as John Locke (1632–1704) did and, in modern times, Willard Van Orman Quine (1908–2000), perhaps the most important postwar American philosopher.[5]

Plato (ca. 428–347 B.C.E.) was the great champion of universals, but the use of the expression *Platonism* to describe a philosophical school in mathematics is comparatively new. It was Paul Bernays, a collaborator of David Hilbert at Göttingen (of whom more later) who at a lecture delivered at Geneva in 1934[6] coined the phrase *mathematical Platonism*, now very widely used. Although mathematical Platonism is not such a precise doctrine as Plato's, because different mathematicians adopt slightly different variants of it, I shall very briefly review Plato's ideas, of which we have already had some advance knowledge (p. 96).

With the myth of the *Cave* in the seventh book of the *Republic*, Plato attempted to demonstrate that the perceptual world is merely a world of appearances: for the prisoners in the cave, with their backs to the entrance, their only "reality" is the shadows of the events outside the cave. This could be a useful parable to illustrate the relation between our percepts and the world "out there," which we have already accepted as an unavoidable concept. But this is not Plato's purport in his parable because he goes much further. What for him "out there" is, is a truly independent world of *universals* or *forms*, inhabited by such things as *beauty* and *mathematical triangles*. Humans have *souls* that are immortal and once dwelt in that world of forms, so that we are born with a *recollection*[7] of them: when we see a triangle, we *recollect* the mathematical triangle to which our soul had been exposed. Likewise, a work of art is more or less beautiful depending on how much the universal *beauty* is attached to it.

Plato enjoyed a long life, and the ideas I have attributed to him are a drastic distillation of the various versions of them that he exposed. But the purpose of my rapid review is that, whether you accept Plato or not, you can construct, as I have shown, a "Platonic" doctrine that is satisfactorily consistent: the material world is contingent, whereas the world of forms, like our souls, which once belonged to it, is eternal. Communication between the two worlds is provided by the soul's capacity for *recollection*. This, as you can see, is *ontological metaphysics* in which you postulate a dual structure of nature.[8]

Plato's doctrine is very consistent if one postulates an eternal soul but, alas, not everybody subscribes to this idea. Plato's pupil, Aristotle (384–322 B.C.E.), accepted the concept of the soul but held that the latter was not capable of existence separate from the body, thus departing from true dualism. It was Plotinus (204–270 C.E.) who revived the concept of the soul, which became a central tenet of Christianism. Christianity, in a different context, gives us a wonderful example of clear thinking in treating a problem of dualism, in this case that of the relation between the Father and the Son. At the beginning of the fourth century, this was a serious theological (and I am afraid also political) problem. The Arians disputed the prevalent notion that Father and Son were of identical substance, or *Homoousious*.

A Church Council met at Nicaea (in Asia Minor) in the summer of 325 C.E. The question I have mentioned may appear to many people today academic, but so passionate its discussion was, that chroniclers say that during that summer you could not buy a loaf of bread in Nicaea without getting engaged in an argument on *Homoousion* (the name given to the doctrine of Homoousious). What I find remarkable is that in order to tackle the problem, one of the questions that the bishops were asked to answer was: "Was there a time when the Son was not?" If Father and Son shared the same substance, the answer to the Nicaean question had to be negative, and this is what the church fathers found: Arianism became a heresy. You may of course not subscribe to the Nicene Creed, but I hope you realize how tough a question you have to answer if you want to assert anything about singularity or plurality of substance. Mathematicians who profess to be Platonists do not appear to consider themselves necessarily bound by such constraints, and it is not always easy to understand what one is talking about. But we shall have to say more about Nicaea later.

MATHEMATICAL PLATONISM

The psychological foundation for Platonism is the awareness, shared by all mathematicians, that the problems and concepts of mathematics exist in some form that is certainly independent of the individual. It is very difficult, however, to define precisely what mathematical Platonism is, since not all its practitioners appear to subscribe exactly to the same doctrine. Because of this I shall now collect a number of statements which I believe are accepted, not necessarily in their totality, by most Platonists. Some of these propositions are not entirely independent, but I nevertheless present them as separate because it is possible that some Platonists might subscribe to one part of a proposition but not to the other.

(1) Mathematics, "if true, is necessarily true, that is, it is true regardless of the contingent details of the physical world."

(2) Mathematical truths are true in all possible worlds.[9]

(3) Mathematical truths were true even before the world existed.[10]

(4) The mathematical world "exists apart from us, because, as all mathematicians agree, its structure is independent of individual perception."[11]

(5) The existence of mathematical reality "is comparable to that of physical reality but distinct from it."[12]

(6) "Our relationship to the mathematical world is exactly the same" as to the outside world.[13]

(7) Our "minds do have some direct access to this Platonic realm through an 'awareness' of mathematical forms." The nature of the communication between our minds and the Platonic world, nevertheless, is a "mystery."[14]

(8) The mathematical world exists independently of the brain and "owes nothing to human creation."

(9) Mathematical truth, because the mathematical world exists independently of human intervention, is *discovered* and *not invented*.[15]

Because of statements like (4) and (5), Platonism in mathematics is also called *realism*, as I have already said, but the converse is not necessarily so, since attempts have been made to present a realist but non-Platonic view of mathematics.[16]

Later in this chapter I shall discuss the views of the greatest mathematical Platonist of the twentieth century, Kurt Gödel, but I shall now consider the various points raised by the above notes.

The Platonic World as Anteceding the World

Conditions (1) and (2), which were asserted by the American philosopher Penelope Maddy, claim that mathematical truths are necessary. Tautologies of course are necessary, but I am sure that such a limited view of necessity would be unacceptable here. Maddy is therefore right in using the "all possible worlds" condition so as to permit the use of the word "necessity" in the wider sense envisaged by Kripke (see p. 41). There are serious problems, however, if the expression "all possible worlds" is understood also to include a nonexisting world as in (3). In many ways this is a necessary extension, because it guarantees that the Platonic mathematical world be entirely independent of the physical one, as required by (5). In a sense, the opposite argument to the Nicaean one is used: if there was a "time" (before the big bang) when the two worlds did not coexist, then they cannot share the same substance. But if this requires that the concept of a triangle be true in the primeval void, we are in trouble, as I shall now discuss.

The extension to "all possible worlds" (including an empty one) can be made in two different ways. One is by using a mathematical proposition in the sense of a rigid designator as on p. 41, and the other is by examining the mathematical proposition entirely independently of any knowledge arising from our world. The first alternative is not satisfactory in the present case because the truth value of a rigid designator as used in world number n depends, as we have seen, on its truth value in our world; and we are therefore in danger of contaminating the mathematical proposition in question with the contingencies of our world. So, in order to satisfy (3), the concept of a mathematical triangle must have been valid in the primeval void *not as examined by our minds* (which cannot be claimed to be independent of the contingencies of our world), but rather as examined by any intelligent mind that we must be able to imagine existed in those inhospitable conditions. But, before the big bang, space did not exist, and a triangle, however you might think about it, at the very minimum (that is, if you try not to visualize the geometrical objects at all as figures) must necessarily entail some spatial relations. Even with the best good will, to countenance a meaning for a triangle in such circumstances, for a mind that had never *experienced* any relations at all is, I believe, impossible.[17] There is of course a simple way out, if we accept that such existence as the mathematical objects enjoyed in that situation, was an existence in the mind of God, in which case the Nicaean question is again answered negatively, identifying now the substance of the mathematical world with that of the deity.

If any Platonic mathematicians wish to put forward such a suggestion, they would achieve instant consistency. But I never heard Platonists like John Lucas or Roger Penrose proposing such a neat solution. Even Paul Davies (1992), who specializes in theistic discussions of nature, does not seem to come loud and clear in support of this proposal. As for Hardy, he was as hardened an atheist as he was a Platonist. (See p. 36.)

I shall later give further reasons why the concept of a preexisting Platonic world is not a good one. As for its theistic version, it seems to me that since it is not overtly entertained by many self-confessed Platonists, it is prudent to try to find other solutions.

The Platonic World as a Parallel Postcreation World

I shall now deal with the points (4) to (6) above, disregarding the question of how the mathematical Platonic world happened to achieve existence. This is not very satisfactory: as regards the material world, the big-bang picture is a *plausible* interpretation of its coming about, but we are asked to believe that at the same time as material particles like electrons and quarks appeared, an entirely different but equally existing world also came about, populated by abstract ideas of triangles, sets, groups, and the like. Whereas the consensus of the individual perceptions of my study table is so far universal (my study, alas, has never been visited by anybody in a state of delirium tremens), the consensus that validates the objectivity of the mathematical world is achieved only by the selected few members of the mathematical community. Undoubtedly, the universality of mathematics is remarkable, but it could be easily explained, not by postulating the independent existence of a world which is difficult to describe, but by merely rejecting (8) and admitting that the "world" of mathematical ideas is a creation of the human mind carefully fitted to nature from whence it receives such objectivity as it possesses. But more about this later.

Communication with the Platonic World

First of all, I shall warn my readers about a type of argument that is often given in this respect and which I believe should be firmly dismissed. Pages and pages have been written with quotes from brilliant mathematicians providing anecdotal evidence for the lightning clarity with which, at the moment of mathematical illumination, mathematicians apprehend the truth of whatever mathematical results they are seeking. This has nothing whatsoever to do with the existence or otherwise of a Platonic world "out there," to be instantly apprehended by the mathematician's mind, but it merely reflects the nature of thought processes in "high" states of the mind. The experience of having discovered a mathematical theorem is exquisite and powerful, one of the most beautiful experiences that I, alas too rarely, have known, but it is nothing more than a psychological state, so that to draw from it ontological conclusions is not acceptable. Arguments based on such psychological events reek of the old-fashioned theological arguments that harnessed the wonderful experiences of life, like the view of the night sky and the smell of a rose garden, in support of theistic belief. If intensity and blinding clarity of human experience is invoked, its counterpart must never be forgotten, that serial killers also experience a sort of ecstasy that makes them feel "good."

We can now safely go back to (7), that is, to the question of how do we manage to communicate with the Platonic world, which I consider as the weakest of all the elements of the Platonic edifice. I take it that we must all agree that this Platonic world has nothing to do with matter: how does information from it then reach the brain? It is difficult to conceive that a nonmaterial object could causally act on a material one. As you remember, belief has to be supported causally; and I have already mentioned that even theologians use this principle as far as possible. It is just not good enough to invoke here a "mystery"—a word that should be expunged from scientific use because it negates scientific discourse. I believe that I see a table because some photons, which I understand how they could be reflected from it, reach my retina, which is connected to the brain via the optic nerve. This is a good and clear causal chain: even if it were wrong, it is an honest attempt at justifying belief. What is the causal chain that can start from the Platonic world and end in my brain?

How can then the Platonic world be apprehended? For it is certainly not open to our senses; hence it must be apprehended by some property of the mind or at least of some minds. The more rational view is that it can be apprehended in the same way as God is apprehended, which means, however, that the apprehension of nature cannot be the subject of natural means, but rather that it requires the gift of faith, or some such gift. This is a very defeatist view about human ability to engage with nature; and after all, even if nature were dual, it is still nature.

No decent theologian would nowadays try to demonstrate the existence of God by appealing to the beauty of the night sky or of a rose. But some scientists appear to be prepared to believe that, the Creator having obscured all his traces by the immensely complex process of evolution, has nevertheless left a visiting card[18] by the underlying (admittedly very beautiful) mathematical structure of the universe. This debases the Creator to the role of NASA, trying to proclaim our existence on Earth to unknown aliens by broadcasting mathematical formulas.

Discovery versus Invention

It is a fundamental tenet of the Platonist *credo* that mathematics is discovered and not invented (see number 9 above), a question which I shall now discuss. Mathematicians are not the only creative people who harbor such views. It is said that when Michelangelo had a Carrara marble block in his studio and carved it, he took the view that he was *discovering* the figure that he sculpted, not *inventing* it. The word "inventing" here is appropriate, although we do not use it now in the context of a work of art. Renaissance artists like Piranesi and Leon Battista Alberti, however, signed some of their works with their names followed by the word "inventor." You can see in the Escorial, the famed Spanish monastery and palace, a memorial to Juan de Herrera as "inventor of the fabric," and so on. I do not claim to know exactly what Michelangelo might have had in his mind,[19] but, whatever it was, one can speculate that he, having a deep understanding of his material, would have been able to "discover" for a given block of marble the right shape of the sculpture that could best be carved within it. When a mathematician says (see proposition number 9) that he "discovers" a theorem rather than invents it, it could likewise be argued that, given the present state of mathematical knowledge, he is able to state a theorem that adequately fits within the latter. It is in this sense, in the sense that every invention has to be adequate to the circumstances, that the inventor has to "discover" what is seemly, that the two activities go together; and it is merely a question of personal emphasis that would lead an inventor to see himself as a discoverer.[20]

To end this part of the discussion, nothing more appropriate than the following query from the late Oxford mathematician George Temple (1981), p. 274: "How do we know that our presumed mathematical discoveries are genuine revelations and not the products of self-delusion?" If you hear the echoes of theological argument here, yes, they are there, because those who avail themselves of quasi-revelatory processes must always have to overcome that doubt, one that Temple knew only too well both as a mathematician and as a monk.

I shall finish this section by considering two arguments about Platonism.

A False Argument in Support of Platonism

A popular argument in defense of Platonism runs roughly as follows. When the brain evolved, say 200000 or 100000 years ago, the tasks that humans had to perform were very simple, so that the ability to do abstract mathematics had no survival value. On the other hand, not a sufficient time has lapsed since then for the brain to have adapted to anything, let alone higher-mathematical thinking. The conclusion appears to be that mathematical ability could not have been created by evolution, and therefore, if this be so, that it must have been acquired by humanity through some form of intercourse with the Platonic world.[21]

Let me first consider the alleged simplicity of some of the tasks that primitive men had to perform in order to survive: it is argued that when ducking a missile, the human brain does not have to integrate Newton's equations of motion. (See note 21.) As a point of fact, this is not true: I have given on p. 148 a reference to recent work that shows that this is precisely what happens when a cricketer is catching the ball when fielding. It is very easy to entertain entirely unrealistic ideas about the simplicity of human or even animal action.

The second mistake in this argument is to ignore that the brain evolved basically as a *universal learning machine*, that is, one most of whose development happens after birth in response to stimuli: humans are not necessarily born with a mathematical ability, but with a *potential to learn* and thus to *acquire* such an ability. In order to recognize this, it is enough to reflect that the alleged evolutionary argument about mathematics would equally well be valid for the ability of the brain to learn Swedish (to which of course no child was exposed 100000 years ago): yet a six-month-old Swedish baby exposed to this language will recognize three sounds of the letter *a*, well beyond the perception of an American baby.[22] It is the early environment of the human after birth that does most of the action.

The third, and perhaps the most profound mistake in this faulty reasoning, is to ignore the fact that it is perfectly possible in the evolutionary process for qualities to emerge that do not have a survival value at all, not because they are not the result of a direct environmental challenge but, rather, because they are a by-product of another feature that the species did develop in their struggle for survival. A very recent and clear example is afforded by the species *Deinococcus radiodurans*, which can resist a dose of radiation several thousand times that which would kill any other living species. This species, however, could never have been exposed to such a dose because such levels of radiation are many times higher than any that could conceivably have ever been experienced on our planet. The reason why this species has this extraordinary property is that they have developed a DNA-repair system in order to sustain severe dehydration, which was the environmental challenge that the species faced; and it is this DNA-repair system that gives them the ability to sustain enormously high levels of radiation.[23] It could be argued along similar lines that a high level of intelligence might, for instance, be a by-product of other characteristics that an individual had to possess to make him or her more attractive to the opposite sex, a proposition that does not appear to be negated by modern experience.

A (Partly) Dubious Argument against Platonism

It has been argued that if mathematical verities always existed in the eternal Platonic world, then it should have been possible for mathematicians in any century to have made amazing discoveries centuries ahead of their historical times.[24] Likewise, why do mathematicians ever make mistakes? If the Platonic world is as real as my table, why have I never met anyone who mistook the latter for a pink elephant (remember the sobriety of my visitors), whereas perfectly sober mathematicians often read their Platonic database wrong?

Let me take this second point first. Let us admit with people like Alain Connes that the Platonic world is perceived just as the ordinary world is perceived. Perception of the ordinary world leads us to natural science, but we proceed slowly through trial and error: it has taken a long time until we learned to look through a telescope, for instance. Platonists could argue that the same is the case with the Platonic world, and that it is only a trained intellect who will learn to decode correctly its signals. This would be as well their defense against the first criticism. However, as regards inventions (or discoveries) ahead of their times, there are some well-known historical examples, perhaps the most notable of which is that of the spinors, to which I shall refer later. (See p. 420.)

The question about mathematical errors, however, is not such a happy one for the Platonists. When William Rowan Hamilton invented his quaternion algebra (see p. 418), he got his picture of rotations entirely wrong. Was it that his communication lines with the Platonic world had noise? It is no more reasonable to say that, in discovering his bit of the Platonic world, Hamilton read the message wrong, than to say that in inventing quaternions Hamilton had to adapt himself to the knowledge of his time, which made it impossible for him to understand how rotations could be made consistent with everything else that he held to be true. But if he used here what he knew, and about this there is not the slightest doubt, it is not very sensible to assume that in part he tapped the Platonic world, and that in part he tapped nature as he knew it. It is far simpler to understand what Hamilton did, as I shall discuss a little later, within the same framework accepted for natural science, as making progress by carefully fitting to the body of nature such cloth as available at the time.

Not Good-bye to Platonism

This has been a very long section, and I have gone into the discussion of many detailed views because the understanding of mathematical Platonism is absolutely essential if you want to develop an informed view on the strength or otherwise of the dualist view of nature. But, alas, we have not yet finished, because the work of the greatest Platonist of our times, Kurt Gödel, has yet to be addressed. We shall soon do that, but, in order to be able to place the discussion in a significant context, we still have to understand a couple of important schools in the philosophy of mathematics, *formalism* and *intuitionism*, which emerged historically in that order. To be able to make the transition toward Gödel smoother, however, I shall discuss the latter school first, but even then, in between the discussion of these two schools, we shall have to learn something about sets and natural numbers, in order to have some actual examples on which different points of view can be rehearsed.

There is one important point that we must learn from the Platonists and that we have to keep in mind: their firm gut feeling that the world of mathematical ideas has some sort of unavoidable necessity. Whatever the grounds for this belief might be, even the hardest-minded monist must take it into account, and I shall come back to this question later.

INTUITIONISM AND MATHEMATICAL PROOF

The names of some of the mathematical schools are somewhat confusing. I imagine that many would think of the Platonists as *idealists* rather than, as they are often called, *realists*. As for *intuitionism*, there is also a trap there. Most people would think of the word "intuition" in the dictionary sense of "the immediate apprehension of an object by the mind without the intervention of any reasoning process"; and although there is a genealogical relation between this concept and the school of *mathematical intuitionism*, there is a lot more in the latter. As we shall see in a moment, *constructivism* might be a more appropriate name to the school, although it is in fact usually applied only to some specific groups within it.[25] For the purposes of this book, since I am not going to get into it in any great detail, we may regard both terms as rough synonyms.

It was Immanuel Kant who analyzed the concept of intuition. *Sense intuition* (which is what the dictionary definition refers to), cannot work on its own, in accordance to him, but it must be supported by some faculty of the mind that is able to provide the framework within which our perceptual field is organized. This framework was for Kant absolute space and absolute time; and he postulated that the mind has a faculty of *pure intuition* that permitted their apprehension. Pure intuition licenses the a priori validity of the perception of space and time, as well as of the axioms of mathematics. Although of course these ideas are no longer tenable in the light of relativity theory, the concept of the *intuition* of a mathematical object has percolated through to modern times.

The leader of the intuitionist school was the Dutch mathematician Luitzen Egburtus Jan Brouwer (1881–1966), and his motivation was a profound distrust of Platonism. For the Platonist, if you wanted to prove that something exists, you did not have to have it in front of your (mental) eyes: it was just enough to know that it belonged to the world of eternal mathematical truths. Brouwer fundamentally rejected this: for a mathematical object to be said to exist, it had to be available for contemplation by the mind; and the only way in which you could achieve this was by providing a procedure to *construct* the given object in a *finite* number of steps. (It is in this sense that the *intuition* of the mathematical object comes in.) Because of this approach, the whole question of mathematical proof came under attack by Brouwer.

The concept of mathematical proof is very complex and I shall only touch upon its surface, but it is important to realize that contrary perhaps to popular opinion, what constitutes a mathematical proof is not anything as clear-cut and precise as one might imagine. This does not mean, of course, that mathematicians are not immensely rigorous. What a proof is, on the other hand, is open to argument, as we shall see in a moment. That the concept of proof varied along the centuries and for different cultures is clear. As early as the fifth century B.C.E. Greek geometry refused to rely on visual grasp, but used instead step-by-step proofs based on axioms.

(This is the Euclidean method, which so much affected Western thought.) Indian geometricians, instead, relied on a figure with relevant auxiliary lines, accompanied by no other comment than the single word "Behold!"[26]

Brouwer took the empiricist position that belief had to be causally grounded. (See p. 376.) What is quite remarkable is that no one realized until then that if you accept this, then a famous logical law, the *law of excluded middle*, must be thrown overboard.[27] To understand this I must state that law: if p is a proposition, such as "unicorns exist," then:

Law of excluded middle: p and not p cannot both be true.

This is a very simple logical result that I hope readers can accept. Even poets believe that this is the way of rational thinking. Thus W. H. Auden (1950, p. 8):

Whether conditioned by God or their neutral structure, still
All men have this common creed, account for it as you will:—
The truth is one and incapable of contradiction;
All knowledge that conflicts with itself is Poetic Fiction.

Beautiful as this stanza is, and convincing as it may sound, it is no more than wishful thinking. If we cannot causally ground "I believe that p," it does not follow that "I believe that not p" be true, *if we insist that for this to be the case this latter proposition must be causally grounded.* For example: if I were to say "I believe that p," as in "I believe that unicorns exist," everyone will know that this is ungrounded, since I cannot have experienced p ("unicorns exist"), and I cannot even invoke reliable historical evidence for it. On the other hand, if I were to say "I believe that not p" (that is, that unicorns do not exist), it is again obvious that I cannot have *experienced* not p. (Which is written $\sim p$ in logical notation. Notice that if I say "There isn't a unicorn in my room," I am not *experiencing* a nonunicorn.)

Brouwer's contention that the law of excluded middle cannot be used because it does not satisfy causal grounding has an enormous importance in mathematics, since it undermines one of the most common methods of proof used in it. This is the demonstration by *reductio ad absurdum* or *reduction to the absurd.* The idea behind this method is very simple.

From table 2 of chapter 3 (p. 52) we know that if the implication $p \Rightarrow q$ is true and the consequent q is also true, the antecedent p can nevertheless be false. (See the third row of the table.) This means that, if we want to prove that p is true, this cannot be done by correctly deriving a true consequent. The trick, instead, is to negate p, and *then* to derive a consequent: $\sim p \Rightarrow q$. If the implication is true (that is, if the reasoning used in forming it is correct), and if I find that q is false, then, from the last row of the table mentioned, ~p must also be false. This means, *if the principle of the excluded middle is accepted as valid*, that p must be true. As you can see, this is the only way in which implication can be used to prove anything as true.

Some readers might wish to have an example of *reductio ad absurdum*, and the following one is just about as simple a case as possible. From the famous Pythagorean theorem, the diagonal of a square of unit side must have a length equal to $\sqrt{2}$, so that this expression should be in some way meaningful, since it must denote a mathematical point of the straight line that contains the diagonal. If you have a calculator, you will find that $\sqrt{2}$ equals roughly 1.4142, and you can also easily work out 355/251 which is equal to 1.4143. Thus, you might think that it would be possible to write $\sqrt{2}$ as some, more accurate, fraction m/n. Naturally, if this were so, and m and n had a common factor, we would simplify them without changing the value of the fraction. So, I shall now prove the following remarkable theorem:

$\sqrt{2} \neq m/n$, *where m and n have no common factor.*

I shall assume the opposite to be true, that is, that

$\sqrt{2} = m/n$, *where m and n have no common factor.*

It is easy to find from this relation that $m^2 = 2n^2$. This means that m^2 must be even, and therefore that m must also be even, because the square of an odd number must always be odd. Thus, we can write $m = 2r$ for some integer r. If we replace this result in $m^2 = 2n^2$, then $4r^2 = 2n^2$, which gives $2r^2 = n^2$. Therefore n^2 is even and, as for m, n is also even. We conclude that m and n must both be even, and thus that they have 2 as a common factor. But by excluded middle this result and the one displayed above cannot both be true. Therefore this displayed result leads to a contradiction and must be false, which proves our theorem.[28]

The conclusion that we can derive from this little theorem is very important. Numbers that can be expressed as fractions are called *rational numbers*. When expressed as decimals, they may entail a finite number of digits as 1/2, which equals 0.5, or an infinite one as 1/3, which equals 0.333333. . . . What we have proved is that $\sqrt{2}$ cannot be a rational number and that it must be expressed as an *infinite* decimal, because any terminating decimal, such as 1.4142, is always rational, being given as 14142 divided by 10000.[29] We have therefore proved that *nonrational numbers*, of which $\sqrt{2}$ is an example, "*exist.*" (This type of reasoning is called an *existential* proof.) I put "exist" in quotes because this proof of the existence of these numbers, which incidentally are called *irrational*, depends on the use of the law of excluded middle and thus, according to Brouwer, is open to suspicion. It is, in fact, as if we had managed to prove *somehow* that the proposition "Unicorns do not exist" is untrue, and thus deduce that unicorns exist, never having actually found one.

As you can see, there is a great deal to be said for Brouwer's insistence in *constructive* rather than *existential* proofs. But if you follow this desideratum, the use of *infinitary* methods, which are methods that require procedures that we must imagine are repeated indefinitely, that is, an *infinite* number of times, must also be proscribed, as Brouwer indeed advocated. However much I like and admire the intellectual probity of the intuitionists, I am afraid that as a practical program for mathematical work, it is just not on. Huge and very useful chunks of mathematics would be left out because there is no alternative to existential and infinitary methods of work; and we would thus be throwing away results that are extremely useful. Having said this, it is often the case that constructive results are obtainable as an alternative to existential proofs, and the attitude of some pure mathematicians that the pursual of such results is unnecessary is somewhat sterile. If you want to apply mathematics, a constructive proof gives you very often a handle that is otherwise unobtainable.

The moral of this story is that if you think that mathematical truth is uniquely defined beyond any possible doubt, you do have a problem. Those who would like the world to be a reflection of some eternal truths might worry, but if you take the more modest line that in this sublunar space nothing is intrinsically *necessary*, and that this is clearly so in natural science, then you might extend mathematics a warm welcome to the club.

SETS AND NATURAL NUMBERS

As I have said, we must get acquainted with another important mathematical school, that of the *formalists*, and in order to try out their methods, I shall require some simple examples that come from the study of sets and of the natural numbers. The theory of sets is fairly modern: it was created in the nineteenth century by the English mathematician and logician George Boole (1815–1864) and the German mathematician Georg Cantor (1845–1918); and it is now recognized as one of the fundamental pillars on which the whole of modern mathematics rests. It is not necessary for us to get involved in much of the detail of set theory, but, even about the little that I shall try to do, it is important for the reader to realize that one cannot improvise on this subject: whatever we do must be done by the book, that is, by the rules that mathematicians use, and it is no good saying that "we could do otherwise." Of course we could, but then we would not be using the word "set" in the way in which it is used in mathematics. So, from now on, I shall use the word "set" in a very specific way, that will not necessarily agree with what some people might like, but it is the very well tried way which mathematicians have found. To start with, whenever I deal with a collection of physical objects, such as apples or eggs, I shall refer to it as an *aggregate*, and although I might from time to time regard an aggregate as a *set*, I shall consider these two concepts as absolutely *distinct*, as it will transpire in the course of our work.

Imagine that mathematician John has an aggregate of three apples, a_1, a_2, a_3, two bananas, b_1, b_2, and one clementine c_1. First of all, we must agree on the reasons why John has differentiated between the fruits of the same species: one apple was a Cox, the other a Granny Smith, and the third a Golden Delicious, and so on for the other fruits. John decides then to write the given aggregate as a *set*, a mathematical set being a mathematical object with strict rules as regards membership and properties. A set is denoted by curly brackets {}, and the *elements* of the set are placed inside it with two conditions: (i) the order of the elements is irrelevant; (ii) *repetition* is not allowed. Thus John's set S_1 will be:

$$S_1 = \{a_1, a_2, a_3, b_1, b_2, c_1\}.$$

Mathematician Jane now comes, observes the same collection of fruit, but decides that she is only interested in distinguishing between the species: for her the three apples are identical in the sense that a Cox and a Granny Smith are just apples. So she writes:

$$S_2 = \{a, a, a, b, b, c\},$$

because the differentiations no longer make sense, *for her*. But, because of the nonrepetition rule, S_2 now has only three elements, which is sensible, because Jane was only interested in how many different species of fruit she had available. Therefore, she writes:

$$S_2 = \{a, b, c\}.$$

We now face a serious philosophical problem: do we perceive sets? Because, although John and Jane perceive the identical aggregate, they both have written down *different* sets. It could be argued that we are more or less in the same situation as when we perceive an ambiguous figure that one moment looks like a rabbit and the next as an old woman, but this won't wash. In the case of the ambiguous figure, there is no choice: you sometimes will see the rabbit, others the old woman, and your changed perception is not deliberately and precisely determined, as in the above case, by your *theoretical* construction as to what it is that you want to *record* as a result of your perception. I thus take the view that sets are not subjects of perception but that they are *abstract constructions of our minds.*

Let us, to support this contention, look at what we have done. We started from aggregates, that is, physical objects, in order to start discussing sets, but as we moved on, the elements of a set became some abstract objects. An "apple," abbreviated *a*, as used by Jane, does not stand for any one given apple. Rather, it stands with respect to "this Cox apple" in the same relation as "redness" stands to "this patch of red." It is an abstract object, a *universal*, and it is no longer available as a possible element of an aggregate.

Realists, of course, will want to claim that sets can be perceived, but you must be careful because the concept of "perception" can be used in two different ways, even by realists. True Platonists, like Kurt Gödel, believe that the concept of an abstract mathematical set exists, not in space and time, but somehow in a Platonic world, within which it is open to "perception" in the elusive, quasi-revelatory sense in which the mind is said to be able to communicate with that world. But some realists believe that when they look at an aggregate, they actually do perceive a set, and that the latter does exist in space and time.[30] I do not have to say that this is a view that I do not entertain because, although John and Jane do *perceive* the same thing, the sets that they recognize are different.

There is another forceful reason that shows that mathematical sets are abstract objects that cannot be perceived (unless one accepts the Platonic concept of some sort of "revelatory" perception). A most important set in mathematics is the *empty set*. I have already said that the number *zero* is a crucially important number, and that without its introduction our mathematics would have been a mess. Likewise for the empty set, conventionally represented with the symbol $\emptyset$. If any reader knows of anyone who has perceived an empty set, I shall be most interested to hear. And it is no good saying that when you remove one by one the six fruits in John's basket you are left with an empty set which you can "observe": what you observe is that you have run out of fruit, and if, instead, you had six eggs in the basket, you would have run out of eggs! Both statements are clearly different and significant, as it is clear if you now have to write down a shopping list: having run out of apples, it would not do to go shopping for eggs. So, an empty *aggregate* of eggs is not the same as an empty *aggregate* of apples, whereas the empty *set* is one and only: it is utter nonsense (within the rules of mathematics) to talk of an empty set of eggs or an empty set of apples.

There is one further problem with the proposal that sets are in space and time and thus perceived. A set that contains a single element, *x*, is written $\{x\}$ and it is called a *singleton*. You can think for example of an aggregate that contains only one apple. If you think that when you perceive an apple, you also perceive the singleton to which it belongs, then you must assume that the apple and the singleton to which it belongs are one and the same thing:[31]

$$x = \{x\}.$$

But since, obviously:

$$x \text{ belongs to } \{x\},$$

then:

$$\{x\} \text{ belongs to } \{x\},$$

which is mathematically incorrect: "To belong" requires to be *inside* the curly brackets, as shown in the penultimate equation. (This is what it *means* to be an element of a set.) Demonstrably, $\{x\}$ is not inside the curly brackets in $\{x\}$, so that the last equation makes no mathematical sense. (And remember that even philosophers are not free to change well-tried mathematical rules.)

Mappings and the Natural Numbers

Given two sets, any rule that associates in some way elements of the two sets is called a *mapping*. One element of a set may be associated with several of the other, but if in each set one and only one element is associated with a unique element of the other, then the mapping is *one to one*. We have to add yet one more condition, because when you are producing a mapping it is not necessary that *all* elements of either set be involved in it: it is possible for some elements not to be mapped at all. When this is not the case, that is when *all* the elements of both sets are involved in the mapping and the mapping is one to one, we then have what is called a *bijection*. (See fig. 1.)

It can be seen at once why these ideas of mappings and bijections are important. If we think along fairly intuitive lines, it is clear from figure 1 that if we can establish a bijection between two sets, then they have the same number of elements.[32] In order not to run into circles in our definitions, one says that if there is a bijection between two sets, then they have the same *cardinality*. Consider the collection of *all* sets (called a *class*) that share the same cardinality, that is, such that a bijection can be established between any pair of them; then the *natural number n* is defined as whatever the sets in the class have in common. This is an abstract definition of the natural numbers and one that is quite useful because, given it, the usual arithmetical rules can be demonstrated. The set of *all* natural

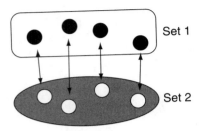

Fig. 1. A bijection, to illustrate the concept of cardinality

numbers 1, 2, 3, . . . can be defined but, as a difference with the sets so far considered, it has an infinite number of elements. This does not prevent this set from being used for counting: this is done by establishing a bijection between the set to be counted and that of the natural numbers. If such a bijection is possible, it is said that the set in question is *denumerable* (or *numerable*). In that case, the set has infinite cardinality, its "number" being conventionally denoted as ω (*omega*).

It is possible to go even further to try to build up directly the natural numbers by means of sets, and the building blocks for this purpose are constructed from the empty set $\varnothing$. The set $\{\varnothing\}$ that contains one element only (the empty set) is associated with the number 1, the set of two elements, $\{\varnothing, \{\varnothing\}\}$, with the number 2, and so on:

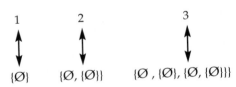

This proposal was made by the Hungarian American mathematician John von Neumann (1903–1957) in 1923, but it must be understood that, although it is the most convenient representation of the natural numbers by sets, it is, unfortunately, not unique. In 1908 the German mathematician Ernst Zermelo (1871–1956) had proposed a rather less transparent alternative, which nevertheless turns out to satisfy all the necessary conditions for the natural numbers. In this representation, the set {} that contains nothing (that is $\varnothing$, which is not written explicitly) is taken to be a single mathematical object and thus associated with the number 1. Then {{}} is the number 2, {{{}}} the number 3 and so on:

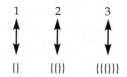

The reader should not worry very much about what the Zermelo nested sets *mean*. The important point, their bijection onto the natural numbers is, however, evident. The existence of these two alternatives, on the other hand, reveals that any attempt at a definition of the natural numbers via sets cannot be satisfactory because the very nature of the numbers differs drastically in both systems. In the von Neumann definition, for instance, it is clear that the number 1 is an element of the number 3, but this is not so for Zermelo. Thus, although set theory is an elegant way of formalizing the idea of the natural numbers, it does not remove the need for knowing in some way what the natural numbers are. If the dream was that pure a priori thought can define the natural numbers, this is the end of such a dream. Or rather one end, because mathematicians do not give up easily. It is because of this reason that considerable work was done to set up arithmetic on an axiomatic basis.

Peano's Axioms of Arithmetic

It was Giuseppe Peano (1858–1932) who from 1889 (when he was a professor of mathematics at a military academy near Turin) until the early years of the twentieth century set up an axiomatic system for arithmetic. He gave five axioms to define the natural numbers (that is numbers like 0, 1, 2, 3, etc.) but the way these axioms are presented in the literature varies widely. I shall follow Bertrand Russell, who is nearest to the original both in time and in spirit, so as to get a better historical perspective.[33] In defining the natural numbers, three primitive ideas (about which later) must be accepted as understood: *zero, number,* and *successor,* and for the time being you may think of *zero* as "0" and of *successor* as the number that follows any natural number when they are given in their natural order. The axioms are:

(1) Zero is a number.
(2) The successor of any number is a number.
(3) No two numbers have the same successor.
(4) Zero is not the successor of any number.
(5) Any property which is valid as a function of zero, and, if valid for a given number holds also for its successor, is also valid as a function of *any* number.

The last axiom is called the *principle of mathematical induction* and it permits us to do in mathematics what is so hard to justify in philosophy, namely to pass to infinity from the validity of a property known to be true only in a finite number of cases. This is so, of course, because the set of natural numbers defined by the Peano postulates contains an infinite number of elements.

The reader must beware that the five axioms above are entirely formal and do not refer to any objects, mathematical or otherwise, in heaven or on Earth. For these axioms to define the natural numbers, they must be *interpreted*, which means that some of the symbols must be identified with mathematical objects that we are prepared to recognize as such. *Zero* must be identified with "0," and the *successor* requires the identification of the number "1" (or "unity"):

> *The successor of 0 is 1.*
> *The successor of n is n +1,*

a definition which entails also that of addition.

The great advantage of a good axiomatic system such as Peano's, is that it transforms arithmetic into a set of rules that can be applied automatically (or more precisely *algorithmically*), without having to enquire about the actual significance of the symbols used. In fact, what it does is give the *syntax* of arithmetic, ignoring its *semantics*. In order to illustrate this, we must drop our instincts about the natural numbers and imagine that we do not know what the unity is: if with the symbol "1" we had denoted for the successor what we normally call "3," then the Peano axioms would work just as well, but the *natural numbers* so defined would not correspond to our intuitive series. Some clever mathematician might then discover that there are other numbers below "1" that should also be considered natural numbers (they would correspond to our 2 and 1).[34]

Having stretched our minds to accept this awesome possibility, how do we know that Peano's "1" is not somehow in the same position as our putative *successor*? The answer is that we don't: we have to accept that Peano's first axiom entails other assumptions about the unity for the whole system to work. Thus the idea that a set of axioms entails self-evident, necessary, truths is not one that we can countenance. Naturally, having lost this chance of bringing in *necessity*, mathematicians were not slow in devising new schemes.

FORMALISM

In discussing natural science we have seen numerous examples which show that no conceptual system that attempts to model nature can be closed, that is, self-contained: in order to save the facts, physical theories have to be adjusted to experiment, and it is only through the interaction between experiment and theory that science can proceed. Pure thought, the a priori, is insufficient. People, however, had greater hopes for the queen of the sciences, mathematics, and it was there that pure thought was expected to reign supreme.

At the end of the nineteenth century two magnificent attempts were made in this direction, basically hoping to treat mathematics as a branch of logic, and thus to formulate rules for working with meaning-free symbols which, once consistency was established, could be *interpreted* as mathematical objects. Bertrand Russell (1872–1970) was the standard-bearer of this movement in Britain, and he joined forces with Alfred North Whitehead (1861–1947) to publish from 1910 to 1913 the most monumental work on the philosophy of mathematics ever attempted, *Principia Mathematica*. Vastly influential as this work was in providing the ideas and the notation that are now the lingua franca in the subject, it is now almost totally unread.

Already in 1903 Russell discovered some paradoxes about sets (see the "catalogues" below) which greatly undermined the foundations of set theory. The latter was not only a cornerstone of Russell's program, but it affected even more seriously the work of his German counterpart, Gottlob Frege (1848–1925), professor of mathematics at the University of Jena. Just when Frege was ready in 1903 to publish the second volume of his *Ground Rules of Arithmetic*, he received a letter from Russell explaining his discovery, and this meant the collapse of his whole program. All he could do was add a disclaimer to his preface, one of the saddest and most noble statements in the history of mathematics, because he calmly accepted the destruction of his work without any bitterness, or any attempt at covering up the unavoidable truth.[35] Russell produced later a very clever (although not entirely satisfactory) way out of his paradox, as I shall show, but the earthquake that made Frege's construction collapse left cracks also in his own.

The desire to found mathematics on an a prioristic basis was nevertheless too strong for attempts to carry out this program to be abandoned. After the publication of the *Principia*, Russell's mathematical energy was exhausted, and the standard passed on to the German mathematician David Hilbert (1862–1943), professor at Göttingen and probably the most influential mathematician of his time, if not of his century. At a Paris mathematical congress in 1900 he presented in a magisterial way the twenty-three most important problems still unsolved, a program that inspired mathematicians up to our own time.

Hilbert's second problem is the one we must be concerned with. And it was not a mean task to present all mathematical concepts and proofs in an entirely formalized way. All mathematics was to be translated in a pure form, as ground rules for the manipulation of signs, without reference to their meaning. If such a finitary analysis showed that no sentence of the form 0 = 1 was provable within the theory, that is, if the theory was found to be free of contradiction, then we would have an absolute *consistency* proof of the theory, and hence of all mathematics. All that remained was to *interpret* the so far meaningless signs, thus identifying them with mathematical objects such as numbers, rules of operation, and so forth.

Hilbert held, in fact, that internal consistency "is the criterion of truth and existence." This meant that, if his program was achieved, then Platonism was unnecessary: "existence" need not be instilled into mathematical objects from a Platonic heaven, but rather defined rationally from the strength of the mathematical edifice. That this program was not as appealing as one might think was recognized by Ludwig Wittgenstein, who quite rightly pointed out that one of the great merits of Lewis Carroll's nonsense is that it is perfectly consistent, though of course meaningless: consistency is not sufficient to justify an ontology.[36]

Hilbert's program was of course music to the ears of the positivists, at that time the leading philosophical school, which held logic to be the fountainhead of secure knowledge. Not surprisingly, the greatest blow to Hilbert's program came from the foremost paladin of Platonism of the century, Gödel. But before we get into this part of the story, I should like to discuss the paradoxes that so much affected Russell's and Frege's work.

SELF-REFERENCE AND ITS PARADOXES

We expect natural language to be unequivocal, but that this is not so has been known for over two thousand years. I have already mentioned that one of the most remarkable properties of the universe is that it is self-referential (as the existence of this book proves), but when language itself becomes self-referential, trouble brews up.[37] The sentence:

John is a liar,

entails no problems. But if I say:

I am a liar,

then meaning breaks down: if the sentence is true, I am lying when uttering it, therefore it is not true that I am a liar. But if I am not a liar, then the sentence is untrue, and so on.[38]

You can see this trouble most clearly as follows.

Call *p* the sentence: "*The sentence q is untrue.*"

No problem here, but change over to:

Call *p* the sentence: "*The sentence p is untrue,*"

and we get caught in an endless mental loop. If "The sentence *p* is untrue" is true, then *p* is true and vice versa, if *p* is untrue, then "the sentence *p* is untrue" is untrue and *p* is true. It is self-reference that leads here to paradoxical recurrence; and we shall soon see that it plays a crucial part in understanding the strategy of one of the most famous results in the philosophy of mathematics in this century, the *Gödel theorem*.

METALANGUAGES

Russell attempted to dissolve paradoxes like the *liar* by inventing a *theory of types*, which essentially involves establishing a hierarchy, so that assertions *about* propositions at one level of the hierarchy must necessarily entail the immediately upper level of it. I shall not get involved in this theory, since it was repeatedly modified by Russell: it is sufficient for my purposes to consider a sequel to it provided by the Polish logician Alfred Tarski (1902–1983),[39] who was professor of mathematics at Berkeley, and whose influence in the development of mathematical logic was vast.

His proposals, like those of Russell, also involve a hierarchy. Natural language provides the first, lower level in the hierarchy, in what we call the *object language*, which entails mainly straightforward statements about things, such as "This table is red." If I say, however, "It is true that 'this table is red,'" I am making an assertion about a proposition in the object language, and I must therefore be in the next level of the hierarchy, the *metalanguage*. In order to avoid confusion, two rules must carefully be followed. First, all sentences must be clearly labeled as belonging to the object language L or to the metalanguage M. Second, the values of true or false cannot be used in L but only in M, in reference to propositions in L.

I shall now show how the last form I gave of the paradox of the liar can be dissolved by this method. Try:

Call p_L the sentence: "*The sentence p_L is untrue.*"

Obviously, we would reach the same stagnant situation as before, but the above line is no longer permitted, because p_L on the left, asserting as it does the truth of a sentence in the object language, must be promoted up to the metalanguage. The correct writing is:

Call p_M the sentence: "The sentence p_L is untrue."

The previous infirmity disappears: p_L untrue entails p_M true and vice versa, so that truth values can now be asserted uniquely. Notice that in the metalanguage I cannot assert p_M to be true or false: a higher order metalanguage is required for this. Thus, any possible paradox at the level of M is avoided, and so on.[40]

All this is impressively clever, but the principle of conservation of human misery requires that we pay for it. The first problem is that, as you can easily see, we have created an infinite recurrence, because we have to step up indefinitely from one level to the next, never being able to stop. This is a problem that was first clearly seen by the Oxford philosopher Francis Herbert Bradley (1846–1924) not just for the truth values, but for any relations to be established between individuals.[41] Such an infinite recurrence, however, is a feature that would not be unwelcome to some Platonists, as creating a staircase to their Platonic heaven.

There is an even greater problem, however, because the whole idea of a hierarchy between the object language and the metalanguage rests on the often unstated assumption that, whatever is asserted in the metalanguage, it leaves the semantics of the object language unchanged; and this is, at best, doubtful. This is a difficult question, and I hope that the following example might go some way toward clarifying it. Consider the sentence in the object language:

You shall have the report yesterday,

which appears to be semantically nonsensical although syntactically correct. It could, for instance, be a mistranslation into English by a Hindi speaker, in which language *kal* means both "yesterday" and "today." But, if I now state in the metalanguage:

The sentence "You shall have the report yesterday" is correct,

then you might imagine that it was the answer to a command such as: "You must produce the report at once." Thus, given that the sentence is said to be correct, its syntax entails that the sentence has to be understood as an *oxymoron*; and this semantic shift has been introduced by the metalanguage assertion about the sentence. A recent literary example of this feature of language is: "He has died tomorrow," used by A. L. Kennedy (1998), p. 88, where the context helps the reader to recognize the sentence as semantically correct.

Since this is an important point, I shall consider a mathematical example. Suppose that, as it was the case for a long time, we reserve the word "number" to indicate positive or negative integers, as well as all rational and irrational numbers. (Rational numbers are fractions, and irrational numbers are numbers such as √2, which cannot be expressed as fractions.) We all know that √(a) = b means that b^2 must equal a. If the latter were −1, however, there is no number the square of which could be −1, since the square or any number, be it positive or negative, must always be positive. So, the sentence

√(−1) *is a number,*

is not just true or false but *meaningless* because √(−1) cannot *refer* to any number within the accepted meaning of "number."[42] This sentence is, or is alleged to be, in the object language of mathematics, but I can now use metalanguage (also called *metamathematics* in this context) to assert:

The proposition "√(−1) is a number*"* is true.

This sentence, however, brings out the symbol √(−1) from the realm of the meaningless to that of the meaningful mathematical signs: it entails that the concept of number, which we started from natural numbers to positive and negative numbers, then rationals (fractions), and then irrationals (like √2), is *now to be extended* so as to admit some new "numbers" *with negative squares.* (These are the *complex numbers,* discussed later in this chapter.) This metamathematical sentence thus signals the formulation of a mathematical program that *modifies the concept of number.* Therefore, the word "number" obeys a different semantics in the two displayed lines shown, and this semantic shift is associated with the use of the metalanguage.

We are now ready to go for the Gödel theorem, but we shall have one more short section, in order to discuss the paradox about sets found by Bertrand Russell, which caused the collapse of Frege's program.

THE CATALOGUE PARADOX

I want to discuss this paradox not just because of its historical interest but because it will reveal a property about sets that, strangely enough, will be important in understanding one of the most remarkable results in quantum mechanics. (See the Aspect experiments in chapter 18.) Let us consider the following straightforward definitions:

Set A = all and only all the catalogues that list themselves.

Set B = all and only all catalogues that do not list themselves.
Set C = set of all catalogues.

It appears self-evident that A and B exhaust all the possible classes of catalogues and that, necessarily,

$$C = A + B.$$

Define however a new catalogue β (beta):

β = *a catalogue of all and only all the catalogues that do not list themselves,*

that is, in principle, a catalogue that lists all and only all catalogues in the set B.

However:

either β lists itself and thus belongs to set A, in which case it contradicts its definition, which should include only catalogues that do not list themselves,

or it does not list itself and thus belongs to set B, in which case it contradicts its definition, because it does not list all the catalogues in set B.

In either case, cannot be a member either of set A or of set B, which means that the neat classification of the set of all catalogues in two disjoint subsets (subsets that do not have any element in common) is not possible:

$$C \neq A+B.$$

This suggestion appears at first sight bizarre. If I have a basket with white and black balls, I can decide, given any ball, to which disjoint aggregate it belongs. But suppose that nature were so malign (or rather that our expectations of her were so naive) that whenever I add a black ball to the basket its color changed to white. The disjointedness of the aggregates would then be meaningless. And if the reader thinks that I am being fanciful, this is precisely what happens in the Aspect experiments that I have mentioned, as we shall see in due course.

The point I want the reader to remember is that, whereas it is very tempting to take for granted that given an aggregate one can always recognize its elements, this is not necessarily the case, about which more later in this chapter. What I shall call the *principle of the stability of aggregates* is a contingent and not a necessary principle.

THE GÖDEL THEOREM: A FIRST SHOT

Now that we know all about self-reference we are ready to dig our teeth into the Gödel theorem, but I must warn the reader that although self-reference plays a part in formulating the *strategy* of the theorem, the theorem itself does not entail it, contrary to the impression that might be created by popular expositions of this work. I must just as well add another warning, namely that more hype has surrounded this theorem than almost any other intellectual construction of the twentieth century, rivaling even relativity in this sense. The importance of Gödel's theorem cannot be denied, but it has to be seen in the right context: it would be, for example as absurd to say that *Daß Kapital* is not an important book, as to claim that it is so entirely by virtue of its impact on economic theory. Likewise, the philosophical importance of Gödel's theorem is far less significant than its effect on a generation of thinkers who had seen the positivist creed crumble in front of their eyes and were looking for new vistas. Alas, just as in the case of Marx, the uses to which the Gödel theorem has been put have not always been beyond reproach. But before getting into all this, let me say something about the man, whose personality, as you will see, was the stuff of which legends are made.

Kurt Friedrich Gödel was born in Brünn (then in Moravia, now Brno in the Czech Republic) in 1906. He was neither of Jewish nor of Czech origin as is sometimes said or implied: both parents were ethnic Germans. Although not a religious man in the ritual sense, he described himself as a theist even in his youth. Gödel entered Vienna University in 1924 to read physics, but switched to mathematics in 1926, and in the next few years he attended the Vienna Circle, the famous cradle of positivism, with whose ideas, nevertheless, he strongly disagreed.

Gödel gained his doctorate in mathematics at the University of Vienna in 1930 and presented the results of his thesis at a conference in Könisberg that year. But it was the next day that this twenty-four-year-old man started one of the most remarkable intellectual events of the century, when he reported, quite casually at the end of a discussion, on his first incompleteness theorem. Fortunately for the success of his ideas, the great Hungarian mathematician John von Neumann was present, and he gave immediate and strong support to Gödel, whereas Zermelo, of set-theory fame, was firmly antagonistic. Von Neumann, who was already a member of Princeton's Institute for Advanced Study, was as enthusiastic about this work as he was influential, and Gödel's paper was published in early 1931 and quickly gained acceptance. On the strength of it, he became a *Privatdozent* at Vienna from 1933 until 1938. During that period, he visited the United States, whereto he finally emigrated in 1940, after marrying Adele Nimbursky (a divorcée) in 1939. With von Neumann's support, he became a member of the Institute for Advanced Study at Princeton, where he remained for the rest of his life.

Gödel was a shy and strange man: he had suffered from rheumatic fever when he was six, was brought up under constant care, and thus became, not without reason, a hypochondriac. Even in his early thirties, in 1934, he suffered a nervous breakdown. He conducted his life on marvelously self-protecting lines: he always answered his letters with meticulous care, but never mailed his replies; and he never failed to make an appointment to anyone who wanted one, but never turned up at the agreed time and place. In his later years he wore an overcoat even in the hot Princeton summers and developed a gradually more aggravated form of paranoia. His death in 1978 was recorded as due to "malnutrition and inanition": he starved himself to death for fear of poisoning.[43]

Let me now sketch the principles of the Gödel theorem, and I expect that the reader will appreciate that whole books are required to deal decently with it, so that I face an almost impossible task: all that I hope is that the reader will get a reasonable feeling about the main ideas. I cannot of course emphasize rigor but, even less can I brush under the carpet the inherent complexities of the work, which must be understood to avoid being mesmerized by siren songs which, by oversimplifying, lead to the use of the theorem for purposes that are not always entirely legitimate.

In order to understand the proof, we first require a technical concept. We must start with a *formal system* (FS) of axioms that embraces the rules of arithmetic (natural numbers as given by Peano's axioms plus addition and multiplication) as well as all the rules of logic. The formal system must be such that by a mechanical application of the axioms and rules of the formal system, true formulas within it are obtained, and only such true formulas. (That is, no untrue results must emerge from the application of the rules, which means that we are assuming for the *moment* the formal system to be *consistent*.) What I have called for simplicity a "mechanical" application of the rules is more correctly expressed by saying that all those true formulas can be derived *algorithmically*, that is, by some sort of computer program which, going through the rules of the formal system FS, churns out true formulas, never getting a wrong one. The reader should appreciate that, by application of the arithmetical and formal rules to the axioms, some of these formulas will entail proofs of theorems. Although not strictly right in logic, I shall often refer to the formulas as *propositions*, since this name will probably convey more of their meaning to the readers.

The exceedingly ingenious first step that Gödel took, was to find a method for assigning a unique number, called the *Gödel number*, to any formula provable (that is derivable) within the system, in such a way that there is a one-to-one mapping between them and their Gödel numbers. This is the most remarkable feature of the system: from the *single* Gödel number (which is an integer), you can completely reconstruct a proposition, like the enunciation of a theorem or its whole proof! Variables can also be coded in the same manner. In order not to interrupt the main thrust of the argument, I shall defer discussion of the Gödel numbers until the next section, where the nature of the formal system will be further clarified.

I shall denote Gödel numbers by letters such as g, h and variable numbers by z. Because of the stated property of the Gödel numbers, a proposition which is applied on a variable number z will be written as $P_g(z)$ where g is a Gödel number. By ranging over g, we have here all possible propositions, and all of them are known and sound, because we must assume that the Gödel numbers have been formed from *all* possible sound algorithms. (*Algorithms* are just repetitive sequences of legitimate mathematical and logical operations that follow one from the other by use of the correct operational rules on a given string of symbols.)

Because Gödel numbers are natural numbers (integers), any given Gödel number can be substituted for any variable number, so that we can consider the somewhat unusual proposition which applies to its own Gödel number: $P_g(g)$.[44] I shall now write some special propositions, and, to make clear that they are propositions, I shall enclose them in square brackets, which have no special meaning except to help in reading. The first one will be:

[$P_g(g)$ *is not provable in the formal system FS*].

This is different from $P_g(g)$ because it is followed by "is not provable in FS," but this is a metamathematical statement (because it is a statement *about* the formal system, which takes the part of the object language), for which there is a code in the Gödel system. Therefore, it must be one of the propositions of the sequence $P_g(z)$ defined above, with the variable number z identified as g, as we have chosen, and for some Gödel number, say h. Thus

[$P_g(g)$ *is not provable in* FS]   *is the proposition*   [$P_h(g)$].

We are almost there! Since the above must be valid for *any* Gödel number, take g equal to h:

[$P_h(h)$ *is not provable in* FS]   *is the proposition*   [$P_h(h)$].

You can now understand the strategy of the proof: we have tried to create a self-referential situation to arrive, as we have seen before in the case of the *Liar*, to some contradiction, as I shall now show. There are two possibilities for $[P_h(h)]$ as read on the right-hand side of the last displayed line, namely that it can be true or false. If it is true, then, from the left-hand side, it is not provable in the formal system. If it is false, then the left-hand side is false and $[P_h(h)]$ is provable. Thus, either the formal system is *incomplete* (because it contains a true statement that is *not provable*) or it is *inconsistent* (because it contains a *false* statement that is *provable*). Therefore, if the formal system is *consistent*, as we had assumed, then it must be *incomplete*.

Gödel's second theorem, of which more a little later, states that the consistency of a formal system cannot be established *within the system*.

I shall now summarize the Gödel theorem, reinforcing at the same time the necessary conditions for it to hold.

FS (the formal system) is an arbitrary formalized and axiomatic theory for which we assume:

(i) *The axioms are recursively denumerable. (See next section.)*
(ii) *Peano's arithmetic theory (denoted with A) is contained in FS.*

The Gödel theorems then state:

(1) *There exists a sentence P_g of the theory which, if FS is consistent, it is not provable in FS although it is true. (I shall call such particular sentence the **Gödel formula**.)*

(2) *If such a sentence P_g exists, then FS is consistent but, because the truth of P_g cannot be proved within the formal system, then the formal system itself cannot be proved to be consistent **within the system**.*

The need for the arithmetic theory "A" to be included arises from the arithmetization of propositions which I have mentioned and that will now be further discussed.[45] But I must first warn the reader that what I have done about the Gödel theorem is no more than a first approximation to the truth: after we understand the technicalities a little better, we shall have, for the sake of decency, to return to the "proof" that I have given. We shall also see in the next section that the conditions that must be satisfied by the formal system must be reinforced. A very important point is that, so far, I have stressed self-referentiality so as to illuminate the strategy of the proof, but we shall see that, in fact, the Gödel theorem does not entail it. Before I go into this, it is worthwhile getting some idea of how the marvelous Gödel numbers work.

GÖDEL NUMBERS

Suppose we want to code the function

$$f(n) = (a+b)^n.$$

Here a and b are supposed to be two arbitrary but fixed numbers, whereas n is understood to be an integer that takes the values 0, 1, 2, and so on. This function satisfies Peano's fifth axiom (p. 389) on the following grounds. It can be calculated for n equal to zero because by mathematical definition any number raised to a zero exponent is unity. Also, if we assume that the form of $(a+b)^n$ is known, then the value of the same function for the exponent $n+1$ is most easy to obtain, by multiplying whatever expression you have obtained for $(a+b)^n$ with $a+b$. Functions with such a property are called *recursive functions*, and they are ideal for *algorithmic computation*. It should be appreciated that, as in Peano's axiom, recursive functions entail a "solution" of the infinite induction problem, since in the above example, for instance, n can range over all the natural numbers (or, if we want to be elegant, we can say that it can range from 1 to ω, the cardinality of the infinite sequence of natural numbers). Because this range of n is denumerable (that is, it can be counted, because it has the cardinality of the natural numbers), the recursive functions generated in the manner described are often referred to as *denumerable recursive functions*.

If we start with the natural numbers, and with all the arithmetical and logical symbols, all the formulas of our formal system FS will take the form of strings of the objects just described; and these strings can be generated recursively, that is, algorithmically. The totality of functions and theorems in the system can then be generated in principle, and such strings can be uniquely denoted by their Gödel numbers, defined as follows.

(1) All the arithmetic operations (+, −, etc.), brackets, and all logical operations as partly listed on table 3 of chapter 3 are given code numbers, 1, 2, 3, and so on. Besides that, we also have codes for statements such as "it is provable" or "it is not provable." Because these statements are not parts of mathematical or logical proofs, they are called *metamathematical*, as we have already seen.

(2) After this is done, a particular string will have a form like, for example, 9, 7, 5, 8, 9, 2, 12.

(3) Once this coding has been obtained, take the prime[46] numbers 2, 3, 5, 7, 11, 13, 17, and so on, and raise them to powers equal to the numbers listed in (2) in that order:

$$g = 2^9 \, 3^7 \, 5^5 \, 7^8 \, 11^9 \, 13^2 \, 17^{12}.$$

This is called a *factorization* of the number g, which is the Gödel number corresponding to the string in question. Factorization merely means that the number has been expressed as a product of factors. In general, a number can be factorized in more than one way: 16 can be written as 2×8 or 4×4. There is a theorem of arithmetic, appropriately called the *fundamental theorem of arithmetic*, that says that the factorization of a number in prime factors is *unique*. For example: there is one and only one factorization of 108 in prime factors: $2^2 \times 3^3$. It follows therefore that g is a unique number coding the given string. Notice that once g is given, its prime factors are *uniquely determined* and computer programs exist that do this job. When this is done, the right-hand side of the number g given in the above example is entirely known, so that from the *single* number g the complete string can be reconstructed. This is an amazingly clever result: if we have all the possible Gödel numbers that have been churned out algorithmically, that is, corresponding to all possible strings of the formal system, then we can reconstruct one by one all those strings.

A very important point is that the Gödel theorem imposes a severe limitation on the formal system treated. It must be such that the true formulas of the system are obtainable by an algorithmic, recursive, method. These true formulas must be recursively denumerable. (This algorithmic recursion, however, will stop short of generating the nonprovable *Gödel formula* itself.) The limitation introduced by the condition of recursive denumerability, which follows from the discussion above, restricts the often alleged very wide significance of the Gödel theorem, since other formal systems exist, which were studied by the American philosopher Barkley Rosser, also entailing nonprovable formulas, but in which both the provable and the nonprovable formulas are not recursively denumerable, and thus cannot be algorithmically generated.[47] We shall see later that this is quite significant.

We must now return to the Gödel proof to tie up a lot of loose ends that I had left in the interest of clarity. This does not mean that I shall become frightfully obscure and rigorous, but that a little decency will now be my main consideration.

A SECOND SHOT AT GÖDEL'S PROOF

Let me recover the last statement that led us to the Gödel's "proof," and remember that the square brackets are only a visual aid:

[$P_h(h)$ is not provable in FS]   *is the proposition*   [$P_h(h)$].

I shall now simplify this statement as:

> [*P is not provable in* FS]   *is the proposition*   [*P*].

On the left here, P started by being a formal string of mathematical and logical symbols. Let me now be more precise about it, and label it as P_{FS}, to make it quite clear that it is a string in the formal system. As such, even if we know that it must be a well-formed string, that is, that it must have been generated by the correct rules, it means nothing because the symbols that appear in it are mere signs deprived of any meaning: all we require of them is that they obey the operation rules of FS. It follows therefore that a true or false value cannot be assigned to those strings, and they cannot be said to be provable or unprovable. All that we can say is, because these strings have been formed by the correct rules, that they are *well-formed strings*. They have, on the other hand, defined Gödel numbers uniquely, and I shall now exploit this.

What we have to do is *interpret* the formal system, and let me explain what this means. In the Peano formal system we had an object called "the successor of 0," and the whole of the *formal* theory can be done never deciding what on earth that is. But in order to *use* the theory, I have to *interpret* "the successor of 0," and identify it with what everybody understands as the number "one." When you interpret the formal system, what you are really doing is *mapping* each formal string P_{FS} with Gödel number g onto an *arithmetical* proposition P_A with the same Gödel number. So, the left-hand side of the last displayed expression is meaningless, unless I substitute P_A for P_{FS} in it. Let us try:

> [P_A *is not provable in* FS]   *is the proposition*   [*P*].

What is the justification for so doing? You might think that you could adduce that the *whole* of the penultimate displayed line could be mapped from the formal system FS onto the arithmetical system A. But this *cannot* be done because in the last line the right-hand side is clearly a proposition *about* the formal system (or more precisely of its arithmetical interpretation) and therefore P on its right must stand for a metamathematical statement:

> [P_A *is not provable in* FS]   *is the proposition*   [P_M].

It should be clear that: (i) it is only now that we have written a line that makes some sort of sense; and (ii) that this line has not been obtained by mapping the original one onto anything: each side of it had to be re-identified *separately*. So, the process by which we have moved from P_{FS} to P_A on the left-hand side is not a genuine mapping. Then, what is it?

The orthodox answer is to say that when we start with a formal system and we treat it in the Hilbert metamathematical way, then the strings of the formal system are, rather than mapped, *mirrored* onto arithmetical propositions. That is, P_{FS} and P_A are regarded almost as different manifestations of the same object, rather than as two different objects related one to the other by a mapping. The reader might get a little bit uneasy at this stage, but this is part of the general strategy within this theory.

As you can see, whereas the idea of self-referentiality was important in order to organize the various steps of the proof, *the Gödel theorem as given by my last displayed line is not self referential*. Therefore, how do we recover the "proof" that we gave before? The basic idea is not very transparent and it goes roughly as follows. Within the Hilbert metamathematical system, an arithmetical proposition that is mapped onto a true metamathematical statement is true. In the last displayed line, we can interpret P_A as mapping onto P_M because both these propositions are "manifestations" of the same proposition P with the same Gödel number. (See the second displayed line in this section.) Therefore, if P_M is true, P_A must also be true. But, clearly, from the last displayed line, if P_M is true, P_A although true, is not provable within the system, so that we have recovered our original version of the Gödel theorem.[48] It is very important to notice that Gödel is here providing us with a new method to ascertain mathematical truth. Although P_A cannot be proved within the formal system, mathematically interpreted, its truth can be determined *metamathematically*.

Let me summarize briefly what is the indubitable significance of Gödel's work. Hilbert had thought that mathematics could proceed from axioms, via rules of operation that are essentially logical functions, to proofs, and thus theorems properly known to be true. The formal system, he expected, would be *consistent* (that is it would not ever entail nonsensical consequences like $1 = 2$) and *complete*, that is, *all* true results could be proved to be so *within* the system. Both these assumptions were destroyed by young Gödel's theorem. Moreover, Hilbert's conception of mathematical truth, as stemming from proofs within a consistent system falls to the ground.

Rather than describing what happened to the world post-Gödel, which I shall do in the next section, I shall now anticipate some ideas that I shall later try to justify. When you use natural language, which is just the language as we use it, we do not expect to be able to justify the truth of a proposition *within* the language. Thus, if I say "There is a table in my room" the only justification I have for uttering that proposition is that there is a table in my room: neither dictionary nor grammar can be substitutes for my table. (As you know, I am not worried about pink-elephant gazing, dreaming, or such nontypical situations: most people know when they are reliable witnesses of the amazing statement that there is a table in my room.) Just as for my table, the whole of natural science is based on the notion that the natural language by which science is transacted is *incomplete*. A language must have *referents* (such as my table), but the referents cannot be *in* the language. (Not all of them, at least.) That theological or mystical language may claim to breach this barrier does not mean that the same can be the case for natural language.

This notion, unfortunately, was not good enough for many mathematicians. The language of mathematics, had to be complete: everything true had to be known to be true *within the realm of mathematics alone*. This conviction, alas, was rudely shaken by Gödel. For what he proved was precisely what any sensible natural scientist would have expected:

> *Just as in natural language there cannot be a complete closed account of why the language is used the way it is, so there cannot be a complete closed account of why mathematical propositions are accepted as true.*

In other words: meaning within a language, be it natural or mathematical, cannot be fully determined by intralinguistic rules, but rather by a connection between the language and reality.[49]

This is very nice and clear, but life is not as simple as all that. The reader must understand that the destruction of Hilbert's program, and with it of the logical positivism that was the fashion of the time, could not take place without some sort of upheaval. The historian Simon Schama, writing of the French Revolution in his book *Citizens*, says that it is a characteristic of revolutions that people develop unreasonable expectations during them. This is precisely what happened in our time. Gödel had proved that we cannot walk without feet, whereupon people decided to levitate—and Gödel himself provided the necessary thermal currents to that effect.[50]

GÖDEL AND PLATONISM

When Gödel presented his work, he was very young and acutely aware that he was swimming against the current created by Hilbert and loved by the positivists. Because of this and of his natural shyness, he discussed his theorems purely as results in mathematical logic, but it is almost certain that right from the beginning he had in mind a Platonist program. He did not go public, nevertheless, until after he was safely ensconced as a member of the Institute for Advanced Studies at Princeton in 1940.[51]

As I have said, Gödel had destroyed Hilbert's notion that mathematical truth equals proof, and he then had two alternatives. The first is the one that I have already mentioned, namely to accept that mathematical language is a natural language which, as all natural languages, requires adequacy to some contingent *nature-dependent* constraints external to the language. The second, which he adopted, was to say that mathematical truth is not contingent but *necessary*, because it derives from the fact that mathematical objects are reflections of universals that form a Platonic world of ideas. The mathematician communicates with this world exactly as the natural scientist does with the natural world. Gödel posited, in fact, that mathematical intuition plays a role in mathematics analogous to that of sense perception in the physical sciences.[52]

Curiously, having debunked Hilbert's philosophy of mathematics, Gödel exploited what remained of it for his own purposes. Hilbert believed that at each level of treatment (in geometry, for instance) gaps were filled that allowed mathematics to reach always a higher level of abstraction: this, as well as the never ending hierarchy of the metalanguages, was for the Platonists a staircase to the Platonic heaven.

THE ANTI-PLATONISTS CONTRA GÖDEL

It is somewhat unfortunate that the main thrust against Gödel's Platonism came from Wittgenstein in work published posthumously and thus necessarily sketchy, being an edition of his extant notes. Partly for this reason, it has often been dismissed as of little significance, and it has even been suggested that Wittgenstein had not understood the delicate technicalities of Gödel's proof. The Canadian philosopher Stuart G. Shanker, fortunately, has published an exhaustive exegesis of Wittgenstein's ideas about the subject by analyzing all the relevant references in his previous writings, and thus producing a new and coherent picture of Wittgenstein's views. And these views present an argument against Platonism which I believe has not yet been bettered. Before I get into battle I should like to acknowledge my debt to Shanker's illuminating article, on which I have based a great deal of this section.

Wittgenstein's remark: "My task is, not to talk about (e.g.) Gödel's proof, but to by-pass it" has often been quoted to prove that Wittgenstein had not appreciated the import of this theorem. What he meant, though, is obvious because for him "Gödel's proof develops a difficulty which must appear in a much more elementary way."[53]

And that there must be a more elementary way is clear, because the one indisputable fact about Gödel is that he closed Hilbert's program full stop. For people like Hilbert, mathematics was the archetype of a priori knowledge and, strangely enough, although both Hilbert and Gödel regarded mathematics as a "paragon of reliability and truth," Hilbert's failure through the Gödel theorem did not affect at all the latter in this view. On the contrary, he used this failure to gain an entry into metalog-ical heights leading to the Platonic world.[54] For Wittgenstein, instead, the Hilbert program to establish the "reliability and truth" of mathematics was a chimera: it was useless to drag the pool searching for a gold coin because that gold coin never existed. And this should have been the end of the story. Gödel, instead, having proved that no gold coin existed in *that* pool, tells us that this is so merely because Hilbert's followers had been searching in the wrong one: what they were looking for was instead to be found in the Platonic world.

I shall try to explain why Wittgenstein considered the Hilbert pro-gram ill conceived. For Hilbert, you start with a formal language, made up of axioms and symbols from which you construct by axiomatic rules, augmented with those of logic, well-formed strings of symbols. Because at this stage those symbols are no more than signs without significance, some flesh has to be put into them by "mirroring" them into arithmetical propositions.[55] For the latter, one can speak of truth or falseness, and such statements are metamathematical statements. I have already advanced (p. 394) Wittgentein's main objection to the concept of metalanguage. For him, this was an altogether false concept because the concept of proof cannot be external to the object language but must instead be firmly embedded in it. Wittgenstein could not have put this more emphatically: "... there can't in any fundamental sense be such a thing as meta-math-ematics."[56] Therefore, since Hilbert's program was based on the use of metamathematics, to prove that it was wrong could not have any episte-mological significance whatsoever.[57] This is all. Wittgenstein feels that we get surprised because we had *philosophical expectations* that were not jus-tified. But mathematics cannot solve philosophical problems.[58]

Let me now expand on the critical argument about metalanguage already mentioned. Take the proposition in the object language: "There is a table in my room" and call this proposition S. To assert that it is true, I must say in the metalanguage "S is true." But in fact, if I want to give the grounds for the assertion S, I must do that in the object language: I have, for instance, to give some causal chain that justifies that belief.

An important point to recognize is that there are two different ways in which the concept of metalanguage can be used. The one that I have described is *within the object language*. But I can also use metalanguage *about the object language*. Thus, if I call T the proposition "There is some-thing in my room," I could run a magnificent metalinguistic proposition: "If 'S is true' then 'T is true'"; that is, I could run a system of rules about how object-language propositions could be used in which there is no commitment as to how such propositions are known to be true or false. In this sense, metalanguage is independent of the object language, but it is in itself a new language, or as Wittgenstein would have put it, a new *language game*. For Wittgenstein, the Hilbert program is really only concerned with the logical grammar of mathematical propositions, that is, with metamathematics *about* the mathematical language, whereas Hilbert added to it a spurious epistemological dimension.[59]

To illustrate the question of the two forms of metamathematics, consider the following example: "The sequence with Gödel number x is a proof of the theorem with Gödel number z." This formula can be understood in two different ways. First, as contingently true or false, in which case the relation between the proof and the theorem is purely external, that is, x denotes the theorem as when I say, pointing at some lines where x is written: "These lines are the proof of z." (It is in this sense that the use of metamathematics is legitimate. It is metamathematics as used not *in* the system but *about* the system.)

In the second way, on the other hand, the formula is taken to be as necessarily true mathematically. It is no longer empirical and descriptive, but it is taken as stipulating the use of the terms "proof" and "theorem" *within* the system.

Gödel's interpretation oscillates between these two meanings of metamathematics. In the second way, we must accept that a metamathematical statement can be *mirrored* by a formula in the calculus of Gödel numbers as the one given, which presents a purely arithmetical relation between a proof and a theorem, via the Gödel numbers x and z. But such formulas are also treated as empirical propositions *about* logical relations in the object language, which are contingently true or false. This "ambiguity underpins Gödel's interpretation of his proof."[60]

In order to understand fully the meaning of Wittgenstein's objections to metamathematics it is important to discuss his views on *conjectures*, which can appear to be counterintuitive. Let me state for this purpose the famous *Goldbach conjecture*. Consider the first few *even* numbers, written as sums of two integers:

$$2 = 1 + 1, 4 = 3 + 1, 6 = 5 + 1, 8 = 5 + 3, 10 = 7 + 3.$$

I could have expressed exactly the same numbers as different sums of pairs of integers, but notice that in every case the numbers used above, 1, 3, 5, 7, are *prime* (see note 46). Vast explorations of similar expressions have been made with modern computers, and so far no even number has ever been found which is not the sum of two primes. Thus, the *Goldbach Conjecture*:[61]

For all x, if x is an even integer, x is the sum of two primes.

This result, however, has never been proved, despite a great deal of work. Now comes the problem. For most people, the proposition stated must be either true or false; that is, either a proof will eventually be produced or a counterexample will be found which will prove it wrong. For Wittgenstein, while such a proof does not exist, the conjecture is neither true nor false but *meaningless*. This seems rather strange because I, for instance, can imagine myself working for years on such a proof. Also, I can imagine what it would mean if I found the proof: I could then confidently take any even integer x, however large, and feed it into a computer, whereupon a sum of two primes equal to x must eventually emerge.

The reader must appreciate that the way in which Wittgenstein uses the word *meaningless* is very specific: the fact that I can *imagine* something does not imply that whatever I imagine is meaningful. Suppose I were to write "Unicorns like eating hot porridge in bed for breakfast." Anyone who has seen the tapestries at the Cluny Museum in Paris can easily imagine a unicorn, and nothing can be more cozily agreeable than visualizing a unicorn in bed having a hot porridge breakfast. And yet, by all normal standards of humanity, "Unicorns like eating porridge in bed for breakfast" is not just false (it could be so, for instance, by virtue of unicorns preferring egg and bacon for breakfast!) it is *meaningless* It is the proof, for Wittgenstein, that gives a *meaning* to a string of otherwise vacuous signs.[62] Try to think, for instance of the meaning of "for all x" in the unproved Goldbach conjecture, and remember that it is not sufficient to imagine what it would be *if it had a meaning*. I hope you can now see the subtle vicious circles against which Wittgenstein fought.[63]

In sum. The crux of the matter is Hilbert's notion of metamathematics. The central point of Gödel's argument is that, assuming that arithmetic is consistent, the metamathematical statement "P_A is unprovable" is true. Given that this statement is merely the metamathematical reflection of "P_A" itself, if "P_A is unprovable" is true, this can only be so because it has been mapped onto the *true* arithmetical proposition "P_A." That is, if "P_A" has been mapped onto a true metamathematical statement, it *must* be true, a result that I have already used in regaining Gödel's result in my second version of its proof. But this is precisely the step which Wittgenstein disputed, not on the basis that what is questionable here is the truth or falsity of "P_A" but, rather, on account of his claim that "P_A" cannot have any meaning at all because *there is no proof of it within the arithmetical system*; and because, in accordance to him, a mathematical proposition not proved is meaningless and its truth (or falsity) cannot in any way be asserted.[64]

Finally, the concept of mirroring which is crucial to Gödel's proof is not acceptable for Wittgenstein, since mathematical propositions are internally tied up to their proofs within a given system, and you cannot export them to another system expecting to keep sense. There are, for example, propositions provable in Euclid's system of geometry that are *false* in another system.[65]

Whatever one's view about the validity of Wittgenstein's arguments, the undeniable conclusion is that the idea that Gödel's theorem provides a cast-iron case for mathematical Platonism is without foundation.

MORE GÖDELIAN DREAMS: HUMAN MIND VERSUS MACHINES

I shall first review very quickly the beginnings of one of the most lively ideas in our recent intellectual history: the question whether computers "think" and the consequential debate on *artificial intelligence*. Although this appears to have nothing to do with the Gödel theorem, you will soon see the connection, real or imaginary.

Alan Turing and Artificial Intelligence

No one would bat an eyelash when exposed to the proposition "airplanes fly." If you think about it, however, it is highly debatable that they do. Certainly, if you call "flying" what birds do, then airplanes do not fly at all. What has happened, of course, is that the word "fly" has changed connotation since heavier-than-air machines have emerged. And it was thanks to this relaxed attitude to the use of language that we actually did learn something about the ways birds do fly: it was only in 1928, when the first successful *glider* "flight" took place, that the falsity was discovered of the theory that birds could soar while hovering because of the air expansion caused by the sun in their hollow wing bones. Biologists, who until then held this theory, had not noticed the existence of thermal currents![66]

I wish that this pragmatic approach to language had been extended to the use of the verb "to think," but here any deviation from orthodoxy (understood of course always as: *me* orthodox, *you* heretic) produces endless argument. Yet the father of the subject, the Cambridge mathematician Alan Mathison Turing (1912–1954), had precisely that prudent attitude in mind when he broached the question of machine thinking in a famous paper of 1950:

> I believe that at the end of the century the use of words and general educated opinion will have altered so much that one will be able to speak of machines thinking without expecting to be contradicted.[67]

I display this sentence because I have never seen it quoted, despite the fact that this must be one of the most cited papers of the century, and that some of its ideas had generated sufficient hot air to keep aloft all those aforementioned gliders. No one, however, could object to the words I have quoted, and yet there are lots of brain-correct people who take offense at the concept that humans might not have an absolute and eternal monopoly on "thinking" in whatever sense that verb be used: they refuse to countenance the possibility that any such process occurs except in human brains, with the slight proviso that it might be only the brains of mathematicians and (some) philosophers that qualify for this purpose. (Of course, this is not necessarily an overt requirement, but when you see what it is to be expected to pass the test, it is difficult to assume otherwise.)

Turing's contributions to mathematical logic are enormously important. Not only did he complete some of Gödel's results, but his own presentation of the Gödel theorem, by means of the *Turing machines*, made the subject powerfully clear, paving at the same time the way for the design of the modern digital computer, in which Turing took a very active part. Unfortunately, in order to concentrate our minds on the essentials, I had to decide not to discuss this work here.[68]

In the spirit of the quotation shown, Turing did not want to get involved in quasi-metaphysical problems entailing the nature of thinking. Therefore, in that paper, he proposed a test which would allow for the suggested semantic shift to take place painlessly: if an observer communicated with a computer in a distant room, and engaged in a dialogue as a result of which she could not recognize that she had had intercourse with a machine, and not with a human, then that machine could be said to be "thinking" or to be "intelligent." This is in fact a sort of Consumer's Association test, which would allow a person to buy whatever level of "intelligence" required; and to treat it in any other way is a complete waste of time.

Such a view of *artificial intelligence* (AI) is described nowadays as *weak AI*: it entails the simulation of intelligent behavior. As artificial flight taught us something useful about birds' flight, so the study of artificial intelligence in the weak sense is useful in generating possible models of the mind, such as the computer neural networks.

People, however, are also interested in a different concept of artificial intelligence, called *strong AI*, and those most interested in it are in fact those who negate its existence, and even the *possibility* of its existence. For strong AI makes an ontological claim, that is, that the type of thinking that goes on in the human brain cannot be qualitatively distinguished from that done by a (not necessarily yet existent) machine. And you will begin to see where we are getting at, because a Platonist must argue that the amazing apprehension by humans of the Platonic world of mathematics cannot possibly be accessible to a mere machine, and thus that the human brain has some properties that no machine could ever have. *Physicalists*, on the other hand, who believe that all that we have in nature is of a single substance, matter, governed by some form of physical laws, advocate strong AI. Although defending forcefully their views, the activities of such people tend to follow the time-honored methods of the physical sciences; in this case, the construction of more, and more sophisticated, computer hard and software to prove their case.

Platonists, on the other hand, like their eponymous hero, tend to distance themselves from experiment, and they probably divide into two classes. First, those who, by the way in which they describe their views, would be more clearly understood (at least by me) if they claimed an immortal soul who does the reading of the Platonic world, as well as the organizing of the brain's thinking processes. Second, there are others who appear to support a soulless view of humanity, but who nevertheless believe that the brain possesses some *physical* properties that no machine could have. As far as I can understand them, it appears to me, roughly, that Roger Penrose belongs to the second and John Lucas to the first category. Both are Oxford thinkers, the first a mathematical physicist and the second a philosopher. Although, as I have described them, their ideas are not identical, they coincide in the view that there is a fundamental difference between human thinking and that of a machine, that this difference is insurmountable, and that no machine could ever breach it.

It is of course quite amazing that human beings could make such a claim for all time, especially since computers are very recent inventions, and since it is impossible to predict their future development. Both Penrose and Lucas, however, are among the most brilliant practitioners of their subjects and of their generation, and although they are obviously sticking their necks out, they do not do that without some reason. The main argument that they use, which was proposed by Lucas in 1961, is based on an ingenious application of the Gödel theorem. Of course, you have to have a certain amount of chutzpah to expect a mathematical theorem to be able to conclude something about the nature of the human mind, but, if you remember that Gödel himself was prepared to derive from his work *onto-logical* conclusions about the duality of nature, they are in good company. Some of my readers, however, might feel that to derive ontological conclusions from a purely logical argument takes us back to the medieval schoolmen who thus "proved" the existence of God. And it is not accidental that Gödel himself produced his own proof to that effect.[69]

In order to simplify a little what can be a frightfully complex discussion, I shall concentrate on the original Lucas's proposal since it shows up very well the main concepts behind these daring ideas.[70]

The Lucas Argument

Lucas, first of all, defines what he means by a machine or computer. It must work following an algorithmic program, and it must therefore be deterministic: what the machine outputs must depend uniquely on the program and the inputs. For his thesis to be tenable that no machine can ever emulate man, his concept of "machine" must be valid *for all time*, and I would not be prepared to bet on Lucas being right in this respect: although forecasting is a valuable activity, prophesying is a bit out of fashion. In any case, I suspect that there are *already* some non-Lucasian systems in operation. (See p. 444.) But let us concede Lucas's definition so that we can enjoy his argument, which is very clever.

So far, we have considered only one formal system, FS but, in principle, many such systems could be devised: call them FS_i with i running over the natural numbers 1, 2, 3, ... Construct a machine M_i that produces algorithmically well-formed strings within the system FS_i, and only such strings. From the Gödel theorem:

> For the machine M_i, if and only if its formal system FS_i is consistent, then g_i is the Gödel number of a sentence true but unprovable in FS_i. (**The Gödel formula**.)

Because of this, Lucas "out-Gödels" the machine M_i since *he* knows that the formula with Gödel number g_i is true, which the machine can't, because the machine cannot prove this. But the Gödel number g_i corresponding to the Gödel formula can be produced algorithmically (as a recursive function). Therefore, a formal system FS_i can be constructed to do this, whence a new machine M_i can be made that can do just as well as Lucas in out-Gödeling M_i.

However, this machine will run into an unprovable formula g_i, which Lucas will know is true. Thus Lucas out-Gödels this new machine. In the same way, you can never build a machine that will be able to prove all the sentences that Lucas knows are true.

All this is very ingenious. Let me now have a look at the other side of the coin.

Contra Lucas

There are dozens of arguments that purport to show up fallacies in Lucas's argument, and I shall try to do my best with the simplest one.[71]

Lucas's argument depends on the formal system FS_i being consistent, so that, in order to out-Gödel M_i, he must first ascertain whether FS_i is consistent or not. However, it is not difficult to prove[72] that the set C of all i for which FS_i is consistent is *not recursive*, that is, that it cannot be generated algorithmically and therefore that it cannot be produced by a "Lucas machine." Therefore, if Lucas wants to out-Gödel a machine and thus prove that no machine is equivalent to Lucas, he must accept that he himself is not a Lucas machine, which is the very thesis that he wishes to prove. He has thus got into a vicious circle.

There is another argument that shows that Lucas puts too much reliance on the original form of the Gödel theorem. For we have seen (p. 402) that formal systems were constructed by Rosser in which nonprovable formulas arise; but that these formal systems are such that the set of all true formulas in them is not recursively denumerable, so that it could not be generated by what Lucas defines as a machine. Therefore, if Lucas engages in his game with Rosser, he can out-Gödel Rosser just as before, although Rosser cannot be a machine! Of course, for this game to work, it is necessary that Rosser plays by the same rules that Lucas, at least implicitly, requires for his machines, namely that they answer "yes" or "no" and *nothing else* to the question: can you prove the following formula? Which remark suggests one more problem for Lucas.

Suppose that the machine, now in the original Gödel-Lucas game, were allowed to say, when faced with the Gödel formula: "No, I cannot prove it, but I know that it is true." It is perfectly possible for the machine to be programmed to do this, since the Gödel number of the unprovable Gödel formula can be obtained recursively. Suppose now that the machine is in a room separate from Lucas, and that he does not know whether he is dealing with a machine or not. Once he gets the above answer, he walks to the machine room, finds that it was a machine that made the statement, and he then says: "The difference between me and the machine, however, is that *I* understand what I am saying." But this entails postulating that, at variance with the Turing test, equality of performance is not sufficient to accept that "thinking" takes place, which is precisely the point that Lucas had set about to prove with his Gödel game. Thus, the likelihood that Lucas's argument can prove that there is a difference between human and machine "thinking" is somewhat remote.

The reader will be glad to know, I am sure, that we have now finished the hard work of the chapter. This does not mean that new ideas are now exhausted: on the contrary, I hope that we shall see that a very sensible view of mathematics can be constructed, which in a way is a synthesis of some of the opposing views that I have described. Before we get into this, it would be helpful to have a look at a case in which one can see how new mathematical "facts" came into being.

A CASE STUDY: COMPLEX NUMBERS

Complex numbers, which I shall define in a moment, were first used in 1545 by the Italian mathematician Geronimo Cardano (1501–1576), also referred to as Cardan. It is interesting to read what one of the most famous mathematicians of the nineteenth century, Felix Klein (1939), p. 56, has to say about their history (his italics):

> *imaginary numbers made their own way into arithmetic calculation without the approval, and even against the desires of individual mathematicians, and obtained wider circulation only gradually and to the extent to which they showed themselves useful.*

Notice that Klein's account does not tally very well with a Platonic perspective, which would have expected so many first-class mathematicians to read correctly the Platonic oracles. Rather, it confirms the view that, as the more pedestrian natural scientists had always done, mathematicians were also adjusting their cloth to the body mathematical that they had to handle.

Cardano had introduced imaginary numbers, as complex numbers were at first called, in the solution of cubic equations, but I shall grossly simplify the problem by dealing with quadratics, that is equations in which the variable appears raised to the power of two.

It has been known for a very long time than an equation such as

$$y = x^2 - 1,$$

can be represented as a curve. In fact, it is easy to see that y equals -1, 0, and 3 for the values of x: 0, ± 1, ± 2, respectively (remember that the square of a negative number is always positive). When you plot these values as shown in figure 2, you get a curve that is called a *parabola*. The particular equation displayed is represented in the figure by the thick line, and the reader can easily verify that it has the values stated.

In exactly the same manner the equation

$$y = x^2 + 1$$

has the values of y : 1, 2, 5 for x: 0, ± 1, ± 2, respectively, and the corresponding parabola is the grey line in the figure. You will now see why I am getting into this simple example. The thick parabola cuts the x axis at two points, ± 1, and these two points, for which the ordinate y of the curve vanishes, are called the *roots* of the equation. You can see that the two parabolas of figure 2 are very much the same thing and yet, whereas one has two roots, the other has none. This is obvious because if we want the right-hand side of the last equation to *vanish* we require

$$x^2 = -1,$$

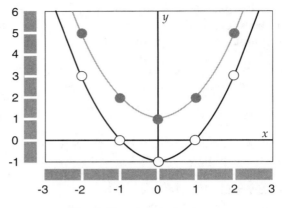

Fig. 2. Two parabolas

which is not possible, because the square of any number, positive or negative, is *always positive*. Speaking historically, mathematics had at this stage two options. One was to forget all about this problem and do nothing. The other was to decide whether the number of roots of a curve was sufficiently important to try to make it independent of the collocation of the corresponding curve. We can see, in fact, that if we move up the black parabola in the figure by two units it exactly coincides with the grey one and thus loses its roots. At this stage mathematical experience, or, if you want, accumulated good common sense, is required to choose the right priorities. What is more important: that a parabola should always have two roots, or that the left-hand side, x^2, in the last equation be always positive?

Platonists would say that at this stage mathematicians discovered one of the eternal truths in the Platonic empyrean, namely that a *number* exists with the astonishing property that its square is –1. I would say that at this stage mathematics took a pragmatic decision. This is based on the fact that the number of roots of curves is heavily related to their shapes, and that the shapes of curves entail some important *invariant* properties (as above, with respect to the translation of the black parabola); and therefore, if a curve has two roots, it should still have two roots after translation. The only way to achieve this is to *invent* a new family of numbers, of which the simplest one (in the same sense in which 1 can be thought of as the simplest integer) is called i (like the symbol for unity, 1, this could have been anything), with the property[73]

$$i^2 = -1.$$

This is the same as to say that the equation $y = x^2+1$ has now two roots, ±i. (We are postulating here that, as for ordinary numbers, the square of i does not change when preceded by a negative sign, and you can see that when i^2 is substituted for x^2 in $x^2 + 1$ you get $-1 + 1$, which is zero.) The number i is called the *imaginary unit*, and its product with any real number b is called an *imaginary number ib*. If you add a real number a to an imaginary number ib you get an expression $a+ib$, called a complex number.

The important thing here is to realize that the use of the word *number* has now *changed*. Previously, numbers were the natural numbers (0, 1, 2, 3 . . .), the negative numbers, the rational numbers (fractions), and the irrational numbers (like √2). All these numbers are now called *real*, and all of them are such that their squares are positive. The complex numbers have squares that may be negative, and because of this many people refused for years to call them *numbers*. Finally, it was proved that, except for this "anomaly," they had all the other properties of real numbers, that is, that they multiply and add up, and so on, by the same rules. So it became to be accepted, not without fighting, that they had a right to be called "numbers."

There is, however, a serious ontological difference. All the real numbers can be visualized as points of a straight line (or, more properly, *mapped* onto points of the line), as long as we take an origin, and agree that one half-line from it, say to the right, will be positive, and the other negative (as we have done in fig. 2). Complex numbers, however, cannot be visualized in (or mapped onto) ordinary physical three-dimensional space. This is what many people found abhorrent about them, until it finally was accepted that numbers need not be associated with any realization, but had to be used as entities with some stated rules of operation. Such Platonists as were involved in this, had therefore to recognize that their first reading of the Platonic world was no good at all, and that they had to change their intellectual spectacles. This is so very much like what happened to scientists after Copernicus, that skeptical people might well believe that the Platonic story is no more than an agreeable but confusing metaphor.

Although I have said that complex numbers cannot be realized in real physical space, many readers will be familiar with a graphical representation of complex numbers by means of the so-called Argand plane, and may be puzzled by my assertion, since this plane appears to be just as good a figure as it was used, for instance, in Cartesian geometry (of which fig. 2 is an example). I show an Argand diagram in figure 3, but I warn the reader that the use of the Argand plane is fraught with danger: whatever you do in it has *nothing whatever* to do (not at least *directly*) with what happens in physical space.

That the Argand plane may be a convenient way to get some visual representation of the complex numbers, but that it has nevertheless nothing to do with physical space, is illustrated by one of the most remark-

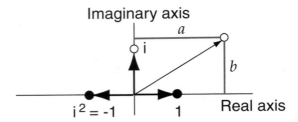

Fig. 3. The Argand plane

The reader is warned that whatever happens in this plane may not bear any relation to the properties of physical space. When you multiply the real unit 1 with the imaginary unit i, you get of course i, which is conventionally represented on the vertical axis, called the *imaginary axis*. You may thus infer from the figure that i acting on a vector rotates it by 90°. This, however, is not true for the same operation in real space, for which the corresponding angle is 180°. (See text.)

able mistakes ever made in mathematics. William Rowan Hamilton (1805–1865) was a mathematical prodigy, Astronomer Royal of Ireland at age twenty-two and knighted before he was thirty. His contributions to mathematical physics in his century are second to none. Hamilton tried for many years to extend the concept of complex numbers, so as to be able to have more than one imaginary unit (that is an unit with −1 as its square).

Hamilton was a very powerful mathematician, and he realized that there were certain algebraic constraints that had to be satisfied if more than one imaginary unit was introduced. It was not possible for him to fulfil them, hard as he tried, when only two such units were used. Then, in a famous day in mathematical history, 18 July 1843, inspiration came to him in a flash, and he saw at once in his mind's eye that three units would do perfectly, all of them, of course, squaring up to −1. For him to realize that this was right, he had to see how these three units would multiply with each other, and part of his inspirational flash was to find what these multiplication rules had to be, and how they would satisfy the algebraic constraints that he knew existed.

Because he now had four units, counting the real one, Hamilton called the new objects *quaternions*. Many people would say that he *invented* them, but the Platonists must believe that on 18 July 1843 in the morning, when strolling in Dublin with his lady wife, Hamilton's Platonic spectacles were exceedingly lucid so as to read correctly the necessary corner of the Platonic world and thus *discover* the quaternions.

If you believe that, you must also believe that his Platonic eyes were not all that reliable, as I shall now show. Of course, I do not expect the reader to try to visualize what quaternions look like, but you do not need to do that to understand the gist of this extraordinary story.[74] The point is this: the same day when he invented (discovered?) quaternions, Hamilton realized that they are related to rotations and, just like it "appears" for i in figure 2, he decided that the imaginary unit, when acting on a vector, rotates it by 90°. This is a result which even today is quoted and which is *wrong*: the correct angle is in fact 180°. I cannot enter here into the reasons why this is the case, but I can easily explain why Hamilton got it wrong. Very roughly we can argue as follows.

Because i^2 equals -1, i^2 acting on a vector changes its sign. (See fig. 2, but I now imagine that we are entirely in the real axis and in a *physical* real axis.) If we repeat this operation once more, we have $(i^2)^2$ acting on the vector and since the bracket is -1 and its square is 1 then i^4 is unity, which should leave the vector, unchanged. Now, if with Hamilton we take i to correspond to a rotation by 90°, then i^4, which is repeating this rotation four times, is a rotation by 360°, which should not change anything. (Rotating anything by 360° is merely bringing back the object to exactly the same position, so that nothing should be changed.)

You can see that all this is so eminently sensible that nothing could go wrong. And yet it is rubbish: the rotation in *physical space* entailed by i is 180°, so that i^2, which is −1, is a rotation by 360°, which should not change anything, (because it brings any object to its precise original position), whereas I am now saying that a rotation by 360° multiplies the object on which it acts by −1, instead of leaving it invariant. You realize that even if Hamilton had had a Platonic telescope to communicate with the Platonic world and thus read the correct result, he would never have believed it, because for him it would have been impossible to countenance the possibility of a rotation by 360° changing the sign of anything. (The Platonic world may be eternal and necessary, as Gödel believed, but this is not so wonderful if we have to accept that its reading by mathematicians, even as clever as Hamilton, is *contingent*. This is incontrovertible here, since Hamilton's "reading" was profoundly affected by his preconceptions, which are of course contingent.)

Be it as it may, the correct result was obtained even before Hamilton's nightmare, in 1840, by a French banker and mathematician, Benjamin Olinde Rodrigues (1795–1851). He was not perturbed by the horrible result just mentioned, which associates a factor of −1 with the rotation by 360°, because, as a difference with Hamilton, he never used the Argand diagram and worked entirely in physical space.

How could we have a rotation by 360° changing the sign of anything? I illustrate how this could happen by using the famous Möbius strip in figure 4, where you can see that when I rotate a point by 360°, it does not return to its original position, but that it rather goes to the "other side" of the strip. (I use quotation marks because the strip actually has only one side: without the twist you cannot move continuously from one side of the band to the other, but with the twist this is now possible, as illustrated just now. And it now takes 720° to return precisely to the same point.) So, if when I rotate the point by 360° it changes sign, it does not matter, because it has not returned to its original position. If I now do the same thing again, there is a second change of sign under the second rotation by 360°, thus canceling the first sign change, and the point returns to its original position not having changed in sign at all.

At the time when Rodrigues got the right rotation rules, no physical object was known to exist that changes sign under a rotation by 360°, and it is no wonder that Hamilton was so misled in his work. It took a long time, until 1913, when the French mathematician Elie Cartan (1869–1951) invented some mathematical objects similar to vectors, but which possess this property. They are now called *spinors*, having been rediscovered (I used the word carefully, since we are talking about physics where discoveries *are* made) by Wolfgang Pauli in 1927. Pauli (1900–1958) discovered that the electron had to be described not by an ordinary vector but by an object that changes sign on a 360° rotation. These objects corresponded to what classically would have been described as the internal rotation or *spin* of the electron around its own axis, whence the name *spinor*.

What we have learned from this story is that, at crucial moments in the history of mathematics, decisions have to be taken that depend on a feeling of what is really important, and what can be sacrificed, in the process of mathematical development. Naturally, inventions made on this basis are not entirely arbitrary. If complex numbers are to be used, for instance, one has to make sure that they do satisfy at least some of the major properties that numbers are expected to have, which provides solid guidance in proposing new definitions. The mathematician is very much like Michelangelo when "discovering" the statue inside the marble block.

Fig. 4. The Möbius strip

A rotation by 360° does not return a point to its original position on the surface of the strip. A second such rotation is required for this to happen.

Invention is possible, but nature, the marble block, and the preexisting mathematical structure, are all constraints to be respected: this is what differentiates the inventor from the charlatan. But in this interrogation of nature, in mathematics as in natural science, mistakes can be made, as Hamilton's example shows; and it is nature that in the end prevails.

One final remark. The question whether $\sqrt{(-1)}$ exists or not is not one that worries anybody these days. One happily writes i for it and that is the end of it. This was not so always, as Klein averred. Even the great Leibniz thought that "imaginary numbers were really something quite foolish."[75] At the time of Gauss there was worry about the "true metaphysics" of complex numbers, since it was not possible to have perceptual intuitions of them. It was Hamilton in 1837 who showed that, in the same way as fractionary numbers could be understood as pairs of reals, so it was possible to do the same for complex numbers; and it was possible in this way to achieve an intuition of their *algebraic structure*.

We now have to face the overwhelming question: did $\sqrt{(-1)}$ exist in the fifteenth century, say, before Cardano? Imagine that someone had then written the formula

$$\sqrt{(-1)} = i.$$

There is no question whatsoever that by all the language conventions of the period such a formula would have been regarded as *meaningless*. This is what Wittgenstein meant when he referred to unproved mathematical formulas. For Wittgenstein, mathematics was language and language was usage, and if that line was meaningless for those who read it in the fifteenth century, it was then meaningless full stop.

For the Platonists, instead, my hypothetical fifteenth-century mathematician was like King Solomon, unable to recognize the timber of the (future) Holy Cross without the help of the Queen of Sheba: but the Holy Cross *was* there! Platonists would even claim that the above formula was true even before the big bang. I leave this to the reader: you pay your money and you make your choice.

Many will feel, however, that this is a very good example of how the concept of truth cannot be applied to an isolated proposition, but rather that it requires the latter to be embedded within an appropriate mesh of interrelated concepts. And it is only within this mesh of entrenched mathematical concepts that true or false mean anything. Quine once wrote: "Any statement can be held true come what may, if we make drastic enough adjustments elsewhere in the system."[76] As you can see, we arrive at a point where mathematical truth works very much like truth in science, which is firmly tied up to an existent and constantly evolving mesh. What was true in the fifteenth century need not be true now.

AN INTERLUDE

Looking back at the work of this chapter it is clear that two major but opposing views of mathematics remain in the field. One is the Platonic view that sees mathematical truth as eternal and necessary. The other is the approach that I have illustrated through Wittgenstein's views, far more empirically oriented, since it regards mathematics as a form of language and thus contingent. I shall try to save some features of either system. The Platonists' feeling of mathematical certainty and necessity cannot be totally ignored. Neither can we overlook the fact that there is a deep relation between mathematics and nature; and we must try to explore how it arises.

I had better reveal my intentions at this stage. I shall try to support the view that there are really two forms of mathematics, although we never differentiate between them. One is fundamentally empirical, unequivocally tied up to nature, from which it comes via our rational system, the evolution of which we already understand. Once you have the first form of mathematics, rules arise that can be woven into formal systems. This permits a development of the second form of mathematics, as a game. Whereas the first form of mathematics is accountable to nature, the second is primarily accountable to a mesh of axioms and rules by which it operates. This does not mean that the two forms of mathematics do not have contact or overlap: on the contrary, there is a lively two-way trade between the two. The reader will immediately grasp, because the rules of game-mathematics are somewhat fixed (although they must from time to time be affected by the two-way trade mentioned), that this encourages the feeling of necessity that informs the Platonist's thinking: you might even accept, if you want, that there is some sort of world of mathematical ideas, very much like the Platonic one, but this world is the creation of our own mind and it is firmly in nature, from which it originated via empirical mathematics.

Of course, I must start my work with the latter. The reader might have noticed that, when I try to understand what is empirical and contingent, I try to construct a scenario in a world sufficiently different from our own, so that what we normally take for granted is no longer valid. So: the kingpin of the whole of mathematics is the concept of natural numbers, which crucially depends on that of counting. Therefore, let us get ourselves thoroughly confused with situations where counting does not work.

And so that you do not think me perverse, I shall tell you a story related by none less than the same Alfred North Whitehead (1861–1947) who was Russell's collaborator in *Principia Mathematica*. And yet, as regards counting, he threw a wrench into the works, by retelling a wonderful theological story about the already known to us Council of Nicaea of 325 C.E. He quotes: "When the bishops took their places on their thrones, they were 318; when they rose up to be called over, it appeared that they were 319; so that they never could make the number come right, and whenever they approached the last of the series, he immediately turned into the likeness of his next neighbor." And then he comments: "It is perfectly possible to imagine a universe in which any act of counting by a being in it annihilated some members of the class counted during the time and only during the time of its continuance."[77] What was good for Whitehead is good enough for me, and, accordingly, I am going to ask a monstrous question.[78]

IS THERE A LARGEST INTEGER?

Most if not all mathematicians would rule out of court the possibility of a largest natural number, say N, because it is "logically necessary" that another number $M = N + 1$ exists, that must of course be larger than N. If you look back at the Peano axioms (p. 389), the successor ($n+1$) of any number n exists, and therefore the set of natural numbers is never ending, its cardinality ω (omega) being infinity. End of story: the question asked was stupid. Answering sensible questions, alas, is easy: it is when answering "stupid" questions that one can really learn; so I shall have a go.

It is said by many, but not by all, philosophers, that $2 + 2 = 4$ is a logically necessary statement. Kant recognized that it depended on our understanding of the unity 1, and this he consigned to metaphysics, via the concept of *category*. (This being some necessary a priori tool of the human rational system.) The problem with this view is that the category invoked appears somehow to transcend nature, whereas it is important to realize that the concept of "one" or "unity" must be empirically justified.

I will therefore look at this in a different way from Kant's. As I have repeatedly said, our minds are the result of the evolutionary process, not only phylo- but also ontogenetically, and the very concepts that we take for granted, such as the principle of sufficient reason, reflect the process of learning, whereby regularities are processed by the mind, and entrenched in some form of rules, which I have called meta-physical. This is the form of the anthropic principle which I proposed in chapter 4. Let us see how this works for the concept of number.

This concept is firmly linked to that of set, and the latter to that of an aggregate of objects. Imagine now that the world were such that, as in the case of the Nicaean bishops, aggregates of a certain size were inherently unstable. Suppose for instance that every time you tried to add a single object to an aggregate of one hundred objects the aggregate exploded. (This, in fact, is what actually happens if you have an aggregate of plutonium atoms, admittedly for a much larger number, which is called the *critical mass*.)

We humans have obtained our idea of number because aggregates are *stable*: we can add elements to them without perturbing the rest of the aggregate; and I shall call this the *principle of the stability of aggregates*. Because of this stability, we can establish a bijection of one aggregate onto another, and thus obtain the concept of cardinality of an aggregate or, in plain language, of how to *count* it.[79] In the strange Nicaean world that I am now describing this could not be done, except perhaps for very small aggregates, and thus, if any concept of number were to emerge among the people living in that environment, it would be one that would terminate at some sort of maximum integer, since talking of any higher numbers would make no empirical sense whatsoever. (Remember that a sandy beach, or the sea, which gave people from Virgil onward the classical example of infinity, could not exist in that world.)

The reader might easily think: even if I were never to experience an aggregate of more than N elements, I would nevertheless invent some sort of Peano arithmetic, and *my* numbers would form an infinite set. After all, I have never counted an aggregate of a billion billion billions, but I can *imagine* what it means. I agree that any of my readers *transplanted* to the Nicaean world would certainly succeed in doing that. But that is irrelevant, because the natives of the Nicaean world would have had an entirely different phylo- and ontogenetic development, and *their* concept of number, if any, would be totally different from ours.

On using a terminology with which the reader will by now be familiar, we can say that the proposed *principle of the stability of aggregates* has allowed the human mind in the evolutionary process to "entrench" the natural numbers. Physical laws, I have asserted, are never *proved* but they are certainly "entrenched," although the situation in mathematics, I believe, is somewhat different.

As I have already advanced, I take the view that there are two *activities*, both of which we call "mathematics." The first one is empirical, the expression $2 + 1 = 3$, being, say, two fingers plus one finger, equal three fingers, in the sense that aggregates have been formed, mapped onto other aggregates, and the stability principle has been used under valid conditions (that is, the aggregates used are "good"). Let me call this version "empirical mathematics." There is now a second version, in which formal rules used in empirical mathematics are extracted, and, say, simulacra of the natural numbers are constructed, as Peano overtly did. (Unfortunately, these simulacra are written in exactly the same notation as in empirical mathematics.) Once this is done, a game is played, in which as many results as possible are derived by the stated rules as applied to the simulacra. Let me call this activity "game mathematics." Of course, in both activities entrenchment works as validating the concepts used in either case. In empirical mathematics, entrenchment entails the adequacy of the concepts as regards facticity, whereas in game mathematics entrenchment is done in relation to the mesh of rules accepted.[80]

There are several good examples in the history of mathematics which show that the two mathematical activities described above were actually pursued. Euclidean geometry was started by Greek and Egyptian surveyors in its empirical form, but it was used by Euclid as "game mathematics." The proof that a Euclidean line is a simulacrum follows from the fact that, when non-Euclidean geometries were discovered, this simulacrum was recognized in one of these geometries as what one calls a "circle" in empirical mathematics!

Quaternions provide another good example. Hamilton quite deliberately defined them in a way that was purely algebraic and thus "game-like," in the sense that he had to satisfy some formal rules which, although derived from simulations of empirical mathematics, were sufficiently removed from it so as to prevent Hamilton from a successful empirical entrenchment of the quaternions themselves. I believe that it is not by chance that it was Rodrigues who successfully entrenched quaternions as rotations: he had taught for the École Polytechnique in Paris, a quasi-military establishment in which mathematics was practiced in association with surveying, and thus kept always closely in touch with empirical necessities.

If you accept that a lot of mathematics is in fact "game mathematics," a great many things become clear. Within the mathematical "game" (or formal system) everything is played by rules that are given logical necessity. A pure mathematician who has never been subjected to the necessity of entrenching in relation to empirical facts (Hardy was such a one: cricket was probably as near as he ever got to nature) is therefore apt to believe that everything in mathematics is not contingent, and he can thus easily fall into Platonism.

One of the very agreeable dividends of the idea that I am proposing, is that the concept of "existence" in "game mathematics" becomes much less troublesome, because we can see that "existence" here is not "existence in the contingent world," which is a much harder problem. (By the "contingent world" I do not mean the world "out there," about which science and mathematics can say very little or nothing, but the world, which I had called before nature$_2$, that we project as well as we can from our sense impressions guided by our cognitive system.)

The answer to our apparently absurd, but nevertheless very important, original question is now obvious. The rules of arithmetic have been entrenched for finite aggregates only, so whereas it is true that for the simulacra of natural numbers as used in game-mathematics we must have an infinite number, on the basis that $M = N + 1$ must "exist" for all N, however large, there is no reason why this should necessarily be the case for "empirical mathematics." The world could be such that, for N larger than a certain value, stable aggregates do not exist, so that the very concept of natural number would break down eventually. Even this instability need not come in the way of game mathematics, because we do know in fact that the total number of particles in the universe, although exceedingly large, is finite, so that aggregates are not available for which the appropriate very large cardinalities can be empirically realized, although they "exist" within the game.

Before I close down, I must dispel any notion that "game mathematics" might possibly be an intellectually suspect activity. On the contrary, it is immensely fruitful for many reasons. Often a "mathematical game" becomes eventually empirically entrenched, and it thus acquires great practical importance. (This was the case, for instance, when Rodrigues successfully entrenched quaternions as rotations, a fact which, given the huge weight of Hamilton in mathematics, took almost half a century to be recognized.[81]) Also, "game mathematics" allows us to work with models that, although entailing meta-physical elements, provide results that can perfectly well be entrenched by mapping onto nature. An example is the concept of the continuum, along a straight line for example. Here you need your infinity, and, although this is a clearly meta-physical concept, it allows us to use calculus, which is a well-entrenched tool. The problem is, of course, why can this be entrenched? I believe that the reason is that the rules used in the "mathematical game" are carefully obtained so as to mesh with those of "entrenched empirical mathematics," which grew in such a way as to be adequate to our contingent world.

Going back to numbers: I take the view that when we recognize the property that two aggregates may be put into one-to-one correspondence, this is not a result of any logical necessity, but rather a contingent fact of the world in which we live, in which our intelligence was developed. From this recognition we arrive at the concept of a cardinal natural number (as that which two aggregates in one-to-one correspondence have in common) and at empirical mathematics, in which one should accept that, if as it is now believed, the total number of particles in the universe is finite, then the largest cardinal number of any physical set (that is any set that could be realized from physical aggregates of objects) is likewise finite. But we can also use the rules that emerge from entrenched empirical mathematics in order to deal with imagined sets, now unrestricted in size, and what I have called game-mathematics appears.[82]

Going back to the question of counting in the case of a Nicaean-type unstable set, I had better stress that, whatever happened in Nicaea, what I had in mind is not a question of perception at all. The crux here is *not the possibility of perhaps better being able to count a tricky aggregate*: the question is that the very concept of counting would never have been a part of our mental set-up if the whole of our phylo- and ontogenetic history had been entirely based on experience of such unstable aggregates.

Finally, a parable might be useful to illustrate the relation between the two mathematical activities described, and if not, it might at least provide some relief in this long and heavy chapter. But, before going into it, I must remind the reader that empirical and game mathematics must not be regarded as completely separate watertight pigeonholes: such a wall as separates these two activities is, to continue the metaphor, semipermeable, and happily allows a two-way trade between them.

A TALE OF TWO WORLDS

Chessworld was a town divided into sixty-four square blocks in an array of eight by eight of them, and its inhabitants were divided into twelve castes: pawns, knights, bishops, rooks, kings, and queens, each of them duplicated in two colors, black and white. Each caste was trained to move from block to block obeying certain strict rules, which happen to coincide with the game which *we* call chess, but which was unknown to the inhabitants of the town. All they knew was that they must follow the rules.

When a pawn wanted to arrive to a particular block, he did so in accordance to the rules. Thus, for this people, what they did was not a game, but a *practical* form of finding their bearings. Although, obviously, they happened to know all the rules of what we call *chess*, they did not play this game at all.

At a certain later time somebody in this town invented a checkerboard and chess pieces, and people started playing the game of chess. But what they did, although it obeyed the ancestral rules that allowed them before to find their way through the town, had *no empirical content whatsoever*. Since the game inventor was a distant and vague historical figure, they believed that chess was a divine game of eternal validity, and that it was a reflection of the wisdom of their gods.

At an even later stage, somebody discovered that she could use the checkerboard before embarking on a perambulation about the town, in order to minimize the number of steps she had to take to arrive at her destination. The chess game was henceforth used as a *model* and, naturally, it fitted perfectly the empirical exigencies to which the dwellers of Chessworld were subjected.

Chessworld had thus split into two. One world was governed entirely by empirical contingent rules. The other, was an abstract apparently eternal world, floating in the imagination of its inhabitants, and manifested by a game whose rules, within the game, had the force of necessity. It took the Chessworldians, however, a not inconsiderable time to realize that this second world was a mere manifestation of the first, and that it was not an entirely independent creation, antedating even that of their own town.

End of the story: you have been warned.

MATHEMATICS AS AN EMPIRICAL SCIENCE

I have already mentioned Wittgenstein's considerable influence in anchoring mathematics within the natural world, as being one form of a natural language. Many authors, however, have also forcefully supported the rejection of mathematics as a priori knowledge. Both Imre Lakatos (1922–1974) and Hilary Putnam (1926–) rejected the view that, as a difference with natural science, mathematics is a priori. For them, it was *quasi-empirical*.[83]

I shall indulge in a few quotations in order to reassure the reader that my views of mathematics as an empirical science are far from unique. Take this: mathematical knowledge is "fallible, corrigible, tentative and evolving as is every other kind of human knowledge" (Hersh 1985, p. 10).

An explanation why we can perform mental arithmetic, for instance, "cannot be found in mathematics and logic. The reason is that the laws of physics 'happen to' permit the existence of physical models for the operations of arithmetic such as addition, subtraction and multiplication. If they did not, these familiar operations would be non-computable functions" (Deutsch 1985 *b*).

Finally, I should like to refer to the work of the American mathematician Gregory Chaitin, whose fundamental results started even in his teenage years, and whose work deserves a far more complete treatment than I can give it in this chapter. His major contributions relate to the study of systems in which the amount of data relevant for the results is very large, which now goes under the name of *complexity theory*. The very important fact is that Chaitin embarked on a profound analysis and extension of the Gödel theorems, and instead of being lured by the siren-song of Platonism, he came out as a convinced empiricist. He was able to study the Gödel theorems from the point of view of information-handling, that is, what is called *information theory*. Any axiomatic system used will have a given complexity, linked to the information content of the system. Thus, within that system, it is never possible to prove that a given object has an information content greater than that of the system, although infinite many such objects exist with that property. Thus, "physics provides a means to expose the limitations of formal axiomatic systems." Also: "mathematics is perhaps more akin to physics than mathematicians have been willing to admit." Even more recently, he writes: "I believe that elementary number theory and the rest of mathematics should be pursued more in the spirit of experimental science, and that we should be willing to adopt new principles. I believe that Euclid's statement that an axiom is a self-evident truth is a big mistake."[84] So, as you can see, not the whole world has gone Platonic.

Are There Empirical Mathematical Proofs?

The idea that mathematical truth might not be a priori or even necessary is one that would be found by many people, I am afraid, almost offensive, and if that is the case, I am sorry. Worse, I think, is the possibility that mathematical proof might be, some times at least, empirical.[85] The situation appears to be rapidly changing, however, owing to the advent of computer proofs, of which I shall give one, the most famous, example.

It had long been suspected that every planar map (such as a geographical atlas) could be colored with only four colors, ensuring nevertheless that no frontier would be equally colored on both sides. Although no exception to this rule was known, no proof was obtained despite many years of work, and this became known as the four-color problem. In the 1970s Kenneth Appel and Wolfgang Haken, with the assistance of John Koch, who did most of the computer programming, produced the proof of the conjecture. This comprised very roughly three different stages. First, a complex algebraic proof was obtained, as a result of which it turned out that the analysis of the problem would depend on discovering the value of a certain integer that would decide the classification of all the possible cases.[86] Second, this integer was guessed on probabilistic arguments, and the guess was then checked by computer. The last and third stage, however, required the analysis of thousands of alternatives: if all of them satisfied the four-color conjecture, then the theorem was "proved." This turned out to be the case. For the first time mathematics was faced with a clear instance of a result discovered a posteriori, very much as the result of a physical experiment.

I seem to remember the comment of a distinguished mathematician when informed that the four-color problem had thus been "proved": "so, after all, it turned out not to be an interesting problem!" Much as I sympathize with him, I believe that there are better ways to bury your head in the sand. Whether you like it or not, with more powerful computers and software, a lot of this is likely to happen in the future.

THE UNREASONABLE WORRIES ON THE EFFECTIVENESS
OF MATHEMATICS IN THE NATURAL SCIENCES

Eugene Paul Wigner (1902–1995) was a Hungarian-born mathematical physicist who spent the larger part of his working life in the United States and whose influence on modern physics was colossal, both in breadth and in depth. He was responsible for showing how the abstract mathematical subject of group theory could be applied to most branches of physics; and since (almost) everything I know about that subject comes from him, and since I have earned such living as I could on the strength of that knowledge, I somewhat feel like an ingrate in defying the judgement of the master, but there it is. On 11 May 1959 Wigner gave a lecture entitled *The Unreasonable Effectiveness of Mathematics in the Natural Sciences*, which, given his great intellectual authority, caused a huge stir in the scientific world, the ripples of which still throb.[87]

I shall reproduce two quotations from this paper. "The first point is that the enormous usefulness of mathematics in the natural sciences is something bordering on the mysterious and that there is no rational explanation for it." "The miracle of the appropriateness of the language of mathematics for the formulation of the laws of physics is a wonderful gift which we neither understand nor deserve." Now, however much the words "mystery" and "miracle" might help in selling popular science books, when I see them used outside their proper theological context, I feel for my intellectual bazooka. For their use in science or philosophy is the abdication of the rational way of thinking that must inform a natural scientist qua scientist. (Others, please remember that scientists and mathematicians and Simon Altmann are just as likely as anybody else to talk rubbish when meandering outside their subjects.)

I believe that discussion of the role of mathematics is hampered by a rather diffuse idea of what mathematics really is about. Fortunately, the great French mathematician Jean Dieudonné (1971) has summarized wonderfully this role for us:

> mathematics is the logical study of *relationships* among certain entities, and not the study of their nature.

This is the key to our understanding of the "Wigner problem." For practically the whole work of a natural scientist is to discover relations, to study relations, to classify relations. Everything, from the discovery of Kepler's laws to that of a penicillin, entails one of these activities; and taxonomy is the ultimate example where a systematic study of relations is practically the whole subject.

But you cannot have anything to describe unless you have a language, and, given Dieudonné's assertion, whenever a scientist is opening his or her mouth, mathematics must be spoken in one form or another. And it will be mathematics that is already known, or mathematics that is invented on the spot for the purpose: it is in this latter way that mathematics was mainly created and empirically entrenched. Naturally, scientists will never miss an opportunity of borrowing some concepts from "game mathematics" whenever convenient. Interestingly, this requires for them a knowledge of what is available in that game. Most often, however, scientists have no time or patience to find this out, so that they reinvent the necessary concepts, which then become entrenched as part of the empirically based mathematics.

Let me give a few examples. When Newton was excogitating his second law, as we discussed in chapter 6, he required a *relation* between the positions and the times of passage of a particle at an initial and at a nearby point. No one had done that for him, and he invented the concept of instantaneous velocity, which formed the basis of his calculus. And this was the theory that dominated mathematics for well over two centuries. No wonder it worked in physics, since it had been tailor-made for it.

In the year 1858, the Cambridge mathematician Arthur Cayley (1821–1895), while studying systems of equations, invented a new mathematical object called a *matrix*. It is not important for us to know in detail what these things are, but the one feature I want to stress is this. If you multiply two numbers, they *commute*: 4×3 is equal to 3×4. Cayley was able to define a multiplication rule for two matrices A and B, but they do not commute: AB is not the same as BA. Good and important as matrices were, they remained for quite a long time a mathematical curiosity; and even good mathematical physicists knew nothing about them. When quantum mechanics was being created, one of the most important papers was that of the Göttingen physicist Werner Heisenberg in 1925, which gave the equations of the new mechanics. He needed some quantities the products of which would not commute, and he happily introduced them, without enquiring whether they "existed" in mathematics. He did not have any doubt that, since empirically he had to have them, then they were mathematically "possible." What he was doing, of course, was discovering an empirical relation and providing the mathematical language to fit it. Three months later his Göttingen colleagues Max Born and Pascual Jordan published a paper in which they start with the "remark that this rule [Heisenberg's product rule] is nothing else that the well-known law of multiplication of matrices."[88] And they then rehearsed the rules about the use of matrices that went back to Cayley. No wonder that matrices are useful in quantum mechanics, since they were virtually reinvented in order to fit with the relations that the study of this subject requires.

In 1926, again during the study of quantum mechanics, the Cambridge mathematician Paul Adrien Maurice Dirac found that he needed a very strange "function" in order to represent the motion of the electron. This function had to vanish for all points of the straight line (I shall call this the *base*), except at a single point. This is quite possible: you can imagine a function looking like the cross section of a mountain and require it to become thinner and thinner until it becomes a mathematical line perpendicular to the base. So far so good: but Dirac faced a constraint that was demonstrably absurd. If you imagine the cross section of that mountain, it will have some area, but when it becomes a straight line, that area must vanish, since a straight line has only one dimension and not two, as areas must have. Yet Dirac required that his infinitely "thin" function had to have an area equal to the unity. At the time, this was mathematical nonsense; that is, functions with such a property could not "exist." Yet Dirac, than whom few physicists have been more inventive or more courageous, stuck to his guns and defined what all the world now knows as *Dirac's delta function*. The interesting thing is that mathematicians did not go around saying "This man is mad, he talks about something that I cannot perceive with my Platonic spectacles." They just went on to work to invent a new theory of functions which would admit Dirac's delta function as a respectable denizen. This work was not trivial: it took almost a quarter of a century and two of the most distinguished mathematicians of their time, the French Laurent Schwartz and the Russian I. Gelfand. The theory they produced is called *distribution theory*, and although it was created to make an honest woman of quantum mechanics, the delta functions were soon recognized to be very useful even in engineering applications. However, the original mathematical presentation was as difficult as it was elegant; and it was the job of the Oxford mathematician George Temple to rework it in a more accessible form. You can see here that you move from empirical mathematics to game-mathematics, and that when the latter theory is really found to be useful, even the way it is presented changes, to adapt itself to the needs of the consumers. No wonder that it works in practice: if it hadn't, it would have been forgotten between the dusty covers of some obscure mathematical tome.

Wigner's own subject, group theory, is the archetypal example of a mathematical tool to study relationships. If you glance back at figure 2, you will notice that the parabolas in it have what is called *mirror symmetry*. This means that if you *reflect* the point with coordinate x in a parabola through the vertical axis until it reaches the value $-x$, then the corresponding point of the parabola does not change in value. When x changes from $+1$ to -1, for instance, y for the gray parabola maintains the same value $+2$ (see fig. 2), which gives it its characteristic mirror symmetry. The interesting thing is that this result can immediately be obtained from the equation of the parabola,

$$y = x^2 - 1,$$

which of course does not change when x is changed into $-x$, since the square is always positive. The important idea here is that a given type of algebraic equation will correspond to curves of some particular symmetry. Naturally, the symmetries that may appear for more variables can be more complicated than mere mirror reflections, and the great virtue of *group theory* is that it allows you to classify all such possible symmetries by studying the set of symmetry operations of a curve. These sets have some special properties which define them as *groups*, whence the name of the theory. Even when you have a single particle moving in empty space, you have a symmetry, owing to the fact that any position of the particle is physically equivalent to any other. Also, if you have a system with many identical particles, permuting two such particles must again lead to a physically equivalent state (the energy, for instance, cannot change under such a permutation, because the particles are identical). Obviously, symmetries are nothing else than relations; and it is no wonder that group theory plays such a crucial part in the study of nature, since it deals with a great many of the relations that one has to study.

I rest my case, and I have thus finished what I wanted to say in this chapter. Before closing it, I must confess that I have tried to shelter the reader from some of the more thorny philosophical problems in mathematics, concerned with the concept of the infinity, and in doing so I have seriously neglected some very important ideas. The reader who is interested will obtain some, I am afraid, very partial redress for this failure in the next section.[89]

A TAIL END: CHASING THE INFINITY

Clearly, the concept of infinity is a very fertile ground for existential thoughts. In what way does the infinity exist? People like Brouwer preferred to abolish it, substituting for it finitary processes in which the number of terms involved could be taken as large as possible. On the other hand, if we take as an example the natural numbers 1, 2, 3, . . . , and use Peano's successor axiom, we can take it that the set so defined will never cease to increase the number of its elements. This is the simplest example of an infinite set, and the cardinality of the set of natural numbers so defined is called ω (omega), as we have already mentioned. We have also seen that any set that can be placed into a bijection with the set of natural numbers is said to be *denumerable*, and that its cardinality is ω. This cardinality, although the set is denumerable, is still infinity, and it is also called $\aleph_0$ (read *aleph null or aleph nought*, aleph being the first letter of the Hebrew alphabet). This symbol was introduced by one of the most original (and most unbalanced) mathematicians of the nineteenth century, Georg Ferdinand Ludwig Philipp Cantor (1845–1918), who was born in St. Petersburg, but early settled in Germany, where he vegetated with a chair at the unfashionable University of Halle. What the reader must notice is that "infinity" is given a name like $\aleph_0$ because Cantor discovered that, just as 1 is not the only number, but 2 follows it, so $\aleph_0$ is not the only "infinity" but it is followed by a second one, $\aleph_1$, and so on. Thus infinities are just as varied as the natural numbers are, and they can be numbered by means of the so-called transfinite numbers, of which $\aleph_0$ is the first. But all this will be much clearer after we do a bit of work.

Suppose that I try to list all the rational numbers. We start with the fractions:

$$\frac{1}{1} \quad \frac{1}{2} \quad \frac{1}{3} \quad \frac{1}{4} \quad \frac{1}{5} \quad \frac{1}{6} \ldots$$

and then repeat similar successive lines with numerators, 2, 3, and so on. In this way we "list" all fractions, and I put "list" in quotation marks because we shall have an infinite number of lines, and each line itself will be infinite. In fact:

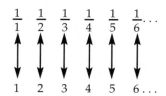

So, because that row alone seems to exhaust all the natural numbers and thus to have the same cardinality as its (infinite) set, $\aleph_0$, then the totality of all fractions might appear to have a larger cardinality than $\aleph_0$. This is not so: all the rational numbers can be counted, that is a bijection between all of them and the set of all natural numbers can be made, showing that both sets have the same cardinality. This requires a little ingenuity and is done in figure 5, in which we must notice the following points. First, the square array, if infinitely continued, both in each line and in the number of lines, exhausts, as already explained, the totality of the rational numbers. Second, if we follow the gray lines of the figure, we can count 1, 2, 3, 4, . . . as we hit each fraction and thus establish the bijection between them and the natural numbers. Thus, the set of all the rational numbers has the same infinite cardinality, $\aleph_0$, as the set of the natural numbers.

Let us now try to count the irrational numbers, for which purpose I shall consider a small subset of them, namely all the irrational numbers between 0 and 1 (as, for instance, 0.15432 . . .). As we know, the irrational numbers must be expressed as a decimal number with an infinite number of digits. Suppose that the set shown in figure 6 were to express the totality of such numbers in the range stated.

This assumption can be demonstrated to be wrong by means of a very ingenious and very general argument invented by Cantor, which is called the *diagonal argument*. Given the alleged array of all irrationals, form a new irrational by replacing the first digit (after the decimal point) of the first number with a different digit a'_1, the second digit by one different from the second digit of the second number, and so on. (The relevant digits to be so replaced are given in gray in figure 6.) In this way the following irrational number is formed:

$$0.\, a'_1\, b'_2\, c'_3\, d'_4\, e'_5 \ldots .$$

This irrational differs from *each* of the entries in the figure, from the first by the first digit, from the second by the second digit, and so on. We have thus proved that, even if I limit myself to a small segment of the real axis, the one between 0 and 1, the irrational numbers in it cannot be denumerated (because I can always insert in this interval new irrationals). This means that they cannot be counted either, in the sense that they cannot be put into a one-to-one mapping with the integers. Because of this, we say that their set is *not denumerable*. The real numbers (integers, rationals, and irrationals) denote points of the straight line or real axis, as we have already seen, and their set forms what is called a *continuum*. In fact, the above argument for the range 0 to 1 can be used for any range, however small, and demonstrates therefore that between two irrational numbers (points of the

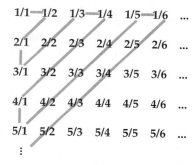

Fig. 5. Counting the rational numbers

You must follow the gray lines starting from the top-left corner.

$$
\begin{array}{llllllllll}
0. & 0 & 0 & 0 & 0 & 0 & 0 & 0 & 0 & \cdots \\
0. & a_1 & a_2 & a_3 & a_4 & a_5 & a_6 & a_7 & a_8 & \cdots \\
0. & b_1 & b_2 & b_3 & b_4 & b_5 & b_6 & b_7 & b_8 & \cdots \\
0. & c_1 & c_2 & c_3 & c_4 & c_5 & c_6 & c_7 & c_8 & \cdots \\
0. & d_1 & d_2 & d_3 & d_4 & d_5 & d_6 & d_7 & d_8 & \cdots \\
\cdots \\
1. & 0 & 0 & 0 & 0 & 0 & 0 & 0 & 0 & \cdots
\end{array}
$$

**Fig. 6. Cantor's diagonal argument:
the irrational numbers are not denumerable**

line) there must always be an infinite number of other irrationals. So, the set of all irrational numbers forms an infinite set, the cardinality of which cannot be $\aleph_0$, since the set is not denumerable, and their cardinality is denoted with the symbol $\aleph_1$, although c is also used to remind ourselves that it is the cardinality of the continuum. That there is no possible cardinality between these two is called the *continuum hypothesis*.

CODA

We have met once again in this chapter one of the major topics of this book, the struggle between contingency and necessity. Everything that depends on the properties of an undivided nature must be contingent, as nature itself is. It is only if we can get away from that sphere to some other one, like a Platonic world (or a theistic realm), that necessities might appear. (As always, I disregard such necessities as arise purely from linguistic conventions.) One of the most cogent arguments for mathematical necessity ever presented came from Gödel himself through his theorems, but there is no question that at least a verdict of *not proven* must be given to this attempt. This does not mean, of course, that of their own free will people are not entitled to support the credo of Platonism, but that their use for this purpose of intellectual props, the validity of which is far from established, is to be deprecated. We have seen in this chapter that even the most simple and most essential concept of natural numbers depends on the property that I have called *aggregate stability*, which is clearly contingent. Thus, the work of this chapter goes quite a long way in rejecting the notion of natural duality or, more properly, in confining it purely where it belongs, to theological discourse.

One of the most interesting things that happened in this chapter is the Lucas-Penrose use of the Gödel theorem in order to show up some properties of human thinking. There are two quests involved in their search. One is to show that it will never be possible to construct a machine that will be able to perform as well as a human. The second, which is a consequence of the first, is that the brain is *inherently* different from a machine, either because it entails some sort of nonphysical activity or, as Penrose thinks, because it depends on some alleged quantum effects that, for some unexplained reason, cannot be available to computers.

Gödel's theorem in any form is most unlikely, as I have shown, to produce a serious positive answer to those suggestions, but let me try and retrieve what I think is the essence of the discussion. If the argument in question were valid, all that it could *undoubtedly* show is that human thinking cannot be algorithmic. That is, that one does not reason always from *a* to *b* by using logical and mathematical steps that are recursively arranged (that is, step-by-logical-step). But we know only too well that this must be the case: when Hamilton invented quaternions, for instance, there was no way in which he could deduce, from what he knew of complex numbers, the steps that he should take in order to create the new algebra. He, after a lot of ruminating about the subject, *guessed* the answer. His mental process was clearly not algorithmic!

Why is it then that we have to get into all the complexity of the Gödel-Lucas-Penrose argument to discover what we already know? I can only guess:[90] if you are a Platonist, the steps that could be considered as departing from the straight and narrow of algorithmic thinking are merely processes of Platonic illumination. Hamilton did not *invent* quaternions, he *discovered* them, and his sudden flash of inspiration was no more than a fruitful reading of the Platonic world. Because such illuminations are taken for granted by the Platonists, their arguments must be concerned with run-of-the-mill logical thinking.

Occam's razor may have become blunted lately, but any sensible use of it would indicate that Platonism, in this particular context, is a heavy price to pay if its object is to discover what for most people must be a fairly obvious property of the human mind. As for the alleged implications as regards machines, the whole argument hinges on what Lucas, quite carefully, defines as a machine, namely one that could never do anything except algorithmic processing. That machines will *never* be designed that can surmount this restriction is not a question for philosophical or even technological argument, or for a well-grounded guess, but must be taken as mere unsupported futurology. Only time will tell.[91]

Let me remind the reader that the proposal I have made about the two mathematical activities that I have described combines the best of the two main antagonistic views of mathematics. On the one hand, we have an area of mathematical activity that is empirically oriented and where entrenchment is done on the basis of saving the facts. On the other hand, the game-activity of mathematics operates on fixed rules, and provides the element of necessity that is given by entrenchment with respect to the mesh of established rules and theorems. Thus, the whole subject is woven into a large mesh, parts of which are more empirically oriented and others more formal; this mesh is very much like the science mesh which I have described before and of which it is, of course, an integral part. This closely resembles a Platonic world but human-made. Of course, the possibility of the Platonic world being a human creation is abhorrent to Platonists, as conspiring against the eternal validity that they claim for mathematical verities. *Chaqu'un à son goût.*

There is one more question I must handle, because some of my readers may feel somewhat cheated that after reading such a long chapter they might still find themselves at sea as regards what a mathematical proof really is. It might be a bit disheartening to be answered: "A mathematical proof is what mathematicians are prepared to accept as such," but this somewhat dry statement has at least the merit of agreeing with praxis. Remember my Tate Gallery philosopher, with whom Wittgenstein would have agreed: " 'This man is a student of mathematics.' How do you know?—'All the people in this room are mathematicians; only such people have been admitted.' " And again: "mathematics is a MOTLEY of techniques of proof.—And upon this is based its manifold applicability and its importance."[92]

Finally, I should like to stress that the Lucas-Penrose attempt to link the Gödel theorem with mathematical proof, and thus to thinking, is not without inherent weaknesses. Because the "proof" provided by Lucas's machine is only a *formal* proof, that is, one in which the logical and mathematical rules are recursively (or algorithmically) used. From what I have just said, the idea that there are no other forms of acceptable mathematical proofs is just a pipe dream: the Gödel proof *itself* is *not formal*, because it requires the use of the concept of *mirroring*, which is just as informal as anything can be. Even without this feature, to extend the very limited form of logical truth with which Gödel is concerned to the whole realm of human truth is no more than wishful thinking. If people *need* to entertain such thoughts (for which I would be the last to object), the best thing they can do is to go back to theological discourse.

NOTES

1. I am grateful to Gerry Altmann for this information. An account of Jeffrey Elman's neural networks may be found in his Altmann (1997).
2. This is roughly what Popper (1972) calls the third world.
3. As an example of this confusion, we shall soon see that the greatest of the mathe-

matical dualists of this century, if not of this millennium, was Kurt Gödel and he is nevertheless often labeled as an *idealist*. (See for instance Yourgrau 1991.) The reader has already been warned that pigeonholing or ism-using is not always a profitable activity.

4. That the truth of a belief is not sufficient justification for it, as Plato had held (*Theaetetus*, 202 C), was first pointed out by Edmund Gettier (1963), a remark that I have learned from Maddy (1990), pp. 36–37. Gettier provides a critique of knowledge as justified true belief.

5. Locke (1690), Bk 3, Ch. 3, § 1; Quine (1964), pp. 1–19. Quine's example is famous: When I say that Ted and Ed are two white dogs, it might be believed that I am committing myself to the universal "whiteness." But for my phrase to be true all that is required is that there be one white dog called Ted and another white dog called Ed.

6. Published in Bernays (1935).

7. For immortality, see Plato's *Phaedo*. Recollection is called *anamnesis* by Plato. See *Phaedrus* 249. Also *Meno* 81: "The soul, then, as being immortal, and having been born again many times, and having seen all things that exist, whether in this world or in the world below, has knowledge of them all."

8. The reader might find it useful to compare with my previous use of the expression *meta-physics*, which did not entail an *explicit* ontology but purely a *nomology*. It was the American philosopher Charles Chihara (1982) who first pointed out the distinction between ontological and mythological Platonism.

9. See Maddy (1990), p. 160, for (1) and (2).

10. I heard John Lucas stating this at a meeting in Oxford in 1995 when he claimed that 2+2 = 4 was true even before the big bang.

11. See Connes in Changeux et al. (1995), p. 56. This book contains interesting dialogues between a Platonist mathematician and a skeptical biologist (Changeux).

12. See Connes in Changeux et al. (1995), p. 205. The separateness of the Platonic world with respect to the physical one appears also to be displayed by Roger Penrose in his fig. 8.1 of Penrose (1994), p. 414.

13. See Connes in Changeux et al. (1995), p. 56.

14. See Penrose (1994), pp. 50 and 414 for the last two quotations. All mathematicians are "in contact with the *same* externally existing Platonic world" (Penrose 1989, p. 428) which endows this world with objectivity as claimed in (4). See also Penrose (1997), fig. 1.3.

15. See Connes in Changeux et al. (1995), pp. 151 and 89, respectively, for the last two assertions. See also p. 89: the "preexisting mathematical world is logically prior to any human intervention." Many of the above notes are contained in the following sentence of one of the most famous British mathematicians: "I believe that mathematical reality lies outside us, that our function is to discover or *observe* it, and that the theorems which we prove, and which we describe grandiloquently as our 'creations,' are simply notes of our observations" Hardy (1992), pp. 123–24.

16. See Hellman (1989).

17. It could be argued that the Platonic world is concerned with *relations*, not with objects. If one could successfully claim that relations have a *necessary* meaning (why?), then it is these relations that are so to speak stored in the Platonic world, in preparation for the world to exist. One would then have to argue that, these relations being necessary, any created world had to carry their imprint. It is difficult, however, to defend such view of necessity without a theistic position.

As regards the earlier question of mathematical propositions as rigid designators, Kripke (1971, p. 78) takes for granted that the proposition "the square root of 25 is in fact the number 5" is *necessarily* true; and this being so in our world can then be taken as a rigid designator. This, however, begs the question: why is arithmetic as it is? Suppose that our own world were such that, whenever you joined together two aggregates of objects, say two eggs and two oranges, an extra object always appeared. If intelligent beings in such a case used *our* arithmetic, they would find counting either impossible or nonsensical. More about such anomalies later, but note that my "absurd" scenario for aggregates is in fact realized in the microworld as we know it. (See p. 549).

18. I am aware, of course, that numerous scriptural passages encourage this interpretation. When John the Baptist sent two of his disciples "to Jesus, saying, Art thou he that should come? or look we for another?" (Luke 7:19), Jesus answered: "tell John . . . that the blind see, the lame walk, the lepers are cleansed."

19. There is some evidence for my story, however, from one of his sonnets, no. 15 in the Guasti edition, which is quoted by Vasari:

> Non ha l'ottimo artista alcun concetto
> Ch'un marmo solo in sè non circonscriva
> Col suo soverchio; e solo a quello arriva
> La man che ubbidische l'intelletto.

A prose translation is roughly as follows: "An excellent artist never has any conception that a piece of marble does not by itself enclose within its discard, and it is only to it that the hand arrives that obeys the intellect."

20. For Wittgenstein "the mathematician is not a discoverer: he is an inventor." Wittgenstein (1978), p. 111, 2.

21. See Penrose (1994), p. 148, who favors this argument, as does Davies (1992), p. 152. See also p. 155 for the erroneous claim that the ducking of a missile does not require the solution of Newton's equation, contrary to the next paragraph in the text.

22. I owe this example to a lecture by Colin Blakemore.

23. Aldhous (1995).

24. Changeux in Changeux et al. (1995), p. 33.

25. A very much related school is also the one called *verificationism*, for which a mathematical statement is said to be true if and only if it has been constructively proved. Its most noted figure is the Oxford philosopher Michael Dummett (1925–). (See for instance Dummett 1977.)

26. Hankel (1874), p. 205.

27. This historical fact was noticed by Popper (1972), p. 128.

28. This proof appears in clear detail in Book X of Euclid's *Elements* (ca. 300 B.C.E.). See van der Waerden (1974), p. 110, but it was almost certainly known to the Pythagoreans around 500 B.C.E. The discovery, however, caused a great deal of worry to the Pythagoreans, who had a mystical feeling for the natural numbers and their ratios, and apparently the members of the sect were sworn to secrecy about the existence of the irrationals. There is a legend that Hippasus of Metapontum broke his oath and was drowned as a result.

29. In between two rational numbers there are always irrational numbers: 14142/10000 and 14143/10000, for instance, embrace between them $\sqrt{2}$. Have we *discovered* these new numbers? Wittgenstein (1974), p. 372, answers this question with a resounding negative: we have *constructed* them.

30. See Maddy (1990), chap. 2; her views are vigorously contested by Chihara (1982). Gödel's view that sets are not in space and time is clearly stated in Gödel (1983). See also Chihara (1990), p. 15.

31. This assumption is made by Maddy (1990), p. 153 and it is strongly criticized by Chihara (1990), p. 201.

32. Even before their seventh month babies can correctly map an aggregate of either two or three sounds to one of the same number of objects on a screen, as discussed by Mehler et al. (1990), p. 139. This mapping ability is probably the way in which the concept of number is learned. Counting, for instance, is nothing else than mapping the names of numbers (or their sounds) onto aggregates (such as fingers!).

33. See Peano (1908), p. 27, or Russell (1919), p. 5. The presentation of Peano's work is ferociously terse, and not made any easier by the fact that it is written in Interlingua.

34. Russell (1919) gives some even more horrid examples of this situation.

35. "Hardly anything more unfortunate can befall a scientific writer than to have one of the foundations of his edifice shaken after the work is finished. This was the position I was placed in by a letter of Mr. Bertrand Russell, just when the printing of this volume was nearing its completion" (Geach et al. 1970, p. 234).

36. Of course, the lack of an ontology would be welcomed by hard-core formalists. Hilbert's quotation appears in a letter to Frege, (Frege 1980, p. 42, cited by Shanker 1988, p. 171), where Wittgenstein's views on this question are also discussed. Even Gödel openly criticized the view that took consistency as the criterion of existence, as quoted by Feferman (1988), p. 102. A good account of the Hilbert program may be found in Smorynski (1977).

37. These two aspects of self-reference are probably not independent. Because the world is self-referential through our use of language, the latter is constrained: in natural language, it is impossible to "get out of the world" to talk about language. Thus at some stage the speaking mouth must bite its own tail, and this is were paradoxes arise.

38. Notice that for the purpose of these examples I use the word "liar" in a restricted meaning, to qualify a person who *always* lies. The *paradox of the liar* here described is also called the *Epimenides*, since it was first uttered by Saint Paul (Titus 1:12) as follows: "One of themselves, even a prophet of their own [who has been identified as Epimenides], said, The Cretans are always liars." In the next verse he goes on to assert: "This witness is true." Although this has often been alleged to show how poor a logician Saint Paul was, this is not necessarily the case: see note 40.

39. This is done by Russell himself in chapter 4 of Russell (1940).

40. This argument, with due acknowledgement to Tarski, was also given by Gödel in lectures he delivered at Princeton in 1933–34. See Feferman (1988), p. 104. See also Russell (1940), chap. 4. Notice, incidentally that Saint Paul's second statement, in Titus 1:13, if interpreted in the metalanguage, absolves him from logical naivety. And if saints are not granted the use of metalanguages, who is?

41. See Bradley (1893).

42. Many mathematicians would claim that the object language of mathematics is like a syntax without semantics, the symbols used being signs that need not signify, but that must be related by the stated rules of mathematics. (See the second epigraph to this chapter.) Any well-formed string of symbols could thus be considered as "meaningful." It is usual, however, to require the mathematical signs to be capable of interpretation, that is, that they be mirrored by mathematical objects. Within this interpretation, the sentence I have given in the object language of mathematics does not obey these rules because, *at this stage*, there is no number that could be represented by $\sqrt{(-1)}$. This symbol was as meaningless at the times discussed, as the statement "a positive number smaller than zero" is now.

43. See Dawson Jr. (1988a) for a biographical sketch of Gödel and Wang (1987) for an exhaustive treatment of his life and thoughts, written by a disciple.

44. Strictly speaking, I should make a difference between $P_g(g)$, which is a formula, and its *name* $P_g(g)$, which is the symbol used to designate the formula. Remember, however, that we are here at an impressionistic level of Gödel's proof. In our second shot later, metalanguage will automatically sort this out, since in it we only have names of the formulas of the object language, and not the formulas themselves.

45. The sketch of the proof which I have given is partly inspired in Mostowski (1952). For a rigorous but clear proof of Gödel's theorems see chapter 2 of Enderton (1972). DeLong (1971) contains a detailed discussion, as is also the case with Hunter (1971). A simple and illuminating, but not very rigorous treatment appears in Nagel et al. (1958). Hofstadter (1979) contains a discursive treatment of the theorem and of much else, but despite the huge popularity of this work, I do not find it always understandable. Three extremely useful articles are Fitzpatrick (1966), Bencerraf (1967), and Lacey et al. (1968). Hersh (1997), pp. 311–15

contains a simple proof of the Gödel theorem, and Kolata (1982) gives some simple examples of problems, apparently mathematically accessible, but that cannot be proved.

46. Prime numbers are integers that cannot be divided by any other integer, except themselves, and the unity, to give an integral result. This is the case for 3 but not for 4, because the latter divided by 2 gives 2.

47. Such formal systems were constructed in Rosser (1937). See also Rosser (1936).

48. See Shanker (1988*b*), p. 221, for a discussion of the assertions which I have quoted.

49. See Shanker (1988*b*), p. 211.

50. The reader might detect here a contradiction about my previously expressed skepticism about revolutions (p. 240). But I have already said that Galileo's was more of a theological than a scientific revolution. Revolutions tend to be noted more by the impact discoveries have on the culture of the period, than on the mainline pursuits of science or mathematics. It is an undeniable fact that whereas the Gödel theorem has been hailed by many as the most remarkable thing since the invention of sliced bread, it has had a minimal impact on the later development of mathematics during the twentieth century. (Although, through Turing, it much influenced the concept of computation.)

51. See Feferman (1988), in particular p. 111. John Dawson shows that Gödel was interested not so much in the falsification of Hilbert's program but rather in revitalizing Platonism in the middle of the positivistic mood that dominated philosophy in the 1930s. (See Dawson Jr. 1988*b*.) He clearly wanted to exploit the hierarchy of languages. In fact, instead of drawing interest away from Hilbert's metamathematics, his work had the opposite result.

52. Gödel (1983), p. 484: "I don't see any reason why we should have less confidence in this kind of perception, i.e., in mathematical intuition, than in sense perception." Feferman (1988), p. 102 asserts that "Gödel openly criticized the view which took consistency as the criterion of existence." And (p. 107) he quotes Gödel supporting the "concept of 'objective mathematical truth' as *opposed* to that of 'demonstrability.' " As regards Gödel's Platonism, it has to be said for him that it had at least the consistency (although not necessarily the increased verisimilitude) of stemming from a firmly theistic position. Hao Wang, who reports firsthand from conversations with him, says that "G's conception of metaphysics as first philosophy includes centrally the concepts of God and soul" (Wang 1987, p. 161). For further discussion of mathematical truth see Dales et al. (1998).

53. Wittgenstein (1978) is the work mentioned in the text, and Shanker's paper is Shanker (1988*b*). For the first Wittgenstein quotation see Wittgenstein (1978), VII §19, p. 383. The second is quoted in Shanker (1988*b*), p. 181.

54. See Shanker (1988*b*), p. 183.

55. The somewhat vague concept of mirroring is not one that Wittgenstein would have accepted. For him the relation between a sentence p and the system S to which it belongs is indisoluble. You cannot assert that p belongs to a system S. This must show itself. "Understanding p means understanding its system. If p appears to go over from one system into another, then p has, in reality, changed its sense" (Wittgenstein 1975, p. 153).

56. Wittgenstein (1975), p. 153, claims also that there cannot be a proof of provability, because it would have to rest on entirely different principles from those of the putative proof. Wittgenstein denies that there can be a hierarchy of proofs, and thus that metamathematics can exist in any fundamental sense.

57. Shanker (1988*b*), p. 193. For Wittgenstein, Gödel's theorem is not a refutation of Hilbert's program but a *reductio ad absurdum* of it because you only refute what *can* be true, whereas Hilbert's program was inherently wrong. (Shanker 1988*b*, p. 182.)

58. See Shanker (1988*b*), p. 184.

59. For Wittgenstein, Hilbert's metamathematics is just mathematics. (See Wittgenstein 1967*a*, p. 136.) I can invent a game in which I play with the rules of chess. "*In that case I have yet another game and not a metagame*" (Wittgenstein 1967*a*, p. 121). Every step in the so-called metatheory of chess corresponds to a move in the game, "and the whole difference

consists only in the physical movement of a piece of wood" (Wittgenstein 1975, p. 327). See also Shanker (1988*b*), pp. 212 and 178.

60. Shanker (1988*b*), pp. 217–18.

61. Christian Goldbach (1690–1764) proposed this famous conjecture in a letter to Euler in 1742.

62. For Wittgenstein "the proof changes the grammar of our language, changes our concepts. It makes new connexions, and it creates the concept of these connexions. (It does not establish that they are there; they do not exist until it makes them)" (Wittgenstein 1978, III § 31, p. 166). Thus, for him (see Shanker 1988*b*, p. 219), it is incoherent to assert that "*P* is unprovable and true." Mathematical conjectures are therefore nothing else than ill-formed expressions. (Shanker 1988*b*, p. 176.) Wittgenstein's view of conjectures has recently received support from the work of Gregory Chaitin, who has proved that some conjectures might not be susceptible of either proof or disproof. See Chaitin (1988).

63. This question of certain mathematical symbols being meaningless within a given system is a difficult one, and the reader might find helpful a later example (p. 421) of how a meaningless symbol, $\sqrt{(-1)}$, can be made to acquire meaning within a given mathematical system.

64. See Shanker (1988*b*), pp. 221–22.

65. Wittgenstein (1978), I, Appendix III, § 7, p. 118.

66. See Whitby (1996), p. 57.

67. Turing (1950), p. 442.

68. A very detailed yet accessible account is given in Penrose (1989) and Penrose (1994). The interested reader will find a critique of Penrose's work in Searle (1995) most useful; it contains a very clear and concise account of the Gödel theorem. See also Searle (1999). Searle (1990) is still a very useful introduction to artificial intelligence and to his famous Chinese room, criticized in Churchland et al. (1990). Alan Turing was one of the most remarkable British theorists of the century and his biography by the Oxford mathematician Andrew Hodges (1983) is essential reading.

69. See Wang (1987), p. 195.

70. The original Lucas paper is Lucas (1961) but it was expanded in Lucas (1970*b*). This work attracted dozens of rebuffs, of which I shall mention a few later, and Lucas's rejoinders appear in Lucas (1968) and Lucas (1996). For Penrose's work, see Penrose (1989), Penrose (1990), and Penrose (1994).

71. I follow here the very useful papers by Krajewski (1993) and Bowie (1982).

72. We want to prove that the set C of all i for which FS_i is consistent is not recursive. From the Gödel theorem we know that if and only if FS_i is consistent there is a Gödel number g_i of a proposition that is true but unprovable in FS_i. So, it is sufficient to prove that the set of all such g_i is not recursive. If such a set were recursive, then it would entail a *consistent* formal system FS_k. (Consistent because all the formulas with Gödel number g_i are true.) Therefore, g_k, which if the system is recursive must not be provable, is, from Gödel theorem unprovable, which is a contradiction. Thus C cannot be recursive.

73. The symbol i was introduced in 1777 by Leonhard Euler (1707–1783). He was born in Basel but lived most of his life in St. Petersburg, except for twenty-five years at Berlin at the court of Frederick the Great, from which he finally returned to St. Petersburg. The definition given in the text had no more than heuristic value until the concept of real number was extended to include complex numbers, for which proper formal mathematical rules of operations had to be defined.

74. Very simple accounts of this remarkable saga appear in Altmann (1992), chap. 2, and Altmann (1989). The reader will later in this chapter notice that the ambiguity reported in these references about the date of death of Rodrigues has been resolved. I have recently checked his death certificate at the Père Lachaise cemetery in Paris: it is 17 December 1851, not 1850 as sometimes quoted.

75. Klein (1939), p. 75.

76. Quine (1951), p. 43. The reader will recognize here my previous description of the scientific mesh, now enlarged to include mathematics and even logic. Quine, in fact, claimed, as I do, the same status for mathematics and logic as that for empirical science.

77. Whitehead (1911), p. 881. I owe this reference to Whitrow (1961), p. 142. Whitehead's quotation comes from Stanley (1907), p. 188. The historical basis for the legend is slim: Eustachius of Antioch gave a figure of 200 or 270 for the number of bishops, and Anasthasius, the only eye witness, writing in 347 C.E., mentions "about 300." The number 318 is theologically significant because it establishes a causal link with the Old Testament, being the number of servitors of Abraham (Gen. 14:14), and because it contains in its Hebrew spelling a reference both to the Holy Cross (through a letter shaped like a T) and to the consonants in the name of Jesus Christ. The earliest reference to the legend was written in Arabic in the tenth century by Severus of Aschmunaïn, a Copt, in a slightly different way: 319 sat in their thrones but only 318 exited, the extra throne being occupied by Jesus. (See Aubineau 1966.)

78. The same question is raised in Wittgenstein (1978), II § 24, p. 133.

79. I have already given evidence in note 32, that even very young babies can establish such bijections.

80. This dual view of mathematics was implicit in Wittgenstein: "I should like to be able to describe how it comes about that mathematics appears to us now as the natural history of the domain of numbers, now again as a collection of rules" (Wittgenstein 1978, IV, §13, p. 230).

81. Even longer; the Oxford English Dictionary, second edition, for instance, still carries Hamilton's wrong definition of quaternions. I have been promised, though, that the third edition will be right in this respect.

82. This concept of game-mathematics is very much what Wittgenstein proposed, particularly in Wittgenstein (1967b) and Wittgenstein (1978).

83. See, for instance, Lakatos (1978), p. 30, Lakatos (1981), and Putnam (1979). An important empiricist in mathematics is the American philosopher Philip Kitcher, who rejects a priorism and Platonism, and equates mathematics with natural science. Kitcher (1984) contains a wealth of material, especially on Platonism. See also Kitcher (1987). A useful review of mathematical empiricism is Galloway (1992).

84. For a simple acount of complexity theory see Lewin (1993) and Gell-Mann (1995). Also Tymoczko (1986b). See Chaitin (1982), pp. 949 and 942, respectively, for his first two quotations, and Chaitin (1995), p. 42 for the last one.

85. This question was rehearsed by Putnam (1979), chap. 4.

86. The details of the proof in a simplified manner appear in Appel et al. (1977). Tymoczko (1986b) gives a very good discussion of the philosophical implications.

87. This lecture, which was first published in 1960 in *Communications in Pure and Applied Mathematics* was reprinted in Wigner (1967), pp. 222–37. The quotations that follow are from pp. 223 and 237, respectively.

88. Born et al. (1925), p. 859.

89. A good and readable account of the infinity may be found in Barrow (1993).

90. My guess is not entirely without basis, since I once asked the same question from Professor Penrose at the end of a lecture (14.10.1995, Rewley House, Oxford), and his answer indicated that this was not a point worthy of serious consideration to him.

91. Herbert Simon (1996), p. 98, for instance, reports on a program called BACON which on being presented with the data historically available to Kepler rediscovered Kepler's law, a clear instance of what people would normally accept as creative thinking. I have repeatedly heard excellent arguments that "proved" restrictions on computer abilities "for all time," which were broken the moment that an unpredicted technology appeared in the market.

92. See Wittgenstein (1978), V § 50, p. 300 and III § 46, p. 176, respectively.

PEEPING TOM PEEPING AT THE MICROWORLD

I can safely say that nobody understands quantum mechanics. . . .
Richard Feynman (1992), p. 129.

results must be reported without somebody saying what they would like the results to have been . . . it is necessary for the very existence of science that minds exist which do not allow that nature must satisfy some preconceived conditions.
Richard Feynman (1992), p. 148.

In the year 1555, the Venetian painter Jacopo Tintoretto (1518–1594) produced a picture whose extraordinary composition was many decades ahead of its time. *Susanna at the Bath*, which now hangs at the Kunsthistorisches Museum in Vienna, goes right to the core of that most despised of activities, *voyeurism*; and places it where it belongs, firmly at the center of art (and, of course, of science). The two elders are there as they should be,[1] resolute in their determination to have nothing whatsoever to do with a *Playboy* centerfold: what they look at must be the *real* thing. Important as the two peeping Toms are, it is Tintoretto who claims the center stage for himself as chief voyeur. Not for him a *Venus pudica* coyly protected from view by merciful foliage, nor a demure perspective that sets the watching eye of the painter at a prudent indeterminate distance, his Susanna is not just a vast figure on the foreground: she almost stands out of the frame to stress that the artist is there, the grand voyeur, nearly at touching distance, as much a part of the scene as the elders of the tribe. What we see, he is saying, is what there is. This *is* reality, not just reality recollected in tranquillity: the artist here is no longer a narrator as in the great Byzantine-medieval tradition, but a witness, if not a participant.

Move away in the same gallery a few paces to the Pieter Bruegel the Elder (1525/30–1569) room, than which nature has hardly ever produced anything more elevating, and have a look at the *Children Games* by the crafty old master, painted five years after *Susanna*. And what makes this picture as remarkable as the Tintoretto in an entirely opposite way is that the children are there, playing their games, happily and totally uninhibited, because *they are unobserved*. The artist is saying "I am not there, I made sure that I did not interfere with my subjects, that is why their games are so *real*."

Notice that I have used twice the word *real*, and in two contrary ways. Pieter Bruegel the Elder is the classical physicist dealing with the macroscopic world. The less you disturb it the better, and this is no problem because you "do not have to be in the picture" to describe the macroworld. If you want, you can, like Bruegel, appeal to memory, because these are things you have grown up with, and they are part of that familiar everyday life—that has helped phylogenetically to model your perceptual and rational apparatus—about which you have a well-entrenched vocabulary. You are never in any serious trouble as to what there is out there to describe.

Tintoretto the voyeur, is instead the quantum physicist, for whom reality cannot be taken for granted because she must search for the secret body of nature. To know what there is, the physicist cannot keep at a distance: unless she talks about what she can "see," she is in danger of inventing a "reality" that does not exist at all. Nothing must be assumed to exist without solid, firsthand, voyeuristic knowledge.

My parable must be understood, however, as parables are, with a pinch of salt. What I am trying to convey is a sort of a principle of maximum prudence, to be used when first navigating the uncharted waters of the microworld. Naturally, as one gains in knowledge and self-confidence (and let us face it, when nothing else works!), some relaxation of the initial austerity might be in order.

Life, alas, is not as simple as art can be, and those who were engaged in the great adventure of discovery that started almost precisely with the beginning of the twentieth century, did not analyze what they were doing in the way that I have described: to continue my parable, they often went to visit Susanna at her bath with a folded copy of *Playboy* in their hands. Of course, most of what they had to see they saw, mostly well, but muddled they were, and muddle they created from which we still suffer; but they also created the most accurately tested physical theory ever known to man: quantum mechanics. No amount of astonishingly accurate testing stopped people, nevertheless, from frequently clinging to their badly creased centerfolds, and we cannot complain about that: wisdom could not exist without a robust backcloth of diffidence and mistrust. And in any case, if scientists from time to time prefer to avert their eyes from the plains of nature and look yonder upon the hills, they are after all made of the same metal as poets: humankind cannot bear very much reality, not at least without a struggle.

Readers of this book, when preparing to study the microworld, should realize that there are problems with their intellectual baggage. As part of the evolutionary process from which we have evolved, our rational system acquired certain principles to organize experience, which I have called meta-physical, such as the principle of sufficient reason and that of causality. They cannot be *proved* to be true, neither do they have any claim to logical necessity. Nevertheless, they are endowed for us with what I have called *empirical necessity*: in describing the macroscopic world it would be bordering on lunacy not to be guided by such principles. But, because they evolved with our species as part of our interaction with the macroscopic world, to assume that they can be carried through unsullied into the microworld is as unjustified as it is unwise.

This is not all: our language has been fitted to the natural world; and for this purpose, the natural world has been the world of day-to-day experience, the macroscopic world. The reader will remember the strange word *grue* that gave us so much work in chapter 7. The important feature that we discovered was that, however strange this word looks, it is just as acceptable a predicate as *green* or *blue*; and that the only reason why we do not use it, is that it has not been (and it could not have been) *entrenched*. The whole of our language depends on a hierarchy of entrenchments and what we have seen is that *grue* clashes with a better-entrenched concept, that of the causal timescale and must thus be discarded.

So, in this voyage of discovery on which we are embarking, we must be prepared to throw overboard suspected baggage: our great meta-physical principles, and our very vocabulary, must be subject to utmost scrutiny.

I repeat that, of course, this is not what happened in the historical development of the subject. It is a feature of humanity that people want to "understand," which in very many cases means that whatever is new must be hung from old and trusted intellectual pegs. (A respectful way of saying that during nature's striptease people revert from time to time to the sterile contemplation of their *Playboy* centerfolds.) Einstein once said that common sense is the set of ideas that one has acquired by the time one reaches eighteen, and, unfortunately, the desire to keep that set of ideas as the bedrock on which to construct "understanding" is overwhelming (it was so, even for Einstein himself!). Macroscopic common sense thus used may become a hallucination, and such "understanding" as might be derived from it, comforting to the wearer as it might be, can be a curse of civilized intercourse. It is in *this* sense of "understanding" that Feynman, in my first epigraph to this chapter, is not only right but will *for ever* be right.

I have already said that the birth of quantum mechanics is almost precisely coincident with that of the twentieth century. But it was a period of some five years in the middle of the 1920s that saw a veritable explosion of the subject. New ideas and results appeared so quickly that, when I briefly review the history of that period in chapter 17, I shall have to discuss the work by the *day* of completion. But what is amazing is not only the pace but also the level of the work. I would not be surprised if something like 90 percent of the scientists who *applied* quantum mechanics during the century worked entirely by using the principles discovered during those few years. Undoubtedly, those who led the world at that time were intellectual giants whose names will remain for ever. And yet, some of them failed to pay sufficient attention to the very simple cautionary remarks above, or were unaware of their existence, and went on regardless, often wallowing in their prejudices: whenever nature conflicted with the latter, something had to be wrong, some "mystery" or "paradox" lurked. We must not complain about that: it is thanks to the fact that the grumblers produced the most ingenious arguments against what they saw as insupportable results, that the subject was mercilessly scrutinized. Every time, however, such principles of quantum mechanics as were questioned emerged unscathed. I wish I could say triumphant, because, as often as the doubters were proved wrong, defeat gave them renewed strength to further their attacks.

All this soul-searching is to the good, because it has forced theorists to analyze the principles used in ferocious detail, and we can thus hope one day to come to a situation in which there is a reasonable consensus within the scientific community as to the correct interpretation of quantum mechanics. (There is no problem of course about it *working*: as I said, it has been experimentally checked to a greater level of accuracy than ever before attempted in physics. Also, the applications of the accurate calculations that can be carried out with the theory range over huge areas, from atoms to transistors, from the design of pharmaceuticals to the understanding of DNA, and much more.)

Because of this situation, I shall not attempt to discuss in this and the next chapter the conflicting views about the significance of the principles of quantum mechanics, although a few deep waves will from time to time hit the surface. I shall try to cash in on my remarks about the dubious validity in the microworld of our meta-physical principles, and of our macroscopically entrenched vocabulary, to present a view of the way in which the microworld might be treated that is reasonably consistent, and probably similar to that implicitly supported by most people working on the subject. The reader must appreciate, however, that quantum mechanics is a very hard subject, and that in order to produce an understandable account, I shall have to concentrate on what is most essential. Even then,

I shall have to consider sufficiently restrictive conditions so as to deal with problems that are within reach. So, I cannot expect to achieve anything like complete generality.

In doing this work, I shall not only depart from the historical development of the subject, but I shall also avoid making detailed reference to those who created it, and to their ideas, which is the only way to save the reader from becoming hopelessly confused. Once we have built some foundations, we shall examine them as far as we can for cracks, whether naturally existent, or whether imagined by excessive clinging to old prejudices. But I shall leave this for later chapters.

The historically counterfactual approach in this and the following chapter does not mean that I shall ignore altogether the history of quantum mechanics, which would be injudicious because, more than in any other subject in science, the role of philosophical argument was here most important, if not always entirely creditable. We shall have a good go at this history in chapter 17, where I shall provide detailed references to the work sketched in the present chapter and the next. So, if the reader finds some of the assertions made in these preliminary two chapters not sufficiently well supported, the discussion in chapter 17 should serve to consolidate ideas, by revealing the way in which they originated.

Although I shall present my account of the ideas of quantum mechanics, in this and in the following four chapters, with what might appear to be a degree of finality, this does not mean that I claim the same for my views. I would not be understandable, however, unless I gave some priority to the need for coherence over that for critical completeness. The reader must understand that, despite this air of finality, my account is necessarily tentative. It is impossible to be entirely accurate at the simple level at which I am working. The treatment that follows must thus be understood as a first approximation to the subject: all that I expect to achieve is to encourage a hope that a coherent view of quantum mechanics is possible that does not entail the "mysteries" with which this subject is so often endowed in popular discussions.

We are now ready to make a start with the problems that, when we began to peep into the microcosm, caused the collapse of the cherished ideas that until then had been successful in the macroscopic world.

QUANTA AND THE INFIRMITIES OF WAVES AND PARTICLES

In some ways, the world picture had not changed very much from the days of the atomic theory of Democritus (460–370 B.C.E.) until the beginning of the last century. If you break a rock into pieces, you ultimately reach elementary particles, which the Greeks called *atoms*, but which nineteenth-century people recognized could be objects like *electrons*. The important thing, though, is that except for their electrical charges, one then thought of electrons in a way not too different from tiny billiard balls. When particles oscillate, as for instance on the surface of water, they create *waves* (as I shall further discuss in a moment) which are completely distinct from the particles themselves, and these waves can carry energy which, as a difference from matter, can be changed at will by quantities as small as we please. These three ideas, which were the foundation stones of classical physics, that the world consists of particles, or of waves, and that the energy can always vary continuously—that is, in steps as small as we might wish to choose—were challenged by experimental results obtained in the first twenty years or so of the twentieth century, as I shall now show.

Waves

I should like first to discuss a little the relation between oscillations and waves. The archetypal oscillator is of course the pendulum, as shown in figure 1. It is well known since Galileo that the *period T* of the pendulum, that is, the shortest time it takes for the bob, starting from the initial position A, to return to it, depends only on the length of the pendulum. This period has been represented in figure 1 by assuming that the horizontal plane shown (a sheet of paper) has been moving *from left to right* with uniform velocity, in order to provide a scale on that plane for the time *t*. There is some device like a pen that can mark the position of the bob on the paper, and the positions A, B, and C of the bob are marked on the graph with the same letters primed, further primes indicating registrations at a later time.

If the pendulum were moving in a fairly viscous medium, its *transversal* movement would be propagated along a line *in space*, disposed exactly like the time axis of the figure and called the *line of propagation*. Thus, it would generate a *transversal wave*, looking exactly like the one in figure 1 (see fig. 3, chapter 11), but the corresponding periodicity (now in space rather than time) would be called the *wavelength* λ (*lambda*). Remember nevertheless the crucial difference with figure 1: each point of the wave is now the *perturbation* itself that constitutes the wave. This wave is said to be *polarized*, which merely means that the perturbation, transversal to the line of propagation, always belongs to the same plane. In fact, if the bob of the pendulum were charged, it would emit an electromagnetic wave of the type discussed in chapter 11. A useful way of

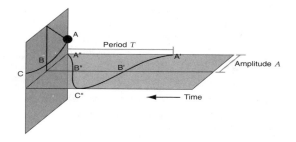

Fig. 1. The pendulum and waves

The horizontal plane moves from left to right. Its central line is taken to be zero. Values above and below it are positive and negative respectively. Although the amplitude is the largest value of the perturbation, we shall sometimes use this word to denote any other values of the latter.

expressing the period, which in most cases would be of the order of millionths of a second—meaning that there are millions of complete oscillations per second—is precisely in this latter form, which is called the *frequency* ν (*nu*):

$$\frac{1}{T} = frequency = number\ of\ oscillations\ in\ 1\ second = ν.$$

It is clear that if we have waves they must, at least within this classical picture, be generated by some oscillatory process, and it is very important for us to get a clear picture of the role that the energy plays in this case. The energy of the pendulum during its oscillation must be constant because of the principle of energy conservation, which is valid if we ignore small frictional effects at the suspension point and with the surrounding air. We can vary the energy, however, by raising the point A from which the bob is released (it is pretty obvious, for instance, that the higher this point is, the greater will be the velocity when the bob traverses the lowest point of its trajectory at B).

There are three important things that we must remember about this energy. First, it must depend on the amplitude *A* (see the figure), since this amplitude will be larger the more we lift the bob at A. Second, the energy of our oscillator can be altered by whatever quantity we want, however small, since this merely entails a correspondingly small change of the initial position of the bob. Third, since *A* can take a negative value (see the figure), the energy cannot just be proportional to it as such, since energies must necessarily be positive. It turns out, in fact, that:

The energy of a wave is proportional to A^2 (amplitude squared).

(Remember that squares, even of negative numbers, are always positive.)

There is one other property of waves that we must recall from chapter 11 (p. 258), namely that they can *interfere*, that is that the *amplitudes* of two waves can be added in such a way that along a line in space their resultant will be periodically zero (destructive interference) or a maximum (constructive interference). Remember that in interference it is the amplitudes themselves that count (because they can be negative as well as positive and thus add or subtract) and not their squares, although the latter correspond to the energies of the waves.

As we have seen in chapter 11, things were pretty clear by the end of the nineteenth century: light was a wave, as demonstrated by phenomena of interference. Other known radiations, such as electron beams, were particles because such interference was not found. Finally, no form of energy was believed to exist that could not be varied continuously. All this changed drastically in a matter of years.

Quanta

The new world was born in October 1900 in Berlin, where Max Karl Ernst Ludwig Planck (1858–1947) had recently been elected to a chair of physics. I shall not enter into the details of the discovery[2] but shall merely give its context. For some years, people had done good experiments to understand the distribution of energy as a function of the temperature emitted by a particular type of radiant body called a *blackbody*. As the experimental results improved, Planck was able to obtain empirically a formula that reproduced extremely well the data; but, being a theoretician, he set himself the ambitious task of deriving this formula from some well-chosen first principles. Of course, since radiation was involved, it was natural to assume that there was some sort of an aggregate of oscillators with different frequencies v, and it was then fairly straightforward to find the desired distribution by using the well-proven methods of thermodynamics. In doing this, Planck found two unexpected features. First, the amplitudes of the oscillators, which classically would determine their energies, did not matter at all: only their frequencies were relevant. But something even stranger appeared. By what Planck later referred to as "an act of desperation," he had to assume that the oscillators did not continuously vary in energy within the aggregate, but that their energies always came in multiples of some fixed quantity, which he called an "energy element."[3] Thus the concept of *quanta* as *discrete* quantities of energy (later extended for other variables) was born, although this word was not introduced until 1905, as we shall see.

But this is not all. I have already said that, amazingly, amplitudes had nothing to do with the energy. Remember that Planck was studying radiation of frequency ν emitted by what he (rightly) assumed was an aggregate of oscillators. Their *quantized* energies, which Planck had to assume in order to reproduce the experimental results, had to be precise multiples of some "energy element" which itself depended only on the frequency ν. Thus, Planck's hypothesis comprises two steps. First, the "energy element" (later called *quantum*) is

energy = constant times frequency or E = h ν, (**Planck's relation**),

where the stated constant is denoted with h, and it is universally called *Planck's constant*. Second, the total energy radiated can only be an integral multiple of the one given above, that is, it must be of the form $nh\nu$, for integral n (that is, 1, 2, 3, ...). Such integral numbers were later recognized as determining in a similar way the values of other variables, and were called *quantum numbers*.

Planck was able to work out from the experimental results the value of his constant h, which is tiny. Thus, the discrete steps that it originates in a system with macroscopically large energy would be so small that such energies would appear to vary continuously.

How Light, the Archetypal Wave, Turned Out to Be Corpuscular and How the Word *Quanta* Entered Physics

Sir Isaac Newton (1642–1727) held that light was corpuscular, but the work of his contemporary, the Dutchman Christiaan Huyghens (1629–1695), on waves and the phenomenon of interference, convinced people that light was a wave. And the world quietly rested on the safety of this knowledge until 1905. This was Einstein's *annus mirabilis*. In March of that year he analyzed Planck's work, in a paper that he published a few months later, and which was to win for him the Nobel Prize in 1922, not for his work on relativity, but for that on the photoelectric effect.

In that paper, Einstein examined and extended Planck's work, and found that, in order to derive other properties of the blackbody radiation, it was useful to assume, not only that the oscillators were quantized in energy, as Planck had postulated, but that light itself consisted of "energy quanta" $E = h\nu$. And this was the first time that the word "quanta" was used in modern physics.[4] Einstein was only too aware of how daring this proposal was, breaking as it did with the accepted wisdom of generations, and he referred to this hypothesis as *heuristic*, a word that he included even in the paper's title. I do not know whether Einstein ever heard of Cardinal Bellarmino and of instrumentalism, but I am sure that he did not require the Inquisition's pressure to abhor the reality of his own creation. There is no question that he considered this idea of quanta as provisional, as he stated even in 1911 at a Solvay conference.[5] And, despite the amazing success of his idea, Einstein expected all his life that one good day the "real" thing would reveal itself to the world. But that day never came.

And now back to Einstein's paper. What made it astonishingly successful is that Einstein was able to use his idea of *light quanta* as particles in order to explain and predict some amazing properties of the so-called photoelectric effect. It was Sir Joseph John Thomson (1856–1940), the discoverer of the electron, who realized in 1899 that when ultraviolet light is made to shine on a metal surface, electrons are emitted by the latter. Philipp Lenard (1862–1947) was able in 1902 to measure the energy of those electrons for different frequencies of the incident light and found an inexplicable result. Considered as a wave, the energy of light must be regarded as proportional to the amplitude squared of the wave (see p. 451), which is the *intensity* of the beam. Nevertheless, the energy of the electrons torn from the metal had nothing to do with this intensity, but depended *only on the frequency of the light*. What particular form this dependency took, he was not able to determine.

Now comes Einstein who deploys a daring but obvious stratagem to solve the conundrum. The light particles have energy $E = h\nu$, he postulated "heuristically," on using Planck's relation. When they collide with the electrons in the metal, they can transfer to them the whole of that energy. But the electrons have to waste a bit of it in order to escape from the metal, so that the energy of the stripped electrons must be

$$E = h\,\nu - P,$$

P being the energy lost in the escape.[6] The important thing here is that, as Lenard had shown, the intensity of the light, couterintuitively, does not affect the energy of the photoelectrons. But, as you can see, Einstein went further than Lenard, now predicting the precise relation between the energy of the photoelectrons and the frequency of the incident light. In the 1910s Robert Andrews Millikan (1868–1953) conducted some very careful studies of the photoelectric effect at the University of Chicago, and he published results in 1916 that corroborated perfectly Einstein's prediction. Hence the Nobel Prize for the latter: even Bellarmino could not have denied that the damned hypothesis did work.

It took a good ten years after Millikan's work for the poor light quanta to be properly christened by the American physical chemist Gilbert Newton Lewis (1875–1946) as *photons*, which, even then, he referred to as "this hypothetical new atom."[7]

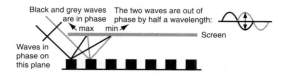

Fig. 2 Diffraction by a periodic structure (Diffraction grating)

The distance between successive lines (represented with black squares) must be very small, comparable to the wavelength of the incident radiation.

How Electrons, the Archetypal Particles, Turned Out to Behave Like Waves

It will be useful to start by reviewing the interference of waves, which, as I have said, was until the turn of the century the crucial test for distinguishing between waves and particles. I want to extend it now to cover the phenomenon of *diffraction*, for which purpose I refer the reader to figure 2.

I represent in the figure a periodic structure that may be obtained by printing reflecting lines on a surface, which is shown sideways, the reflecting lines being represented by the squares shown. If we shine on this grating light perfectly in phase, that is, such that the waves of all different incident rays register exactly, they will be out of phase by the time they arrive on the screen because, as it should be clear from the picture, the paths that they traverse are different in length. They will, therefore, no longer register; and their relative displacement will depend on the length traversed by each ray. Thus, there will be points at which the waves will arrive at the screen in phase, so that they will interfere constructively giving a maximum, whereas at other points of the screen there will be a minimum (actually a nil value), as shown in the figure. Obviously, this pattern will be repeated periodically on the screen, and this is called a *diffraction pattern*. It can be shown that the distance between two consecutive maxima depends on the distance between the consecutive lines of the grating and the wavelength of the incident light. Thus, the latter may be determined from simple geometry.

We can now go back to our history. As we have seen, Einstein had written the energy of the photon of frequency ν as $E=h\nu$. Louis de Broglie (1892–1987), working in Paris for his doctorate, published a paper in 1923 in which he proposed that this relation be extended for other particles, including the electron.[8] What he was then suggesting, of course, was that the electron, particle as it was well known to be, could also behave like a wave satisfying the relation mentioned. In 1924 he incorporated this argument in his thesis, which was read by Einstein, and apparently also by Max Born (1882–1970) of Göttingen.

Just about at the same time, some experiments were being conducted at the laboratories of the American Telephone and Telegraph Company (later the Bell Telephone Company) in New York, under the direction of Clinton Joseph Davisson (1881–1958)[9] in which electrons, all with precise velocity v, were made to impinge on a nickel or platinum plate. On measuring the distribution of electrons coming out of the surface in different directions, strange patterns appeared, with some maxima along certain directions. It was at first thought that these maxima might be due to the vibrations of the atoms of the material and, since Max Born was the expert on these phenomena, he was consulted. Born immediately recognized that the effect might be a manifestation of the electron waves predicted by de Broglie, and he suggested this to his colleague at Göttingen, James Franck. As a result of this, a young student of Franck's, Walter Elsasser (1904–1991) looked into the Bell Laboratory work, and he suggested that their results were indeed a manifestation of de Broglie's electron waves.[10] The atoms of the material, regularly spaced, acted as diffraction centers, and he suggested that experiments performed with a single crystal would show strong interference effects, until then obscured, because the samples so far used were an aggregate of many small crystals, so that the regular spacings of the latter were somewhat confused.

At the beginning of 1927, C. J. Davisson and Lester Halbert Germer (1896–1971) actually performed the experiment with single nickel crystals, and did verify precisely the expectations of quantum mechanics: the electron behaved as waves diffracted by the regular structure of the crystal. Soon, they received the Nobel Prize for this work.[11]

As I have said, the equally spaced atoms of the crystal act as a diffraction grating, and since the distance between those atoms was known, Davisson and Germer could quickly determine the wavelength λ to be associated with a beam of electrons of mass m and velocity v. They actually found the relation

$$mv = h / \lambda, \qquad (de\ Broglie's\ relation),$$

which de Broglie himself had guessed in his thesis, whence its given name. So, four years after de Broglie's first paper, the fact that electrons can behave like waves was confirmed, and since then the same has been proved for much heavier "particles," such as neutrons and even atoms.

THE UNCERTAINTY PRINCIPLE

We have now reached an extraordinary situation that breaks down the barriers between particles and waves. To a wave of frequency v, like light, for instance, we must associate a particle of energy E given by

$$E = h\,\nu, \qquad (\textit{Planck's relation}).$$

Conversely, with a beam of particles of mass m and velocity v we must associate a wave whose wavelength λ is given by

$$p = h / \lambda, \qquad (\textit{de Broglie's relation}),$$

where, for convenience, we have written the product mv as the *momentum p*.[12] This is an important quantity in mechanics because, for instance, when two particles collide, their *total momentum remains constant*. In other words, any momentum one particle loses is gained by the other.

We appear to have reached an intolerable situation in which the same object may appear to us as a particle or as a wave, and this worry is one that even now people do not cease to express. Thus, in 1996: "One of the most counterintuitive consequences of quantum mechanics is that particles can behave like waves."[13] But this conundrum is only counterintuitive with respect to the "commonsense" approach already criticized: there is no reason whatsoever why one should expect the well-known ideas of particles and waves, entrenched as they have been only in the macroscopic world, to survive in the microworld. On the contrary, the experimentally observed particle-wave duality must be welcome as a reminder by nature that we must not be pitifully naive.

This does not mean that we should not investigate why, for instance, a microscopic particle does not behave like a billiard ball. We must ask ourselves for this purpose why we can confidently talk about macroscopic bodies, even when they are not observed. Is the moon there when we do not look at it? Of course, and we need not appeal to other observers to endure a vigil while we sleep. We know that we can apply the laws of classical mechanics to the moon, and therefore that if we determine its position and velocity at some time (I am simplifying a little), we can go to sleep, and we can predict exactly where it will be when we wake up. Even more, we can satisfy ourselves that the moon follows a precise and computable *trajectory*, that is, a curve with precise position and velocity (or momentum) at each point of it, so that at every point of the trajectory we can predict the next point, *however near*. Thus the trajectory is *continuous*, which is what licenses us to assume that, between observations, the moon is *there*, and it is there with properties that depend *continuously* on those it had when we ceased observing it. In order to be able to generalize this later, I shall call this trajectory a *history*, since it unfolds in time.

Could we similarly define a trajectory for the electron? Let us try, and I shall use here an argument first proposed by the young Werner Carl Heisenberg (1901–1976) in 1927, although I shall simplify it for my purposes.[14] Imagine that in order to establish the electron trajectory I want first to measure the electron position (after all, we also measure the moon position). I need a ruler, and this must be one with very fine divisions: there is nothing I can use for this purpose, except a light beam of sufficiently small wavelength λ (which takes the place, of course, of the spacing of the ruler). Now, with a ruler divided in millimeters the best accuracy I can get is 1 mm (maybe a little better, but we can ignore such possible refinements). So, the error with which I can measure x, which I shall denote with Δx, will be $\Delta x = \lambda$. The trouble starts when I measure the momentum p, because I cannot choose freely the error in its measurement. In fact, the light beam of wavelength λ impinging on the electron is made up of photons with momentum h/λ (see the de Broglie relation above). On colliding with the electron, those photons will necessarily have altered its momentum by, possibly, as much as that amount. (On collision, momentum is transferred, but remember that it is conserved.) The error Δp thus introduced in the momentum will be $\Delta p = h/\lambda$. It is useful to consider the product of the two errors so far found:

$$\Delta x = \lambda, \quad \Delta p = h/\lambda, \quad \text{so that} \quad \Delta x \Delta p = h. \quad \textbf{\textit{(Uncertainty relation.)}}$$

Many people think of the process that I have described as entailing a perturbation of the system examined, introduced by the macroscopic measuring instrument. I shall have to challenge this view later (see next section), but let me say for the time being that such perturbations are also known in the macroworld. Whenever I take my temperature—which I must do with a thermometer—at, say, 30° C, I am *reducing* a little, by contact with it, my original temperature of, say, 37° C. In this example, the error introduced is minuscule, but if I want to measure in the same manner the temperature of one drop of my blood, the error could be significant. In the macroworld, however, you can always reduce such errors. You can achieve this in my example either by using a larger amount of blood, or by employing a thermocouple of extremely small mass.

What is so remarkable and new about the uncertainty relation is that two errors of independent variables are *coupled* and in such a way that if one diminishes the other must *necessarily* increase. Notice, for instance, in the uncertainty relation that if Δx is *halved*, then Δp must be *doubled* in order to keep the value of their product. It follows that we have two alternatives in trying to describe our microsystem.

I can take Δx very small, in which case the system will be localized and appear very much as a particle. But then Δp will be very large, so that I could not possibly determine anything resembling a trajectory for the "particle." In other words, I know very well where the "particle" *is*, but I lose any idea of *where it is going*. (Because, its momentum being very blurred, we hardly know anything about its velocity.)

The second alternative is to measure the momentum very accurately, in which case the "particle" will have a very large error in position. Thus it will be *delocalized*, this meaning that its position could be anywhere in a very large region. The object will thus behave like a wave, because waves are delocalized, covering a large region of space, but they do have a very clear direction (given in this case by the precise value of p measured).

The net result is that microscopic "particles" are not particles at all, and that they cannot be described by trajectories, as follows from the two alternatives discussed. They will sometimes appear as particles (with great uncertainties as to the direction of their "motion") and sometimes as very delocalized waves. Because these objects are neither particles nor waves, the only sensible thing to do is to invent a portmanteau word (as *grue* was) in order to describe them. (Remember that *grue* was the color of any object that was green in case a certain condition was satisfied and blue in case that condition was not satisfied.) So, let us introduce the word *wavicle*:[15]

> *A wavicle is a physical object that as an element of certain phenomena manifests itself as a wave, but as an element of other mutually exclusive phenomena manifests itself as a particle.*

It is clear that my first warning as to the passage from the macro to the microworld was worthwhile: our macroscopically entrenched vocabulary has to be altered. The physical objects of the microworld are neither particles nor waves: they are *wavicles*. The recognition of this fact immediately dissolves the apparent paradox entailed by the physically observed duality of waves and particles.

Before we move on, I must warn the reader that in chapter 17 we shall have to understand why young Heisenberg was in tears after trying to convince Niels Bohr of the validity of the argument here sketched to obtain the uncertainty relation.

TRAJECTORIES, HISTORIES, AND COMPLEMENTARITY

Going back to the moon: from the previous discussion, our ontological postulate that the moon is there when we do not observe it, is based on the fact that it has a continuous, well-determined trajectory. But we must realize that the very concept of trajectory, however well attested, is based on the principle of uniformity or continuity of nature; and that this is a meta-physical principle that has no more than macroscopic entrenchment. For wavicles, we cannot have trajectories but only *histories*, a concept that I shall try to explain in this section.

And this brings us back to Heisenberg's argument for the uncertainty principle. In it, we started with an unobserved particle, then perturbed by the act of measurement, either of its position or of its velocity. This is not a good picture, because if we conclude, as we did, that particles do not exist as such in the microworld, it is inconsistent to postulate an "unobserved particle." Remember that in order to speak of the unobserved moon, the concept of trajectory is required, and that this concept is based on principles that have no valid entrenchment in the microworld.

So, the Heisenberg argument, however convincing, is not good enough. (This does not mean that the uncertainty principle must be discarded; on the contrary, having discovered it in a suggestive but nevertheless dubious manner, it means that a better proof must be sought, which in fact was successfully done, as I shall discuss a little later.) If we think of wavicles at all as elements of reality, they must be indissolubly associated with the phenomena (such as measuring arrangements) whereby they are manifested.

Although wavicles cannot have trajectories, they can have *histories*. Imagine an electron gun that issues electrons well separated in time. I fire the gun and then observe a flash on a screen. Because I can be sure that there was one and only one electron involved in this phenomenon, I can talk about a *history* for this electron from the gun to the screen, but I must not imagine that those two points are joined by anything like a continuous trajectory.

In a way, a history is as closely associated to a wavicle as a trajectory is to a particle. For a classical particle, however, vigilance may be suspended because the trajectory is sufficiently robust to sustain the belief that the particle is somewhere along it, even if unobserved. For a wavicle, instead, unless there is a history, there is nothing which we can get hold of in order to sustain any form of belief about the wavicle itself. A history, therefore, must always entail some macroscopic events that start and terminate the history, and by which the *whole* phenomenon may be considered as the manifestation of a wavicle.

We may now come to the enunciation of the *complementary principle*. Wavicles are inseparable elements of the phenomena that manifest them, and these phenomena belong to one of two mutually exclusive sets, entailing two different types of histories, in which the wavicle is manifested either as localized (particle-like) or unlocalized (wavelike). The phenomena here discussed usually entail operations that can be described as measuring operations, in which some macroscopic pointers or dials will be used to determine values of some physical variables. They may also be, however, events that in some way localize the wavicle.

It is most important to realize that a history must always be considered *as a whole*. Half a history is meaningless: the history of a wavicle must be a *complete registered phenomenon*, not just a bit of it. This is so because it is only when a phenomenon is completed and registered that one can examine what sort of manifestation has taken place.

I hope that the reader will remember that the present account bears no relation to the historical development of the subject, and that what I call the complementary principle may or may not contain much in common with what is usually enunciated under that name. Likewise, the somewhat simplified concept of a *history* here introduced must not be confused with other more technical definitions of *quantum histories*, with which I shall not be concerned.

Now for a little bit of philosophy. For some people, as I have more than once intimated, philosophy entails fitting ideas into pigeonholes. The question might thus be raised at this stage whether, in describing as cautiously as I have done our relation to the microscopic world, I have abandoned any hope of fleshing out the latter with any form of reality—and that, instead, I have considered only the reactions of an observer to it. There is a tendency, in fact, to overestimate the influence of the observer on quantum phenomena. Admittedly, we must have histories, and the histories *that we can discuss* must necessarily entail observations. Despite the obvious fact that quantum processes must have existed for millions of years before any observer existed, it is often claimed that quantum mechanics is not concerned with the "real thing," but only with the way the latter appears to an observer, who always interferes with it during the act of observation.[16] In other words, it might be thought that we have renounced *ontology*, and that we only deal with the *epistemology* of the microworld.

I must issue at this stage a serious health warning. Some people appear to think that nothing is ontological that is not based on "objects" of the same nature as macroscopic objects: if we talk of particles, yes, we have an ontology, but if we talk of wavicles, we only have an epistemology. I am not going to say that this is nonsense because people are free to use words in their favorite way, and to try and canvass support for their semantics. But I should like to claim in as forceful a way as possible that such an approach is to be strongly deprecated, as being based on what Einstein (who nevertheless fell into his own trap) recognized as the misconceived use of common sense. That we cannot ascribe comfortably familiar properties to wavicles, does not mean that we cannot claim that they exist, at least in the same way in which the unobserved moon exists: even in the latter case we do require meta-physical assumptions.

Another source of frustration for the reader will be that I have given some rules which, although somehow justified, might appear to be a dogmatic attempt at stopping argument. Yes, the rules are produced in a fairly ad hoc manner, but remember that we are now bereft of the wonderful principles that we acquired phylogenetically and that I have called *meta-physical.* They have been in the making for thousands of years: no wonder that they fit comfortably to us like a second skin. We are now trying to construct a new set of concepts and a new language, and we have not had much more that fifty years to do so: we must expect the new suit to feel a little tight under the armpits.

All that the new rules attempt to do is to provide a language that saves the facts without contradiction. I shall discuss, for example, the rule that incomplete histories are meaningless. Imagine I do not obey the rule when studying a phenomenon of interference of electrons, and that I consider what happens *before* the electron hits the screen. Because the electron is charged, and because the electron is undoubtedly unlocalized at that stage, I have in this unfinished "history" a charge distribution spread out over all space. If I terminate the "history" with the electron hitting the screen, I am forced to require that all the charge be instantaneously transported in such a way as to condense at the point on the screen where the electron flash was observed. Such an effect on a charge distribution would be absurd because, in accordance with well-tested physical laws, movement of charge would produce radiation, which is not observed. Let alone the fact that instantaneous movement breaks also basic and well-entrenched laws. What I had proposed is that the wavicle history has no meaning until the electron hits the screen, whereupon the wavicle is manifested in a particle-like event. But the "part" of the history *before* that event cannot be discussed per se: it is an incomplete history.

You can see why such a rule is important. One of the features that a good language must have is the avoidance of time-wasting with statements that are inherently contradictory: we have learned as children to avoid saying that a surface is both red and green all over. We have to learn new similar rules. Of course, they are unfamiliar. But they are not "mysterious": they are the rules that, it appears, nature imposes on us if we want to engage in rational intercourse with it.

I shall now describe a couple of experiments that will allow us to try out our new concepts, and which will be very useful in order to acquire new ideas to move forward in our exploration of the microworld.

THE TWO-SLIT EXPERIMENT. INTERFERENCE

This experiment is a classical source of confusion and dismay about the treacherous behavior of photons, but we need not worry about this. First, we know that photons do not behave like classical particles, and, second, we must always calmly adhere to Richard Feynman's injunction in the second epigraph to this chapter, which, needless to say, was not always followed by the great and the good.

The two-slit experiment is very simple in principle, and it is sketchily illustrated in figure 3. I shall assume, for reasons that will presently become clear, that the light source is *sparse*, sending well-separated photons, one at a time, and that I have a device that permits me to shut either slit at will, very quickly with respect to the interval between photons. I shall consider three experiments, the first of which is *not* illustrated in figure 3. First, we have only one slit open for a sufficiently long time to collect, say 10000 photons on the screen, a number which I shall denote with N. I shall obtain one wide "line" with more photons at the center than at the edges, as shown in figure 4, where the total number of photons counted would of course be N. When you look at this curve and you compare it with figure 1 of chapter 13 (p. 348), it is clear that we have a probability distribution for the photons along the width of the "line."

We now perform a second experiment, illustrated on the left of figure 3 and involving exactly the same number N of photons as before, but in which the two slits are alternately opened. The timing is so well chosen that whenever a photon reaches the slits one and only one of them is open. Each slit is opened exactly 5000 times, so that, as before, the total number of photons that reaches the screen is 10000. You will then find two "lines," and each of them will have a distribution of photons exactly like in figure 4, except that the total number of photons in each one will be half the original number. This is a perfectly normal result, and if you are in the habit of having expectations, and if you think of photons as "particles," this is exactly what you would have expected: the results that would have been produced by each separate slit are simply added up. But remember that the two slits are never open at the same time.

In the third experiment, the two slits are simultaneously open, and as before *one and only one photon* is in the apparatus at any one time. One would expect, therefore, that the fact that the two slits are open would not make any difference with the second experiment, a photon going through either the first or the second slit, just as before. Under this assumption, these two events would be *mutually exclusive*, so that their probabilities would simply add up (see p. 356), precisely as in the second experiment.

The experiment with only one slit open at a time. The N photons are equally distributed between the two lines shown, each of which looks like fig. 4.

Appearance of the screen when the two slits are simultaneously open. The N photons are uniformly distributed over the "lines" shown.

Fig. 3. The two-slit experiment, first with only one slit open at a time and then with the two slits simultaneously open

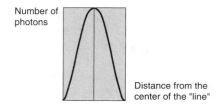

Number of photons

Distance from the center of the "line"

Fig. 4. The distribution of photons as a function of the distance from the center of the fringe, given in some convenient units

This, alas, is not at all what happens. What you now observe on the screen is a series of "lines," exactly like in the diffraction pattern of a wave. The only way in which one can understand the maxima (center of the "lines") and the minima (center of the spaces in between them) is by assuming that there are waves that hit the screen, so that when they are in phase, their amplitudes add up (maxima), and when they are out of phase, subtract, leading to the minima.

We must thus conclude that in the first and second experiments, wavicles are manifested as particles, whereas in the third one they are manifested as waves. This is perfectly in agreement with our complementarity principle, but we must try to understand why this change of manifestation arises. The reason is very simple. Remember that in the first and second experiment any photon that reaches the screen does so having passed through a perfectly well determined open slit. The wavicle history, therefore, is one in which the photon has been *localized*, and cannot therefore manifest itself as a wave: a localized wavicle must be manifested in a particle-like phenomenon. In the third experiment, the two slits are simultaneously open and the wavicle is delocalized: its history is such that when the photon finally appears localized on the screen, *we cannot say whether it has gone through one or the other slit.* We know experimentally that this is so, because whenever we can actually identify the slit through which the photon has gone, as in the second experiment, interference does not occur.[17]

There are two questions that we must discuss here. If instead of thinking, as I urge, quite neutrally about wavicles, one clings to the view of the photon as a particle, one can produce an apparent paradox when the two slits are open because a particle must necessarily pass either through one or the other. People have lost sleep over this "paradox." If we think of wavicles, it is clear that this requirement is totally unnecessary, and that one must not cover up the deficiencies of one's thinking with accusations that nature is mysterious. Nature is what she is.

The second question is this. Those who want to cling to some form of classical corpuscular theory at all costs, might claim that what appears to be wavelike interference is merely the result of some ingenious interactions between the photons in the region between the slits and the screen. But this cannot be so because we are using a sparse source, which guarantees that only a single photon is in the apparatus at any one time. Therefore, *each photon must interfere with itself* because no other photon is there with which to play. This, in fact, makes no difference whatsoever to the experimentally observed interference with normal sources.[18]

We have learned a few things. When the history of a wavicle is particle-like, then the particles are distributed in accordance with certain *probabilities that add up*. But when the wavicles manifest themselves through wavelike histories, these *probabilities do not add up*. Instead, the wavicles appear to be associated with waves *the amplitudes of which add up*, thus causing interference.

I shall now discuss a highly ingenious new experiment that appears to have most counterintuitive results.

DELAYED CHOICE EXPERIMENT

John Archibald Wheeler (1911–), of the University of Texas at Austin, proposed a very remarkable experiment with even more remarkable consequences.[19] I hope, however, that none of my readers will lose sleep over it, but it is a very good test of clear thinking.

The half-mirrors shown in figure 5 are carefully silvered, so that precisely half the photons that impinge on them go through, and half are reflected. The photon hits the first half-mirror and it is *either* reflected (black path) *or* transmitted (gray path) so that it is detected either in the upper or lower detector, but this is the case *only* if the half-mirror M is not in place. The thickness of this mirror is carefully chosen so that the transmitted wave (from the black path) plus the reflected wave (from the gray path) interfere *destructively* so that the intensity of the wave on *b* vanishes. In *a*, instead, the transmitted wave (from the gray path) plus the reflected wave (from the black path) interfere *constructively*. Therefore, when M is inserted, only the top detector will register.

When M is not inserted, the photon behaves like a particle and it follows either the black or the gray path, but when M is inserted, the photon behaves like a wave. It cannot be stated which path it follows, because, since *b* is extinguished, both a transmitted (black line) and a reflected wave (gray line) must exist there, whereas, on the other hand, a signal is detected in *a*. Thus, a wave must exist both in *a* and in *b*.

Wheeler argues that by inserting the special mirror M at the time *t* we

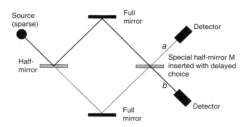

Fig. 5. Delayed choice experiment

The source emits well-separated photons, one at a time, and the mirror M is inserted only *after* the photon has hit the first half-mirror. This is the delay that gives this experiment its name. The insertion of the mirror M can be omitted or repeated many times at convenient intervals.

change the history of the photon *prior* to that time, until when the photon had to behave like a particle, and was thus allowed to follow either the black or the gray path. When M is inserted, the possibility of the photon "having been" in a given path (black or gray) disappears. This *retroaction* appears as a reversed causal relation in which the effect alters the nature of the cause. Even more, Wheeler has shown that this type of experiment effected on light coming from a distant star could "retroact" by millions of years. This observation led Wheeler to his idea of a *participatory universe*, whose history is modified by the observer.

I shall now attempt an interpretation of this ingenious experiment on the lines so far developed: what "exists" "out there" are neither particles nor waves but wavicles, and wavicles evolve through *histories*. In the delayed choice experiment, we can observe the same wavicles manifested by two complementary histories, wavelike or particle-like. The delayed choice does not change the past history, it changes the way in which *the one prevailing history* is manifested: the Lord's Prayer and the Paternoster are identically the same prayer in two different languages.

It will help to remember that a phenomenon is not a phenomenon (or a history is not a history) until it is completed and registered. (See the enunciation of the complementary principle on p. 460.) This means that until we introduce the mirror M, there is *no history at all*: I have said that half a history is meaningless. A history must be completed. It is only because we stick to the intuitive but dangerous idea of a partial history that retroaction appears to take place. You cannot retroact because there is nothing about which you can properly discourse until the "effect" takes place and a history is completed one way or the other. This is quite different from measuring, for instance, the position of the moon. Because of continuity, we can extrapolate backward in time, that is retroact, but this cannot be done for a wavicle.

Counterfactual Infirmities

The delayed choice experiment splendidly illustrates how one can fall into easy traps by the unreliable use of counterfactuals in quantum mechanics.[20] Whereas in classical physics we can happily use counterfactuals such as "If I had measured the position of the moon yesterday at 1 A.M., it would have been such and such," you cannot do so, in general, in quantum mechanics. To say "If I had not inserted the mirror M, the wavicle would have been either on the black or the gray path" is meaningless. All you can say is "I am not inserting the mirror M, and therefore the wavicle is either in the black or the gray path." In the first statement, we are describing an unfulfilled history, which is meaningless, as a difference with the second statement.

This is a subtle point that requires further analysis. Basically, the problem is this: counterfactuals require *contextual stability*. If I say "If I had placed this salt in water, it would have dissolved," I can trust that the context assumed remains valid, even if the negated event were to take place. If I had reasons to suspect that water would turn into ice whenever I put salt in it (which given an adequate refrigerator is perfectly possible), my counterfactual would have been not only useless but positively misleading. Thus, in the alleged quantum mechanical counterfactual, the *context* is profoundly sensitive to whether the possibly negated event does or does not take place. You have been warned. And you will see in chapter 18 another, opposite, pathological case in which a counterfactual must *necessarily* be true.

I am afraid that the reader might at this stage be thoroughly fed up, because I have stated a series of strange rules which may appear to mean nothing at all. Please, remember that we are trying to learn how we can discourse about the microworld in a way in which experiments do make sense without getting involved in contradiction. This is not an easy task, because, in order to make progress, we have to jettison some precious intellectual baggage that we (and our species) have acquired in intercourse with the macroworld. As shown in the first part of the book, even that intellectual baggage is not obvious and requires serious analysis. No one can expect fifty years of new thinking to compare with the intuitions acquired over 5000 generations. If you reflect about it, the progress in such a short time, however painful, is marvelous. That the new ontology that emerges is not obligingly familiar must be utterly disregarded: if you want to see palm trees, please do not go to Lapland.

MEASUREMENTS AND EIGENSTATES

The left-hand side of figure 3 (p. 463) illustrates the two different types of measurement used in quantum mechanics, *preparative* and *destructive*. When one slit at a time is open, the position of the wavicle as it goes through the slit is known, but the wavicle itself remains extant after the measurement (*preparative measurement*). When the wavicle reaches the screen, it becomes localized, and the position of the corresponding flash on the screen can be measured. This is a *destructive measurement* because the wavicle is no longer available for further experiment: it has been absorbed by the measuring device.

Preparative measurements are required in order to be able to follow the further evolution of a wavicle during an experiment, and it must be appreciated that they are *Procrustean*. That is, the measuring apparatus (like the slit in our case) acts as a sieve that singles out one of the possible values that a variable can take and allows only the wavicles with that particular value to emerge from the measuring device (as it happens with the wavicles that come out of the slit).

We shall discuss the question of what is the *state* of a wavicle in the next chapter, but it is not difficult to accept that the preparative measuring device allows only wavicles to emerge that are in a state in which the variable measured has a very specific value. Thus, all the wavicles coming out of the slit have (within some margin of error) the same value of the *transversal coordinate* in the plane of the slit. Likewise, there are devices that will single out wavicles with the same energy value.

In general, a preparative measurement commences a new history of the particle, thus establishing a profound time asymmetry: the history of the particle *before* that measurement cannot be related to its history *after*. This is so because only one of several successive histories may be considered at any one time, phenomena entailed in successive histories being in general unrelated.[21] Thus, previous to passing through the slit, the wavicles could have been in states with a wide range of values of the transversal coordinate (as shown by the fact that, if it is the other slit that is instead open, they will emerge out of it). But, when they emerge from only one open slit, their state changes to that with the value of the transversal coordinate corresponding to *that* slit. In order to be able to represent this, let me introduce the symbol ψ to denote the state of the wavicle, whatever that state means (see chapter 16). What I have just said means that, on measurement, the original unlocalized state ψ changes into a new localized state ψ':

Measurement of position: changes ψ into ψ'.

Not all measurements change the state in that way. If after you have changed ψ into ψ' through some measurement, you *immediately* repeat the same measurement, the state ψ' will remain unchanged, as I show in figure 6.

I illustrate in the figure the preparative measurement of the x,y coordinates of a wavicle, by means of a "sieve" screen. If we call ψ the state of the wavicle after the measurement, this state remains unchanged when the coordinate is immediately measured again. The state ψ is called an *eigenstate* of this coordinate, and the value of the coordinate itself is called the *eigenvalue*:

> *Second measurement of x,y : leaves ψ (eigenstate) unchanged.*
> *The eigenvalue is the value x,y of the coordinates.*

VARIABLES AND OPERATORS

Quantum mechanics has a very rich and interesting mathematical structure, which I have to ignore in this book. This is not altogether a bad thing, since people can get a little mesmerized by the mathematics in this subject and lose sight of its underlying philosophy. There are some essentially mathematical ideas that I cannot entirely avoid, however, and the reader will have to be a little tolerant because a half-told story is never fully satisfactory. I hope, nevertheless, that the reader will be able to imagine that what we are now going to deal with must, if completed, form a formidable apparatus for the study of quantum systems.

We must recognize that if I discuss a certain history, some variables might make no sense at all with respect to it. In figure 6, for instance, in the region between the two vertical screens, the transversal *momentum* has no meaning whatsoever, because the transversal *position* has been *exactly* measured. A given history, on the other hand, will entail the time evolution of some initial state ψ; and in order to make sure that we are licensed to use the variable x, say, we must assume that this variable has been measured. From the discussion in the last section we have:

> *Measurement of x acting on ψ changes ψ into ψ'.*

I shall take the first timid steps to flesh this out into a mathematical notation, and this is an example where the use of an appropriate notation produces vitally important new ideas. First of all, I shall say that the dash in ψ' is an *operator* that changes ψ into another function. Now, everybody has met instances of the use of operators, without knowing: the dagger in Joe Bloggs† is an operator that changes the name of a person into that of

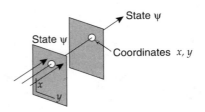

Fig. 6. Eigenstates

The state ψ (notice the change of notation from the text) does not change after measurement.

the same person after death. And the surd in $\sqrt{(2)}$ is an operator that changes 2 into 1.4142. . . .

The next step is to recognize that the *prime* symbol ' used for the operator in the displayed equation is not very convenient because of its being too generic: if you read the symbol ψ', you do not have the faintest idea of which was the variable whose measurement entailed the change of ψ into ψ'. I shall therefore use the following convention: to each variable, such as x, I shall assign an operator written with the same letter in bold, x, so that $x\psi$ will mean exactly the same as ψ', that is, the state in which the system remains after the variable x is measured.

In the same way, the operator for the momentum p is p, and likewise for any other variable. The only exception, for historical reasons, is the operator for the energy, which is written H and is called the *Hamiltonian*. With this notation we have

$$x\psi = \text{transform of the state } \psi \text{ after the variable } x \text{ is measured.}$$

If we write

$$x\psi = \psi,$$

it means that the state is unchanged, so that ψ is an eigenstate of x.

The important point that the reader must appreciate is that, although this appears to be a mere matter of notation, the theory of quantum mechanics allows us to determine the *mathematical form* of the operators corresponding to the variables, which means that equations such as the last one become mathematical equations that can be solved by standard mathematical methods. This is the crucial point, and it is the one that we have to skip because it would require too much mathematics, and would in any case contribute little to our understanding of the philosophical principles of the subject.

There are a couple of mathematical details that I must nevertheless give. The first is quite amazing: whatever ψ is which represents the state (which will turn out, of course, to be a mathematical function), its meaning is such that

ψ and $c\psi$, for any constant c denote identically the same state.

I cannot go into the reasons for this result at this stage, but the reader will find a discussion on p. 505 (note 3). It follows that in the definition $x\psi = \psi$ of an eigenstate, which I had displayed before, the right-hand side may be multiplied by any constant: $x\psi = c\psi$. And it can be proved from the mathematics of the subject (see the above-mentioned note) that the constant in the last equation *must* be chosen to be equal to the eigenvalue of the variable measured. For convenience, I shall denote eigenvalues with the same letter as the variable but in capitals. Therefore,

$$x\psi = X\psi.$$

This is called the *eigenvalue equation*: X and ψ are respectively the *eigenvalue* and the *eigenfunction* for the measurement of the variable x, whose operator is X. Once this operator is given mathematically, which is done once for all from quantum theory, the mathematical equation above can be solved to obtain both the eigenvalue X and the eigenfunction ψ, which is of course a result of much practical importance. I hope that the reader will accept for the time being the few mathematical results that I had to assume. So, we can now come to a most interesting point.

Commutation of Operators

I have already given one social operator affecting poor Joe Bloggs, the †. Another is equally simple and, in most cases, far more agreeable: Mrs. Joe Bloggs. The operator Mrs., of course, changes the male Joe Bloggs into a female, his wife. There is one little problem, because although the operator Mrs. is correctly given on the left, (like x in the previous equation), the dagger, traditionally, is tagged at the end. This is very inconvenient so that I shall write †Joe Bloggs, instead of the more common Joe Bloggs†. We now have two operators that can be applied singly or jointly on the name "Joe Bloggs," and we shall see that they do not commute:

> Mrs. †Joe Bloggs = *the wife of the late Joe Bloggs.*
> †Mrs. Joe Bloggs = *the late wife of Joe Bloggs.*

We could write:

$$Mrs.\dagger \neq \dagger Mrs.$$

Notice that the symbols of the operators act on whatever they act *from right to left*. In Mrs. †Joe Bloggs, Joe Bloggs is first dead, and then "Mrs." operates. In †Mrs. Joe Bloggs, Joe Bloggs is first replaced by his wife, and then she is dead.

COMMUTATION AND THE UNCERTAINTY PRINCIPLE

All this is most important, and it will give us a very good example of how, with a minimal amount of mathematics, one can understand the physical meaning of the lack of commutation of operators. All we have to remember is that we have introduced the operators to represent measurement operations. Take the eigenvalue equations of the two operators p and x, corresponding to the momentum and position respectively:

$$p \; \psi = P \; \psi,$$
$$x \; \psi = X \; \psi.$$

Consider now the meaning of $x \, p \, \psi$. Remembering, as we have seen in the last section, that the operators operate from right to left, it means that p is first measured on state ψ (result: eigenvalue P) and that then x is measured on the resulting state. Instead, $p \, x \, \psi$ means that x is measured first, and then p is measured on the resulting state $x \, \psi$. Because we know that the measurement of x changes the value of the momentum (see p. 458) of the state ψ, the previous value P found for it cannot now be obtained. The two measurements do not commute, and neither do their operators:

$$x \, p \neq p \, x.$$

Because in quantum mechanics, as I have said, one is able to obtain the mathematical form of the two operators that appear here, it is very easy to show mathematically that they cannot commute. This is equivalent to saying that the measurement of one variable introduces an error in the other one, thus bypassing Heisenberg's gamma-ray microscope argument. Moreover, from the mathematics of the subject, the complete form of the uncertainty principle may be derived.

It is not too unplausible to accept that if two operators commute, which means that their measurements do not interfere one with the other, then the corresponding variables must have simultaneously precise values. So, if an operator commutes with the energy operator (the Hamiltonian), it must have a precise value for as long as the energy has a precise value. The principle of conservation of energy states that this is the case while the system is isolated. Therefore, any operator that commutes with the Hamiltonian will keep a precise value for as long as the energy is constant, that is for as long as the system is isolated. For this reason, such operators are called *constants of the motion*, *motion* here being understood as the evolution of the system in isolation.

THE UNCERTAINTY PRINCIPLE
AND PROBABILITY DISTRIBUTIONS

We must look again at the uncertainty principle to try to understand better the delicate meaning of such expressions as Δx used in its enunciation. Consider a set of electrons all with precisely the same momentum, going through a single slit, very much like in the first case of the double slit experiment of figure 3. The slit may be taken as a preparative measurement of the transversal coordinate x, whereas the screen may be considered as a destructive measurement of position. As we have seen in figure 4, we get a *probability distribution* for x on the screen, which we represent again on the left of figure 7.

Let me call the set of electrons incident on the slit, all of them with uniform properties (they had identical momentum), an *assemblage*. By this I mean what in statistics I have called a sample, but with this proviso: a statistical sample entails an element of diversity, however carefully you choose coincidence of some properties, but an *assemblage* will be an aggregate of particles prepared all of them with identical properties. It must not be confused with an *ensemble*, of which more later.

When you have probability distributions of the type displayed in the figure, the *half-width* of the distribution curve (given by the segments shown at the bottom of fig. 7) is often used to express the error of the data. (Clearly, if this half-width is very large, it means that the results are very scattered and thus not very precise.)

We can now go back to our experiment. Imagine that we have measured x on our assemblage of electrons as explained. We have then obtained the probability distribution shown on the left of the figure, and the value of Δx as explained. Imagine now that, instead of doing the experiment this way, we divide the electrons in two channels as they come out from the slit, and that in the first channel we do exactly the work already explained, whereas on the second channel we measure the

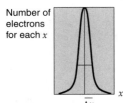
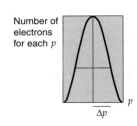

Fig. 7. Probability distributions and the uncertainty principle

The variables x and p are measured with respect to the centers of the respective distributions.

momentum p, and we get the probability distribution shown on the right of the figure. Because the momentum of the electrons must have been altered as they go through the slit, and because Δx *is* very small (since all the electrons go through the very narrow slit), Δp *must* be sufficiently large so that the uncertainty relation be satisfied.

There are a few questions that I must now address, arising from the above discussion. First, whereas in classical mechanics x and p are independent variables (which means that their values are entirely unrelated), this is not quite so in quantum mechanics. The two probability distributions of figure 7 are firmly coupled by the uncertainty relation. (It can in fact be shown mathematically that given one, the other follows.)

The next question is far more delicate. Having seen figure 7, you may think that the Δ's in the uncertainty principle should be obtained from the study of assemblages as suggested in that figure, just as errors, classically, are determined from probability distributions on samples. This is not so: it is important to recognize that the uncertainties involved in that principle can be defined *for a single particle* rather that for a particle as a member of an assemblage. Once these uncertainties are so understood, it does follow that if assemblages are studied, they will behave as explained, but, conversely, that they are not required in order to define the uncertainties or errors. This is another case of a profound distinction in the way in which simple concepts from the macroworld translate into quantum mechanics: I shall explain in the next chapter why it is claimed that assemblages may be dispensed with when discussing statistical properties normally understood through samples, a point still strongly disputed by some theorists.

CODA

I have followed in this chapter what you could call a minimalist strategy. I have accepted that such elements of reality as might exist in the microworld cannot be manifested to us by means of the tools that we had for the macroworld: both the principles that we have used for the latter and our vocabulary must be considered suspect and/or incomplete. Neither particles nor waves may be used in the microworld, and they must be replaced by wavicles. Because trajectories cannot exist in the microworld, I have replaced them by the alternative but more elusive concept of histories. Wavicles, however, cannot be isolated from their histories and, most important, histories must be considered in their entirety: nothing can cause more trouble than the admission of incomplete histories. It is this principle that eliminates the apparent paradox associated with delayed choice experiments.

Another important concept that we learned is that variables must be considered via their corresponding operators, which idea, because the latter do not commute, allows us to understand the uncertainty relations.

Finally, I should like to stress that the history-wavicle concept entails an ontology: we postulate that these are the objects that are "out there" as elements of the microworld. The fact that histories must be determined through macroscopic events does not mean that we only have an epistemology. Wavicles are just as good candidates for an ontology as anything can be. This ontology is meta-physical in the sense that the wavicles manifest themselves to us via their histories, which entail macroscopic physical events. This meta-physics is a little more remote, but nevertheless no less salubrious than the macroscopic meta-physics entailed, for instance, by the use of the principle of continuity in macrophysics. (Remember the unobserved moon!)

NOTES

1. Old Testament Apochrypha, *Book of Susanna*. The totally different perspective of the otherwise similar Tintoretto's *Susanna* at the Louvre (ca. 1550), makes that of the Vienna version more significant.

2. The interested reader could do no better than consult Pais (1982), pp. 364–72.

3. See Pais (1982), p. 370 for Planck's desperation. *Energieelement* appears for the first time in Planck (1900), p. 242.

4. His Nobel Prize paper is Einstein (1905a). Quanta appear in it in various forms: *Energiequanten* (p. 133), *Elementarquanta* (p. 136), *Lichtenergiequanten* (p. 147).

5. Pais (1982), p. 383.

6. I have oversimplified the explanation. Clearly, on collision with the metal electrons, the light quanta will transfer *at most* their total energy. For some electrons emitted, the

energy received will be less: hence the energy quoted in my formula is not that of all the emitted electrons, but only of the most energetic ones, as can be observed experimentally.

7. Lewis (1926).

8. de Broglie (1923).

9. See Davisson et al. (1921), (1923).

10. Elsasser (1925).

11. Davisson et al. (1927) fail to acknowledge Walter Elsasser's important *prior* contribution, in particular his suggestion that single crystals be used, of which they were aware. They go to some length, instead, in describing an accident as a result of which nickel crystals were grown on the surface of the plate used, which they claim was the reason why experiments on single crystals were performed.

12. I must warn the reader that I am being grossly careless here. Although the relation between the velocity and the momentum appears entirely trivial, differing only by a factor m, this is not so. This relation is trivial if and only if the particle is in free space. An electron in an atom, or for that matter in most *real* situations, is far from free, and although the relation between p and v is *formally* the same, the mass is no longer just a number but what is called a *tensor*. (An applied force will accelerate a free particle along its own direction, but this is not so when the particle is bound. Thus, whereas Newton's law, force equals mass times acceleration, works well for a free particle with the mass as a number, this is no longer true for bound particles: the acceleration cannot be parallel to the force, thus requiring a redefinition of the mass.) This, unfortunately, is a question not made transparent in most elementary books on mechanics, and even less so in some philosophical works, causing serious problems.

13. Pfau (1996).

14. Heisenberg (1927). Heisenberg's imagined experiment requires a microscope to measure the electron position, and because the wavelength of the radiation used must be very small, this is called the *gamma-ray microscope*, gamma rays satisfying that condition. A more detailed treatment of Heisenberg's gamma microscope than he provided, however, must be discussed (Jammer 1974, p. 64).

15. The word *wavicle* was coined by Eddington (1928), p. 201 but, unfortunately, it has not been much used. It is my impression that it is not sufficiently distinctive, in the sense that people who hear or read it for the first time might think of objects such as wavelets instead. The problem is that the formation of Eddington's portmanteau word is defective, because the second moiety of the composite must never be a suffix, since suffixes do not have *equal status* with the prefix to which they are attached. The word *particle* derives from the Latin *particula*, itself a diminutive of *pars*, so that -*icle* is a clear diminutive suffix, which conveys the reading of *small wave* for *wavicle*. I am thus afraid that Eddington made a serious philological mistake in coining this latter word. The correct portmanteau should have been *partave*, and I think that if Eddington had chosen it, it would have been far more used. It is too late, alas, to propose new coinage and, *faute de mieux*, I shall stick to *wavicle*.

16. This view is not entirely sound, of course. As I have said, we are concerned with histories, and a history cannot be analyzed in separate parts such as "the real thing" on one hand, and the "measurement" (always entailing some macroscopic registration), on the other. For better or for worse, a history is just one whole: it is for this reason that I have insisted that half a history is nonsense. Remember that what I am doing is trying to construct a language that empirically steers clear from nonsense, and that I claim that, if such a language were possible, then the objects of the language, such as histories or wavicles, may safely be regarded as manifestations of elements of reality, endowed with an ontological status. This is, after all, what has happened in developing the macroscopically-entrenched language: even stars must be surmised from observations rather than immediate intuitions.

The reader must realize that the definition of *histories* that I have used is but a pale reflection of the real thing. (See Omnès 1994, p. 104 and Gell-Mann et al. 1990.)

17. The reader might think that I am contradicting myself in considering only a part of the wavicle history. This is not so. In the first two experiments, the wavicle history really *begins* at the slits, since the wavicle has been localized there, and ends at the screen. In the third experiment, the history goes *from the light source to the screen*. Whenever you have anything like a measurement of position (as localization is) or a measurement of momentum, you finish a previous history and commence a new one.

Thoughts experiments were early proposed in which special detectors that would not stop the "particle" were placed in one of the slits in order to determine which way the "particle" goes, hoping that interference would nevertheless be observed. It was argued, contrariwise, that the uncertainty principle would, on localization of the particle by the detector, blow up its momentum and thus render interference impossible. Scully et al. (1991) proposed experiments with heavy particles (atoms, rather than photons) so that ingenious detectors could be designed that would have a negligible effect on the momentum. They argued that, although the uncertainty principle would now not play a part, interference would be eliminated by the correlation of the particles with the detector. The latter thus is the relevant feature, rather than the operation of the uncertainty principle.

The correlations mentioned may be described as due to the *entanglement* (see p. 542) between the particles of the beam and those of the detector, or as due to *decoherence* of the beam. (See p. 560.) This experiment has now been conducted (Dürr et al. 1998), and the proposals discussed have been fully verified. All this is in agreement with my previous remark that the uncertainty principle should not be understood as the result of a disturbance of an "unobserved particle." This question is discussed in detail by Brown and Redhead (1981).

18. Early experiments were done by Sir G. I. Taylor (1909), who took up to three months to register one photographic plate, so little light he used. More recently, very accurate experiments have been done by Jánossy et al. (1958). Because the velocity of light is 300000 km per second, we know that a photon travels 3m in one millionth of a second. These authors have used a source that produces 10000 photons per second; hence the time interval between photons is a tenth of a millisecond, in which time a photon moves 300 m. By taking the relevant length of the instrument as 14 m, they thus ensured that only one photon could be in that length at any one time. Fiendish thing to do, because even looking at the instrument would produce upsetting thermals, so that remote control of an underground instrument was required.

19. Wheeler (1983). See also Selleri (1989), pp. 114–18.

20. The reader may wish to refer to the treatment of counterfactuals on p. 62. See also d'Espagnat (1983), chap. 12.

21. The reader should abstain from drawing premature conclusions that quantum mechanics is inherently time-asymmetrical. The question of the arrow of time in quantum mechanics is very delicate and will be discussed further in chapter 19.

QUANTUM STATES

W e saw in chapter 6 that a *state* is defined as a set of self-predicting variables. This means that these variables are such that a predictive equation is found that, from the values of those variables at an initial time, determines the values of the same variables at any later time. The variables so defined are called *state variables*. We also saw that in order to simplify the predictive equation, it is often worthwhile to replace the state variables by some function of them, which is called a *state function*. Of course, the state function must be well chosen, so as to satisfy the proposed simplification. Once the state variables or the state function is known, the state is fully specified: to know the state variables (or the state function) is to know the state. Naturally, we must expect some changes in order to define the state function in quantum mechanics, but the basic principles of the definition will remain: no scientific knowledge is serious knowledge that is not predictive.

THE STATE FUNCTION

In order to define the state in the manner described, we must choose possible state variables in a sensible way and, to have a practical example of how this is done, let us assume that we want to deal with electrons. We find here our first hurdle, because, as a difference with Newton's definition of state (chapter 6), we cannot choose both the position x and the momentum p (mass times velocity) to be starting values of the presumed state variables. The reason for this is that, as discussed in chapter 15, the uncertainty principle precludes these two variables to be treated as independent. Obviously, presumptive state variables must be independent, since otherwise you are piling into the definition of your state redundant and thus useless information.

As the reader probably remembers, finding state variables is not a mean task, because of the trial-and-error circular process involved, and this search requires plausible rather than logically deductive argument. In our case, such plausible arguments as I can use, since I do not want to get into the mathematics of the subject, must necessarily be crude. I hope that the reader will accept that to see through a glass darkly is better than to see nothing at all: so please, do not get bogged down by the imprecision and hand-waving that follows.

Let us look again at the two-slit experiment which, when done with electrons, behaves identically as with photons. When the electrons go through one slit at a time, we know their initial coordinate, call it x, but what we observe on the screen is a probability distribution, as shown on the left of figure 7 of chapter 15 (p. 474), which I repeat here. (Fig. 1.)

Since we have to accept that in many cases our initial conditions will not be entirely sharp, owing to the operation of the uncertainty principle, we might think that all that happens in quantum mechanics is that, rather than dealing with x and p as such, we must deal with their probability distributions. As I just have said, however, these are not independent. So, let us assume that we jettison the redundant distribution in p and that we only leave the one for x. We can see that if the state function were only to carry the information in figure 1, we would be in a pickle, because there is nothing whatever in that figure to tell us in which direction the electron is moving, and without that information, there is no way in which we could predict its motion.

Fortunately, we can get a clue from the two-slit experiment as to how things might move forward. Figure 1 corresponds to the case in that experiment when the wavicle manifests itself as particle-like. When the two slits are open and there is interference—and thus wavelike behavior —we saw that the probabilities *do not add up* but that there must appear the amplitudes of some wave, call it $\psi(x)$, which do. Of course, if we were to know

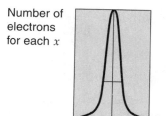

Number of
electrons
for each x

Fig. 1. Probability distribution for the position *x*

The horizontal line shown is well determined from mathematical principles, and it is called the half-width of the probability distribution.

this wave, because wave motion depends on the time and a wave always moves in a specific direction, then we would be able to add the missing element to the probability distribution of figure 1: we would know how this probability distribution varies with the time, that is, its rate of change, and this will give us the missing direction of motion in that figure.[1]

So with a little bit of optimism and a great deal of faith[2] we can imagine that what we need as the state function is a wave, call it $\psi(xt)$, because it now depends on the time, *such that it allows us to recover the probability distribution in x*. This wave will also do two things for us: it will automatically carry the rate of change of the probability distribution in x, and its amplitude will add up in accordance to the rules of interference, thus agreeing with experiment. (See fig. 2 where I assume, as I shall show in a moment, that given the wave $\psi(xt)$ it is possible to derive the probability distribution associated with it. Please notice also that I have made no attempt to make this figure correspond to any real physical situation.)

First of all, we must ensure that the function $\psi(xt)$, whatever its mathematical form, sustains the vital predictive equation without which no function could be a state function. Of course, to get this equation is a major task of judicious guessing, which I cannot here reproduce, but this job was done by Schrödinger (see chapter 17). He found the predictive equation (as the Newton equation was in classical mechanics) that given $\psi(xt)$ determines $\psi(xt')$, thus verifying that $\psi(xt)$ is the state function. Of course, the detailed form of $\psi(xt)$ has to be found in each case depending on the circumstances, and the Schrödinger equation itself will be further discussed in the next section.

The state function remains nevertheless a concept that is not so easy to grasp. Consider, however, the first case of the two-slit experiment when only one slit is open at any one time. What we observe on the screen are probability distributions, and this is what the state function

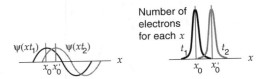

Fig. 2. A wave determines the change of the corresponding probability distribution with respect to the time

From the two probability distributions shown on the right of the figure, the rate of change of the probability as a function of the time can in principle be determined. Therefore, this rate of change can be determined from the single wave $\psi(xt)$ depicted on the left of the figure at two different times. We assume for each of them, as discussed in the text, that $\psi(xt)$ determines the probability distribution in x.

must be able to predict. This is quite sensible, since from $\psi(xt)$ at the initial time (when the photons or electrons go through the slit), we cannot expect to obtain precise values of x, but only a probability distribution for them. So this probability distribution must somehow be obtainable from $\psi(xt)$ which, as we have seen above, must be some sort of a wave. The probability distribution that we must find cannot be given by the amplitude of that wave, because in the interference experiment the amplitudes add up but not the probabilities. We know, on the other hand, that the amplitude squared $\{\psi(xt)\}^2$ of a wave denotes its intensity, which, it is possible to imagine, might relate to wavicle numbers, and thus (through appropriate ratios) to frequencies or probabilities. It is thus sensible to continue our guessing game by interpreting $\{\psi(xt)\}^2$ as the *probability* of finding the wavicles at the point x at the time t.

Notice that we have now satisfied our requirement that the presumptive state function should allow us to derive appropriate probability distributions. Because this state function $\psi(xt)$ is in general a wave, it is most often referred to as the *wave function*.

There are several technical points that need a little tightening up, which will be done in the notes, but the reader would be probably well advised to return to them at a later stage.[3]

THE SCHRÖDINGER EQUATIONS

We have discussed eigenvalue equations on pp. 470–71, and I shall move forward from that idea. The energy operator is called the *Hamiltonian* and it is represented with the symbol H. More precisely, we should write $H(xt)$ to indicate that the energy operator depends on the position of the wavicle (it cannot of course depend also on its momentum), and that it varies with the time. (I can for instance switch on at a certain time a magnetic field that will change the energy.) If the energy operator is time independent, in which case I shall write it as $H(x)$, then the energy can have a value that is constant in the time. This is its eigenvalue E, and we can write the eigenvalue equation

$$H(x)\ \psi(xt) = E\ \psi(xt).$$

The reader might be surprised that I write the state function, which is the eigenfunction of a time-independent Hamiltonian operator, as depending on the time. But waves are just like that: they must always depend on the time because they are always on the move. (See fig. 2.) Since the Hamiltonian is constant with respect to the time, the value of the latter should not affect anything that can be observed at different times, and for this to be the case all that we require is that the probability $\{\psi(xt)\}^2$ be *time independent*.[4] States for which this is the case are called *stationary states*, and the equation displayed above is the *Schrödinger equation for stationary states*. Such states occur for instance for the electrons that surround the nuclei of atoms (as long as nothing like magnetic fields are switched on and off, so that the Hamiltonian is time independent) and are crucial in our understanding of the way in which atoms emit light.

I must now come to the most important Schrödinger equation, for the case when the energy or Hamiltonian is time dependent, which is the acid test to verify that $\psi(xt)$ is self-predicting and that it therefore is a correct state function. Schrödinger obtained his *Schrödinger equation with the time* on reasoning by analogy with some known problems of classical mechanics, and what he found was:

$$H(xt)\ \psi(xt) = constant\ times\ the\ rate\ of\ change\ of\ \psi(xt).$$

It is easy to check from this equation that, given the state function at the time t_1, the same function can be derived at any later time. In fact, from the above equation, the rate of change of the state function is the left-hand side (computed at the time t_1) divided by the constant, the value of which is actually given by the Schrödinger equation. The rate of change of ψ is the change of this function in, say, one second. Therefore, at the end of one second, the value of $\psi(xt)$ will be the original one at t_1 plus the found rate of change. From the new value of the state function at the end of the first second, its value at the end of the second second is similarly computed, and so on: in this way the state function may be predicted at any desired time.[5]

In practice, this work would have to be refined, because we have used the rates of change to compute increments after one second, and this timescale may be too coarse, since the rates of change obtained from the left-hand side of the last equation are really valid instantaneously (that is, for instance, at the precise time t_1), and are thus somewhat inaccurate over a large interval. But the work described can be conducted for microsecond intervals or even smaller, so as to obtain the desired accuracy.

It must be appreciated that the Schrödinger equation with the time is *deterministic*. This is a *necessary* formal requirement, because, as we have seen, the definition of state requires a deterministic predictive equation. So, quantum mechanics is not as haphazard as some people think. On the other hand, once the state function is determined, what we obtain are probability distributions, and not exact values of the variable we are working with (the coordinate x in our case).

THE MEANING OF THE PROBABILITIES

We used two different types of aggregates in discussing probabilities. If I toss a coin a finite number of times, say one hundred, this aggregate of tosses is called a *sample*, and in general the frequency of heads over it will not be the probability $1/2$. I can *imagine* an arbitrarily large number of repetitions of my single toss, all in identical conditions with the same coin, and the frequency over this aggregate, which is called the population, may be taken, within certain severe limitations discussed in chapter 13, to be the probability in question. Such *imagined infinitely large aggregates* are called in physics *ensembles*. They are purely theoretical constructions, used in order to be able to correlate probabilities with frequencies, albeit in an entirely nonexperimental way. An ensemble can never be constructed in practice because it is an imagined aggregate, but, like a mathematical point, it is a model that can be mapped by appropriately chosen aggregates of actual events.

All the elements of an ensemble are absolutely identical, but this is not necessarily so in discussing samples. In quantum mechanics we shall use a special type of sample to discuss probabilities, which, as a difference with an ensemble, is meant to be experimentally constructed. Let me try to describe how we must understand the probabilities that are built into the state function. When we say that the probability of finding the electron at x is $\{\psi(xt)\}^2$ (see, however, note 3 for a more precise statement), this means that if we prepare a large aggregate of electrons *all of which have been subject to exactly the same preparative measurement as the electron in question*, then, when we measure the coordinate x over all the elements of this aggregate, the frequency distribution of x over the latter is the probability we are talking about. Such an aggregate of *identically prepared* wavicles is called an *assemblage*.

The fact that we invoke an assemblage in order to interpret the probability entailed by the state function (very much as ensembles are invoked in classical physics) does not mean, conversely, that the state function of a single wavicle can only be defined for that wavicle *as an element of some assemblage*. That is, it is not necessarily right to assume that $\psi(xt)$ has no meaning for a single wavicle but only within an assemblage, as a statistical distribution over it, although this is a view held by many people.[6] This idea is not entirely convincing: in the double-slit experiment there is interference even when we have a *single wavicle* in the apparatus at any one time, so that we should be able to assign a physical meaning to the state of such a single wavicle. Indeed, recent work claims that even the experimental determination of the state function for a *single wavicle* is possible, but this assertion is not free from criticism.[7]

Although consensus on this point cannot yet be considered universal, I shall take the state function ψ as defined for a single wavicle, and ψ^2 to be the probability of finding the wavicle at some point (or, more in general, in some particular eigenstate). The probability here must be understood in the sense in which we discussed it in chapter 13, p. 360. We saw there that, following Kolmogorov, a probability can be defined in a formal way just as for instance a volume is defined, and that, just as the volume, it can be given an intrinsic meaning for a single entity without reference to such things as sequences of events. All that is required is that some formal mathematical rules be satisfied. The important thing is that one can work with this quantity just as with any other defined quantities, and it is only when it is required to *interpret* it, that assemblages must be invoked, such that the frequency of the eigenstates over them are identified with the probabilities.[8] I shall soon discuss some examples that will make the significance of this interpretation clearer.

STATIONARY STATES AND QUANTIZATION

By the time quantum mechanics was developed there were two major experimental results that required the acceptance of energy quantization. One, was the fact that, as Planck and Einstein had recognized, the photon states in the case of the so-called blackbody radiation formed a discrete set. The other was the existence of a huge amount of data on atomic spectra. If you have a sodium atom, as for instance in ordinary table salt, it does not emit any light at all, of course, but if you put it into a flame, it does. The light it emits, if properly analyzed by an instrument called a *spectrometer*, shows very clear lines, the frequencies (p. 454) of which can be measured and are a sort of fingerprint for the sodium atom itself.

It was Niels Bohr who in 1913 explained this phenomenon, already sketched in chapter 11, p. 288. The electrons outside the nuclei of the atom can adopt stationary energy states (that is, states in which in principle they can remain indefinitely), and these states are quantized. At low temperatures, the electrons are in low-energy states, but if the atom is excited (as when put into a flame), higher energy levels become occupied. By some mechanism that Bohr did not explain, but which was later discussed by Einstein, an electron in an excited state falls down to another one lower in energy, and the energy difference is emitted as a photon, that is, light. Because the energy difference between those two states is given by the Planck's relation as $h\nu$, and because this energy is the one that is passed on to the photon, this argument gives the frequency ν (*nu*) of the emitted light. Whole books of experimental results on atomic spectra were instantly explained by this hypothesis which, as we can see, rests on the assumption of stationary energy states in the atom, with quantized energies. It was because Bohr's ad hoc hypothesis could now be derived perfectly well from first principles in quantum theory that people were immediately willing to accept the general tenets of the theory, however hard it was to give up cherished classical principles.

Since energy quantization is such an important phenomenon, I shall try to produce a simple example of how it can be explained within quantum mechanics. For this purpose, I shall now give a little more detail about what a blackbody is supposed to be. We can imagine that we have a fairly large black cavity with a very small hole. Any photon from outside that hits the hole will be captured by the cavity, which is why we call this a blackbody, black always meaning that no light is reflected by the object. (Whereas a green object, say, reflects radiation of the frequency that corresponds to green.) As a result of this process, the cavity will be full of photons after a time. We can now imagine that those photons will have wave functions corresponding to all sorts of frequencies. We also know that these photons cannot get out of the cavity (otherwise no blackness!) so that their wave functions must lead to zero probabilities for the photons to come out. I illustrate this condition in figure 3.

Because the photon cannot exit from the region, it means that the probability of finding it outside that region must be zero. This in turn means that the wave function itself must vanish at the boundary of the region. Therefore, its half-wavelength multiplied by an arbitrary *integer n* must equal the length of the region. In figure 3 this factor is given as 3, but it could quite clearly be any integer. From the condition that the wavelength is *quantized* by the relation that $n\lambda/2$ be equal to the length, say l of the region, it is easy to show that the frequency of the wave, and hence its energy must be quantized. (Remember the Planck relation $E = h\nu$.)[9]

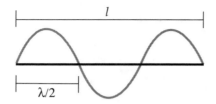

**Fig. 3. The wave function of a photon
confined within a finite region of space**

The thick line represents the region in question.

Electrons in atoms are not confined to a finite region as in the example above, but they can in principle move away from the nucleus up to infinity. Since the electrons are in *bound states*, however, that is states that are not free to leave the atom, the probability of them reaching an infinite separation from the nucleus must be zero. It can be proved that, as in the example of figure 3, it is the corresponding condition that the wave function must vanish at the boundary (in this case infinity) that forces the quantization of the atomic states.

EIGENSTATES AND THE PRINCIPLE OF SUPERPOSITION

I shall now discuss a most important principle of quantum mechanics that will allow us to describe a quantum state that is not an eigenstate. I shall provide for this purpose an example concerning polarized photons, which I shall discuss in a way that will lead us later on, in chapter 18, to a simple description of one of the most remarkable experiments of the last century, concerning the so-called Bell inequalities.

Most people are these days familiar with the phenomenon of polarization: if you take two polaroid spectacles and place the second perpendicular to the first, light will not go through. What happens is this. Light, as we know, is a transversal wave (p. 257), but in an ordinary light beam, the *electric vector* changes direction for different positions along the beam (to which it is always perpendicular), whereas in a *polarized beam*, it is constantly in the same direction. In my example, the first spectacle polarizes the light, which means that it eliminates (absorbs) all electric vectors except in one direction. When this polarized beam hits the second polarizer, which absorbs all light except that with its electric vector perpendicular to that of the first polarizer, there is of course nothing that can go through.

In the laboratory, the simple polarizing plate of the spectacles is most

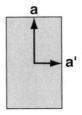

Fig. 4. Symbolic representation of an analyzer plate

The two perpendicular directions of polarization of the two emerging channels are given, but this picture does not show that the emerging channels are well separated, as is the case in practice. (Since the beams polarized along a and a' are perpendicular to the arrows shown, they appear in this symbolic representation to be coincident, which they are not.) Notice also that the analyzer is not symmetrical with respect to its plane, having only one entry channel on one side and two exit ones on the other.

often replaced by a more powerful device, called an *analyzer plate*. This is a crystal with the remarkable property that the incident light emerges through two *well-separated* channels, both of them polarized along directions that are one perpendicular to the other. I illustrate this in figure 4 in a convenient but purely symbolic way, because I make no attempt to represent in it the all-important fact that the two channels are well separated. All that I want to stress is that there are two perpendicular emergent directions of polarization, labelled **a** and **a'**, respectively. (The reader should keep in mind that when describing analyzers I always add a prime to the letter denoting a vector, to indicate another one *perpendicular* to it, a convention which will be much used in chapter 18.)

Of course, if I have a polarized beam with polarization direction **u**, all the photons in it can be said to be in a state of polarization denoted by that vector, $\psi(\mathbf{u})$, since they all emerge out of the channel polarized in the **u** direction. We are now ready to consider what we mean by photons in an eigenstate $\psi(\mathbf{u})$, and we do this in figure 5, very much as in figure 6 of chapter 15 (p. 470).

If we select from the first analyzer the channel polarized along the **u** direction, the state of the photons in that channel will not be disturbed by the second analyzer, since all of them will emerge from it polarized along **a**, which is parallel to **u**. The photons emerging from the first analyzer will thus be in an eigenstate which we write as $\psi(\mathbf{u})$, but which could equally well be labeled as $\psi(\mathbf{a})$, in terms of one of the two possible eigenstates of the second analyzer. Let us now see what happens when the photons impinging on the second analyzer are not in an eigenstate of it, which we illustrate in figure 6.

All that happens in figure 6 in comparison with figure 5, is that the second analyzer is rotated as shown with respect to the first one. We can

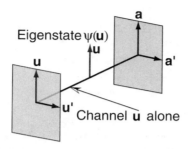

Fig. 5. A polarization eigenstate

safely expect the photon beam incident on the second analyzer to be still in the same state $\psi(\mathbf{u})$ as before, but this can no longer be identified with the state $\psi(\mathbf{a})$ that pertains to the photons exiting from the **a** channel of the second analyzer. In fact, when the experiment is performed with only one photon at a time arriving to the second analyzer, the remarkable result is found that a photon emerges either through channel **a** or channel **a'** *at random*. Just like when tossing coins, it is impossible to predict the outcome of an isolated event, but the probabilities of the outcomes, that is, the proportion of the total number of the incident photons that emerge through either channel, depends only on the orientation of **u** relative to the second analyzer.

Let me call $n(\mathbf{ua})$ and $n(\mathbf{ua'})$ the respective probabilities (or frequencies) of the photon emerging through either channel of the second analyzer and let me call $\{\mathbf{ua}\}$ and $\{\mathbf{ua'}\}$ the *angles* between the vectors **u** and **a**, and **u** and **a'**, respectively. The experimental result is that the probabilities are some function of the corresponding angles. This function is very well known in mathematics where it is denoted with the symbol "cos." In many simple calculators, if you enter the angle and then press the cos key, you will get the value of the function. (You will find that it is one and zero for 0° and 90° respectively, but you need know nothing about the cos function, except that it can be obtained from a calculator.) The results (which are in fact predicted by quantum theory) are:

$$n(\mathbf{ua}) = \cos^2\{\mathbf{ua}\}, \quad n(\mathbf{ua'}) = \cos^2\{\mathbf{ua'}\}.$$

Notice that a bracket such as (**ua**) on the left of these equations *is not an angle* but that it merely denotes a pair of parameters. It is easy to learn how to recognize angles in my notation: they are always indicated by braces.[10] Notice also that both equations are precisely of the same form, because the number of photons emerging in any direction is the square of the cos of the angle between that direction and the one that gives the polarization of the photon.

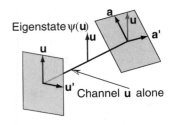

Fig. 6. The state ψ(u) in terms of the eigenstates of the second analyzer

The planes of the two analyzers are parallel.

From our discussion on the eigenstates, an incident beam of photons in a state ψ(u) could be in one or other of the two eigenstates ψ(a) and ψ(a′) of the second analyzer, in which case all the photons will emerge through one and only one channel. This is why we have an eigenstate, which is very easy to represent since ψ(u) will be either ψ(a) or ψ(a′). How can be represent the state ψ(u) when, as in the example of figure 6, it does not identify with either of the two possible eigenstates? There is a very simple rule which is used in quantum mechanics and that can be, at least to some extent, justified mathematically. Let me call ψ(u) the *unknown state*, because, until the experiment is carried out with the second analyzer, we need know nothing about its state of polarization (that is, we can assume that the details of the first analyzer are unknown). This unknown state is then written in terms of the possible eigenstates that can be the result of the experiment in a combination of the following form:

$$\psi(\mathbf{u}) = a\psi(\mathbf{a}) + b\ \psi(\mathbf{a'}),$$

where a and b are some constants. These constants are interpreted as follows: their squares give the corresponding probabilities for the corresponding eigenstates to emerge as a result of the experiment. In other words, if the isolated wavicle is in the state ψ(u), it will continue in that state until an experiment is performed, in which case the state of the wavicle will jump at random into one of the possible eigenstates of the experiment. If this experiment is repeated over an assemblage of identically prepared wavicles, each eigenstate will appear with a probability (frequency) equal to the square of the coefficient corresponding to that eigenstate in the expansion of ψ(u). On comparing this statement with our previous result on the probabilities, it is immediate to see that a and b are cos{**ua**} and cos{**ua′**}, respectively.

The prescription that I have given, which is called the *superposition principle*, can, as I have said, be grounded on mathematical principles, but it must be appreciated that it is, in any case, a fair description of the experimental results. Its interpretation, however, is delicate and, as we shall see in the next chapter, it has not been accepted without a great deal of pain and anguish. Let me discuss a little the most important features of the superposition principle.

Superposition principles have also been used in classical physics. Maxwell and Boltzmann had discovered that a gas is a *mixture* of molecules with different speeds (p. 365), and this mixture could be represented as a sum like the one in the last equation but with very many terms, each corresponding to a given velocity, with the coefficients so chosen that their squares give the *proportion* of molecules with that velocity in the mixture. Those proportions are in fact nothing else than the probabilities for the corresponding velocities, and those authors have actually given expressions that would allow their calculation. So, the situation appears to be very similar; and yet it is entirely different. In the Maxwell-Boltzmann case we have a *mixture*, and the probabilities are *actual probabilities*, that is, nothing else than the *proportions* with which each of the components appear in the mixture.

In the quantum mechanical superposition principle, the situation is completely changed in two ways: to start with, we do not have a mixture at all. This can be verified in figure 6 by rotating the second analyzer until **a** coincides with **u**, whereupon all photons will emerge in identically the same state. (As we, in any case, know they were, since the assemblage was formed by extracting all the photons from a single channel of the first polarizer and thus in identical polarization states.) Therefore, we do not have a mixture but a *pure state*. Once this is understood we can notice the second fundamental difference. The probabilities that appear in the quantum mechanical superposition principle, as the squares of the coefficients therein, are *potential probabilities*. By this expression we mean that they are the probabilities that determine the pure state in the sense that, if this pure state is now measured, that is, if the possible outcomes for the emerging eigenstates are determined, then these potential probabilities give the frequencies with which each of those eigenstates *will* appear.

Notice the dual role of the probabilities in the quantum mechanical case. Through the coefficients, they *determine* the pure state of a single wavicle. (*Each* photon will have a state fully specified by the last displayed equation.) But in order to experimentally observe them, we need an assemblage of wavicles, all resulting from the same preparative measurement; and when the second measurement is made on the whole assemblage, then the various eigenstates will appear with frequencies equal to those probabilities.

The concept that we have developed of a pure state that is not an eigenstate, as a superposition, is not easy to grasp, and some of the founding fathers themselves refused to countenance it, as we shall see in the next chapter. The problem is that if we want to have some intuition of what the state $\psi(\mathbf{u})$ means in the last displayed equation, we are led to think that this photon is partly in the state $\psi(\mathbf{a})$ and partly in the state $\psi(\mathbf{a}')$, these states being weighted by some sort of ghostly probabilities. Of course, this type of intuitive thinking can lead to an equally ghostly ontology. We should instead accept a prudent one, in which unknown states exist, in the sense that they can be determined through observations that permit the state to be manifested. I shall have to say a great deal more about this, nevertheless, in chapter 17. Even in this chapter, I shall soon return to the problem with a practical example, which I hope will help the reader to acquire a better understanding of the meaning of the superposition of states.

Reduction of the Wave Function

The reader should notice that randomness has reared its head, almost unnoticed, in the above discussion. And I refuse to complete the cliché by qualifying this head as ugly: our job is not to judge nature but to understand it. If nature needs randomness, so be it. Having said this, I am duty bound to point out what appears to be a contradiction, which will indeed be revealed to be so in chapter 19. The evolution of quantum mechanical states is ruled by the Schrödinger equation with the time, and this equation is *deterministic*. (It could not be otherwise, since it is only a deterministic equation that licenses the use of the word "state.") So the unknown pure state is happily evolving deterministically in the time, as some combination of potential eigenstates, each of which must also evolve in the same manner deterministically. This happy deterministic evolution, which the Schrödinger equation predicts for *all time*, continues as such *only until the state is measured* (in our example, until it hits the analyzer: but see the next subsection), whereupon it changes *at random* into one of the states appearing in the combination. In other words, starting with a number of eigenstates (two in our example, but it could be any number, even infinity), when the measurement is effected, all those states disappear except one, obviously breaching determinism. It is usual for this reason to refer to this situation as the *collapse of the wave function*, but I shall prefer to use the less calamitous name of *the wave function reduction*, a problem to which I shall return in more detail in chapter 19.

Superposition, Reduction, Measurements, and Histories

Imagine a photon in a state $\psi(\mathbf{u})$ given as before as a superposition of the eigenstates $\psi(\mathbf{a})$ and $\psi(\mathbf{a'})$ of the analyzer; also as before we shall assume a single photon in the apparatus at any one time. For convenience, I will take $\mathbf{u}$ to make the same angle with $\mathbf{a}$ as with $\mathbf{a'}$. It is easy to believe that when this photon hits the analyzer plate, its state will instantaneously reduce, that is, that the photon will jump at random into one or the other of the two eigenstates, which will thus have been measured in the *preparative* sense described on p. 468. This is not necessarily so, as I shall illustrate in figure 7, in which we introduce a second analyzer on the right, rotated by 180° about the $\mathbf{a}$ axis with respect to the first. This is so as to present the polarised channels $\mathbf{a}$ and $\mathbf{a'}$ as *entry* channels for the photon that is in between the two analyzers. Once they go through the analyzer, those channels join in a single exit channel $\mathbf{e}$.

The surprising experimental result is that this channel is *polarized precisely as the incident channel* was, that is, along the direction $\mathbf{u}$. Most important, this being the case, this channel is in a *pure state and not in a mixture*. As I have said, it is tempting to believe that the reduction of the wave function (from a pure state into a mixture) takes place when the microscopic system (photon beam) interacts with a macroscopic object (analyzer). From this point of view, one would conclude that the state function has reduced once it goes through the first analyzer, and therefore that in the region between the two analyzers we have a mixture, half the photons being in one beam polarized along $\mathbf{a}$, and the other half along the other beam, polarized along $\mathbf{a'}$. (See note 10.) This experiment, however, shows clearly that this *cannot be right* because, if this were so, the emergent beam would also be a mixture of photons polarized along those directions, whereas it is a pure state, still described by the original superposition state. (That the emergent beam would be a mixture of photons polarized along $\mathbf{a}$ and $\mathbf{a'}$ follows from the fact that, if in the central region of the figure the photons were already in the eigenstates of polarization, parallel to $\mathbf{a}$ and $\mathbf{a'}$ of the analyzer, then as eigenstates do, they would go unchanged through the analyzer.)

How and when, we must ask ourselves, does the state function really reduce? In order to understand this, consider the first analyzer as an instrument to measure the polarization state of the photons. Have we measured this state once the photons emerge on the right of this analyzer? The answer is: no. We do not know (that is we do not measure) the polarization state of a photon *until we ascertain in which channel the photon is*. We can do this in a variety of ways. The simplest way is to put one photon counter along each of the channels. If we do this, we shall find

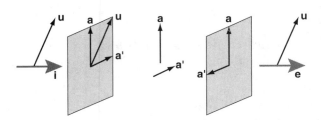

**Fig. 7. Verification that interaction of a wave polarized along u
with an analyzer does not in itself cause the reduction
of the superposition state ψ(u)**

that both counters produce exactly the same readings, thus verifying that the *potential* probabilities of 1/2 per channel have transformed themselves into *actual* probabilities, therefore creating a mixture rather than a pure state. (The value 1/2 arises because both channels are equiprobable, and the sum of their probabilities must be unity.) Another way, as already done in figure 5, would be to block off one of the channels, say the one parallel to **a'**, whereupon you would find that half the photons emerge from the second polarizer parallel to **a**.

The net result of this story is that the operation of measuring is more delicate than we might have thought. It not only involves an interaction between the microscopic system and a macroscopic one but, also, *it is not completed until some macroscopic counter or pointer has been read or registered.* And this should make it easier to understand my previous requirement that a history is not a history until it is completed and registered. Measurements will be further discussed in chapter 19.

Hidden Variables

The breakdown of determinism entailed by the reduction of the wave function might be thought to be susceptible to cure by postulating a fundamental revision of the concepts underlying the superposition principle. I have said that although the state

$$\psi(\mathbf{u}) = a\ \psi(\mathbf{a}) + b\ \psi(\mathbf{a}'),$$

when measured, yields photons in either of the two states on the right-hand side, it is not a mixture of them. Assume that the description of the beam of photons in the state ψ(**u**) as being all in identically the same state with polarization vector **u**, is *incomplete*, and that the apparently identical photons are in fact a mixture of two types. That is, assume that there is a variable, as yet unknown, that can take only one of two values, say 1 and -1, distributed among the photons in actual proportions a^2 and b^2, respectively. Assume further that photons that possess the value 1 of this variable are determined to change into the state $\psi(\mathbf{a})$ and that those with the value -1 must change into $\psi(\mathbf{a}')$. With these simple assumptions, the superposition principle now becomes identical with the classical description of a mixture and determinism is fully restored, the postulated variable *determining* in each case the outcome of the process in question.

The variable that is thus introduced is appropriately called a *hidden variable*, and the propounders of the existence of such variables very often appear to try to claim credibility for them from the fact that the introduction of hidden entities has a not negligible pedigree in the development of science. Atoms, for instance, were used as a sort of hidden variable during the nineteenth century.

Before we listen any further to the siren songs, let us reflect upon this. Yes, suppose that the state $\psi(\mathbf{u})$ is a mixture of photons, a^2 with hidden variable 1 and b^2 with hidden variable -1. But we know that a and b are determined by the position of $\mathbf{u}$ relative to $\mathbf{a}$ and $\mathbf{a}'$ in the analyzer. So, by merely rotating the analyzer we must change the composition of the mixture! In other words, the hidden variables cannot be so hidden: they must be able to interact with the polarizer so as to change value instantaneously in accordance to the orientation of the latter. This requirement that hidden variables be *nonlocal*, that is, that they be subject to *action at a distance*, makes them of course far less attractive as theoretical constructions. Dogmatism in discussing quantum theory, however, is always premature, and a healthy but critical spirit of inquiry about hidden variables should not be dismissed. I shall say something further about this in chapter 19.[11]

MORE ABOUT THE PHYSICAL MEANING OF THE SUPERPOSITION PRINCIPLE

It is very tempting at this stage to believe that the somewhat ghostly states postulated by the superposition principle are purely verbal props with no physical content. In order to dispel this possible misconception, I shall provide a practical example. It was accepted in the nineteenth century that the molecule of benzene, a very well known chemical, contains six carbon and six hydrogen atoms. This was a bit of a puzzle, because it was also known that atoms have definite *valencies*, which are integers that determine the numbers of atoms with which they can link. Thus, it was recognized that hydrogen and carbon have fixed valencies of one and four, respectively. It was difficult to produce a picture, as it is normally done in chemistry, with atoms joined together by lines (representing *bonds*), that can join six carbon and six hydrogen atoms while satisfying their valencies. It was the German chemist Friedrich Kekulé (1829–1896) who, in a notorious flash of inspiration, solved the problem in 1865, although his solution created a new conundrum.

If we look at the left-hand side of figure 8, the hydrogen atoms in it satisfy their valency of one, but to do so for the valency of four for the carbon atoms requires the assumption that they are alternatively linked by *double bonds*. Once Kekulé hit on this idea, he soon realized that there was another way in which the double bonds could be arranged, as shown on the right-hand side of the figure. This means that the liquid that we call benzene in a bottle must be a *mixture* of the two forms shown in the figure. Even in the nineteenth century, however, chemists were pretty good at replacing atoms in a molecule by other atoms (of the same valency) and the dichlorobenzene compound, in which two contiguous hydrogen atoms are replaced by chlorine atoms, was well known. The great question then arose: if benzene is a mixture of its two Kekulé structures, how is it that dichlorobenzene does not appear in the two forms shown in figure 9? In fact, since in one case the bond across the two chlorines is single, and in the other case it is double, chemical experience indicated that they would have different physical properties and thus be susceptible of separation from the "mixture" resulting from the chlorination of benzene. But no evidence whatsoever was ever found to confirm, or even to suggest, the existence of those two forms of the compound.

Almost as soon as quantum mechanics was developed, the Kekulé conundrum was resolved. The two Kekulé structures in figure 8 may be regarded as eigenstates of the molecule. The latter is properly described by the superposition principle as a superposition of those eigenstates, *but it is not a mixture of them*. The eigenstates have only a potential and not an actual probability, and thus they would not lead to two different forms of dichlorobenzene. The latter (fig. 9) is itself given by a superposition, but not as a mixture, of its two possible forms. What happens is that, within the superposition principle, all six bonds in benzene are of exactly the same nature, neither single, nor double, but of a superposition of these two types of bond. Even more evidence is now available, since physical methods exist that permit the very accurate determination of the geometry of the molecule.

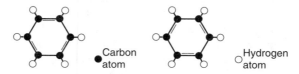

Fig. 8. Kekulé structures of the benzene molecule

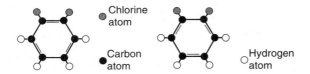

Fig. 9. The two possible forms of dichlorobenzene compatible with the Kekulé structures but which were never found

If the classical chemical structure with three double bonds existed at all, those bonds being double and thus stronger, would be *shorter*, whereas within minute margins of error modern experiments show that the six bonds in benzene are all of precisely the *same length*. And these experiments can now be done with one single molecule at a time, so that there is no possibility that the measurements provide some sort of an average. The evidence for this type of success of the superposition principle is now overwhelming.

I hope that the reader can see with this example that the difference between the quantum mechanical superposition and a classical mixture is not just a form of words. Without the somewhat elusive concept of superposition, even simple chemical facts would not be understandable. And it is the superposition principle which is the basis for the design of quantum computers.

STATIONARY STATES AND QUANTUM JUMPS

I have already discussed in this chapter the electron states in atoms and showed that they must be quantized. An electron, in general, will be able to assume one of the many quantized states in the atom, but, in normal conditions, it will lodge in the state with the lowest energy available. If the atom is excited, the electron will itself be excited to a higher energy state. These states are of course stationary states, but nevertheless the electron cannot stay in one of the higher states forever and falls down to a state of lower energy, emitting a photon with energy equal to the energy difference between the two states.

I have discussed in chapter 11, p. 290, the mechanism whereby the electron is jogged out of the higher energy state by interaction with the photon field. The jump from the higher to any lower state is a *random process*, and this is another case when quantum mechanics abandons determinism. We can nevertheless relate this phenomenon to the reduction of the wave function, which we know is not deterministic. This is so because, when the electron is excited, its state can be written as a superposition of all the energy states of the electron in the given atom, each state with its corresponding coefficient, the square of which gives the probability of the electron falling into that state. It is the interaction with the photon field (as invoked in chapter 11) that now causes the reduction of the wave function, as a result of which after a certain (very short) time, one and only one of the states appears in the reduction. In this case, of course, certain restrictions must be obeyed by the emerging state, such as being lower in energy than the initial one. We can see from this that the reduction of the wave function can take place, not only as a result of measurements, but also in what appears to be a spontaneous way, as a result of the interaction between the wavicle and some field.

IDENTICAL PARTICLES AND THE PAULI PRINCIPLE

I shall handle this problem with a certain abandon at first, although such hand waving as I shall indulge in will be so subtle as to be almost imperceptible. Once we learn a few things, I shall go back to the beginning and expose a certain amount of dishonesty in my arguments. In order to follow a more traditional terminology, I shall resume the use of the word "particle" as a substitute for the more rigorous "wavicle," but this is no worry at all, and will not entail any later mea culpa.

Consider a time-independent Hamiltonian $H(x)$. The variable x here is the coordinate of one single particle, say an electron. Suppose now that I have two electrons. I shall have to differentiate somehow between the two, and I shall use the fairly obvious notation $H(x_1 x_2)$ which I shall abbreviate as $H(12)$, and similarly for the wave function $\psi(12)$. Electrons are identical, so that I can permute them without altering the Hamiltonian (that is, without changing the energy).

I am afraid that in order to get what we need, I shall have to be a bit more sophisticated, and I shall invent a *permutation operator* P which when acting on anything (an operator, a wave function, etc.) permutes the electrons 1 and 2. For instance, when I apply this operator on the Hamiltonian $H(12)$, it swaps the labels of the two particles, but *in this particular case* it cannot introduce any change, as I have just said:

$$P\,H(12) = H(21) = H(12).$$

I shall now prove that this operator commutes with the Hamiltonian, by acting with it on the Hamiltonian operator, now abbreviated as H, itself acting on the wave function ψ. The operator P acting on $H\psi$ must act on each factor separately, but we have just seen that $P\,H$ is the same as H. Therefore,

$$P\,(H\psi) = (P\,H)\,(P\,\psi) = H\,P\,\psi,$$

where I have added parentheses as a help to read the equations, but they must be discarded when analyzing the result.

When dealing with operators, one very often introduces an *operand* like the ψ here, in order to understand how things work but, because we know that this operand is quite arbitrary, we can get rid of it at the end of the work to obtain an *operator relation* valid for any operand. Thus

$$P\,H = HP,$$

so that *P* commutes with the Hamiltonian. This entails (see p. 473) that *P* is a *constant of the motion*. That is, if we consider an isolated system and this operator has a certain eigenvalue at the start, it will keep this eigenvalue for as long as the system remains isolated.

This, I hope, is all very clear despite the little feat of mathematics performed, but the reader may wonder what on earth the eigenvalues of such an abstract operator as *P* might be. If $\psi(12)$ is its eigenfunction, they must of course satisfy the eigenvalue equation :

$$P\,\psi(12) = p\,\psi(12).$$

(Compare with p. 471; please, do not confuse the eigenvalue *p* here with the momentum.)

It is fairly simple to prove that the eigenvalue *p* in the above equation can have only one or the other of the two values ± 1.[12] We can now feed this value back into the eigenvalue equation, to obtain:

$$P\,\psi(12) = \pm\,\psi(12).$$

Remember now that these eigenvalues must remain the same for as long as the system is isolated, because *P* is a constant of the motion. We thus have two possibilities:

$$P\,\psi(12) = \psi(12) = +1\,\psi(12), \quad \text{or} \quad P\,\psi(12) = -\,\psi(12) = -1\,\psi(12).$$

In the first case, we say that the wave function is symmetrical with respect to permutation of the particles, and in the second that it is *antisymmetrical*. The latter case is the most common one, and the particles for which this relation is valid are called *fermions*, of which electrons are an example. The former case is obeyed for instance by photons, and the corresponding particles are called *bosons*. The classification of particles in fermions and bosons was first done by Wolfgang Pauli (1900–1958), in accordance with the results just given, which he derived, but the category to which each particle belongs was obtained by him at that time from experimental evidence.

The *Pauli principle* is a most important and very simple property of fermions. Consider $\psi(12)$, the state function of a two-electron system, and imagine that the two electrons were in two identical states (stationary states, for instance, of the same atom). On permuting the particles, nothing significant about the state could possibly change, since the particles are identical, and all that you are doing is swapping them between two identical states. On the other hand, because they are fermions, we know that the state function should change sign. For the state function to remain the same and yet to change sign is impossible, unless the wave function $\psi(12)$ *vanishes*. (Zero and its negative are identical!) Thus *two electrons can never occupy identical states*, which is the *Pauli Principle*.[13] The physical significance of this principle is enormous: if you have an atom with eighty electrons, all of them would sit in the lowest energy level if electrons were bosons, but Pauli requires that the electrons distribute themselves among levels of gradually increasing energy, until their total number is exhausted. Naturally, this causes profound changes in the properties of atoms and, even more, of solids.

All this, I hope, is just as clear as a little bit of formula bashing permits, but I must now come clean. I have made an assumption which is far from obvious: I have been talking of electron 1 and electron 2, as if we were able to come with a brush and paint labels on the particles, whereas even in chapter 12 (p. 315) we found that labeling molecules in a classical problem is not a minor task. We managed in that case because we could distinguish the particles by their trajectories. Since the latter do not exist for quantum mechanical particles, all that I have done is based on the assumption that at least under certain conditions we can label them, which is extremely difficult to justify. I shall try.

BACK TO PHILOSOPHY: IDENTITY AND NAMING

I now have to redo in a little more detail some of the work of chapter 12 on identity, since things become much harder in quantum mechanics. Identity, alas, is one of the great problems of philosophy, and I must warn the reader that what I shall connote with this word in this book is not necessarily what philosophers mean by it.[14] When we say that Tom and Sam are *identical* twins, we understand that there are two separate individuals; hence the possibility of naming them. In a strong use of the word *identical*, on the other hand, philosophers would understand identical twins to be a *single* individual, and hence deny the possibility of separate names. (The reader must realize that philosophers, theorists as they are, would construe the expression "identical twins" to signify that they do not differ in a single molecule or even in a single instant of their entire histories. And please reflect that if their histories are identical, they could not even stand in two different places! No wonder that Saint Thomas Aquinas permitted souls, otherwise indistinct, to have different histories.[15])

This connotation of the word *identical* goes back probably to Nicholas of Cusa (1401–1464) but, as I have done in chapter 12, this is not the meaning I use, of which more later. I must first refer to what is perhaps the most famous principle concerning the problem of identity (this word understood in the philosophical sense just discussed), the *Principle of the Identity of Indiscernibles* due to Gottfried Wilhelm, Freiherr von Leibniz (1646–1716), and enunciated in his posthumously published *Monadology*:

> there are never in nature two beings that are perfectly alike and in which it would not be possible to find a difference that is internal or founded on an intrinsic denomination.[16]

This is an extension of Leibniz's *principle of sufficient reason*. If distinct things could be exactly alike, then there would not be a sufficient reason for any differentiating relationship that one bears to another.[17] Two particles, say, with exactly the same intrinsic properties (charge, mass, state, etc.) would therefore have to be *identical* (in the philosophical sense described above) and thus would have to be the same particle.[18] If two electrons, which in any case share the same mass, charge, and any other intrinsic attributes they might have, were to be in the same state, then they would be indiscernible and thus identical; and they would therefore be the same electron, which is not possible. This is what Weyl calls the Leibniz-Pauli exclusion principle, as a philosophical denial that two electrons can occupy the same state.[19] One cannot take very seriously the possibility of a physical principle to be derived from philosophical considerations alone, and, in fact, if Leibniz's argument were all that we need for the Pauli principle, then, why is it not obeyed by bosons?

So, we must return to the problem of identity and naming in order to clear up the formal work done before and to justify, when to do so is possible, the use of labels for the elementary particles. To start with, I shall redefine some of the terms that I shall use, since there is a lot of confusion in the physical literature between the words *identical* and *indistinguishable*, both being used almost as synonyms. I shall not use the word *identical* in the Nicholas of Cusa or Leibniz sense, but rather in its dictionary meaning from the *Oxford English Dictionary* (2d ed.):

> Agreeing entirely in material, constitution, properties, qualities, or meaning: said of two or more things which are equal parts of one uniform whole, individual examples of one species, or copies of one type,

so that any one of them may, for all purposes, or for the purposes contemplated,
be substituted for any other. [My emphasis.]

Just as I have done in chapter 12, identical particles will be understood to share all *intrinsic properties*, such as charge, mass, spin, and so on. (These are the *essential properties* of the scholastic philosophers pertaining to their *substance*.) On the other hand, they may differ in *extrinsic* or *relational* properties (position, velocities, and states, which correspond to the scholastic *accidental properties*).[20]

It will be a bit more difficult to understand what I mean by *indistinguishable* until we have an example, although of course I shall start from the dictionary definition that if two things are indistinguishable, they cannot be recognized as being different, so that they cannot be given different names or labels. But a little more discussion will be necessary. On the other hand, I shall always use the words *indistinguishable* and *indiscernible* as exact synonyms.

In classical mechanics, we started from identical particles, all with the same intrinsic properties and thus exchangeable *for our purposes*. This is in accordance with the above dictionary definition, but, in order to be more precise, I shall restrict such purposes to those related to keeping quantities like energies and probability distributions unaltered by those exchanges. We were able to label classical particles by consideration of their trajectories, but now in quantum mechanics, trajectories are no longer significant. How can we then label identical particles?

If you think about the work I have done, such labeling is essential in order to make sense of permutations, and permutations are operations that are effected on the Hamiltonian. Also, these operations are constants of the motion, so that I do not have to apply them to the Hamiltonian as it changes throughout time, but I can take it to be given at some initial time. At that initial time, I can assume the electrons (or whatever particles appear in the Hamiltonian; but keep an eye on the future, please) to have been localized. I can thus use those initial coordinates to label the particles, which allows us properly to use the permutations as I did. Notice that the lack of trajectories does not preclude labeling as long as some initial localization is possible, which allows the classification of permutations (as symmetrical or antisymmetrical) for all later times. This solves the problem, and we can therefore understand how fermions and bosons arise, but this is not yet the end of the story.

The question is this. In order to label my identical particles I had to appeal to our ability to localize them for some particular instant in time. This, alas, might not be universally possible. Photons, for instance, are strange particles because they have no mass and they travel always at the velocity of light: the only way to localize them is to destroy them, as when getting a light beam impinging on a fluorescent plate. Otherwise, one would have to consider photons belonging to different light beams on which there is of course a high degree of transversal localization.[21]

Such transversal localization, however, is obtained by macroscopic means, such as collimators or lasers. Because of this, it is important to consider the situation of a photon "gas," as can be found for instance inside a blackbody cavity, as we have seen. Although in a "gas" of photons I might be able to talk about the total number of them in the gas (the cardinal number of the aggregate), I cannot count the photons: counting is naming and this might not be possible just because I cannot keep photon 1 fixed in one place while I look for photon 2. So, I could easily "label" my first photon thinking that it is my second one. Because of this, the ordinal number of an aggregate of photons is not necessarily meaningful. Photons might thus be *indistinguishable,* and this remark qualifies the meaning of indistinguishability that I had started to define before: if there is no way in which I can name or label identical particles, then they are *indistinguishable,* and this is a concept that classically would be extremely difficult to grasp since the required restriction is impossible.[22]

I shall now examine the consequences that indistinguishability would confer to a set of particles, should particles with such property exist at all. I must first remind the reader that what we had found before was that when a permutation P is applied on a Hamiltonian, it either multiplies it by +1 (fermions) or -1 (bosons). This is all that we could discover, and it was left to Pauli to assert the empirical rule that electrons are fermions (and thus that no two of them can occupy the same quantum state). Consider now a Hamiltonian for two particles, which I shall write as $H(12)$. If the two particles, as in the case of the photons, cannot be labeled, it means that P acting on such a Hamiltonian does "nothing" because the labels 1 and 2 cannot be discriminated. Doing "nothing," however, is not meaningless. Just like adding "nothing" to a number gives us the wonderful "zero," a permutation that does "nothing" gives us what we call the *identity permutation.* Because, as a difference with all the other permutations, the identity permutation leaves the Hamiltonian entirely unchanged (or as one says, *identically invariant*), it might be thought that the possibility for it to be multiplied by -1 cannot arise, and therefore that indistinguishable particles must necessarily be bosons. This, however, is wrong: indistinguishable particles can be either fermions or bosons, as it was the case with distinguishable ones.[23]

The reason why I have gone into all this detail is that I hope that the reader now grasps that in principle two identical particles may or may not be indistinguishable. Whether just a situation really arises in nature is still an open question, although some philosophers appear to support the view that photons are actually indistinguishable. In view of this, it appears prudent to keep a distinction between "identical" and "indistinguishable" particles, a distinction that is unfortunately blurred in the physical literature.

Let us blow up, since we are here, another old philosophical canard. The principle of impenetrability of matter is a hallowed one: two different particles cannot occupy the same point in space at the same time. Is this now true? Think about two electrons with identical wave functions: they would have the same probability of being found at any point in space, and this is what in the new physics would be the nearest one can get to the definition of interpenetrability. But the Pauli principle forbids two electrons to be found in identical states, so that the old wives' tale is true for them.[24] Bosons, on the other hand, do not obey Pauli, which means that two bosons can interpenetrate to their hearts content. (The question of course is that, if indistinguishable bosons existed, their individuality would be so tenuous that the fact that they may interpenetrate would not be very significant.)

CODA

We have seen in this chapter that the state of a quantum particle may be defined in a way that is not as different from the classical case as one might imagine, in the sense that the state function (or wave function) must satisfy a deterministic equation, the Schrödinger equation, in the time. Eigenstates, which are the states in which a given variable has a precise value, are not too difficult to understand. What entails more mental strain is the use of the principle of superposition to give an arbitrary state as a combination of eigenstates because this combination is not in any way a mixture. What happens is that, when such a quantum state is measured, the superposed wave function reduces at random into that of one of the eigenstates. There are two problems here, one a very serious one, which is the meaning of the reduction of the wave function, because it entails a breakdown of the deterministic evolution of the state function under the Schrödinger equation. I shall discuss this question in chapter 19. The other problem is more philosophical, and it entails the meaning of the superposed wave function, in particular in the case when macroscopic bodies might be considered. This is the famous Schrödinger's cat problem, which will entertain us for a while in chapter 17.

I hope also that the reader will have appreciated that the old philosophical problem of identity is of serious importance in quantum mechanics and that, conversely, the latter provides a new example of what we mean when we say that two particles are indistinguishable or indiscernible. The reader will have noticed, I hope, that by stepping out of the philosophical use of the word "identical," I have avoided getting into the thorny metaphysical problems of the nature of individuality, which are behind the work of Leibniz and his followers.

We are now ready to return in the next chapter to a more historical and critical account of the main principles of quantum mechanics.

NOTES

1. Notice that we are returning here to the Newton situation just as closely as the constraints of quantum mechanics permit. In the Newton case, the state variables were x and its rate of change (instantaneous velocity). We are now replacing x by its probability distribution, and its rate of change by the rate of change of its probability distribution.

2. Optimism and faith are here under complete control: if falsely held, no predictive equation will ever appear for the presumptive state function, and we soon shall have to renounce sin, the devil, and the flesh.

3. There are three major problems. The first one relates to the fact that the wave function must be *complex*. Let me explain this. As we know (p. 416), the imaginary unit i is such that i^2 equals -1. Otherwise, a *complex number* $a+ib$ behaves like any ordinary sum, adding and multiplying by the ordinary rules. As we shall see, the *complex conjugate* of $a+ib$, which is defined as $a-ib$, and represented with the symbol $(a+ib)^*$, is very useful, because the product of any complex number with its complex conjugate is always real and positive. The proof is very simple: $(a+ib)(a+ib)^*$ is the same as $(a+ib)(a-ib)$ which is the sum of four terms: a^2, $-aib$, iba, $-i^2b^2$. The second and third term cancel, and the fourth is $+b^2$. The result of the product is thus a^2+b^2, which is necessarily real and positive, as asserted. We can now understand what happens. First of all, the state function must be complex because, in fact, it must carry within itself two pieces of information, one the probability distribution in x, and the other its rate of change. A complex function is ideal to cope with this, since it already has two parts, one real and the other imaginary. On the other hand, if $\psi(xt)$ is complex, its square $\{\psi(xt)\}^2$ will in general be a complex number, whereas a probability must necessarily be a positive real number. This difficulty can easily be remedied, on replacing $\{\psi(xt)\}^2$ by $\psi(xt)\{\psi(xt)\}^*$, which must always be real. (The reader who is not used to these tricks must appreciate that, although this appears to be somewhat arbitrary, it is nevertheless a sensible way of adjusting the mathematics used to the physical constraints encountered.)

The second technical point is that, even with the above correction, $\psi(xt)\{\psi(xt)\}^*$ cannot be defined as the probability of finding the particle at the point x because such probabilities for a continuous variable must vanish. (Only one favorable case out of an infinite number of possible outcomes.) What that product really is is a *probability density*, such that if l is a small length that contains the point x, then $\psi(xt)\{\psi(xt)\}^*$ times l is the probability of finding the particle in l.

The third problem follows at once from the above. Assume l to be a small length, and then add up $\psi(xt)\{\psi(xt)\}^*$ times l over the whole of the x axis. The result should be unity, since the probability of finding the wavicle *somewhere* along the axis is a certainty. Nevertheless, there is nothing in the definition of $\psi(xt)$ that ensures that this will be the case. Therefore, one revises slightly the relation between the state function and probabilities. Instead of saying that $\psi(xt)\{\psi(xt)\}^*$ times l is the probability of finding the wavicle in the small interval l, we say that this probability is:

$$\psi(xt)\ \{\psi(xt)\}^*\ \textit{times l divided by}$$
$$[\textit{the sum of } \psi(xt)\ \{\psi(xt)\}^*\ \textit{times l, over the whole length of the x axis}].$$

It is now clear that if we sum this expression over the whole length of the x axis, this sum affects only the numerator (that is, whatever is before "divided"), and thus that the result of the sum is necessarily unity.

A remarkable property of the state function $\psi(xt)$ follows from this redefinition. Because $\psi(xt)$ appears in this expression both before and after the word "divided," *it may be multiplied by any arbitrary constant without altering its physical meaning*, that is the stated probability. This is a situation that is entirely new in quantum mechanics. The energy, which is a state function in classical mechanics could not possibly be allowed to be multiplied by an arbitrary constant (if I could do this, I could cut my electricity bill in a most satisfying way). Functions that may be multiplied by arbitrary constants are called in mathematics *rays*. Whereas a vector has a direction and a length, the ray has only a direction.

The reader must beware of statements that the state function cannot be fully determined because it may always be multiplied by an arbitrary constant. This is incorrect. The wave function is fully determined as the quantity that it is, which is a ray, and rays do not have lengths. We do not say that the area of this page is not fully determined because we do not give its color: the full specification of different physical quantities must be made with regard to their nature.

4. More properly, from note 3, it is $\psi(xt)\{\psi(xt)\}^*$ that should be time independent.

5. All this work assumes, of course, that one knows the mathematical form of the Hamiltonian operator—which I have not discussed—and hence that the result of its action on the state function on the left of the Schrödinger equation may be found mathematically.

6. A very early example of this approach is the book by Kemble (1937), but this view has many supporters, few more compelling than Peres (1993). For a strong criticism of the views of this author, however, see Primas (1990), p. 60.

7. See Aharonov et al. (1993*a*), Aharonov et al. (1993*b*), and Anandan (1993). The validity of this work is queried in Stamp (1995), p. 130, and Ghose et al. (1995). These last authors accept, nevertheless, that this work shows that quantum mechanics can make meaningful empirical statements about single systems. I shall discuss the question of the significance of the state function further in chapter 17. Zeitlinger (1990) discussses neutron diffraction experiments in which no more than a single neutron is present at any one time in the interferometer and concludes (p. 20), as I have suggested in the text, that the wave function has to be associated with each individual neutron and not just with an assemblage.

8. As discussed in chapter 13, the passage from the formal probabilities so defined to experimental frequencies, although straightforward in practice, entails theoretical problems that are not universally considered as fully resolved at the present time.

9. You have to obtain a relation between the wavelength λ of a wave and its frequency ν. This is easy. In a time equal to the period T, by the definition of the latter, the wave advances a length λ. The velocity v of the wave is therefore λ/T. Since the frequency ν is the inverse of T, v equals $\lambda\nu$, whence ν is v/λ. The boundary condition in the text requires $\lambda / 2$ to be of the form l, the length of the region, divided by an integer n. Therefore ν must be equal to $vn/2l$. On introducing this result into Planck's relation the energy E equals $h\nu n/2l$ and must therefore be an integral multiple of the constant $h\nu/2l$.

10. The reader should notice that, as is standard in mathematics, $\cos^2\{ua\}$ means $[\cos\{ua\}]^2$. From the values of the cos that I have given, it is clear that $n(ua)$ is unity when $\{ua\}$ is $0°$ and that it vanishes when this angle is $90°$, exactly as one would expect from the case discussed in fig. 5. When the angle $\{ua\}$ is $45°$ (in which case $\{ua'\}$ also has this value), it is easy to prove that both probabilities equal $1/2$.

11. Hidden variables had a very checkered history. In one of the most influential books ever written on quantum mechanics, von Neumann (1932*a*), pp. 209, 305–24, had constructed a proof that hidden variables were incompatible with quantum theory. Given the gigantic authority of that author, his result was accepted as gospel truth but, more than thirty years

later, Bell (1966) found an unjustified assumption in the proof. Bell's results have been refined since then, and the consensus is that there is no theoretical impossibility for the existence of such variables. As we shall see in chapter 18, however, they must be nonlocal, that is, they must be influenced by actions at a distance as already exemplified in the text.

12. To find this eigenvalue I shall perform a little trick, which is to apply P on both sides of the equation

$$P\psi(12) = p\psi(12).$$

When I do this I must realize that, on the right-hand side, P has nothing to do with the constant (the eigenvalue), and must only operate on the function:

$$P^2\psi(12) = pP\psi(12) = p\{p\psi(12)\} = p^2\psi(12).$$

On the other hand, on writing explicitly the left-hand side of the above equation, and then using the definition of the operator P, we obtain

$$P^2\psi(12) = P\{P\,\psi(12)\} = P\{\psi(21)\} = \psi(12).$$

On comparing the right-hand side of these two equations, it follows at once that p^2 must be unity, whence p must be ± 1.

13. Wolfgang Pauli was hardly 25 when he discovered his famous principle (Pauli 1925), which immediately revolutionized our understanding of atomic structures and spectra. Pauli was born in Vienna and was the godson of Ernest Mach, but he studied at Munich where he did his doctorate with Sommerfeld. At the time of his enunciation of the principle, he was a mere *Privatdozent* at Hamburg, where the *Pauli effect* was soon discovered: it was enough for Pauli to enter a laboratory for some apparatus to fail. (It was later said that even his presence in a given town was experimentally ominous.)

14. The standard reference on identity is Hirsch (1982), and chapter 11 of van Fraassen (1991) is a beautiful essay on the philosophy of identity in quantum mechanics.

15. Rescher, in Leibniz (1991), p. 66, quotes Nicholas of Cusa: "there cannot be several things exactly the same (*aequalia*), for in that case there would not be several things, but the same thing itself."

16. Leibniz (1991), Section 9, p. 62. I cannot commend sufficiently Rescher's edition, which collates with the text of the *Monadology* numerous quotations from Leibniz's other works, that are invaluable to understand his ideas.

17. Leibniz (1991), p. 66.

18. French et al. (1998) considered the possibility that, among the attributes to be shared by two indiscernibles, relational properties may have been excluded by Leibniz. This is what they call the *strong principle of the identity of indiscernibles*. This would lead to a definition of identity that permits two particles, differing in a relational property, to be identical and yet separate, which does not seem to agree with Leibniz's numerous quotations on the subject in the Rescher edition of the *Monadology*, neither with Weyl's reading of the principle, as discussed in the text. See also French (1989*a*), (1989*b*), (1989*c*), and Redhead et al. (1991), (1992). The article of da Costa et al. (1992) contains a very useful and comprehensive review of the problem of identity in quantum mechanics. See also Dalla Chiara et al. (1985), (1993), and French et al. (1995).

19. Weyl (1949), p. 247.

20. As in chapter 12, I follow here Jauch (1968), p. 275. See also the Dante quotation given as an epigraph to this book.

21. I am much indebted here to correspondence with Professor Peter Higgs, whose view is that whereas the word *identical* is a good one to use, in the sense that I have explained, the term *indistinguishable* is not so, since in accordance to him, all identical particles, photons included, may be considered distinguishable. I feel, however, that there is a vast difference

between two photons belonging to two different beams, which have been created by macro-scopic means, and two photons in a photon "gas." Thus, it appears prudent to leave open the possibility that genuinely indistinguishable particles might exist. Berry et al. (1997), in a very important paper, use the word "indistinguishable" rather than "identical," but Professor Berry has kindly written to me indicating that this use might be tentative.

22. This, however, is what d'Alembert had envisaged when he suggested that the dis-tinguishability of two probabilistic states was a matter for empirical assessment. (See p. 343.) A good discussion of the problem of counting photons was given by Toraldo di Francia (1978). See also Dalla Chiara (1985), who rejects the possibility of assigning ordinal numbers (and thus labels) to photons within a set, a constraint that is extended by Redhead et al. (1991) to *all* quantum "particles."

23. The reason for this situation is subtle. When we consider the wave functions, the identity permutation must not necessarily leave them identically invariant. What must remain so are the probability distributions $\psi\psi^*$. But it is clear that for this to be the case, the wave functions may be multiplied by -1 just as well as by +1. These two cases correspond respectively to fermions and bosons. Nevertheless, a difference arises for indistinguishable particles. Consider two distinguishable particles 1 and 2, which may have wave functions corresponding to two distinct states. For convenience, I shall dispense with the letter ψ for these two wave functions and denote them with the letters a and b, respectively. The terms $a(1)$ and $a(2)$ will indicate that the particles 1 and 2, respectively, are both in the state a. The pair of particles may be found in four different states:

$$a(1)\,a(2), \qquad b(1)\,b(2), \qquad a(1)\,b(2) + a(2)\,b(1), \qquad a(1)\,b(2) - a(2)\,b(1).$$

The first three states are clearly symmetrical with respect to interchange of the two par-ticles (like bosons are) whereas the last state is antisymmetrical (like fermions). When the particles are distinguishable, so that the labels 1 and 2 are meaningful, bosons have three states (the first three) and fermions only one (the last). When the particles are indistin-guishable, the single antisymmetrical state cannot be written because it would be identically zero. Thus, if indistinguishable particles are fermions, this cannot be so by virtue of the exis-tence of the antisymmetrical state discussed; the antisymmetry of the wave function with respect to the identical permutation (which is the only one possible) must instead be intrinsic, that is not due to the formation of some composite wave function. Such intrinsic antisymmetries are not unknown in quantum mechanics: the wave function of the spin of an electron, for instance, changes sign when the function is rotated by the identical rotation, the rotation by 2π. (See p. 419.)

24. My statement that two electrons cannot interpenetrate has to be strongly qualified. The question is that the state of the electron is determined, not only by its ordinary space coordinates, but also by an internal coordinate, which is called the *spin*. So what the Pauli principle requires is that no two electrons share the same probabilities for the space and spin coordinates. Therefore, the classical principle of impenetrability holds only *if the spin coor-dinate* is disregarded. In fact, two electrons of different spins automatically satisfy the Pauli principle, and they can perfectly well interpenetrate as far as the space coordinates are con-cerned. This is most important in practice, since it allows two electrons to build up a high negative charge density in the same region of space between two positive nuclei, thus holding them together. The statement in the text about the impossibility of interpenetra-bility for two electrons holds true only when the electrons have the same value of the spin, in which case the Pauli principle forbids them to share the same spatial wave function. The reader will notice that the famous principle of impenetrability does not work even for elec-trons, if impenetrability is restricted, as it was undoubtedly the case in the past, to space coordinates only.

THE GREAT QUANTUM MUDDLE

It was you who, not speechless with shock but finding the right language. . . . did most to prevent a panic . . .
Wynstan Hugh Auden (1948)

The fact that an adequate philosophical presentation has been so long de-layed is no doubt caused by the fact that Niels Bohr brain-washed a whole generation of theorists into thinking that the job was done 50 years ago.
Murray Gell-Mann (1979), p. 29.

We are faced with what I shall call the great quantum muddle. . . . Heisenberg's famous formulae . . . have been habitually misinterpreted by those quantum the-orists who said that these formulae can be interpreted as determining some upper limits to the precision of our measurements . . . we have to be able (and are able) *to make measurements which are far more precise.*
Karl Popper (1967), pp. 18, 20.

I find this [Popper's] view of the quantum theory to be rubbish of a most stimulating kind.
David Mermin (1983), p. 655.

The time has come for us to do a bit of thinking about how the inter-pretation of quantum mechanics arose, and in order to do this we must look at the intellectual history of the subject. Let me say at the outset that few other topics in the history of science provoked so much hostility *among the very people who developed them.* We already know about the suf-ferings of Galileo and of Boltzmann, but the new feature here is that, as we shall see, some of those who most contributed to the subject most hated it. Naturally, it required a great deal of nerve to keep going while such a bitter struggle took place, and it was the strong and stubborn per-sonality of Niels Bohr that saved the day: I should like to celebrate this achievement by adopting for him the words that W. H. Auden addressed to T. S. Eliot, in a poem from which I quote in the first epigraph.

In saying this, I am nevertheless acutely aware of Niels Bohr's limitations. Finding the right language could hardly be said to have come naturally to him.[1] He was a good linguist, but of few men it could more appropriately have been said that they had the ability to be unintelligible in so many languages. He could sometimes be unbelievably confusing if not desperately irritating. I remember a lecture he gave at the University of London in 1950, or thereabout, and on a scale of 1 to 10 for lecturing proficiency, he won in my view a comfortable −3. I also believe that a good deal of his philosophical approach was misleading, and that his philosophical understanding was in fact meager, as I shall discuss later.[2] I nevertheless think that Gell-Mann, whom I quote in the second epigraph, is too harsh on him. You cannot expect a fireman to appreciate with a keen eye the architecture of the house that he is defending from a blaze; had it not been for the immense scientific authority that Bohr enjoyed, Einstein's continuous forays against the ramparts of quantum mechanics could have ended in destroying the very citadel. More than once Bohr was able within hours to counteract Einstein's complex arguments—not a mean achievement. The results of quantum mechanics, however, were so amazingly important (whole areas of physics and chemistry were cleared up in a few years) that most physicists, happily engaged in doing the good work, were content to leave the giants well alone to tilt against their windmills. And it must be admitted that in his own confusing way, that so much exasperated his opponents, Bohr had a knack of often hitting, if not the bull's-eye of the subject, just as near it as it was possible to aim with the intuitive methods at his disposal.

If the physicist Bohr was weak at philosophy, one must at least admit that he did not try to teach the philosophers' grandmothers how to suck eggs. The philosopher Popper, not noted for his intellectual humility, did not have such compunction, and he went head-on to teach the physicists how they should renounce muddle and do their job *properly*. I am afraid that in so doing he ran enthusiastically from one howler to another: his misunderstanding of quantum mechanics was encyclopedic, and Mermin is the kindest of men in finding any stimulation in his gibberish. So much for the exegesis of my epigraphs. We could now start, but I should like to say a few words about Bohr because he occupies a strange position in our story. As you will see, he did not discover any of the major results of quantum mechanics, and, in fact, during the first decade or so when the subject was developed, he never engaged in any serious application of it. So how was it that he could single-handedly have assumed the intellectual leadership of quantum mechanics? So much so, that the most usual interpretation of quantum theory in use is still called the *Copenhagen interpretation*, after his home town. (As to be expected from Bohr's propensities, what is called the Copenhagen interpretation, far from being a clear-cut, well-defined system, is rather a collection of views and rules around a basic theme.)[3]

The reader interested in Bohr's life, work, and ideas can do no better than consult the wonderful book on him by Abraham Pais.[4] Niels Henrik David Bohr (1885–1962) was born into a well-off family of Copenhagen. Bohr's father, Christian, was rector of the University of Copenhagen when Niels was twenty, just a couple of years after the latter started his studies there. In 1912, one year after Ernest (later Lord) Rutherford discovered the atomic nucleus in Manchester, Bohr went to work in his department. At that time, there was a vast amount of data on the spectrum of atoms, which were well known to be composed of discrete lines of specific frequencies, but there was no explanation of how these spectra came about. It was the great triumph of young Bohr that he produced a theory (later proved to be untenable) which proposed that the electrons in atoms were held in orbits about the nucleus, very much like the planets around the Sun. These orbits were stable and had definite, discrete (and therefore *quantized*) energies; and the spectral lines were produced by electrons jumping from one to another of these states. At a stroke, by applying the quantum idea of Planck and Einstein, Bohr was able to explain volumes of data on the atomic spectra; and this work achieved such quick recognition that he was awarded the Nobel Prize for it in 1922, at the same time as Einstein received his for his 1905 work on the photoelectric effect. Already, in 1921, Bohr had become the first head of the Institute of Theoretical Physics in Copenhagen, which soon became the Mecca for the obligatory pilgrimage of everyone with quantum mechanical ambitions.

Not only was Bohr a welcoming and generous (if somewhat tediously demanding) host at the institute, but in 1932 he was granted by the Carlsberg Foundation the use for life of a palace in Copenhagen, where he could entertain like a prince: he was a most important man and again at the cutting edge of physics, contributing at that time new work on nuclear theory. No wonder that people paid attention to what he said. We can now start our review, although I must warn the reader that I shall be highly selective and refer only to such work as is immediately relevant to the interpretation and the philosophy of the few quantal problems that I was able to cover in the two previous chapters.[5]

THE BEGINNINGS

The somewhat ad hoc quantum theory of atoms introduced by Niels Bohr, although destined to become a dead end, had the strong support of one of the ablest mathematical physicists of the time, Arnold Sommerfeld (1868–1951) of Munich, who developed it exhaustively, inventing at the same time a number of mathematical methods that later were crucial in the development of quantum theory. In the summer of 1925, Werner Carl Heisenberg (1901–1976), who had recently completed his Ph.D. with Sommerfeld, had a bad attack of hay fever and went to Heligoland, in the North Sea, to recover. There, in just a few weeks, he invented quantum mechanics and sent his paper to the publishers on 29 July.[6] It is not necessary for me to go into the details of this work, although it was here that Heisenberg introduced the concept of a quantum mechanical operator; he also showed that in many cases these operators do not commute. The work was presented in a way that was unfamiliar at that time because he introduced some mathematical concepts, arising from the lack of commutation, which appeared strange.

Max Born (1882–1970), professor of theoretical physics at Göttingen, had another graduate student, Pascual Jordan (1902–1980), at twenty-two even younger than Heisenberg. It took Born only eight days to recognize that the strange mathematics of Heisenberg was none else than the well-known matrix calculus invented in 1858 by the Cambridge mathematician Arthur Cayley (1821–1895).[7] A few more days of mathematics with Jordan, and the work was completed. Their paper was received by the editors on 27 September 1925, less than two months after Heisenberg's.[8]

Important as Heisenberg's original idea was, it was the work of Erwin Schrödinger (1887–1961) that really sold wave mechanics to the physicists. To start with, he produced a presentation of quantum mechanics on mathematical lines that were far more familiar than those adopted by Heisenberg and which connected more easily with what people knew of classical mechanics. No less important was the fact that in a series of four papers Schrödinger solved many of the major problems of atomic and molecular spectra which had puzzled people for decades, obtaining results that could immediately be compared with experiment. In that remarkable year of 1925 Schrödinger, then at Zurich, was, at thirty-seven, a middle-aged man in comparison with the youngsters in the field; and this is not without significance, because he later proved to be unable to accept the consequences of his own discovery, as I shall show. Also, he came not from Germany but rather Vienna, where he already had done work on the theory of color which, although very important indeed, had not given him much in the way of academic kudos.[9]

If Schrödinger was no longer in the prime of his youth in 1925, he found other means to pump up his mental energy. Just before Christmas a vacation was necessary, and he left Zurich for the Alps. But this was a vacation in more than one way because he left his wife in town and took with him a lady friend. Who she was, it is not known, but she must have been as patient during the day as efficacious at night, since Schrödinger worked furiously in an outburst of almost adolescent energy and returned to Zurich on 9 January 1926 with his discovery done and his paper written. If you are worried about Mrs. S., you should know that they remained a devoted couple until the end of their lives, although Schrödinger kept falling in love repeatedly, especially with underage girls. The only real tragedy in all this is that, when so inspired, he wrote the most atrocious love poetry. In any case, Mrs S., as I understand, was not left entirely to her own devices in Zurich; Hermann Weyl, (1885– 1955), one of the greatest mathematicians of his time (whose very own sherry glass I was once allowed to sip from in Manhattan), providing adequate company.

Less than two weeks after his return to Zurich, Schrödinger communicated on 27 January 1926[10] the first of the four papers that dominated the subject for decades; and the fourth one was received five months later on 21 June. Schrödinger's work was an adaptation of the mathematical techniques that had been used classically in the study of waves in fluids. The picture he had in mind was this: he assumed that the electron charge, rather than being pointlike as classically assumed, was extended over a large region of space with variable density. He could treat this variable-charge distribution by methods similar to those that had been used for fluids, and in his fourth paper, he interpreted the wave function ψ, which he had introduced, in terms of this charge density, as follows:

$$(point\ charge\ of\ the\ electron)\ times\ \psi\psi^* = charge\ density.$$

(Please look up p. 505 for the meaning of the complex conjugate ψ^*.) This interpretation, however, was soon discarded, as I shall now discuss.

THE INTERPRETATION OF THE WAVE FUNCTION

Although there was no difficulty in solving the mathematical equations proposed by Schrödinger for the wave function, the question of its interpretation caused a great deal of trouble and it still is somewhat controversial.[11] On 25 June 1926, four days after Schrödinger's fourth paper, Born communicated the paper in which he interpreted $\psi\psi^*$ as a probability density, as explained on p. 505.[12] In his Nobel lecture, delivered in 1954 just four months before Einstein's death, Max Born stated that he had founded his statistical interpretation on the one given by Einstein for light quanta (photons) in 1905.[13] In order to relate particles with waves, Einstein had already interpreted the square of the amplitude of the optical wave as the probability density for the occurrence of photons.[14] This is somewhat ironical, given the skeptical attitude toward the role of probability that Einstein held, as we shall see later.

As soon as Born formulated his probabilistic interpretation of the wave function, an element of schizophrenia affected the scientific community, people holding opposite views which were occasionally supported, not always at different times, by the same person. (The reader must notice that all this was happening well before Kolmogorov, as discussed on p. 360, allowed for the use of probability just as any other variable, its definition no longer requiring the awkward consideration of sequences of similar events, a feature that I exploited in discussing the wave function on p. 485.)

In accordance to the Born hypothesis, quantum mechanics predicts the probability of the result of a measurement performed on a single system. Born, however, espoused more than once the statistical interpretation, which I shall call the Einstein hypothesis, since the latter was its most distinguished exponent. Quantum mechanics, this hypothesis claims, merely predicts the relative frequencies of the results of measurements performed on an assemblage of identically prepared systems.[15]

The reader must beware of statements such as the one that I have just made, which do not really amount to much. Because, as we have seen in the last chapter, and as it is in any case obvious when probabilities are used, any statement made about a probability must ultimately be tested over some sequence of adequately chosen events. The crucial distinction for a statistical formulation of quantum mechanics is the insistence that *the state function of a single system is meaningless*; and Einstein himself produced such a statistical interpretation, to which he adhered throughout his life, at the Solvay Congress of 1927. In 1936 he made absolutely clear what he meant by it: "The ψ function does not in any way describe a condition which could be that of a single system: it relates rather to many systems, to 'an ensemble of systems' in the sense of statistical mechanics."[16]

Despite the enormous and well-proven physical intuition that Einstein possessed, and despite the fact that his statistical interpretation is still supported by many distinguished people,[17] I take the view that it must be rejected. First, there is nothing in the nature of the state function itself that requires this restriction of its meaning. It is one thing to say that its predictions must be tested statistically, and another to say that it does not apply to a single system, when we know that the Schrödinger equation with the time is in fact applied, and mathematically solved, for a single particle. Second, it is difficult to understand within the Einstein hypothesis what we mean when the state is defined as a superposition of states, as in the case of the benzene molecule. (See p. 496.) It is perfectly possible in modern laboratories to deal with a single molecule at a time, and there is no evidence whatsoever, for instance, that a single Kekulé structure can exist. Thus, unless the wave function can be defined for a single molecule, an adequate description of a benzene molecule would not be possible. Third, I mentioned (p. 485) recent work that suggests that the experimental determination of the state function of a single particle might be achieved.

A line of retreat often taken as regards the interpretation of the wave function is that ψ does not represent the physical system, but only our *knowledge* of it; and it appears possible that at the time when Born proposed his interpretation, he was prepared to hold this view.[18] I shall discuss this attitude a little later, when dealing with Bohr's philosophy.

One more gambit as regards the state function. As you remember, since Cardinal Bellarmino, a way out of difficulties with theories about which people do not want to stick their necks out, is to adopt an *instrumentalist* position, and it is no wonder that this strategy was strongly used in quantum mechanics. I shall quote the Hungarian American mathematical physicist Eugene Wigner: "The observation results are the true 'reality' which underlie quantum mechanics. The state vector does not represent 'reality.' It is a calculational tool."[19] This appears to be gospel truth, but let us try this test. Change "quantum" into "classical" and "state vector" into "state function called energy." If you use Wigner in this modified form, his words apply perfectly well to the energy: when you go to a water mill, you do not "observe" anything like energy. All that you observe is *motion*. Yet from ancestral times people have "measured" the energy of a water mill: an experienced miller would say "This waterfall gives me the power of two horses." It could be argued indeed that energy is merely a bookkeeping tool (as indeed it is when used to calculate my electricity bill), but very few classical physicists would have denied the "reality" of energy.

The fact is that what is "real" in the physicist's description of nature is not amenable to obvious instant definition. (Although in the next chapter we shall try!) Starting on the negative side, I would take the line that ontological excommunications and benisons should never be accepted as valid elements of discourse in asserting what is and what is not "real." On the positive side, instead, tests of objectivity and invariance are useful. Ultimately, entrenchment in the scientific mesh will license the use of the predicate "real." This might appear to be a somewhat skeptical view, but if you think about this a little, you will see that it is not. Most scientists, for instance (but not all philosophers), are prepared to accept happily that electrons are "real," and they appear to understand what they mean by this, just as Mach in his own time understood what he meant by claiming that atoms were not "real." Reality by decree, alas, is but a dream: beware of false prophets. If alleged instrumentalism in any subject succeeds for a long time and on a wide range, you can bet your bottom dollar that people will be prepared to predicate as "real" the objects behind such instrumentalist inquiries. And remember: the meaning of a word is its use.

WHY HEISENBERG CRIED: BOHR'S INTERPRETATION OF QUANTUM MECHANICS

At the end of 1926 Heisenberg is in Copenhagen with Bohr. Endless and most often frustrating discussions with the master bring despair to young Werner. Thank goodness, the great man needs a rest and goes on a skiing holiday to Norway. The time is February 1927. Before the end of the month Heisenberg had discovered the uncertainty principle and, wisely, had written a long letter about it to his friend, one of the most precocious young men of the quantum gang, the Austrian Wolfgang Pauli (1900–1958), who at the time was at the University of Hamburg. Heisenberg needed to confide in someone who would not drown him in a flood of quasi-metaphysical worries. The middle of March came, and Bohr with it, and with him trouble. As you might remember, Heisenberg described a hypothetical microscope with which to measure the position of the electron. His optics, alas, was not very sound, and Bohr quickly put him right. And then day after day after day of discussion arose, ten solid, murderous days. Heisenberg probably did not understand at the time what it was that really worried Bohr: "I remember that it [the discussion] ended with my breaking out in tears because I just couldn't stand this pressure from Bohr."[20]

And yet Bohr was right, and it was only when I heard for the first time about Heisenberg's sobs that I understood why this was so. Heisenberg's picture was that of a particle "out there," and of an experiment that measures the particle and in so doing "interferes" with it. Bohr's view instead was that you should not speak of a particle "out there," but rather describe the whole complex event as *one phenomenon*. The "particle" cannot be interfered with, because before we measure there is nothing like a particle of which we can legitimately say that it is "out there." In the terminology I have used in the last chapter, we have one and only one history, and it is only of this completed and registered history that it makes sense to speak. Heisenberg was splitting the history, in two parts, first particle and then measurement, and this should not be done.[21]

It is almost unbelievable that Bohr went on and on for some ten days and that what I have written in a dozen lines did not come across to Heisenberg. But it is my view that few people could beat Bohr at repeating the same idea incessantly never making it clearer. And yet, *he was right*. Fortunately, the impasse was somehow patched up, with the master finally allowing Heisenberg to publish, and the latter's crucial paper on the uncertainty relations reaching the editors on 23 March 1926.

EINSTEIN, PROBABILITY, AND THE BREAKDOWN OF DETERMINISM

We have seen that it was Einstein, in his 1905 paper on the Brownian motion, who made Boltzmann's probabilistic ideas finally acceptable to the scientific community. (See p. 322.) Even then, probability was at this stage a classical concept: we treated the movement of molecules in a liquid as random because we had no access to all the details of their motion. The latter was ruled by classical Newtonian laws, and it was our *ignorance* of the precise values of the huge number of variables involved that entailed the resulting statistics. The great change came with the work that Einstein did in 1916 on emission of light by atoms. Although originally based on Bohr's theory of the atom, it could immediately be applied to the new quantum mechanics. Electrons in excited Schrödinger's stationary states interacted with the electromagnetic field (see p. 290) and thus fell down to lower energy states, emitting a photon with the corresponding energy difference. The important point is that this transition was entirely at random. That is, the electron jumps down from the excited state at *unpredictable* times. Also, most important, the direction of the emitted photon was random, contrary to causality, and this could easily be observed.

It is this type of phenomenon that entails a fundamental departure of quantum mechanics from determinism or causality. We know, however, that to expect causality to survive the passage from the macroworld to microphysics is wishful thinking. Men like Heisenberg were prepared to concede this,[22] but the older Schrödinger was not. He visited Copenhagen in September 1926 when Heisenberg was there, and they had heated arguments over this problem. Heisenberg reports that Schrödinger ended the debate by exclaiming: "If we are going to stick to this damned quantum-jumping, then I regret that I ever had anything to do with quantum theory."[23]

As for the man most responsible for pointing out the breakdown of determinism, Einstein, he never accepted it. "I would be very unhappy to renounce complete causality," he wrote to Born in 1920.[24] The whole world knows his famous dictum "God does not plays with dice,"[25] and he never, to the end of his life, abandoned this line. Even in 1948, writing again to Born, he says: "I have become an evil renegade who does not wish physics to be based on probabilities."[26] In pondering over this dicta, I urge my readers to reflect that even the greatest of men can sometimes be very silly. Where was the *evidence* for Einstein's pronouncements, and what has anybody's *wish* to do with what the world is or is not like? And if there was no evidence, wherefrom comes the philosophical urgency to extend a principle, such as that of causality, well away from its proper domain of entrenchment? Unless, of course, you firmly believe that what you know is eternally, universally, and necessarily valid because you have gained such knowledge by some process akin to Platonic illumination. To which illusion, I grant, Einstein had a greater right than most humans to succumb.

I should perhaps give a warning as to the meaning of the breakdown of causality in quantum mechanics. I want to stress that this is not due to unavoidable ignorance of any necessary initial conditions, a view which is often entertained. Thus, Heisenberg, in the paper in which he introduced the uncertainty principle:

> In the sharp formulation of the law of causality, "If we know the present, then we can predict the future," it is not the consequence but the premise that is false. As a matter of principle we cannot know all determining elements of the present.[27]

This is not quite right; *it is not that we cannot know the present sufficiently well*: we know the present as well as the present can be known. No, the problem is that whatever information we might have about the present, it is not information that can allow us to predict the future except in a probabilistic way. Nature is not an oyster that denies access to the pearl that would give us the future. That pearl *does not exist*, at least in the sense of the word "exist" that can be used in natural science.

I cannot end this section without a brief mention of one of the phenomena that most dramatically showed the fundamental truth of quantum mechanics and of the unavoidable existence of randomness in nature. This is the α-decay of radioactive atoms. It had been known for a fairly long time that some radioactive nuclei emit α particles, which are helium nuclei. This emission is random: no amount of heating or cooling or anything else would alter the rate of emission, and the particles come out at entirely unpredictable intervals. It was George Gamow (1904–1968) at Göttingen who in August 1928 explained the phenomenon. The α particles inside the nucleus have a wave function $\psi(x)$ that gives the probability of finding the particles at any value of their position x. Although this probability will be very small outside the nucleus, it is not exactly zero, and this means therefore that the α particle has some chance of getting out of the radioactive nucleus, which is in fact what one observes. (This is appropriately called the *tunnel effect*.) Because this effect is governed by a probability function, its occurrence is entirely random, and such radioactive phenomena have in fact been used to generate sequences of random numbers.

QUANTUM MECHANICS À LA POPPER

When I was a child, the image of the drunken Helot had a lot to do in keeping me on the straight and narrow; it is in this spirit that I shall entertain my readers with Popper's activities in eliminating every possible misconception from quantum mechanics. Popper produced various proposals to this effect, and the reader of *The Logic of Scientific Discovery* has to deal with a veritable palimpsest of assertions which were stated in the German original, and which later on even Popper admitted were wrong. This did not deter him from formulating a grand vision of quantum mechanics, in which the dualism particle-wave was rejected, and the uncertainty principle was bypassed, by dint of Popper producing precise position and precise momentum measurements of the "particle" *but not at the same time*. (This, of course, ignores the gist of the uncertainty relation: for a classical Newtonian particle those quantities are required at the identical time, and it was the importance of Heisenberg's argument that if you knew precisely the momentum at a given time, then a measurement of position *at that time* would alter the momentum.) Alternatively, if he manages to "produce" simultaneous values of the two variables, Popper uses Newtonian techniques that presuppose continuity and thus a trajectory.[28] Of course, there is no experimental evidence whatsoever that continuity can be used in the microworld. What Popper is doing is nothing else than postulating metaphysically the existence of continuous trajectories. This position is described by many philosophers as *realist*, a realist presumably being in this context (as in mathematical Platonism) a person who claims that things that they imagine are "real." The "great quantum muddle" for Popper was the physicists' misconception that the wave function, which in accordance to him had no more than a statistical meaning over an ensemble, could be applied to a single particle.[29]

BOHR: COMPLEMENTARITY AND ALL THAT

John Stewart Bell (1928–1990) was a British theorist who did some of the most important work of the century on the foundations of quantum mechanics, which I shall discuss in chapters 18 and 19. I was fortunate in attending some of his lectures, and, great man as he was, I felt that he went too far in his distrust of Bohr, especially when claiming that he never understood what the latter meant by complementarity. The problem, to a large extent, is that Bohr extended his idea of complementarity to cover so much, life itself, that it is sometimes difficult to get to the crux of his thinking. And he elevated the concept to the level of a comprehensive, philosophical doctrine, which obscured its important physical significance. Yet, the concept that there are no waves or particles, but objects that manifest themselves through mutually exclusive phenomena that show up either particle-like or wavelike behavior; and that these phenomena are *complementary*, in the sense that it is through their joint not concurrent consideration that the object is apprehended, is correct. (The reader will notice that even I became infected by Bohr's proclivities in producing a long and convoluted sentence, although this is nothing compared with the master. Indeed, I doubt very much that Bohr would have approved of my description of complementarity, since he would have objected to the intrusive word "object," as you will see in a moment.[30])

The concept of complementarity was presented to the world by Bohr in a lecture he gave at Como, Italy, on 16 September 1927.[31] I must admit that this was a novel idea. Bohr (unfortunately) realized that there was a lot of mileage in it, and he went on endlessly discussing applications of complementarity. In so doing, he created a philosophical system of sorts that in my view was founded on a misconception.

The uncertainty principle had shown that it was necessary to relate the act of observation to anything that could be said about the fundamental "particles." Accordingly Bohr, like many of his followers, took the view that they were firmly avoiding any ontological commitment, and that they were dealing purely with the epistemology of physics. I shall give a couple of quotations to show what he had in mind.

> There is no quantum world. There is only an abstract quantum physical description. It is wrong to think that the task of physics is to find out how nature is. Physics concerns what we can say about nature.

> Our task is not to penetrate into the essence of things, the meaning of which we don't know anyway, but rather to develop concepts which allow us to talk in a productive way about phenomena in nature.[32]

These statements might sound appealing on first reading, but there is an inherent vagueness that makes them of very little use and that, I think, led many people astray who in any case were skeptical of Bohr's views. In the first quotation, we run into the problem of what "nature" means in it. If by "nature" you mean what I called in chapter 8 "nature$_1$," that is the

Kantian *Ding-an-sich*, the statement is trivially true. If, on the other hand, "nature" is meant to be what I had called "nature$_2$," that is, the nature we experience through our perceptual and rational system, then Bohr would have to explain what he means *by saying things about nature at all*. It will be easier to illustrate what I have in mind through a negative example, by stating a sentence that we must rule out: "Lengths are objective elements of nature." We know in fact that relativity theory demonstrates that the concept of length is observer-dependent and thus not objective. And here we have a clear example that physics attempts to do what Bohr refuses to accept. Because in negating that statement we are trying to explore, not just what we can say about nature, but what can be said in a way that is objective or invariant. It is perfectly adequate to claim that special relativity, in claiming a special invariant status for the spacetime interval (p. 278), is making as much of an ontological statement as I do when I claim that there is a table in my room. That this ontology is of the sublunar species, that is, that it does not attempt the impossible in breaking into "the essence of things" in the Kantian sense, goes without saying.

There is also an inherent contradiction in regarding the complementarity principle purely as an epistemological approach. Quite rightly, Bohr insists on a holistic approach to phenomena, rather than to allow their partition in "observed object" and "measuring apparatus and registration of observations." But he also insists, equally rightly, that we require two complementary phenomena, to discuss *what*? Just the phenomena, in the extravagant waste of saliva to which Bohr was prone? Surely, if we assert that the two complementary phenomena are coupled in some way, it is because we need them to identify what they are coupled about, and this is the underlying *object*, the wavicle. To say that this is only an epistemology of the wavicle would only make sense if one extended this skeptical approach to the whole of nature (nature$_2$), eliminating all ontology from our discourse. But if this is not claimed, then to say that complementarity gives you only the epistemology of the wavicle implies that there is still some clever ontology that could be assayed, in order to introduce objects with better ontological credentials than wavicles.

This is no idle speculation. The clever ontological program just suggested has been carried out under the leadership of the American David Bohm (1917–1992), who, as a relict of Senator McCarthy's days, was professor of theoretical physics at the University of London. Precisely as I have said, he claimed mere epistemological status for Bohr's ideas and proceeded to invent an amazing "ontology." Needless to say, I do not agree with this approach, although it is immensely clever, but I shall try to express my worries about it more fully in chapter 19.[33] Bohm was a very powerful theorist, whose work in solid state put him at the top of his generation, and his orthodox contributions to quantum mechanics have left a permanent mark. No one who knew him could have failed to be attracted by his personality. His own quantum mechanical ideas were strongly influenced by the most charismatic physicist of the century, Einstein, with whom he had been in contact during his Princeton days. So, I should say a few words about the Einstein-Bohr relationship.

EINSTEIN VERSUS BOHR: FIRST ROUNDS OF THE DOGFIGHT

Already in 1920, when Bohr visited Einstein in Berlin, they had argued about the wave-particle duality. Although electron diffraction was still to come, the problem was already there, and Bohr expected a clean break with classical physics, whereas Einstein had a more conservative view; their respective positions were fairly fixed since that time, and quite opposite.[34] This controversy could have remained latent (or at least merely epistolary) but for a most fortunate circumstance. The chemical company Solvay, which had made a fortune out of caustic soda, had started in 1911 a series of international conferences at Brussels, and it was they that provided a public forum for the discussions between Einstein and Bohr.

The fifth Solvay conference in October 1927, and the sixth in 1930, came at a time when the problems of interpretation of quantum mechanics were most urgent, and gave an opportunity to Einstein and Bohr to fight the battle of the giants in front of a world audience.[35] The pattern in each case was the same. Einstein would produce some experiment that majestically demonstrated quantum mechanics to be wrong, whereupon Bohr would grow almost hysterical, spend a largely sleepless night, and by the next morning produce a marvelously detailed analysis that blew up Einstein's argument. It is amazing how Bohr could manage to put his finger on the weak point of Einstein's lucubrations in such a short time and in such hair-raising circumstances. General opinion accepted that Bohr got the best out of the exchanges, and this gave confidence to a generation of theorists to just get on with their work and not to worry. But Einstein never gave up. (See chapter 18 for the next round of the fight.)

THE SUPERPOSITION PRINCIPLE AND THE IGNORANCE INTERPRETATION

I have already mentioned that one of the ways in which people tried to make palatable the breakdown of determinism was the *ignorance interpre-*

tation: it is only because we do not know precisely the initial conditions, it was alleged, that accurate prediction is not possible. We have already seen that this is not a sound argument, and where it becomes most dubious is when it is applied to the superposition principle. If ϕ_1 and ϕ_2 are the only two possible eigenstates of a system, when that system is not in either of them, its wave function, as already stated, is

$$\psi = a\,\phi_1 + b\phi_2.$$

We have also seen (p. 491) that this state is not a mixture of the two eigenstates. One way of explaining the situation is this. If we had measured the state of the system, then it would have been either in ϕ_1 or in ϕ_2. But we have not measured that state, so that it is because of our ignorance that we have to use the state ψ. The ignorance interpretation has a long history in the treatment of probability, but its use in the superposition principle is a far cry from saying that we believe that the odds against Black Oyster in tomorrow's Kentucky Derby are five to one, in which case we accept that there are factors that we necessarily ignore (such as the weather), which might decide the outcome, and which an experienced punter tries to estimate when placing his bet.

The situation in the superposition principle is entirely different. When we use it, as we had done on p. 496, to give the state of a benzene molecule, we are not reasoning from any lack of information. On the contrary, we are doing so from what, within the present theory of quantum mechanics, is the sum total of available knowledge. Naturally, a new theory might come but, meanwhile, the ignorance interpretation is not an interpretation at all, it is a *cri de coeur* for a *different* theory. Within the context in which we are working, the superposition principle gives a *complete* description of the state. In the case of benzene we saw that, otherwise, compounds would exist that have never been found, and that the theory gives correctly the geometry of even a single molecule.

The question of ignorance has been so much exploited in quantum mechanics and has created so much confusion that it might be helpful to consider a classical situation where ignorance genuinely matters.

The Classical Fruits of Ignorance

Jane lives in Wales, but her husband John, aged ninety-eight, is in New York. Suppose that in New York the life expectancy of a man of this age is six months, this meaning that half the number of people of this age die in this interval. Jane telephones her husband on 1 January and finds that he is alive. Afterward, she does not hear from John for a few months, and in late May she begins to worry about him; and because she knows the statistics, she reckons that John is as likely to be dead as to be alive. Jane is very rational, and, being aware that her husband has a one in two chance of being dead, she thinks half the time about him as alive and half the time as dead. Of course, Jane never thinks of her husband as being in a superposition of the dead and alive states because she knows that there is nothing potential about the probabilities with which she reckons: her husband is a member of a sample, the elements of which, at that particular time, are not all in an identical state, half of them being dead and the other half alive. On 25 May Jane telephones her husband and she hears that John is alive and well. During the next few days Jane thinks of her husband as alive, but on 30 May at 9:00 A.M. she receives a letter. When she opens the letter, she discovers that her husband had died on 28 May at 3:00 P.M. I shall now propose a couple of questions which are if not ridiculous at least obvious, but the reader will find in the next section that almost all of them, in a quantum context, were supposed to enshrine extraordinarily deep ideas.

First question. As far as Jane's knowledge went, Joe was still alive at 8:59 A.M. on 30 May, but he was dead at 9:00 A.M., when she opened the letter. Since this is a sequence of events entirely in Jane's mind, owing to her state of ignorance, can it be claimed that there is some relation between Jane's reading the letter and her husband becoming a dead person? This is of course complete nonsense. The second question, instead, is not as obvious as it appears: at what time did Jane become a widow? There are legal problems, such as powers of attorney conditional on the life of John and exercised by Jane after 3:00 P.M. on 28 May, which would in effect require a decision on this question, and which might have to be settled in a court of law. If we want to be pedantic, however, we would say that Jane became an ontological widow at 3:00 P.M. on 28 May, and an epistemological one at 9:00 A.M. on 30 May.

Be it as it may, classical effects of ignorance are not serious, but much has been claimed for this being the case in quantum mechanics. There are some problems as regards the use of the superposition principle when we want to apply it in its quantal form to macroscopic bodies or, even worse, to systems that have to be treated partly quantum mechanically and partly classically. Such problems as might exist have forcibly been brought to the attention of the public by Schrödinger's feline companion, to be discussed in the next section.

SCHRÖDINGER'S CHESHIRE CAT

As all the world knows, the Cheshire cat had the amazing faculty of vanishing while leaving behind nothing more than its grin; I thus hope that the reader will not be too puzzled if no cat of any description will be visible on these pages. Obviously, neither Schrödinger nor we were or are interested in cats, at least qua theorists. What Schrödinger wanted to do, *franc-tireur* as he was, was to discredit the superposition principle as a description of a state and, incidentally, to create terror and despondency about the use of quantum mechanics. All very creditable because without vociferous objections scientific ideas can never become adequately entrenched. Schrödinger was of course amazingly clever in bringing a cat as an archetype of a macroscopic object into his imagined experiment because this made his case far more poignant. But it also made it a little more emotional and unphysical, so that I shall prefer to present Schrödinger's case to my readers in a catless form, preserving nevertheless all the features needed for the discussion.

I shall first describe the experimental set up that we shall need, which is very simple (fig. 1). I have an analyzer like the one depicted in chapter 16, figure 4 (p. 488) in which **a** and **a'** are the (perpendicular) directions of polarization of the two emerging channels. I have a single incident photon, with polarization vector **u** at precisely 45° to the direction of polarization of the channels. This means (see note 10, p. 506) that the probability of the photon emerging through either channel is precisely 1/2. Therefore, if I use the superposition principle, the wave function of the photon before it hits the analyzer is of the form

$$\psi(\mathbf{u}) = a\psi(\mathbf{a}) + b\psi(\mathbf{a'}),$$

where the square of the coefficients a and b, which give the probability of the photon emerging in each corresponding channel, is 1/2.

We place on each channel a photon detector, and the two detectors are connected to a single digital counter which displays "0" or "1" when the detector on **a** or **a'**, respectively, has registered a photon. (I assume that the photons arrive on the analyzer one at a time, and that the beam is so sparse that, however long the experiment, there is never more than one single photon involved.) The analyzer and the two photon detectors are inside a black box with a hole through which the polarized photon gets in. As you will see in a moment, this is not quite, but almost, what Schrödinger had in mind, cat or no cat. But to understand precisely what Schrödinger was about, it will be very useful to consider first this case of the *almost* Schrödinger Cheshire cat.

In the region between the analyzer and the detectors, the photon can

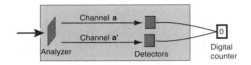

Fig. 1. Schrödinger's Cheshire cat (Almost)

The digital counter is represented showing a "0," corresponding to the photon in channel **a**. If the active channel is **a'**, instead, the counter displays "1."

be in either one of two states, corresponding to the two possible channels, and we shall simply label these states with the letters **a** or **a'**, as $\psi(\mathbf{a})$ or $\psi(\mathbf{a'})$ respectively. By the superposition principle, we write the photon state as

$$\psi_{photon} = a\,\psi(\mathbf{a}) + b\,\psi(\mathbf{a'}),$$

where, as before, a^2 and b^2 equal precisely $1/2$. This means the following: when I measure the photon state (which I do by reading the digital counter) the probability of getting in the counter either 0, for the state $\psi(\mathbf{a})$, or 1, for the state $\psi(\mathbf{a'})$ is precisely $1/2$. This is the meaning of the quantum mechanical description of the pure state of the photon in the situation described. I am not saying that the photon is in either of the two states (that is, in either of the two channels), but that its state is described by means of those two states. (This is important and please remember: I am not allowed to think of the photon in that state as being in either $\psi(\mathbf{a})$ or $\psi(\mathbf{a'})$ because I have no means of knowing which is the active channel until the measurement is completed.) Since I am invoking potential probabilities, in order to check this statement, I have to repeat the basic experiment on injecting into the black box an assemblage of photons, all identically polarized. I shall not be able to predict the outcome for each photon: this will be 0 or 1 at random. But for the assemblage, I shall find that both these entries will appear in equal numbers.

I am of course interpreting the phenomenon within the Copenhagen picture, so that already Schrödinger would not have been entirely happy. But never mind him, he will have his say in a moment. Before we allow him to worry us, it is important to reflect on which is the *history* that we are dealing with. This is very simple; it is all in the black box, that is, it starts with the photon at the pinhole and ends with it at one of the detectors. This is as it should be, the history starting and ending with a measurement on the wavicle (or some form of localization of it), and it is this history which is represented by the given wave function.

Bohr, quite wisely, given the limited means at his disposal at the time when he had to fight dragons, never allowed classical histories to be

mixed with quantal ones. What I mean by this is that he always considered that the measuring instruments must be treated classically.[36] In our
example, this refers to the digital counter, since it is this that registers and
thus finalizes the measurement, and it is for this reason that I have put
the counter *outside the box*. Notice that I have separated things neatly:
whatever is quantum is inside the box: this is where the quantum history
begins and ends. Outside the box is all classical.

Now Schrödinger comes to spoil the fun, even if there is not a cat in
sight within a mile. He insists that we must put the counter *inside* the
black box (fig. 2). This is not daft. What Schrödinger is saying is that, if
quantum mechanics is any good, it should be able to deal also with classical systems. Each elementary particle of the counter must obey
quantum laws, and thus the whole counter must be treatable by means of
some wave function, however horrible. Although Bohr did not consider
it necessary to worry about such cases, Schrödinger was right in trying
this out, since in modern physics one must be able to deal even with large
systems by quantum mechanical means. (Cosmologists, for instance, like
to think of the wave function of the whole universe!)

Let me denote with $\psi(0)$ the combined wave function for the photon
going through channel **a** and the counter being in the state "0," and similarly for the state function $\psi(1)$. (These functions entail, of course, the
states of the corresponding detectors.) The combined system, by the
superposition principle, will have a wave function

$$\psi_{photon\ +\ counter} = a\,\psi(0) + b\,\psi(1),$$

with the same values of a and b as before. Notice, however, that the history corresponding to this phenomenon has not been closed, *because the
digital counter has not been measured, that is, its reading has not been registered.*
I can close the history in various ways. To please Schrödinger, let us
assume that the black box has a lid and that we finish the experiment, that
is, complete the phenomenon, by lifting the lid and looking at the counter.
Now comes the Schrödinger rub: until I do that, the counter is neither in
state "0" nor in state "1," but it is in a state that is described in terms of
those two states by the state function shown above. This, Schrödinger
says, is nonsense, because even if we have not observed the counter, we
know that it must be in one or the other of the two states, so that the
quantum mechanical description clashes with "reality."[37] No one can conceive of a macroscopic counter which is not showing "0" or "1" but that
is partly in either state, just as Jane could not think of John in a superposition state of being dead or alive. But there is a great difference with the
poor widow. She did not set up a quantum mechanical wave function for
her husband, whereas this is precisely what we are now doing for

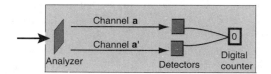

Fig. 2. Schrödinger's Cheshire cat (The real, albeit catless thing)

everything inside the black box, *including* the macroscopic counter. What is even worse, we pretend that the mere positioning of the counter inside the box will change its behavior from that of figure 1 (where it was classical) and, therefore, that we must prudently consider the significance of a quantum mechanical superposition state for it. (The reader may begin to smell a rat here, but no cats yet: let us relax and play Schrödinger's bizarre game, since something useful will come out at the end.)

A great deal of mystification has been created in this and similar examples by the apparent importance of a human observer in doing such innocent things as here lifting the lid and peering inside. (This is very much the same as in my previous story of Jane reading the letter from New York and, just as in that case, any conception that this action might be significant in determining the event studied is highly suspicious: but remember that we are playing an entirely quantum mechanical game now, even if we know that the counter weighs a ton.) Contrary to much opinion in this respect, I take the view that all you want here is to complete the history properly, and whether the registration is done by human or computer intervention is irrelevant. In most practical cases computers are used for this purpose, and in figure 2 we could perfectly well assume that we have a lead coming out of the box connecting the counter to a secure computer, that is, one in which I shall not be able to output the data until a password is keyed in. If this is so, Schrödinger would claim that until the password is entered and the output read (which will be simply 1 or 0), the use of the superposition principle would force us to accept that the counter is in the limbo state of neither displaying 0 nor displaying 1.

Let us reflect about Schrödinger's claim. The counter is in the quantum mechanical state given by our last displayed wave function until a measurement is completed. When this is so, the wave function is *reduced*, and the system jumps at random into one or the other of the two possible eigenstates, $\psi(0)$ or $\psi(1)$. However, once this happens, all that remains is a classical history: the photon and with it the trail of the superposition wave function is gone. But, because the whole thing happens inside the black box, Schrödinger forces us to extend the use of the quantum state well past its sell-by date, which is the instant when the wave function reduction was completed. At that moment a classical history starts, and I should be able to apply the principles of continuity of classical physics to the state of the counter. The fact that I am not reading it is, of course, classically irrelevant from the point of view of my certainty that, whether observed or not, the counter must read either 0 or 1, just as I know that the moon is there even when I do not look at it. This is what you do in the laboratory in millions of cases. We can see what happens here going back to the computer output. This is, of course, registered by the computer the moment that the wave function is reduced, that is, the moment the counter shows a reading. Computers habitually enter not only the data captured but also the time at which this has happened. If before entering the password in the computer I go for a two-week vacation to the South of France, I am most unlikely to entertain thoughts of the counter being in a limbo superpositional state. I know that when I come back to the laboratory and I read the output, I shall know exactly the precise time at and after which the counter read 0, say.

The reader may think that I have fully answered Schrödinger's case, but this is not quite so. All I have done is isolate its tender point. What I have said is that, whether I put the counter inside the black box or not, whatever happens in figure 2 must be what happened in figure 1, that is, that the photon wave function reduces the moment the photon hits a detector. But this is not the whole story. What Schrödinger wants us to prove is that *the combined photon-counter wave function also reduces instantaneously at that moment*. That it reduces, we know experimentally from figure 1, but why this reduction takes place, we do not know. If we can prove why this reduction takes place and how, then, whether we operate in the way of figure 1 or of figure 2 cannot make a scrap of difference. But we must show that even for this combined function, the reduction will take place.

This is a difficult problem because, as I already advanced on p. 492, the reduction of the wave function does not fall easily within quantum theory. The problem is that the Schrödinger equation with the time is deterministic, so that the wave function is happily evolving in accordance to it until it hits the detector, whereupon a nondeterministic random process takes place, and this process is not explainable by the use of that deterministic equation. This is what we have to investigate if we want to be able to cohabit adequately with Schrödinger and his cats.

But, I repeat, this is the only problem with Schrödinger, cat or no cat. If the existence of this reduction can be demonstrated for a macroscopic body like the detector plus counter (which experimentally we know must be the case), then there is no question whatever that even inside the box we need not consider the possibility of the counter ever being in a superposition state. This problem will be further discussed in chapter 19 (p. 560), when the saga of the Schrödinger cat will finally be completed.

Which reminds me that I must reluctantly reveal how cats come into this story. Instead of a counter, Schrödinger places in the black box a vial which, when the **a** detector registers a photon, explodes spreading hydro-cyanic acid onto a cat that has also been put in the box before the experiment starts. Thus the state "0" of the counter is "cat dead" whereas the state "1" is "cat alive." Hence, before the lid is opened, the cat is neither dead nor alive but in a superposition state, Schrödinger tongue-in-cheek claimed! As you can see, whether you cruelly use a cat this way, or humanitarially spend a couple of dollars on a digital counter, does not make a scrap of difference to the physics or to the philosophy of the problem. The amazing difference, however, is the impact that this thought experiment has had on the general public: Schrödinger created a veritable cat industry. I am absolutely sure that no one would ever have written a book called *The Schrödinger counter*![38]

CODA

I have tried in this chapter to give some account of the ideas and the work of the major contributors to the invention of quantum mechanics, but I left a serious gap in my brief account which I must now fill. It would be wrong not to mention in this chapter Dirac's outstanding contributions, which placed him easily among the top of the youngsters who were responsible for progress in the subject. Paul Adrien Maurice Dirac (1902–1984) was born and educated at Bristol, where he graduated in engineering. Fortunately, he went to Cambridge in 1923 where he heard about the work of Heisenberg and Schrödinger, and he immediately made important contributions to the mathematics of quantum mechanics. His lasting monument, made public in January 1928, was his discovery of the relativistic version of the Schrödinger equation. This was one of the major events of the century in physics, since it opened the way to an entirely new world of ideas. Dirac, for instance, predicted the existence of a particle like the electron but with a positive charge, which was later found and named the *positron*. Unfortunately, I could not discuss these developments because, at the simple mathematical level which I have maintained, I could not have done much more in this subject than deal with banalities.

The reader must notice that, of the various infirmities of quantum mechanics so far discussed, the only serious problem that remains is the reduction of the wave function, which will be discussed in chapter 19. On the other hand, Einstein's reluctance to accept quantum mechanics as a complete theory provoked a number of attacks on it, by him and by his followers, which we shall discuss in great detail in chapter 18. The theoretical and experimental treatment of the alleged incompleteness resulted, in actual fact, in the most extraordinary success for standard quantum theory, a success that has important philosophical implications for our present understanding of nature.

NOTES

1. Einstein, in a letter of 20 March 1954 quoted by Pais (1991), p. 178, wrote about Bohr: "He utters his opinions as one perpetually groping and never like one who believes he is in possession of definite truth."

2. It has been claimed that he was much influenced by Hegel and specially Kierkegaard (see Faye et al. 1994, p. 144), but though the Danish philosopher is from time to time a very good read, his approach is not the obvious best introduction toward a philosophy of science. Pais (1991), pp. 423–24, who knew Bohr well, denies Kierkegaard's influence on Bohr, and implies that the latter had never read philosophers such as Hume or Kant, which would explain his philosophical naivete. He also reports (p. 421) that Bohr's philosophical pretensions were ridiculed by the Copenhagen philosophers.

3. Every writer on the philosophy and history of quantum mechanics owes a debt of gratitude to Max Jammer for his painstaking and perceptive record of the subject (Jammer 1974). Fortunately also, there is an excellent modern reformulation of the Copenhagen interpretation by Omnès (1994). See also Omnès (1999a), (1999b).

4. Pais (1991).

5. A very comprehensive review may be found in the book already mentioned by Jammer (1974), but chapter 12 of Pais (1986) contains a brief and very useful account of the development of the subject.

6. Heisenberg (1925).

7. Born (1962), p. 4.

8. Born et al. (1925); see also Born et al. (1926).

9. Biographical details on Schrödinger may be found in Moore (1989).

10. Schrödinger (1926).

11. The early work on this subject is described in detail in Jammer (1974), pp. 440–69.

12. Born (1926).

13. Einstein (1905a).

14. This lecture appears in Born (1962), and the reference to Einstein is given on p. 7. Pais 1986, p. 259, however, doubts this story. Notice that $\psi\psi^*$ is ψ^2 for ψ real.

15. See Jammer (1974), p. 441.

16. See Jammer (1974), p. 440. Notice that Einstein is not making the necessary distinction between ensemble and assemblage. (See p. 484.)

17. Edwin Crawford Kemble (Kemble 1937) and even von Neumann (although not explicitly) followed the statistical interpretation. (Jammer 1974, p. 443). Hartle (1968) claims that the quantum "state is not an objective property of an individual system, but is that information, obtained from a knowledge of how the system was prepared, which can be used for making predictions about future measurements. . . . The 'reduction of the wavepacket' does take place in the consciousness of the observer, not because of any unique physical process which takes place there, but only because the state is a construct of the observer and not an objective property of the physical system." This is not a view that I support in the text, however. Belinfante (1975), p. 8, asserts that state functions are properties of ensembles and the ensemble description is used also by d'Espagnat (1976), pp. 15–16, 21.

It is useful to understand that the Copenhagen interpretation means different things to different people. Ballentine (1970) claims that within the Copenhagen interpretation *"a pure state provides a complete and exhaustive description of an individual system."* But this is contested both by Stapp (1972) and by Peres (1993). The Copenhagen interpretation, in accordance to Peres, is entirely the opposite: it is pragmatic and does not accept the absolute or objective meaning of the state function. Stapp's interpretation of Copenhagen coincides with that of Peres but it is diametrically opposed to Ballentine. Peres, however, writes (p. 23) that there is "no real conflict between Ballentine and Stapp, except that one of them calls Copenhagen interpretation what the other considers as the exact opposite of the Copenhagen interpretation." I leave it to the reader to ponder over this.

18. See Jammer (1974), p. 43. Also, note 17.

19. Wigner (1983), p. 288.

20. Jammer (1974), p. 65. See also Pais (1991), pp. 304–306.

21. Bohr (1961), p. 73, requires that the word "phenomenon" be severely restricted in meaning, and that it be used only to refer to "observations obtained under circumstances, whose description includes an account of the whole experimental arrangement." Also (pp. 63–64): 'I warned especially against phrases, often found in the physical literature, such as "disturbing of phenomena by observation" or "creating physical attributes to atomic objects by measurement." Such phrases are . . . apt to cause confusion.' Finally (p. 73): "every atomic phenomenon is closed in the sense that its observation is based on registrations obtained by suitable amplification devices . . . the quantum mechanical formalism permits well-defined applications referring only to such closed phenomena."

22. Thus Heisenberg (1930), p. 63: "To co-ordinate a definite cause to a definite effect has sense only when both can be observed without introducing a foreign element disturbing their interrelation. The law of causality, because of its very nature, can only be defined for isolated systems, and in atomic physics even approximately isolated systems cannot be observed."

23. Heisenberg (1955).

24. Quoted by Pais (1991), p. 192.

25. Stated in a letter to Max Born 4 December 1926 (Einstein 1971, p. 91), although what he says is: "I, at any rate, am convinced that *He* ['the "old one" '] is not playing at dice." Other references to dice-playing also appear on pp. 149 an 199. This dictum is also quoted as *"Gott würfelt nicht."*

26. Jammer (1974), p. 189.

27. Heisenberg (1927). I have used the more reliable translation of this sentence in Pais (1991), p. 306, rather than the one in Wheeler et al. (1983), p. 83.

28. In Popper (1967) he proposes to measure the two positions of a particle at two times. From the positions and the time difference you can work out the velocity and then the momentum "with an accuracy as great as you like." For this to be true, the time interval must be sufficiently small, and continuity must exist, to permit a set of successive values of the average velocity to converge to a precise value for a very small time interval. This is what Newton assumed, as discussed in chapter 6, but continuity is or is not a property of nature: it is contingent, and anyone who claims it has to provide evidence to that effect. In Newton's case, the evidence that continuity may be used is overwhelming. Not only is such evidence lacking in the microworld, but to expect continuity in it is almost absurd A rock may be treated as a continuous body, but already when it is crushed into sand, continuity (literally) breaks down. What can you expect when you go into the much smaller constituents of the sand grain? In any case, it defies imagination to see Popper using in quantum mechanics *exactly* the same procedures used by Newton in classical mechanics, which anyone who has read my modest chapter 6 would recognize as such.

29. Popper (1967). See also Popper (1982), in particular p. 148, where the argument given here in the text is substantially repeated. Popper's views were strongly attacked by his former

collaborator Paul Feyerabend (1968, 1969), while providing a very useful (and sympathetic) account of Bohr's ideas. For Popper's "realism" see for instance, Gibbins (1987), pp. 80–81.

30. A very clear critique of Bohr's ideas is given in chapter 4 of Gibbins (1987).

31. Pais (1991), p. 311. The lecture was published in Bohr (1928).

32. Pais (1991), pp. 426–27 and 23, respectively.

33. For an authoritative statement of Bohm's ideas, his posthumous book Bohm et al. (1993) is the best source. Already its title gives the game away, and chapter 2, called "Ontological versus epistemological interpretations of quantum theory" shows how Bohr left open a fatal trap for people to fall into: the dangerous idea that his interpretation of quantum mechanics ruled out an ontology.

34. See Jammer (1974), p. 121.

35. See Pais (1991), pp. 316 and 427, respectively, for these two conferences.

36. Bohr (1961), p. 50: "the *measuring instruments* . . . serve to define, in classical terms, the conditions under which the phenomena appear."

37. The reader should recall my remark about why we know that the moon is there when we do not observe it: it is because for macroscopic bodies the principle of continuity can be used to describe a trajectory. Schrödinger's clever ruse is to force us to abandon that principle by coupling the macroscopic counter with the microscopic photon.

38. The cat jumped into the world in Schrödinger (1935). Instead of the polarized photon beam, Schrödinger had a lump of radioactive material and a detector for the emitted radiation. Lockwood (1989), p. 198 claims that Schrödinger understands the superposition principle in the *ignorance interpretation*. This, applied to the cat, would lead to an absurd conclusion because we cannot conceive of the cat as partly alive and partly dead. But, because this macroscopic absurdity comes from the quantum mechanical description of radioactive decay, then Schrödinger claimed that this description must be regarded as suspect or at least incomplete, as Einstein, Podolsky, and Rosen had done in a paper that we shall discuss in the next chapter. (Schrödinger's paper was a sequel to that one: thus far reached Einstein's hand!)

IS THE QUANTAL DESCRIPTION OF REALITY COMPLETE?

Einstein attacks quantum theory

New York Times, 4 May 1935, heading on p. 11.

First, and those of us who are inspired by Einstein would like this best, quantum mechanics may be wrong in sufficiently critical situations. Perhaps Nature is not so queer as quantum mechanics.

John Bell (1987*b*), p. 154.

results must be reported without somebody saying what they would like the results to have been.

Richard Feynman (1992), p. 148.

The reader would do well in comparing John Bell's desires with the sober rule of conduct suggested by Feynman. Feynman knew only too well what had to be said, because the interpretation of quantum mechanics has been obscured by charismatic but prophetic figures, who allowed their instincts to prevail over their powerful minds. Einstein was the head prophet, and David Bohm and John Bell his most distinguished apostles.

The story I shall tell in this chapter is astonishing. First, Einstein and some of his collaborators at Princeton produced a "counterexample" which was taken, especially by philosophers, to signify that some flaw had been found in quantum theory; consequently, this work was quite inappropriately referred to as the "EPR paradox," after the names of its authors, Albert Einstein, Boris Podolsky, and Nathan Rosen. This, however, was *not a counterexample*, as Einstein himself knew. In a counterexample of a theory, the principles of the theory must be used to derive a wrong result, whereas EPR arrived at a result that *they* considered unacceptable, but *that required the use of principles not valid within standard quantum theory*. This, however, did not deter anybody from trying to go further from the thought "experiment" discussed by EPR, to one that could be done in the laboratory, to prove once and for all that the quantum theory was, if not wrong, at least seriously incomplete.

David Bohm was the first to propose such an experiment, and then John Bell did a marvelous piece of mathematical analysis, arriving at some inequalities which, if disobeyed experimentally, would have led to the breakdown of quantum mechanics. The experiments were eventually done with exquisite attention to detail and extraordinary accuracy; and the expected breakdown of quantum mechanics did not take place. (Having expressed before lukewarm feelings about Popper's falsifiability, I cannot find a more perfect example in history of an attempt to put this concept into effect.) Such a resounding victory for quantum theory would, you think, have silenced the opposition. But not at all. Bell, Bohm, and many others continued with their dark mutterings that nature could not be what nature was in accordance to the maliferous theory. And here we now are, still looking for the holy grail. But this will be the story of the next chapter.

THE EINSTEIN, PODOLSKY, AND ROSEN SO-CALLED PARADOX

I shall first discuss the main ideas concerning this thought experiment. We start with two identical "particles" (I shall continue with the illegal use of this word), say two electrons, which, either have been created simultaneously by some process, or have been allowed to interact for a time, after which they fly away in separate directions. Each particle, of course, will have its own momentum and position at any one time, which will be denoted with p_1, x_1, p_2, x_2, respectively. (Remember that bold letters indicate the operators corresponding to the given variables.) Because the particles form a combined system as a result of their initial interaction, it is appropriate to consider their combined total momentum P and their distance X:

$$P = p_1 + p_2, \quad X = x_1 - x_2.$$

These two operators commute, as it is very easy to prove:[1]

$$P X = X P.$$

It is important to observe, however, that x_1 does not commute with P: for this to be the case it would have to commute with each of the two operators in it, whereas it does not commute with p_1 (see p. 472).

Before I go any further, I must mention a result of quantum theory that is as important as it is counterintuitive. Because P and X commute, they can have simultaneous, precise values, which you might think could be measured by simultaneously determining p_1, x_1, p_2, and x_2. This is not so, for the simple reason that p_1 and x_1, which do not commute, cannot be measured simultaneously. Thus, we must accept that the precise values which the eigenvalues of P and X must assume (because of their commutation) must be measurable by some direct process, not involving measuring the momenta and positions of the individual particles. However counterintuitive this result is (and I suppose that it is for this reason that it is not mentioned by Einstein et al. 1935), it is a crucial consequence of the basic principles of quantum mechanics.

So, we assume that at a certain initial time we have measured P and X. I can now start with the reasoning of Einstein, Podolsky, and Rosen (famously known as EPR). They first engage in a bit of practical ontology, which appears to be so plausible that they enunciate it as some universal rule of nature:

> Every element of the physical reality must have a counterpart in the physical theory. *We shall call this the condition of completeness.* . . . If, without in any way disturbing a system, we can predict with certainty (i.e. with probability equal to unity) the value of a physical quantity, then there exists an element of physical reality corresponding to this physical quantity.[2]

I shall try to carry on within the EPR spirit, so that the argument of their paper can be understood, but I must warn the reader not to be too impressed by the above words of wisdom: notice that you are mixing here a nomological rule with a point-of-fact requirement, which is that of *not in any way disturbing the system*. Thus, the apparent universality and necessity with which the rule is given is seriously impaired by the contingent requirement that the system be assumed not to be in any way disturbed. This is a point of fact, the validity of which in an actual experiment must be left to the experimental result, rather than be posited beforehand. But let us be philosophically naive and follow EPR. Here they come.

As I have said, they start by measuring P and X. Then they measure p_1 (the eigenvalue of p_1) whereupon, from the known eigenvalue P of P, p_2 becomes known "without having disturbed" particle 2. Therefore (in their language), p_2 is an element of reality. Next, they measure x_1 and from the eigenvalue of X, x_2 follows, again "without having disturbed" particle 2. Therefore, x_2 is also an element of reality; thus, both p_2 and x_2 are elements of reality. But quantum mechanics tells us that p_2 and x_2 cannot simultaneously be elements of reality because of the noncommutation of the momentum and position operators of a given particle. Therefore, quantum mechanics is incomplete.

In order to understand what EPR were about, we must consider the previous paragraph in two different ways. If we analyze it from the point of view of quantum theory, it is, of course, unmitigated garbage. The reason is that, as I have shown before, x_1 does not commute with P. This means that, when x_1 is measured in the second EPR step, the eigenvalue P of P changes, whence the value of p_2 derived from P in the first step is no longer valid. The thought experiment, therefore, cannot reveal any contradiction of the rules of quantum mechanics because *those rules are disregarded*.

The reader must appreciate that EPR assume that their condition *"without disturbing"* is one that *they* can decide when it is satisfied, whereas the quantum mechanical commutation rules entail perturbations which they choose either to ignore or to declare unacceptable.

Einstein and his collaborators were of course, clearly aware of this difficulty. Thus, they write:[3]

> We are thus forced to conclude that the quantum-mechanical description of physical reality given by wave functions is not complete.
> One could object to this conclusion on the grounds that our criterion of reality is not sufficiently restrictive. Indeed, one would not arrive at our conclusion if one insisted that two or more physical quantities can be regarded as simultaneous elements of reality *only when they can be simultaneously measured or predicted*. On this point of view, . . . [p_1 and x_2] . . . are not simultaneously real. This makes the reality of . . . [p_1 and x_2] . . . depend upon the process of measurement carried out on the first system, which does not disturb the second system in any way. No reasonable definition of reality could be expected to permit this.

What they are saying here is basically this. If we assume that quantum mechanics is correct, it follows that measurements on particle 1 (which is all that we do in the EPR thought experiment) affect the values of variables of the quite separate particle 2; this entails an *action at a distance* which *they* rule out as impossible. This is the major point that arises from this work, the elucidation of which induced a great deal of both theoretical and experimental work for several decades. Finally, nature spoke and determined that the philosophical necessities invented by Einstein and colleagues were false. But I shall refer to this, and to the question of action at a distance, later in this chapter.

The EPR paper appeared in print on 15 May 1935, and Bohr became very nearly hysterical on its account. As always, he soon recovered and sent a letter to *Nature* on 29 June and then a main paper to *The Physical Review* on 13 July.[4] I shall not try to summarize his arguments, which I find unnecessarily obscure. But if one reads his papers very carefully, the one important weakness of the EPR argument that he discovered is that they rely on the immediacy of the concept of not "in any way disturbing a system," whereas the measurement of x_1, which does not commute with P disturbs the value of p_2 previously derived from the now disturbed value of P. Hence, EPR's contention that measurements on system 1 do not in any way disturb system 2 is not, as a point of fact, correct. That this disturbance entails an apparent action at a distance does not worry Bohr: nature must be relied on to do its own business. On the other hand, this argument shows that the apparently beautiful definition of *physical reality* given by EPR in the first displayed quotation, depends on an *arbitrary* definition of what a disturbance is or is not, rather than entailing a true necessity.

Bohr also objected to EPR's view of "reality." They were still stuck with the concept that there is something "out there" which "exists" independently of any procedure whereby that "existence" can be ascertained. He strengthened his views by moving the emphasis from "measurement" to "phenomenon." The second concept is the one that for Bohr licenses the use of the word "exists." A phenomenon can be a "measurement" as long as all the necessary conditions for the latter are added to the phenomenon itself.[5]

Was EPR a waste of time? Not at all. Although their general conclusion as to the incompleteness of quantum mechanics cannot be sustained, their emphasis on the appearance of action at a distance, inherent in the Copenhagen interpretation, was very useful, even if their distrust turned out eventually to be wrong. Indeed, one of the most extraordinary features of nature found last century, namely that action at a distance must be accepted, resulted from careful attempts to sustain Einstein's contrary views. Nature is *nonlocal* in the sense that, under certain conditions, action taken at one point can have instantaneous consequences on properties defined at another point that can be enormously distant. We shall in fact see that experimental work of the highest degree of precision shows that this is the case. It should be noticed, however, that what we could call the EPR effect (rather that the misnomer of EPR paradox sometimes used) relates to *entangled particles*. In our case, for instance, the particles 1 and 2 are *entangled* through the original interaction that permits the definition of the commuting observables P and X.[6]

I shall now start the work necessary to understand the very important results that proved the EPR doctrine to be experimentally false.

AN INTRODUCTION TO THE NEW RESULTS
ABOUT ACTION AT A DISTANCE

The EPR program was an attempt to show that the description of a state in quantum mechanics was incomplete because it led to a "paradoxical" action at a distance. Ironically, EPR's criticism, based on a thought experiment, led to proposals of experiments that could actually be performed and that would test the EPR hypothesis; and this hypothesis was then proved wrong beyond experimental doubt: action at a distance is in certain conditions (for entangled systems) a feature of nature, and it is thus not paradoxical. EPR and their followers had labeled it "paradoxical" because they expected microscopic nature to obey the same rules that had been well entrenched in macroscopic physics. It must nevertheless be realized that the action at a distance that we shall find is one of the most puzzling features of nature ever discovered. Thus, although these results entirely agreed with orthodox quantum mechanics, such is their significance that they encouraged the expectation that a new quantum mechanics might one day emerge, just as relativistic mechanics did replace Newtonian mechanics. It must be stressed that at present, however, no proposed form of quantum mechanics exists that eliminates the requirement of some form of action at a distance, thus "completing" the quantum mechanical description of nature in the way propounded by Einstein. But more about this later.

The experiments that we shall describe use some properties of polarized photons and polarization analyzers (see chapter 16) which we shall first recall, as well as their interpretation in quantum mechanics within the Copenhagen approach.

Polarized Photons and the Copenhagen and Hidden-Variables Interpretations

I shall first review briefly the relevant features of polarized beams of photons and their description in quantum mechanics within the Copenhagen interpretation, which were discussed in chapter 16. (See fig. 6, p. 490.) Consider a photon corresponding to a light wave polarized along the direction **u**. (This vector denotes the direction of the electric vector of the light wave.) As shown in figure 1, this photon impinges on an analyzer plate, which has the property of allowing any incident light wave to emerge in one of two well-separated output channels polarized in both cases, respectively, along the two perpendicular directions shown by **a** and **a'** in the figure. The proportion of the total number of photons of the incident wave that emerge through the **a** and **a'** channels will be denoted as $n(\mathbf{ua})$ and $n(\mathbf{ua'})$, respectively, and, as discussed in chapter 16, p. 489, they are

$$n(\mathbf{ua}) = \cos^2\{\mathbf{ua}\},$$
$$n(\mathbf{ua'}) = \cos^2\{\mathbf{ua'}\}.$$

Notice that the parentheses (**ua**) and (**ua'**) on these equations *are not angles*: they merely denote pairs of parameters. Angles, on the other hand,

Fig. 1. Polarized light and an analyzer

The vector **u** is the incident beam, and the beams polarized in the **a** and **a'** directions emerge from the analyzer in well-separated channels.

are denoted with braces. The Copenhagen interpretation of these results is as follows. (i) All photons of the light beam are in the *same* physical state: there is *no way* of distinguishing between them, even when we know that they behave differently once they hit the analyzer. (ii) When the identical photons hit the analyzer, they exit *at random* through the **a** or **a'** channel. (iii) Although this is a random, nondeterministic process, the probabilities of the outcomes are well defined, and given by the two equations shown.[7]

It is extremely natural to try to describe the polarization results in an alternative way on introducing so-called hidden variables by means of the following assumptions. (i) Although all the photons of the beam are in the same quantum mechanical state, this state description is incomplete. The photons, it is assumed, possess a variable A which can take one of two values, let us say +1 and -1, that *determine* whether the corresponding photons exit through channel **a** or **a'**, respectively. These variables are not accessible (at present, but perhaps in the future) to experimental observation and are thus called *hidden variables*. Within this interpretation, we must assume that the photons, so far considered to be all in identical states, separate out into two types, say a and a'. Photons of type a have a value +1 of the hidden variable A, which determines them to exit through channel **a**, and likewise those of type a', with value -1 of A, must exit through channel **a'**. (ii) The photon beam is an *actual mixture* of photons of these two types, and the *proportion* of each type in the mixture is given by precisely the same numbers $n(\mathbf{ua})$ and $n(\mathbf{ua'})$ in the previous equations.

These very simple hypotheses thus explain the experimental results without having to invoke a random process in the interaction between the photon and the analyzer. I have already (p. 495) given arguments against the use of hidden variables, but I shall now describe a number of experiments that will allow us, eventually, to discriminate between the Copenhagen and the hidden-variables interpretation.

The First Experiment: Two Parallel Analyzers

All the experiments that follow entail the use of a very special type of light source, which produces *pairs* of photons that propagate in opposite directions but that are both in *identical*[8] polarization states. I must stress that the stated identity of the polarization states is a stronger condition than saying, for instance, that two photons belonging to the same polarized beam are in the same state of polarization. The pairs of photons coming out of the source are, in fact, said to be *entangled*, which means in our case that they behave identically (albeit randomly) in polarization experiments, whereas any two photons of an ordinary beam exhibit *individual* random behavior. (This entanglement is a consequence of some perfectly well established results of quantum mechanics, and that such sources can be constructed is in fact a successful prediction of the theory.)

The first experimental setup that I shall consider consists of a light source as just described, equidistant between two analyzers oriented in a parallel way, as shown in figure 2. Each beam carries one of the photons of each pair produced by the source, and the beams are polarized, of course, as follows from our description of the source, along the same direction, which we shall call **u**.

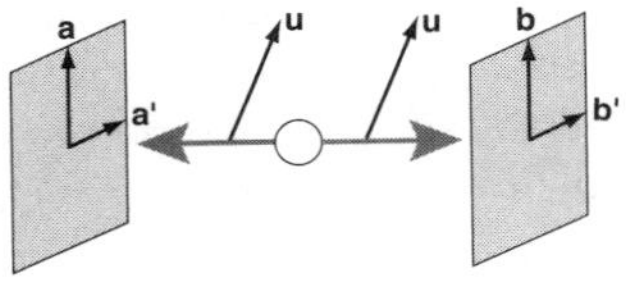

Fig. 2. The parallel analyzers experiment

Photons belonging to the same emitted pair are easily recognized, because they are emitted within an interval of a few nanoseconds, which is well within the resolution time of the photon counters that count the photons emerging in all four channels. We know that when the left photon hits the left analyzer, an output will be registered *at random* along either channel **a** or **a'**. Because of this randomness, one would expect not to find any relation at all between the results registered by the two different analyzers. Instead, the following extremely puzzling result is obtained, which is characteristic of entanglement:

> *Whenever the left-hand photon emerges through channel **a**, the right-hand photon emerges through the parallel channel **b**, and the same happens for channels **a'** and **b'**.*

As we shall see in a moment, this experimental result has very surprising consequences, which we shall now discuss. There are two alternative interpretations that we can use.

Copenhagen Interpretation

The left and right photons, although in the same state, belong to a state of polarization which, as we have discussed in chapter 16, is neither parallel to **a** nor to **a'**. These states of polarization are only realized when the photons hit the analyzers. When the left photon jumps at random into the state **a** on hitting the analyzer, then the experimental result mentioned shows that the right-hand photon must also jump, albeit at random, to precisely the parallel state **b**. We are not allowed to assume that the right-hand photon "remembers" the state of polarization of its sibling (not possible, because this was not specified, further than the orientation **u**, when the pair was created, and because in this interpretation a hidden variable that carries the identical information for both photons is not permitted). This means that whatever happens when the left photon hits its analyzer and jumps[9] to the **a** state, must somehow be transmitted to the right-hand photon and cause it to behave appropriately. This is *action at a distance*. Although we met action at a distance when Newton postulated the gravitational force, the present one is action at a distance with a vengeance, since *it does not depend on the separation between the two analyzers*, as experiments confirm. Moreover, it is *instantaneous*, whereas we have seen that when Newton's action at a distance was made honest by relativity theory, the gravitational effect was found to propagate with a finite speed, namely that of light. On the other hand, remember that this form of action at a distance is very specific since it is only found within a pair of specially prepared *entangled* systems.

Another Counterfactual Infirmity

It is worthwhile noting here a counterfactual infirmity opposite to that encountered on p. 467. If we have two entangled photons, say 1 and 2, and we measure the polarization state of photon 1 as, say, parallel to **a**, then the counterfactual (strictly speaking, a *conditional subjunctive*): *If we were to measure the polarization state of photon 2, it would be parallel to **a***, is necessarily true! However, as I have already warned, the use of counterfactuals in quantum mechanics is not recommended.

The Hidden-Variable Interpretation

The above description is so counterintuitive that it invites the acceptance of a hidden-variable interpretation: we assume that although different pairs of photons may have different types, photons of the same pair must either both be of type *a* and thus both emerge parallel to **a**, or both be type *a'* and thus emerge parallel to **a'**. End of story: no "nonsense" about action at a distance.

It is fairly clear that this experiment cannot discriminate between the two interpretations presented. On the other hand, good solid common sense tells us that the only sensible explanation of this strange result must be along the lines of the hidden-variables hypothesis. Alas, good solid common sense can sometimes be a dangerous tool: and this is one of those times. Because these concepts are so unusual, I feel that the moment has come for me to entertain my readers with a fable.

Interlude: The Story of Lady Mantenida and Lord Maltornato

I found this story in two different versions, registered in the library as Codex EPR and Codex Copenhagen, which I shall now summarize.

Codex EPR

The notorious procuress, Lady Eduina, tries to arrange assignations between two lovers, Lady Mantenida of Polenta and Lord Maltornato of Ravenna. She has for this purpose a large number of pairs of silk stockings, which she perfumes at random. The number is large, and she is quite good at randomizing, so that the chances of extracting a perfumed pair from her press is just as good as precisely fifty-fifty. She sends every day one stocking from the same pair to each lover, and she instructs her clients that when they both receive perfumed stockings on three consecutive days, Lady Mantenida must drug her husband, the Duke of Cornucopia, so that Lord Maltornato can rush from Ravenna to Polenta for the desired assignation. Because the lovers are very stupid, they follow this ploy, until they realize that they can save Lady Eduina's stiff fees by arranging to take the same actions on the first and third Mondays of each month.

This, I am afraid, is a rather dull tale, but notice that the lovers know every day that there is a fifty-fifty chance of a perfumed stocking being in the mail, and that the same chance holds, of course, for finding such a stocking when opening the parcel.

Codex Copenhagen

This must be an earlier version of the same story, with exactly the same three characters, but now Lady Eduina is presented as a powerful and famous witch. She still has the same press with the same stockings, but *all* of them are now perfumed. Before she sends the usual half-pair to each lover, Lady Eduina performs her magic, whereby she transforms the Chanel No. 5 that she had applied to the stockings into an odorless product. When Lady Mantenida opens her parcel, Eduina performs *occasionally* a second bit of magic, and restores the perfume everywhere to full fragrance. Being good at probabilitites, she does this at random and achieves an exact fifty-fifty chance of both lovers receiving odorous stockings.

As you can see, the lovers cannot notice any observable difference between this procedure and the one described in Codex EPR. There are, however, two conceptual differences. First, it does not make sense for them to say that there is a fifty-fifty chance of a perfumed stocking *being in the mail*, since they know full well that while the stockings are en route they are all equally odorless. Second, Lady Mantenida realizes that when she tries the fragrance of the newly opened parcel and Lady Eduina performs her perfume-restoring magic, then this latter action is *instantly* transmitted from Polenta to Ravenna where Lord Maltornato is conducting his olfactory activities. The only problem is that, because of the instantaneous action at a distance now involved, Lady Eduina's fees are even larger.

Question

Given the amorous proclivities of our lovers, and given the fact that it is a truth universally acknowledged that perfumed stockings were actually found from time to time by Lord Maltornato in Lady Eduina's parcels, the reader is asked to decide which account is more plausible. The answer must of course be that of Codex EPR. However: when experiments are performed in the same manner with photons, nature appears to require something like the intervention of Lady Eduina, and it is the second scenario that more adequately resembles the physical facts. Notice incidentally that in the second story, the stockings do have a hidden variable, this being the liquid that has come out of the perfume bottle and which the bewitched lovers cannot smell. But this variable is *nonlocal*: its value at one point (Lady Mantenida's, or rather the Duke of Cornucopia's palace at Polenta) *determines* its value at a distant point in Ravenna. This is precisely what the experiments we shall describe show: a *local* hidden-variable interpretation, in which the values of the hidden variables are unaffected by distant bodies, is incompatible with experience. Obviously, postulating nonlocal hidden variables does not help anyone who wishes to oppose the Copenhagen interpretation, so that the possibility of such variables is not one that is seriously entertained.

End of Tale

Fables are very nice, but they can also be misleading. The reader might think that there is no possible way in which Lady Mantenida and Lord Maltornato can know whether they live in the cozy world of Codex EPR or in the fantastic one of Codex Copenhagen. They cannot, because they are not very good experimentalists, but we can. What we must do is design a rather more involved experiment than the Polenta-Ravenna lovers performed. (Also, the question of action at a distance in the experiments will be somewhat different, of course.)

THE CRUCIAL EXPERIMENT:
TWO NONPARALLEL ANALYZERS. BELL'S INEQUALITY

John Bell had the marvelous idea of designing an experiment which could actually be performed, and which could discriminate between the two possible interpretations discussed. It entails, as before, having a pair of analyzers, but they are rotated in various ways, as shown in figure 3.

It will be easier to start by putting down some simple equations that should be valid when we have some form of a hidden variable, which we shall describe as before by means of photon types. Let us call:

$N(\mathbf{ab'})$ = *proportion of the total number of photons that belong to the appropriate types in order to pass through channels* $\mathbf{a}$ *and* $\mathbf{b'}$ (see the top of fig. 3).

I must make it clear that this number can perfectly well be interpreted as the proportion of the total number of photons that pass on the left through channel $\mathbf{a}$ and on the right through channel $\mathbf{b'}$. The only reason why I use the language of types is that in this way it will be clearer that we are thinking in terms of local hidden variables, of which more later.

Likewise,

$N(\mathbf{ab'c})$ = *proportion of the total number of photons that belong to the appropriate types in order to pass through the channels* $\mathbf{a}$, $\mathbf{b'}$, *and* $\mathbf{c}$ (shown in the two top parts of fig. 3).

We shall have similar notations for other channels. To make things easier, I shall first put down three relations and I shall then explain what

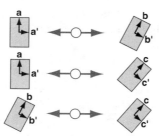

Fig. 3. Rotating analyzers

The analyzers are actually perpendicular to the photon beams represented with the arrows at the center, as it was shown in fig. 2.

they mean. The reader who is not very keen on formulas might be helped by a couple of comments. First, it is sufficient to remember that there are three pairs of channels: **a** and **a'**, **b** and **b'**, **c** and **c'**; although they are illustrated in the figure, it might make things easier not to use the latter at this stage. Second, the reader might puzzle how the left-hand side of the equations to be displayed has been chosen (their right-hand sides will be explained presently). The answer to this query is that this is a typical example of inductive thinking. Bell had in mind what he wanted, and he guessed what he had to put down here in order to get the type of relation that he required. There is nothing more to it, except that this is the type of thinking that requires quite a bit of genius, although once you see it done, it is easy to understand. So let us write down the equations first:

$$N(\mathbf{ab'}) = N(\mathbf{ab'c}) + N(\mathbf{ab'c'}),$$
$$N(\mathbf{bc'}) = N(\mathbf{abc'}) + N(\mathbf{a'bc'}),$$
$$N(\mathbf{ac'}) = N(\mathbf{abc'}) + N(\mathbf{ab'c'}),$$

The first equation says something pretty obvious (but *please* see later), namely that, given the types entailed for the photons corresponding to the channels mentioned on the left of this equation, if we add to our consideration photons that may go through the right-hand analyzer in the orientation corresponding to the channels **c** and **c'**, the photons counted in $N(\mathbf{ab'})$ can, either belong to the type that will result in them going through channel **c**, or that which will result in them going through channel **c'**. Likewise, the second equation results from the same consideration when adding channels **a** and **a'** to channels **b** and **c'**; and the third equation entails adding channels **b** and **b'** to the two channels listed on its left. It is easy to notice that the second term on the right-hand side of the first equation and the first such term on the second one are the two terms that appear on the right-hand side of the last equation for $N(\mathbf{ac'})$. Thus, on adding up the first two equations, the right-hand side will contain those two terms, that is $N(\mathbf{ac'})$, plus two *positive* additional terms. Therefore,

$$N(\mathbf{ab'}) + N(\mathbf{bc'}) \geq N(\mathbf{ac'}).$$

In order to be able to compare with the standard quantal results and with experiment, we have to sharpen our definitions a little. As you will see in a moment, all that we have to do is require that the proportion $N(\mathbf{ab'})$ be calculated for observations in which the counts along the channels **a** and **b'** correspond to the *same pair* of photons, and similarly for the other terms in our last equation. (That is both counters must click within a few nanoseconds of each other, as already explained.) To stress that this is the case, I shall substitute a lowercase n for the capital N in the last equation:

$$n(\mathbf{ab'}) + n(\mathbf{bc'}) \geq n(\mathbf{ac'}). \qquad \textbf{\textit{(Bell inequality)}}$$

This is the famous Bell inequality, but before we see what it does for us, let us go back to the three equations that we wrote for $N(\mathbf{ab'})$, $N(\mathbf{bc'})$, and $N(\mathbf{ac'})$.

The Crux of the Matter

The way in which the Bell inequality was derived is so straightforward that it is difficult to believe that the result might not be true, even if we were to remove such things as hidden variables and types from our language. *But this is not so.* In order to make this quite clear, we shall now discuss those three equations on the experimental assumption that $N(\mathbf{ab'})$, say, means the proportion of the total number of photons that have been counted as passing through the channels **a** and **b'**. Consider again the first of those equations:

$$N(\mathbf{ab'}) = N(\mathbf{ab'c}) + N(\mathbf{ab'c'}).$$

The constancy of **ab'** on both sides means that the aggregate from which the proportion $N(\mathbf{ab'})$ was measured, call it N, divides in two, say $N(\mathbf{c})$ and $N(\mathbf{c'})$. These two are mutually exclusive because photons with the type suitable for **c** can never exit through the perpendicular channel **c'**. If you look at the displayed equation, you may then think that there is nothing that could possibly go wrong, since photons must necessarily belong to one or the other of the two aggregates on the right. If you believe that this is self-evident, look back at the catalogue paradox of chapter 14, p. 395: we there had a case in which the neat division of a main set into two disjoint subsets breaks down. I am not suggesting that the infirmity of our equation is Russellian, but rather that Russell's paradox should alert us to avoid naivety as regards the properties of sets (or, correspondingly, of aggregates). A more significant problem behind the present example is the one that I have called in chapter 14 that of the *stability of aggregates condition* (p. 396). Suppose that the type of a photon that you add to an aggregate N depends on the number of photons already in it, so that photons of the type adequate for **c** can switch over to **c'** under certain conditions. This would make it impossible to guarantee that N divides neatly into $N(\mathbf{c})$ plus $N(\mathbf{c'})$ because the numbers entailed by these two aggregates would become meaningless. It must be appreciated that the alleged stability of the aggregates entailed in the equation displayed requires *locality*. If there is *nonlocality*, in fact, whatever choices that may have been made by the photons for the channels **a** and **b'** in the aggregate N, *may change* once the channels **c** and **c'** are open on the right-hand analyzer. The important meaning of this argument is that, despite its apparently innocent derivation, the Bell inequality is valid if and only if nonlocality does not exist. (It was of course because of this very feature that Bell invented these equations.)

Comparison with Quantum Mechanics

Because of our restriction to entangled pairs, it is very simple to obtain the terms that appear in the Bell inequality displayed previously. Consider $n(\mathbf{ab'})$. The problem is this: from figure 1 and the two equations on p. 540, we know how to calculate the proportion of photons that emerge through either channel of a *single* analyzer, but now we have *two*. Remember however that the count $n(\mathbf{ab'})$ refers to photons that always belong to the same entangled pair, in which the two members of the pair are in identical states of polarization. Thus the photon that emerged through the right-hand-side analyzer in channel **b'** must have approached this analyzer with a polarization direction parallel to **a**, since this was the polarization direction of its sibling. We can therefore construct a figure like figure 1 but now for the right-hand side analyzer, which we do in figure 4, where the vector **a** takes the place of the vector **u** in the former figure, and **a** and **a'** of figure 1 are replaced by **b** and **b'**, respectively. All you have to remember now, from the equations on page 540, is that for any pair of vector parameters (which are the incident and emergent polarization vectors, respectively), you have to form the $\cos^2$ of the angle between the two same vectors. You can therefore write immediately the terms required in the Bell inequality:

$$n(\mathbf{ab'}) = \cos^2\{\mathbf{ab'}\},$$
$$n(\mathbf{bc'}) = \cos^2\{\mathbf{bc'}\},$$
$$n(\mathbf{ac'}) = \cos^2\{\mathbf{ac'}\}.$$

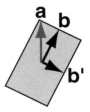

Fig. 4. Orientations for the right-hand-side analyzer

The letter **a** is the direction of polarization of the incident photon and **b** and **b'** are the analyzer's channels.

In order to test the Bell inequality, take the numerical values of the required angles shown in figure 5, which are chosen for convenience. (Any values would do!)

We enter these values in the last three equations, so that the Bell inequality now takes the form:

$$\cos^2 112.5 + \cos^2 112.5 \geq \cos^2 135.$$

It is sufficient to have a small hand calculator to find that the left-hand side here is 0.29, whereas the right-hand side is 0.50, so that quantum mechanics violates the Bell inequality. Therefore, either the experiments show that the Bell inequality is respected, in which case quantum mechanics is wrong, or they show that it is violated, in which case the hypothesis that local hidden variables exist (which is entailed in the Bell inequality) is wrong.

The Experiments

The experiments conducted by Alain Aspect and his collaborators (which I mentioned briefly in chapter 14) at Orsay in France demonstrate well

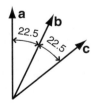

Fig. 5. Relative orientations of the analyzers

The vectors **b'** and **c'** are perpendicular to **b** and **c**, respectively, so that their angles to **a** must be incremented by 90° with respect to those in the figure.

beyond the margins of experimental errors that the Bell inequality is violated and that, instead, quantum mechanics reproduces precisely the experimental results. Thus, the quest started by EPR has resulted in the most clear demonstration possible that the conclusions that EPR tried to obtain are not tenable.

General Comments

I must first refine a little the question of the hidden variables. In trying to "explain" how it is that the photon randomly jumps, from the polarization state it has in the incident beam, to that pertaining to one of the analyzer channels, hidden variables have been tried. So far, we have attached them to the photons, but, alternatively, one could assume that they are associated with the analyzers, and that it is the interaction of their two sets of hidden variables that determines the outcome of the experiment. When this is done, however, nothing much changes in our discussion, except that one has to subsume the analyzer hidden variables into the hidden variable (which I called the type) of the photon.

A few details of the experimental set up must be mentioned. First, a problem arises because the photon counters are not 100 percent efficient, and the errors thereby introduced in the analysis would be too large to permit reaching definitive conclusions. It was found, however, that if altogether four different orientations of the analyzers are introduced, a new inequality similar to the Bell inequality stated arises, and that now the inefficiency of the counters is compensated. The actual definitive experiments were carried out in this manner, and their results are exactly as discussed. A final version of the experiments goes even further. One possible objection to the experiments is that in some unexplained way the analyzers could communicate between each other while the experiment is performed, and thus determine its outcome. To avoid this, the analyzers are rotated while the photons are in flight, and this is done at such a speed that any signal communicating the analyzers would have to surpass the speed of light.[10]

Is Relativity Violated?

We have seen that a remarkable feature of the action at a distance revealed by these experiment is that it is instantaneous, which appears to violate relativity. This is not so. What relativity forbids is sending a *message* with speed faster than that of light. Two observers can easily correlate their behavior instantaneously. Lady Mantenida and Lord Maltornato can agree, for instance, to get up when their alarm clocks sound at 12 noon (they are of course late risers), and there is nothing to admire here. Let us see instead how they could behave within the world of Codex Copenhagen, which, whether we like it or not, is a little more similar to real (quantum mechanical) life than the EPR version.

It is easy to see that they cannot act on the "information" received at a distance. Suppose that they agree that when Lord Maltornato finds a perfumed stocking in his parcel, he should kill Lady Mantenida's husband, the redoubtable Lord Cornucopia. When Lady Mantenida knows that the time is ripe, and she therefore exercises her olfactory activity as vigorously as she can on her stocking moiety, Lady Eduina happens not to revive the perfume, and Lord Maltornato remains both uninformed and guiltless. There is no way in which Lady Mantenida can *cause* any course of action to be taken by Lord Maltornato.

Although action at a distance is imposed upon us by the quantum mechanical experimental evidence, it should be appreciated that it has a limited domain of application, since it only occurs between two entangled particles, that is, particles that either by their initial production or by their initial interaction have formed at some time a single system. Thus, alarming a concept as it is, it is not one that should be confused with action at a distance as it occurred, for instance, in the old Newtonian gravitational theory. It is for this reason that Abner Shimony (1928–) has proposed the name of *passion at a distance* for this interaction.[11] The philosophical status of action at a distance is interesting, however, and deserves a little more comment.

ACTION AT A DISTANCE AND METAPHYSICS

The temptation to trivialize the results of the Aspect experiments by an analogy with my fable of the Codex EPR is very strong and must be avoided. Healey (1989), pp. 54–55, reduces to some extent the interpretation of the experiments to the fact that the two particles are entangled, and thus that the measurement on one leads necessarily to the result on the other. He introduces for this purpose what he calls a *probabilistic disposition*. This is not trivial, insofar as the experiments show that the second particle cannot keep any local memory of its entanglement to the other particle, and thus that this probabilistic disposition cannot be local. But the point I want to discuss here is "the general metaphysical principle that there is no instantaneous action at a distance" enunciated by Healey (p. 50). Although one cannot but sympathize with this sentiment, the fact is that in the experiments discussed here, nature appears to care very little for our metaphysical feelings. And yet the principle proposed by Healey appears to have a true metaphysical, that is a *necessary*, urgency.

I hope that at this stage of the book the reader will be quite sure that I do not oppose metaphysical hypotheses wholesale. So let us have a look at Healey's "general metaphysical principle." Before we do this, we must remember clearly the context of our discourse: we are not here selling socks, a situation for which it is perfectly possible to invoke a whole armory of meta-physical principles more or less well entrenched. We are here instead at the very crossroads of our excursion into the properties of nature, and we are faced with the need to accept a result that is, in all sincerity, repugnant to most of us. But it is there, and if such a counterintuitive result is there, it is extremely dangerous to try to explain it away by principles that are not sufficiently well entrenched *in the context in which our present discourse lies*.

Let me give an example to illustrate why we are in danger. If we say that there is no instantaneous action at a distance, and we want this to be accepted as a meta-physical principle, we are accepting *that the concept of distance is untouchable*. If nature is as strange as it appears to be, one might, however, assume equally strange things about distance. We have lived over two thousand years with a geometry that was based on the premise that the simplest of all geometrical objects is the point: Democritus and Euclid have given us a picture of the world as a bucket of sand. What would happen if it were instead like a bowl of spaghetti, that is, if the simplest geometrical element was a filament rather than a point? Remember that if two objects are at the same point, their distance vanishes, so that what I am saying is that, now, it might be being on the same filament that licenses a vanishing distance. Imagine that entangled particles were on the same filament, and thus at a null distance: the "action at a distance" that we have experienced would become action between two contiguous objects or, even more, the two "distinct" objects would be just manifestations of a single one.

I am not suggesting for a moment that this is the way in which the physical description of nature is going to move. What I am trying to stress here is that to say "there cannot be instantaneous action at a distance" makes sense only if we use the concept of "distance" in the same way as when we apply it to socks, but of necessity our universe of discourse has to move forward. Thus, although the principle appears extremely desirable as a meta-physical principle, its validity cannot be accepted without question, and its validation is just about as complex as the validation of the presently accepted result that instantaneous action at a distance is a property of nature as we know it. The use of this (otherwise excellent) meta-physical principle *in this context* appears to me therefore of no help whatsoever.[12]

It does not matter whether the "filament" scenario which I have created is a "true," or even a "possibly true," picture of nature, but I hope that its formulation throws a warning signal on the dangers of using ready-made metaphysical principles whose application might result in an impoverishment of the range of possibilities that should be considered. We are faced with a very strange situation indeed, and in order to understand it, we shall have to question even the most obvious of our entrenched ideas. At this early stage in the quest, the greatest danger is that of people coming forward and defending their own preferred patch as untouchable. After we all understood the Michelson and Morley experiment, it was Einstein who taught us how to think the unthinkable: time was no longer absolute. Who knows how many cherished absolutes we might have to drop now?

CODA

It might be useful to reflect that the whole of this exercise, like that of the EPR program itself, had one and only one object: to prove the Copenhagen interpretation of quantum mechanics wrong. (Although to accuse the Copenhagen description of being incomplete is marginally more polite than urging that it is wrong, it is precisely the same thing, within the terms of the Copenhagen interpretation itself.) Experiments, more careful and more accurate than ever before, have instead confirmed beyond any reasonable doubt that at least *some form* of the Copenhagen interpretation is not contradicted by experience. You would therefore expect the scientific community to have raised their metaphorical hats to the memory of old Niels Bohr in admiration for his perspicacity: instead, the standing of the Copenhagen interpretation seems to be now lower than ever before! (See chapter 19.) Science, of course, like the rest of life, is not an orderly progression in just one direction. People must be cantankerous, and thank goodness for that: otherwise, we would still be sitting at Aristotle's feet.

NOTES

1. To prove this commutation, call

$$A = (p_1 + p_2)(x_1 - x_2) = p_1 x_1 - p_1 x_2 + p_2 x_1 - p_2 x_2.$$
$$B = (x_1 - x_2)(p_1 + p_2) = x_1 p_1 + x_1 p_2 - x_2 p_1 - x_2 p_2.$$

Remember that p and x do not commute as long as they pertain to the *same* particle, but they do commute for *different* particles. We can therefore write:

$$B = x_1 p_1 + p_2 x_1 - p_1 x_2 - x_2 p_2.$$

Hence:

$$A - B = (p_1 x_1 - x_1 p_1) - (p_2 x_2 - x_2 p_2) = 0,$$

because whatever the value of the brackets here might be, it cannot depend on which is the particle to which they refer.

2. Einstein et al. (1935), p. 777. Notice that EPR used a strange word, *predict*, rather than the word *measure*, that would have fitted more comfortably within the Copenhagen interpretation. This is part of a basic trouble. If EPR wanted to find a contradiction in the quantum mechanical description provided by Copenhagen, they should have stuck to the Copenhagen rules. They do not do that: they wanted to explore their own vision of "reality." For this purpose, they were not concerned with a quantity being actually measured or not, and they felt more comfortable with the rather prophetic "predict."

3. Einstein et al. (1935), p. 780.

4. Bohr (1935a), (1935b).

5. Many authors have discussed the weakness of the EPR argument. (See for instance Murdoch 1994, p. 306.) Einstein himself in a letter to Schrödinger of 19 June 1935 (Murdoch 1994, p. 308) claimed that his argument had not come across well in the paper, and he later presented his own improved version (Einstein 1936, p. 376). For a careful discussion of the ways in which Einstein's views differed from EPR, see Howard (1995). Bohr's answer to the EPR paper is discussed in Beller et al (1994).

6. In order to avoid the EPR problem, Selleri (1990), p. 185, assumes that the measurements are taken on different elements of an assemblage. Podolsky himself, in an unpublished paper (Jammer 1974, p. 192), accepted that quantum mechanics was a correct and complete *statistical theory* of assemblages of like systems. The idea that different variables may be simultaneously measured for different elements of an assemblage is nevertheless one that would contradict in a rather sterile way the important consequences of the uncertainty principle, and must be regarded with the highest suspicion.

7. Notice that this description of the Copenhagen interpretation would not satisfy, for instance, Peres (1993). See further discussion in chapter 19.

8. Strictly speaking, the two photons of each pair differ in polarization state because they are circularly rather than linearly polarized: whereas one photon is circularly polarized clockwise, the other is counterclockwise. This relation is the one that keeps the photons *entangled*, but it does not affect any result from the point of view of the interaction between the photons and the analyzers, and it may be ignored by the reader.

9. This jumping is some sort of a metaphor. Before the photon hits the analyzer, it is in no precise state, so that it can hardly be said to jump to a new one. When it hits the analyzer, its state becomes determined with a probability given by quantum theory.

10. The first person to propose a possible experiment on the lines of EPR was David Bohm (1917–1992), in Bohm (1951), pp. 611–23. Bohm was highly influenced by Einstein and had serious doubts about the Copenhagen interpretation. (He went on to produce a deterministic version of quantum mechanics, which I shall discuss in chapter 19, by introducing *nonlocal* hidden variables, so that even then, Einstein would not have been pleased.) John Stewart Bell (1928–1990), born in Belfast, was a theoretical physicist working at CERN (the European Center for Nuclear Research) in Geneva. He described himself officially as an engineer; quantum theory, on which he was a world leader, was his hobby. His first paper on the inequalities is Bell (1964), and more of his work on this line can be consulted in Bell (1987). It was the paper by Clauser et al. (1969) that first solved the question of how to circumvent photon-counter inefficiencies, thus paving the way for proper experiments. Description of early experiments as well as an excellent discussion of the general theory may be found in Clauser et al. (1978). The definitive experiments that disprove the hypothesis of local hidden variables were performed by Alain Aspect (1947–) and his colleagues at Orsay (Aspect et al. 1981, 1982a), and the final experiment with analyzers set during the photon flight was performed by Aspect et al. (1982b) with analyzers separated by 12 m. Similar experiments, with analyzers at 400 m have recently been performed, totally excluding

any possible interaction between the analyzers. (See Aspect 1999.) Delicate tests to rule out the possibility that the analyzers might in some instances act as photon multipliers have also been performed by Haji-Hassan et al. (1989). (See also Rae 1989.)

11. Shimony (1984). A very important constraint that impedes the transfer of information by means of the Aspect type of experiment is discussed by Peres (1993), p. 290.

12. Howard (1995), in fact, interprets the Aspect experiments as indicating that the two alleged "particles" are not separable and constitute a single object.

WAVE FUNCTION REDUCTION AND DREAM QUANTUM WORLDS

Reams of paper have been wasted on the supposedly weird quantum-mechanical state of the cat, both dead and alive at the same time.
Murray Gell-Mann (1955), p. 153.

I f one looks at quantum mechanics in a sensible way, disregarding preconceptions that largely arise from our macroscopic conditioning, there are only two major conceptual problems requiring attention.[1] One is the question of the reduction of the wave function, which contradicts the deterministic Schrödinger equation with the time, as we have seen in chapter 16, p. 492. The other problem is how do we account for the fact that even macroscopic bodies must be represented by some wave function, whereas they do not behave in the elusive way in which wavicles do. This second problem is the one fruitful result of the Schrödinger's cat pastiche, and it can be much better understood by asking some simple questions: How do we know that the moon is there when we do not look at it? Why when I aim a billiard ball at one of two identical pockets do I never find the ball in a superposition state straddling the two possible results?

These are very hard and intricate problems, and all that I can do is to try to show that, if one addresses them in an unprejudiced way, it is possible to realize that they do not entail any magic or mystery, and that sensible rational explanations are, at least in principle, possible. As regards the moon, I have already provided an answer: as a difference with wavicles, the moon has a *trajectory* that is continuous. Hence, I can predict before going to bed where it is going to be at any time of the night I care to look at it. It is this certainty that sustains our rational (but nevertheless meta-physically grounded) belief[2] that the moon is there even when unobserved. This is more or less what I have said before, but the reader should by now know better than to accept everything in print as true.

Because, in proposing that solution to our question, I have dodged the fundamental problem behind it: how is it that the wave function of the moon does not behave like the wave function of the electron? That is: why has the moon a trajectory at all?

Although there is no universal consensus about the answer to all these questions, as we shall see later on in the chapter, I shall put my money on what I consider to be the most sensible and fruitful proposal so far made: the concept of *decoherence*, which will be the most important idea to emerge from our discussion.

REDUCTION OF THE WAVE FUNCTION. DECOHERENCE

I shall first show in more detail how the reduction of the wave function cannot be explained within the main principles of quantum mechanics, since it contradicts the use of the Schrödinger equation with the time. We have seen (p. 484) that this equation is deterministic, so that given the wave function at the time t, $\psi(t)$, it predicts the value of the same function at a later time t', $\psi(t')$. Let me represent such prediction with an arrow:

$$\psi(t) \to \psi(t').$$

Consider now a wave function representing a superposition state at the time t:

$$\psi(t) = a\psi_1(t) + b\psi_2(t).$$

It will be useful to consider an example, which I illustrate in figure 1. The detectors correspond to the eigenstates ψ_1 and ψ_2, so that if an incident photon is in the region between the diaphragm and the screen at the time t, its wave function is then given by the above superposition.

Reduction of the previous wave function entails that a measurement at the time t' will yield either one or the other of the states $\psi_1(t')$ and $\psi_2(t')$ with probabilities a_2 and b_2, respectively:

$$a\psi_1(t) + b\psi_2(t) \to \psi_1(t'), \text{ probability } a^2, \text{ or}$$
$$a\psi_1(t) + b\psi_2(t) \to \psi_2(t'), \text{ probability } b^2.$$

But this is in contradiction with the deterministic evolution required by the Schrödinger equation with the time:

$$a\psi_1(t) + b\psi_2(t) \to a\psi_1(t') + b\psi_2(t').$$

Fig. 1. An idealized experiment

The opaque screen is supposed to restrict the possible outcomes for the incident wavicle to two.

The reduction of the wave function, as described, must therefore be introduced as an additional postulate,[3] which is of course very unsatisfactory, because this postulate is far from self-evident and requires explanation. This is provided by a mechanism under the name of *decoherence*, which is extremely complex, so that the argument I shall now give has to be taken with a salubrious pinch of salt and merely regarded as a token for the real thing. Let me first of all simplify the notation for the superposed wave function in the region between the diaphragm and the screen, leaving out the time:

$$\psi = a\psi_1 + b\psi_2.$$

What I want to do is compute its square, which should give me the probability of the wavicle being in that state:[4]

$$\psi^2 = a^2\psi_1\,\psi_1 + b^2\psi_2\,\psi_2 + a\,b\psi_1\psi_2 + ba\psi_2\psi_1.$$

Notice now the following facts. The first two terms on the right-hand side here can easily be interpreted as the probabilities of the wavicle being in the states 1 and 2, respectively, but the last two terms (*cross-terms*) cannot be probabilities at all. This is so because the separate wave functions ψ_1 and ψ_2 may be either positive or negative (as the amplitudes of waves are). Thus, their product may be negative as well as positive, whereas probabilities must always be positive. Imagine for a moment that we could find a mechanism that allowed us to cancel those cross-terms. It would follow immediately that the probability of the superposition state ψ is the sum of the two probabilities of the *separate* states ψ_1 and ψ_2. That is, the original superposition state has now become a *mixture* of the two corresponding states. (Remember that superposition states are never mixtures; the latter occur only in classical superpositions.) Naturally, if the wave function reduces to a mixture, then in a single event only one component of the mixture may appear. But this is precisely what the reduction of the wave function means.

So, in order to understand how reduction might come about, we need a mechanism that might cause the cancellation of the cross-terms. We consider for this purpose the huge wave function that represents the wavicle plus its interaction with the environment. (In the photon case, the interaction with the screen, detectors, etc.) What we shall have at each point in space is a huge sum of terms like in the last displayed equation. Because the cross-terms can be either positive or negative, and because of the billions of such terms that one is adding up (corresponding to the interaction of the photon with each particle in the environment), it is plausible to accept that the huge sum of the cross-terms will cancel.

The reader must realize that I am proposing here a very major assumption, somewhat related to the activities of the Buridan Ass Mark II in chapter 13, in the sense that I postulate a random distribution of the huge number of cross-terms, with the probability or frequency of a given positive term being equal to that of its negative one. Thus, in the final sum, every positive cross-term will cancel with its negative counterpart. In the end, what we are doing here is once more introducing an *ergodic hypothesis*, this time as regards the distribution of cross-terms, and it is this ergodic hypothesis that permits us to understand their cancellation.

We have now, on accepting these assumptions, a new concept that can explain the reduction of the wave function. This postulated process is called *decoherence*.[5] The good thing about it is that we can immediately understand why, if I substitute a billiard ball for the photon in figure 1, I shall be able to have a *trajectory* for it. That is, I shall be able to say with certainty, for instance, that the billiard ball will be caught by detector 1: contrary to Schrödinger's cat, I am not tempted to hallucinate, thinking that the ball might have to be represented by a state given as a combination of the two states ψ_1 and ψ_2. This is so, because the huge wave function of the billiard ball will have billions of interactions with the environment in the region between the diaphragm and the screen (with the cue, the baize, and so on). Therefore, it will instantly reduce or, as one says, *decohere*. What is valid for the billiard ball is valid a fortiori for the moon. So we can rest assured that the moon does have a trajectory and thus that it is there when we do not look at it.[6]

A very important idea that emerges from this discussion on decoherence is that, the moment that Schrödinger's cat is poisoned, its wave function, being a large body, will reduce in millifractions of microseconds. We can distastefully but safely leave the cat inside the black box for as long as we want: he is now a classical body like the moon, with a classical trajectory, and we do not need to look at him to know that he is there. Of course, we might be quite ignorant about his state of health, but no more ignorant than Jane was on p. 524 before she opened the fateful letter about her husband.

Even if we do not know what happened to the hapless animal, the important thing is that it will never occur to us that it is in any other state than either alive or dead: at most, we shall be interested in finding out when he died, which is quite trivial. This simple conclusion is wonderfully summarized by Gell-Mann in the epigraph to this chapter.

It is important to understand that decoherence is essentially dependent on coarse graining of the states. At the fine-grained level, the wave function is always deterministic in the Schrödinger sense. It is only when we consider an enormous number of interactions, and we coarse grain them, that decoherence arises. Because coarse graining will always involve, as I have suggested, some probabilistic averaging, it is no wonder that the outcome of a single event entailing reduction is random.

Now that we have some idea about the reduction of the wave function, we can have a good look at the question of measurement, which is quite necessary because, well before people hallucinated about cats, the ideas involved in analyzing measurement evoked truly Platonic shadows.

MEASUREMENT

Consider figure 1 again, now as a rather idealized measurement of the position of the photon, each detector supposedly including a counter that will show 0 when no photon impinges on it and 1 otherwise. The wave functions ψ_1 and ψ_2 will now correspond to the first and second detector, respectively, showing the digit 1. We start with the same superposed wave function for the photon that we had before

$$\psi = a\psi_1 + b\psi_2.$$

But we want to make a measurement, and let us call for this purpose the (macroscopic) wave function of detectors 1 and 2, when they show the digits 1, or 2, χ_1 and χ_2, respectively. The states ψ_1 and ψ_2 must therefore be respectively coupled with the wave functions χ_1 and χ_2. The total wave function that I am measuring is thus[7]

$$\Psi = a\psi_1\chi_1 + b\psi_2\chi_2.$$

When a measurement is completed, this wave function will reduce, only one of its two terms remaining. The doubting Thomases now come and say, wait a moment, this is not the end of it: I have here an endless recurrence; I shall need a camera to photograph the counter, a Xerox machine to copy the photograph, a computer to record the copy of the photograph, and so on. Each step will introduce one new factor to each of the two terms in the above equation. We must have a way to terminate this recurrence, they say. Add the mind of a human observer with wave function Ξ_1 and Ξ_2 corresponding respectively to "observer sees digit 1 in the record of counter 1," and "observer sees digit 1 in the record of counter 2." So, what we get is

$$\Psi = a\psi_1\chi_1 \ldots \Xi_1 + b\psi_2\chi_2 \ldots \Xi_2.$$

It would be entirely appropriate to terminate the recurrence by a further term, this being the mind of God apprehending the mind of the observer but, strangely, the proposers of this analysis assume that the recurrence terminates with the mind of a mere mortal. Thus, it is the mind of the observer that finally determines the reduction of the wave function and closes the measurement.

This approach is wholly unnecessary. When you consider the compound wave function of the particle with the detector, $\psi_1\chi_1$, because of the environmental interaction decoherence supervenes and the wave function of the detector, χ_1, becomes classical. Thus, the detector will have a classical trajectory (most simple, in fact, since in general it will be stationary). Thus, as with the moon, we can look at the detector at any time of our choice: what we see will be what was there in any case. You need as much the mind of the observer to do this, as you need the mind of the observer to decide that the moon is there when you do not observe it.[8]

THE TIME ARROW IN QUANTUM MECHANICS

At the end of chapter 12 we were left with the apparent conclusion that quantum processes had to be responsible for the time arrow. I had mentioned, however, that there is a snag, because the Schrödinger equation with the time is *time reversible*. There is an enormous temptation (and I am afraid that I had built up the reader's expectations in this direction) to believe that the random processes involved in the reduction of the wave function are the key to explain time asymmetry. Such an expectation is easy to entertain: consider a polarized photon hitting an analyzer plate. If we find that it emerges in one direction, say **a'**, we cannot retrodict to a moment before this measurement and say in which state the photon was. It could have been polarized in the **a'** direction, but it could equally well have been in a superposition state.

That this type of process *alone* might "explain" the arrow of time is nevertheless highly doubtful because, if we accept the decoherence theory of wave-function reduction, we know that this theory depends on some sort of ergodic principle in order to cancel *macroscopically* (that is, on

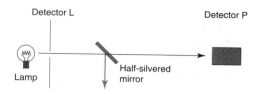

Fig. 2. Alleged irreversibility of measurement

The mirror allows precisely half the photons to be reflected in the direction of the gray arrow. (After Penrose 1989, p. 357.) The detector L does not impede the photon's passage.

coarse graining) certain terms which I have called the cross-terms. So, any hope that we can explain the time asymmetry on purely microscopic grounds is unfounded, and we find ourselves returning to where we were, to the world of macroscopic interactions.

In a way, this is the end of our road, and we have to concede failure. But, because it is very easy to fall into misconceptions, I shall discuss here an example that allegedly reveals the time asymmetry involved in the process of reduction. Let us consider figure 2 for this purpose, and in what follows I shall use the superscript τ to denote that the time has been reversed. I can now ask two questions, one Q in direct time and the other Q^τ in reversed time:

> Q: *Given that L registers a photon, what is the probability for P to do the same? The answer is very clear: 1/2. (Because there is equal probability for the photon to be on the black and the gray paths.)*

> Q^τ: *Given that P registers, what is the probability that L had registered (retrodiction)? Assuming that we do not allow stray photons in the apparatus, the answer is 1.*

Since Q and Q^τ do not coincide, we seem to have discovered a time asymmetry. This discovery, however, is of no significance, and it does not show up any feature of quantum mechanics that makes it differ from classical physics. To show that this is the case, replace the lamp by a gun that fires bullets, and the mirror by a diaphragm that opens and shuts in such a way as to be open precisely half the time. I can ask the questions Q and Q^τ above, and the answers are identical with the previous ones.[9]

The reason why we find ourselves with a result of no interest on our hands is this. If we call two successive events in time A and B, the time asymmetry that we have established is that, if A *evolves* into B, then B^τ does not evolve into A. To show an arrow of time, what we would have liked to have established is that, if the evolution of A to B is *dynamically possible*, then that from B^τ to A^τ is not so. And this does not follow at all from the above example. (Boltzmann's teacup could first be stirred up and could then, by proper dynamic evolution, separate out into tea and milk: it is the failed performance of the latter process that is the interesting time's arrow.)[10]

So, Penrose's conclusions from this thought experiment are not beyond criticism. Besides the references given in the notes, they were also contested by Lawrence Sklar, who has worked longer on this problem than most people, and who concludes that "the basic principle of temporal asymmetry still eludes us." With this, I must reluctantly abandon my arrows of desire (to echo the title of Sklar's paper) and hope that the future will know better.[11] I expect that the reader, on the basis of the brief evidence presented, will understand why I have abstained from adding one more dubious solution to the history of the arrow of time, and have preferred simply to report what the state of the game is at present. It is somewhat sad that, because this is an unsolved problem, solutions are trailed before the public that stir emotion rather than thought.

BOHM'S SO-CALLED ONTOLOGICAL THEORY OF QUANTUM MECHANICS

Before decoherence theory was developed, the reduction of the wave function was not seen to be a natural part of quantum mechanics. Many attempts were made to circumvent it, and two of them, to be discussed in this and the next section, entail a drastic revision of the theory. Bohr had never claimed any ontological status for wavicles: their very use as the building blocks of matter was resented by many, especially Einstein and his followers. David Bohm, as we have seen, came under Einstein's influence at Princeton, and over the years tried to revive a theory, called the *pilot-wave theory*, that had been originally proposed by de Broglie and others, but which he developed well beyond his predecessors.

The fundamental philosophy behind his ideas was that Bohr's preoccupation with not being able to *know* the simultaneous position and momentum of an electron was irrelevant; electrons are there with positions and momenta, even if we cannot know them: ontology takes precedence over epistemology. (Even if Bohm were eventually to be proved right, this approach is extremely dangerous. It reeks of medieval modes of thinking in which inaccessible features of reality were described by arbitrary models. I have already argued against so-called ontologies based on objects postulated just because they are *psychologically* acceptable. Not even Bohm claimed that experiments could ever be performed to determine the simultaneous position and momentum of an electron.)

Be that as it may, Bohm started from his classical electrons, which he managed to make behave classically and deterministically, reproducing nevertheless all the experimental results of quantum mechanics. In order to do this, he had the electrons moving under a force field, which he defined *from the quantal wave function* and called the *quantum potential*. Because the motion of the particles in this field is deterministic (the electron faced with two slits, for instance, would clearly traverse one, *guided* by the quantum potential), superposition states no longer exist as such, so that no reduction of the wave function is required. This is obvious because the wave function no longer exists, except in an indirect way to generate the quantum potential. This approach has of course some attractions, as John Bell emphasized, but it entails several problems.

The first and most serious one is this. It is well known in physics that when fields are introduced in order to consider the evolution of some objects in them, the principle of action and reaction must be obeyed. This is a very clear and very simple physical necessity. As the field acts on the object, the object must act on the field. (A most important branch of quantum mechanics, quantum electrodynamics, studies such interactions between the electrons and the electromagnetic fields in which they move, and it is only by considering them that the energies of the electron states can agree to eight or ten significant figures with the experimental results.) Bohm's quantum potential is *rigid*: although it acts on the electron, it receives no reaction whatever from it.[12] (It can't, because the quantum potential is nothing else than the wave function repackaged: if you were to change the quantum potential, the wave function would have to change, and it is the wave function that guarantees the agreement with quantum mechanics and thus experiment.)

The second problem centers on the claim that Bohm's theory is deterministic, although it uses the wave function, and thus its attendant probabilities, in a repackaged form. I take the view that this being so, it is not appropriate to claim determinism for the theory, and I shall give an example to this effect. Consider a hightech roulette that is governed by a computer. The computer contains a true random number generator (like a radioactive source), so that there is no question that the roulette is random and not deterministic. (Call this *Case A*.) Suppose now that the computer generates the random numbers during a time T (which could be overnight) and stores them in a hacker-proof hard disk. (Call this *Case B*.) The roulette now runs from the hard disk, so that we might be tempted to say that the roulette is deterministic, since its outcomes are all preordained in the disk. This, however, is nonsense. To start with, assuming that the security of the disk is not compromised, there would be no observable difference between *Case A* (manifestly random) and *Case B*. Moreover, we can continuously go from *Case B* to *Case A* by allowing T to diminish gradually until it reaches zero. Would we say that the roulette is deterministic if the numbers are computed during only one second, stored on the disk and then instantly used? Or what about a microsecond?

The fact that Bohm *stores* the probabilities in the quantum potential does not mean that they are not there, just as the random numbers stored in the disk do not cease to be random. Thus, the behavior of Bohm's electrons is far from being manifestly deterministic.[13]

Bohm and the Scientific Method

The introduction of Bohm's version of quantum mechanics is a most interesting example of the scientific method. By no stretch of the imagination can we view this work as an attempt to falsify quantum mechanics, since the theory was proposed in such a way as to coincide with the latter whenever observable results are available. Neither can it be viewed as a refutation of any proposed conjecture. (The Copenhagen interpretation was not refuted but merely bypassed.) Thus, the presentation of this theory was an entirely non-Popperian situation. And yet, it is good science, even if you do not agree with it. People like Bohm and Bell were perfectly entitled to feel uncomfortable about quantum theory as they knew it, and what Bohm was doing was trying to achieve what was, for him, a better *understanding* of processes in the microworld.

I have already said that one person's understanding may be another person's muddle. Science, however, is not a form of psychotherapy. Science demands that people deliver the goods: it is not enough for the propounders of a new theory to say that they feel more comfortable with it. They must prove that they have both consolidated and extended the scientific mesh. If they do, then *their* form of understanding will become *our* form of understanding.

When in the nineteenth century the atomic hypothesis was proposed, a new understanding of heat resulted, and even then it was strongly attacked by many major scientists. But this understanding has now been adopted because without it things like microwave ovens, for instance, could never have been made. The new Bohmian "ontology" appears to be mainly justified by a refusal to accept an ontology that is not a classical one. Admittedly, the Copernican ontology took some time to produce new results, but we now move at a different pace: even in the nineteenth century, the atomic theory was constantly being supported by new experimental evidence. In over twenty years not a single new result has come out of Bohm's heroic proposals. No one can say what the future of the theory will be, but it is unlikely that its fate will be that of a *refuted* conjecture, à la Popper. As time goes on, if nothing new comes out of it, it will become a *failed* conjecture. But even the fact that it was tried and it failed (if it does fail) will have been useful in reinforcing our confidence in whatever consensus on quantum theory eventually emerges.

THE MANY-WORLDS INTERPRETATION

I approach this subject with great trepidation. Many theorists treat it with complete and utter contempt, whereas some very good physicists and a fair number of philosophers regard it as the most complete description of reality yet available. I find it very difficult indeed to give a balanced presentation of the subject, partly because among those who practice it there are different views about what they are doing. So, I will have to present an account which, I have no doubt, will be considered as a misrepresentation of their ideas by many of the theory's practitioners.

With apologies then, I shall go back to Schrödinger's cat in its superposition state inside the black box, and I shall forget the fact that, given decoherence, the cat's wave function must have reduced instantly. Ignoring that situation, the conceptual problem raised by Schrödinger was that we would have to visualize the cat as neither dead nor alive, but in a superposition state. Joe Bloggs, faced with this conundrum, complains that no one has ever perceived a cat as partly dead and partly alive, and that he, Joe Bloggs, is incapable of doing so. Yet, he is wrong, because there are two worlds, each with a cat in a box and a copy of Joe Bloggs. In one world Joe perceives the cat as dead, and in the other he perceives it as alive. Thus we define what it means to perceive a state of superposition, and the nasty reduction of the wave function is not even required. Naturally, we have to go a bit further because Joe Bloggs must perceive the cat as dead or alive in each case with a 50 percent probability, and just two instances are not sufficient to make a probabilistic assertion. So, we have a very large, perhaps infinite, number of universes, and in each of them we have the black box complete with cat, and in each of them there is a Joe Bloggs, perceiving in half the cases a dead cat and in half the cases a live cat. This is all amazingly clever, and although it might appear to be an ontological extravaganza, like all such proposals, it might perfectly well be true.[14]

I repeat that some many-world practitioners would object to my example as not being representative of their views, and I am sorry about that. So, with that proviso, let me go on. It must be understood that the many universes postulated are all just as real as the one that we use for everyday purposes. Each and every one of them obeys the same physical laws. The fact that I am not aware of my infinite copies is neither here nor there: neither am I aware that the earth moves *e pur si muove*.

Now I come to the point that worries me, at least about the particular form of the many-worlds theory that I have chosen to illustrate. Each of the universes postulated in it is real and subject to the laws of physics, so that each copy of Joe Bloggs must have been born, some will have smoked, some will have had lung cancer, and some will have died. So what do we do to compute our probabilities if we do not have one Joe Bloggs in each world? Even more: would it not be possible to imagine a Joe Bloggs that is unique? Let me try an experiment. We have a large square array of very small photon detectors, so that the array looks a bit like a computer screen with each detector a pixel. Claudius fires a photon at the array, and the experiment is designed so that there is equal probability for the photon to reach any pixel. The single photon is now in a superposition state over the huge number of outcomes. $Claudius_1$ sees the photon in pixel number 1, say, $Claudius_2$ in pixel 2, and so on. Claudius is as good a computer engineer as he is ruthless, and he arranges that whenever any pixel registers, *except* pixel 1001, then a bomb will explode. So all copies of Claudius, except $Claudius_{1001}$ will experience a bomb exploding. Now it happens that Claudius was at the same time engaged in an in-vitro fertilization experiment, and that the bomb was arranged to destroy the machine where such procedure takes place. So, in all the universes only one egg is fertilized, and that is the one from which the unique copy of Joe Bloggs will be born. This hapless man is thus unable to partake of the multiuniversal perception of superposition.

Of course, all this might be nonsense, I mean nonsense within the many-worlds doctrine, because it might be assumed that the copies of Joe Bloggs are instantaneously created as and when required by some form of experiment. This might still be possible, but then the meaning of the postulated universes becomes more elusive, because events must take place in them beyond the normal laws of physics. (Indeed, I remember David Bohm suggesting, at a meeting at Balliol College, Oxford, in the sixties that poltergeists might be explained as objects that transfer themselves from our universe to another and then come back to the previous one, *at a different place.*)

So, all I can do is recommend the interested reader to buy the best popular exposition of many worlds available, *The Fabric of Reality*, by David Deutsch, from which they themselves should be able to form their own opinion about Joe Bloggs, the "One and Only."

One final and interesting point about scientific method. The audacious concept of many worlds was introduced by Hugh Everett III in a thesis submitted in 1957 to Princeton University, but despite the revolutionary aspect of his proposals, he made it quite clear that this was not an attempt at contradicting quantum theory (pace Popper), but rather to make it easier to adapt it to the theory of relativity.[15] For better or for worse, he was just trying to expand the scientific mesh.

CODA

We have seen in this chapter that some of the more delicate problems of quantum mechanics can be understood by the concept of decoherence. Not only the reduction of the wave function begins to make sense, but measurement can be understood without having to give the human mind a role that is special compared with that it has in classical physics. The price one has to pay is that some sort of coarse graining coupled with ergodicity must be accepted, and when this is taken into account, any hope of tying up the arrow of time to quantum mechanical behavior dissolves.

The reader will have noticed that the interpretation of quantum mechanics that I have developed in the last five chapters is ontologically parsimonious, as regards both matters material and matters regarding the human mind. I commend this minimalist approach to ontological constructs in a theory, as a prudent way of leaving room for further reinforcements *when new evidence becomes available*. This rule, alas, is not always followed. I have illustrated in this chapter two interpretations far more adventurous in this respect. Interestingly enough, both of them entail nonfalsifiable postulates. In one, intrinsically nonobservable entities, the classical states of electrons, are introduced. In the other, the nonfalsifiable splitting of our world in many copies (the *multiverse*) is introduced. In fact, when Galileo defended a (then) nonfalsifiable rotation of Earth around the Sun, one could even at that time imagine improved experimental conditions that could indeed negate or confirm that rotation. Such similar proposals to falsify the multiverse hypothesis do not appear available at present. What is interesting to remark is that despite these deficits, both new ontologies have enthusiastic supporters, which is a good example that adherence to rigid rules for the scientific method is not necessarily accepted by its practitioners.

POSTIL

We have now finished our work on quantum mechanics, and I should like to explain why we had to spend so much time upon it. It is not that I expected the reader to become knowledgeable about the subject: I wanted to exploit a superb example of how we should approach our understanding of nature. One of the most important ideas in this book is that in order to describe phenomena we need certain normative principles that I called meta-physical, like continuity or regularity, causality, and sufficient reason; and that the use of such principles is justified by their entrenchment. But we have seen when we discussed *grue* that entrenchment is environment-dependent. And this is the rub: whereas those principles had been abundantly entrenched in our interaction with the macroscopic world, we are bereft of similar props in relation to the microworld because, neither as a species, nor as individuals, have we been exposed to it. This requires a detachment on our part in trying to understand this aspect of nature, since we must avoid a naive vision of it tinted by principles valid only in the macroworld. However tempting it is to take the macroscopic world as the "real" thing, and thus our perceptive-rational understanding of it as the basis of all understanding of nature, this is entirely wrong. The macroscopic world is only a crust, so to speak, built upon the underlying quantum world. Lack of recognition that this is so, and that the principles valid for the former cannot be transferred to the latter, produces paradoxes of understanding that the unphilosophical mind reads as paradoxes of nature, as our work on quantum mechanics has often shown.

A great deal of work remains, of course, in order to understand how the at times random behavior of the microworld appears orderly and deterministic when read at the coarse-grained level of macroscopic events. I have only been able to give a glimpse of this with the phenomenon of decoherence and our discussion of ergodicity, but all this is work that still requires further understanding.

So, the great moment has come: we have now done all the major work of the book, and we are ready to tackle in the next chapter the awesome subject that has given its title to it.

NOTES

1. I do not include here the question of action at a distance because, however strange this result is, it does not violate or throw doubt on any of the quantum principles.

2. The concept of trajectory depends on the meta-physical principle of uniformity or regularity discussed in chapter 8, p. 185.

3. This postulate of reduction was first discussed by von Neumann (1932), p. 351, who called it a process of type 1, type 2 being instead the deterministic evolution of the wave function under the Schrödinger equation with the time.

4. Notice that I assume that the stated probability is desired at one particular value of the coordinates, *not* over the whole space. All the probabilities mentioned here must be understood in that sense. Also, I have throughout taken liberties with the probabilities, that should always entail the products of a function times its complex conjugate (see p. 505) rather than the simpler squares that I have used.

5. An important feature of the process described is that it is virtually instantaneous.

This is so because the probability of any cross-terms remaining is so small that the sojourn time of the system in such states is infinitesimally small. The origin of the name *decoherence* is this: a perfect laser beam is *coherent*, that is, it can engage in interference (see p. 258) and its photons are totally delocalized. When the latter localize, thus resembling the reduction of the wave function, the beam has *decohered*.

6. One of the earliest references to decoherence is Hepp (1975); and Zurek (1982) was a very influential paper in getting the theory recognized. (A mathematical but fairly simple review of the subject is his Zurek 1991.) More recent work is due mainly to Gell-Mann and his collaborators. (See Gell-Mann et al. 1990, 1993.) A very simple account appears in Gell-Mann (1995), chap. 11, and Omnès (1994), chap. 7, contains a complete mathematical discussion. See also Omnès (1999 *a*, *b*). Useful compilation of articles appear in Cini et al. (1990) and Wheeler et al. (1983). Lindley (1997), pp. 205 ff., gives a useful qualitative discussion of decoherence. A strong criticism of decoherence as an explanation of wave-function reduction is given by Albert (1992), pp. 91–92.

It is important to realize that it is not necessarily the case that the wave function of a macroscopic object will necessary decohere. Superconductors, for instance, are very well sheltered from environmental interactions, and plasmas containing macroscopic numbers of electrons can in fact exist in quantum mechanical superposition states. (Remember that it is the interaction with a large environment that produces decoherence.) This is a point first recognized by Leggett (1980). See also Legget et al. (1985) and Leggett (1987) for further discussion. New experiments by Friedman et al. (2000) entailing huge numbers of particles satisfy entirely the quantum mechanical superposition principle.

An alternative theory to explain the reduction of the wave function was proposed by Ghirardi et al. (1986, 1988). This is an ingenious proposal: the Schrödinger equation is modified by adding a term that is only significant for macroscopic interactions, and which is of such a form that for such bodies it entails the reduction of the wave function. The advantage of decoherence, however, is that no modification of the Schrödinger equation is required (which involves an arbitrary new constant) and that the whole process invoked is more physical.

7. I am here assuming that the combined wave function of the photon and the detector is the product of the corresponding wave functions, a result that under certain simple conditions can easily be proved in quantum mechanics.

8. The need to invoke the mind of the observer to close down a measurement was first proposed by von Neumann (1932*a*) and later much elaborated by London et al. (1939). It was Wigner who discussed this question in the greatest detail and who, given his great scientific authority, created a huge movement in this direction. A short list of his relevant works is: Wigner (1961, 1963, 1964, 1983). Two recent books that follow this line are: Lockwood (1989) and Squires (1990).

9. I owe this point to Savitt (1995*b*), p. 18.

10. Quite early on, Aharanov et al. (1964) recognized that the apparent time asymmetry I have mentioned was not significant, but their views seem not to have been fully appreciated for many years. Savitt (1995*b*) confirms this result and discusses very clearly Penrose's alleged time asymmetry. This thought experiment is also analyzed in detail in the same volume by Unruh (1995), pp. 46–49, who also showed very clearly (pp. 46–49), *contra* Penrose, how time symmetry does indeed arise, a conclusion also supported by Leggett (1995).

11. See Sklar (1995), p. 192. The key to the problem is probably some strong postulate of ergodicity, but why such a postulate may be used requires explanation.

12. This point was first raised by Anandan et al. (1995). See also Brown et al. (1996).

13. Bohm's original presentation of his theory was given in Bohm (1952). The most complete exposition of his ideas appears in his posthumous book Bohm et al. (1993). Albert (1996) gives a simple discussion in support of the theory, and a good introduction is Cushing (1996).

14. I have adapted the Joe Bloggs example from Papineau (1996), p. 234. He discusses a version of the theory proposed by Lockwood (1996), in which it is the brain that goes into a superposition state of "registering alive" and "registering dead" which would be "like being two different people who know nothing of each other, one of whom sees a live cat and the other a dead cat." I have treated this example in the more physical way in which Deutsch (1997), p. 52, considers copies of frogs experiencing photons. Lockwood, however, would almost certainly not accept my example as representative of the theory (see Lockwood 1989, p. 225), since he refutes this type of interpretation of the wave-function reduction.

15. See Everett (1957). Everett never used the words "many worlds," neither did he refer to a plurality of universes. Rather he described the "universe" as consisting of many "branches." It is now more common to refer to the whole thing as the "multiverse," consisting of many copies of "universes," more or less as a book contains many pages but remains one book. Although no one paid much attention to Everett for some time, it was the combined efforts of John Wheeler and Bryce DeWitt that brought his work to the public's attention. Michael Lockwood has presented a form of the theory based on the multiplicity of minds, which is well discussed in his book, Lockwood (1989). Perhaps the most vigorous and consistent presentation of the multiverse is the one provided by Deutsch (1997).

In mitigation for my inability to present a balanced and accurate view of this subject I shall quote from Lockwood (1989), pp. 229–30: "Everett was nowhere . . . explicit on certain crucial philosophical matters . . .—thus bearing much of the blame for the confusion evident amongst his commentators; nor, it must be said, did he show the least inclination to repudiate the more extravagant formulations of the many-worlds view promoted in the name of DeWitt and others."

A number of authors have found difficulties in the probabilistic interpretation arising from the multiverse theory. Cassinello et al. (1996), p. 1360, points out that, as discussed in chapter 13, the strong law of large numbers entails a probability in order to relate frequencies and probabilities. Because of this, it is impossible to obtain the probabilistic postulate by means of the frequency analysis of infinite copies of a single system, as required in the Everett model.

IS NATURE SUPERNATURAL?

After we came out of the church, we stood talking for some time together of Bishop Berkeley's ingenious sophistry to prove the non-existence of matter, and that every thing in the universe is merely ideal. I observed, that though we are satisfied his doctrine is not true, it is impossible to refute it. I never shall forget the alacrity with which Johnson answered, striking his foot with mighty force against a large stone, till he rebounded from it, "I refute it thus."
James Boswell (1791), p. 333, Saturday, 6 August 1763

You cannot step twice into the same river.
Heraclitus of Ephesus, ca. 513 B.C.E.

The whole business was either a mystery, or else meaningless, and of the two, the meaningless is the more difficult to take.
Patrick White (1967), p. 9.

I abhor a mystery.
Anthony Trollope (1861), chapter 13, line 1

It's true that if you watch the sky-wheel turn for a while you'll see a meteor fall, flame and die. That's not a star worth following; it's just an unlucky rock. Our fates are here on earth. There are no guiding stars.
Salman Rushdie (1995), p. 62

We face at last the overwhelming question: is nature supernatural? This is our final chance to answer it, but if readers expect me to do so, they will be disappointed: it is not my job to tell people what they must think or, even less, what they must believe. All I can do is help the reader look at the evidence required to form a considered opinion, and warn about the various pitfalls to avoid in this process. In any case, the question at hand is hugely metaphysical, and I cannot possibly get into all the philosophical arguments that inform it, since they cover whole libraries of philosophical and theological books. My modest aim is to analyze the way in which the scientific approach to nature that we have so far discussed might cautiously illuminate the problem.

The first thing I want to do is to help the reader shake off some easily acquired but nevertheless naive ideas about the solidity of nature. Read the first epigraph about the robust reaction of Dr. Johnson against Bishop Berkeley's denial of the existence of matter. If you think that the good doctor was right and the bishop a fool, please think again: and I am not concerned with what is or is not "true," but with how we argue in search of what we consider an acceptable "truth." For, if Berkeley's denial of matter was based (as we shall soon see) on his contention that the only things accessible to the mind are *perceptions*, then all that Dr. Johnson's kick gave him evidence of was not a stone but pain in his foot! Of course, you will now say: but this pain must have had a *cause*, and the cause *must* be the stone. But if your mind is not made of matter, as Berkeley and many before and after him have claimed, how can something allegedly material act as a cause of the immaterial?

I shall go back to Berkeley soon but, because it is so important to clear the mind away from deeply ingrained preconceptions, I shall first give you another example coming, not from a literary man (and what a man!), but from one of the greatest of modern philosophers, Gottlob Frege (1848–1925). Frege distinguished between *psychological logic*, which is concerned with what people hold to be true, and *logic* proper, which normalizes truth. Thus, "It is impossible for people . . . to acknowledge an object as being different from itself" is a statement within psychological logic, whereas "Every object is identical with itself" is a *law of logic*.[1]

If you think that this is as clear as pure water, take a bath in the Heraclitus river of my second epigraph, for what Heraclitus is querying there is the very notion of an *object*. You could paraphrase his dictum as "You never perceive the same object twice," because objects *change*. Even my sacrosanct table is not the same as a minute ago: it has acquired a chip at a corner. Is it still the same *object* as before? Is it identical with itself à la Frege? And if you object that the chip is there for everyone to see, what if a single electron evaporated out of the table?

The moment you start thinking in this direction, you will realize that you never perceive objects, but you perceive *events*. It could be argued that we do not perceive a stone falling but, rather, a time series of events that we link together by inventing the concept of an object that changes throughout that series, keeping nevertheless its identity. Instead of many different stones, each perceived "now," I describe my perception as a single stone perceived at different successive times. Even if the stone stands still, I have to invent the concept of *object*. The fact that you are not aware that you are inventing this concept is irrelevant. You might remember the meta-physical *principle of uniformity or regularity of nature.* I have argued that, because it is a principle that we must have acquired during the evolution of our species, it is empirically necessary for our rational and cognitive system, in the sense that our neural network has evolved the better to make use of it, whence the neural network itself is a testimony to the regularities that helped create it during evolution. Thus, those regularities may be considered empirically entrenched, this entrenchment holding purely for that construction of nature which we make by interaction of the latter with our cognitive system.

When you perceive the stone as a single stone falling, rather than as a series of different ones at different times, you are implicitly applying a principle of regularity. That is, it is the postulated *continuous trajectory* in our model that is used to denote the surmised *object.* (Remember how it is that trajectories give identity to classical particles. Also, remember that this is manifestly essential when you deal with moving "objects" of variable mass, like rockets, cars, and planes.) And in quantum mechanics we saw that *histories* were defined by events, and that they in turn are used to reveal the "objects" of quantum mechanics, the wavicles.

So have this in mind: Frege was a mathematician, and he probably thought that he could consider the notion of *object* as a given: surely scientists knew all about them. Thus even his logical law, if it refers to "objects" from "nature," is far from obvious because we never perceive them, but only events. Of course, the situation is different if we think of abstract "objects" like mathematical sets, but then Frege's law is just a rule without empirical content, little more than a linguistic convention. Good old Heraclitus was right to worry. Objects only exist as part of our model making of the nature that we know through its interaction with our rational and cognitive system, which I had called nature$_2$, and thus they "exist" in nature$_3$, the model of nature$_2$. (As always, I say nothing of nature$_1$, the transcendental, inaccessible, thing-in-itself.)

I hope that it is now clear that one has to approach the nature of nature with an open mind, prepared to query even the hardest and surest "facts."

THREE VIEWS OF NATURE

There are almost as many views of nature as there are philosophers, but I shall simplify this matter ferociously and classify the systems that people propose as either *monist* or *dualist* ontologies. A monist ontology views nature as formed of a single substance, be it *immaterial* or *material.* Obviously, the two views corresponding to this disjunction are entirely opposite.

Those who, like Berkeley, deny the reality of matter are *idealists*, whereas those who believe that everything in nature is to be understood on a material basis are called *materialists*. These words, unfortunately, have specific connotations in the ordinary use of language, idealists being the good guys (or otherwise loonies if their ideals do not coincide with yours) and materialists being the people from whom you would not buy a secondhand car. We, however, must use these words in a strictly technical sense, although I shall deal a little with their pejoration when the time comes.

Dualism, on the other hand, has a very long pedigree. Early philosophical systems were in fact polyvalent, in the sense that nature was viewed as comprising many different substances and powers, but in the Western culture the tradition of two substances, one material of which bodies are made, and one immaterial of which our soul partakes, is an ancient one. It was Plato, in the *Phaedo*, who created most vividly the opposition between bodies and immortal souls. Aristotle, on the other hand, accepted the concept of a soul as independent of the material body, and even took its seat to be the heart (the most popular choice nowadays being the brain), but he did not believe in its immortality. Although belief in the soul's immortality is fairly widespread, it does not necessarily entail a fully dualistic ontology. In opposition to Hellenistic thought, Jewish tradition, even in Jesus' time, did not accept the possibility of the soul existing independently of the body, At death, the soul was left in abeyance, "sleeping," confined to the *Sheol* (later translated as *Hades* or even *Hell*), waiting for resurrection, although even the latter was rejected by the Sadducees. On resurrection, the soul would be revived, but never independently of the body. This belief is still at the core of Jewish thought.

It is arguable that even Jesus might have embraced that view, since the Gospels are not transparent in this respect.[2] Doubting Thomas, for instance, testifies that what the disciples were seeing was not an immaterial soul, repugnant to the Pharisaic tradition, but a *real* body with *real* wounds. And the Ascension, as traditionally interpreted, was a *bodily* event and not the spiritual passage of the soul onto the heavenly sphere, so graphically depicted by El Greco in *The burial of Count Orgaz* at Toledo. It was Saint Paul who first preached the immortality of the soul, but the clearest and strongest influence in this direction is that of Saint Augustine of Hippo (354–430). Saint Augustine revived Platonism within Christianity, and from then on, the immortal soul was accepted as an essential concept, although belief in it was not established as a dogma of the church until the Lateran Council of 1513.

Mention of which brings me to a point that must not be forgotten if one wants to understand the history of the concepts that we are discussing. The idea that science and philosophy can express themselves freely in complete independence from religion is very new. Scientists like Galileo and Newton, and philosophers like Berkeley, Descartes, Leibniz, and Kant, operated in a cultural tradition that required them, even forcibly as in the case of Galileo, to present such views as they had within a theistic framework. And it would be naive to doubt that this did not affect their philosophical or even scientific stances.

In trying to assess the value of different views of nature, one must be aware of the fact that a theory that cannot be refuted is not thereby more sustainable. James Boswell was right to claim that, within his own terms, Berkekey's denial of the existence of matter was not refutable, and if the reader nevertheless were to conclude that this doctrine is silly, I would not be too worried, although it is important to have some grounds for that judgement. One other example of an ontology that cannot be disproved is provided by the Creationist constructions addressed to reconciling current scientific evidence—which gives the age of the universe as some ten to fifteen billion years and that of some fossil remains as hundreds of millions—with the Genesis account that the world was created some six thousand years ago. This Creationist hypothesis claims that the world was indeed created about 4000 B.C.E. complete with ancient ruins, fossils, rocks, and galaxies, all with the right amount of aging in order to agree with the currently accepted views. Of course, that creation act is also supposed to have carefully simulated the evolutionary process in order to account for the alleged passage of time. Many people sincerely believe that this is the best picture of nature available, and they hold this partly on the grounds that this belief cannot be refuted, which is absolutely true. Here, unfortunately, one cannot invoke Popper's criterion that if a theory cannot be falsified it cannot be scientific. Because nothing that we can say about the ontology of nature can ever be falsified. We cannot do experiments, for instance, to *prove* that stones exist, despite Dr. Johnson's naive outburst. All we can do in order to test an ontology (besides proposing some general grounds for entertaining it) is to see whether the arguments that lead to it are or are not equally compatible with a variety of other ontologies, the existence of which should weaken belief in the proposed one. For instance, since all that it is required for the Creationist scenario is that creation agrees both with Genesis and with accepted scientific facts, numerous alike but perhaps extravagant scenarios could be postulated for the known large number of planets in the universe, providing them with an instant (and perhaps bizarre) history, archaeology, and paleontology. And if we choose our planets sufficiently far, we shall never be the wiser.

All that I am suggesting is that one should try to devise an ontology such that, given the various hypotheses that lead to it, the resultant ontology is reasonably unique. This is of course a tall order, and I do not claim that it can be achieved: I merely suggest that some movement in that direction is advisable. I do not expect, of course, Creationists to be moved by this plea: the anatomy of belief is very complex. Others, however, might find that a rule of ontological parsimony is a useful guiding light, ontological parsimony being of course nothing else than the good old Occam's razor.

In trying to choose among the possible views of nature the one that you find most satisfying, it is useful to bear in mind one or two guiding lines. The first question one has to decide is whether one believes or not that there is something in nature that is *necessary* as opposed to *contingent*. If you hold the latter view, you accept that things are as they are: you can describe them, but you cannot give reasons why they are not otherwise. (By "things" I mean here the *totality* of things: you might be able to give reasons, of course, why, given *A*, *B* must be the case.) If you believe that there are things in nature that are *necessary*, then you are more likely to be prepared to accept some sort of dualist approach, since the nonnatural or nonmaterial sphere, which is then accepted, may entail necessary truths. Of course, if you look at the evidence from natural science, as we have done in this book, you cannot expect to discover any supernatural entities because that is a contradiction in terms. (Although there is no denying that there are people who manage to allegedly do that, thus creating an immense amount of interest in the public.) As we have seen, nothing studied by natural science is *necessary*, except for some linguistic conventions that are, in any case, conventions enunciated on a contingent substratum. Even the great meta-physical principles that we have used are contingent, and we are allowed to use them only in the macroworld whence they originated. These contingencies persuade us that my table is indeed there, and that this world exists (of which more later). Necessary truths would contrariwise incline us to think that some other "world" is there in which those truths are supreme.

Mathematics has traditionally been a haven for necessary truths and for a Platonic world of ideas allegedly independent of the human mind and of the contingencies of nature. But, again, a naturalistic approach to mathematics, as in chapter 14, shows that there is no clear case at all for such ontological inventions. All this does not mean that the only consistent view of nature possible is a monistic one: people do believe in such things as the deity and the eternal soul. But it would be very unwise to attempt to provide a ground for such beliefs on evidence arising from natural science, of which more later (under the great name of Giordano Bruno).

We are now ready to go back to Bishop Berkeley and various other philosophers who have conformed our views on the ontology of nature, but I must warn the reader that I shall be almost absurdly selective: all I want is to set up the necessary intellectual landscape for our discussion, and if some of the greatest names (like Hume, Kant, or Leibniz) do not appear in the next sections, it is partly because they have already been discussed in this book, however briefly, and partly because I want to concentrate on the absolute minimum of ideas that will give us some understanding of one of the greatest problems in philosophy.

THE ABOLITION OF MATTER

I shall not follow a chronological order, because I want to discuss monism first, in the idealistic form of which Berkeley is an exemplar; but, to understand how Berkeley operated, it is useful to say something about his rival and predecessor John Locke (1632–1704). Locke, educated at Christ Church, Oxford, was a contemporary and friend of Newton, and one of those who inaugurated the age of reason that goes under the name of the Enlightenment. Rather late in life Locke published, in 1690, his most influential book, *Essay Concerning Human Understanding*, which was the masterpiece of English *empiricism*. Locke rejected that humans are born with innate ideas; all ideas, instead, come from experience, through the five senses. He realized, however, that not all the properties of matter that inform our ideas share the same qualities. The color of an object, for instance, is contingent on the light under which we observe it. Thus, he held that matter had *primary* and *secondary* properties, the primary properties being such as extension and motion, and the secondary such as color. And it is here that we take our cue to move on to Berkeley.

The crux is that Berkeley recognized that, if we are prepared to have a skeptical attitude about the so-called secondary properties of matter, we could just as well go the whole hog and be skeptical about matter in toto. George Berkeley (1685–1753) was a very different man from Locke. Not only was he Irish, but he was also born by the time when the Restoration that followed the Commonwealth, and finally the Glorious Revolution of 1688, had weakened in Britain the impetus for the Enlightenment. Also, as a difference with Locke, he published his most influential *Treatise* in 1710, when he was very young, only twenty-five. Installed as bishop of Cloyne in Ireland in 1734, he died at Oxford during a visit to his son.

Berkeley produced three main arguments to reject the *independent* existence of matter, of which I have already briefly mentioned two. (What the qualification of independence means, I shall explain in a moment.) First, matter cannot be perceived through the senses, because "by them we have the knowledge only of our sensations." (As Dr. Johnson had knowledge only of the pain in his foot.) The senses "do not inform us that things exist without our mind, or unperceived."

Second, substance cannot be known by reasoning from sense experiences by means of a logical (and thus for Berkeley *necessary*) deduction. Also, dreams, hallucinations, and so on, cast doubt on the generality of any such reasoning. The supposition of external bodies, moreover, is not necessary for producing ideas in our minds.

Third, one might think that it is at least probable that "there are such things as external bodies that excite their ideas in our minds," but this is false in accordance to Berkeley, because he holds that the physical, if thus postulated, and the mental, must be ontologically distinct, and thus that the physical could not be causally efficacious on the mental. (Berkeley is positing here, of course, that the mental could not possibly be physical, which is no more than an unsupported assumption.)

It follows from these arguments that there is no other form of existence than to be perceived, either by us or "*some other . . . spirit*" (of which more in a moment).[3] It is from here that one of the most famous (but not necessarily true) dicta of philosophy was pronounced by Berkeley: *To be is to be perceived*, often quoted in its Latin form, *esse est percipi*.

I can now tie up a few knots. First, *existence* for Berkeley is not independent of the mind, which is the basis for his rejection of the *independent* existence of matter. Second, we must discuss a little what Berkeley calls "some other spirit," and why this is in fact a kingpin of his doctrine. When I do not look at my tree in my lonely garden, my tree ceases to exist, and it will only recover existence when I look at it again. Thus, the existence of what we normally call objects is full of gaps but, because those gaps are existentially and observationally meaningless, I can fill them up in any manner I want. While my tree is unobserved and thus nonexistent, it is equally valid from the point of view of my ideas that it be transformed into a statue of Margaret Thatcher, or into a mute cow, or a wooden bench, or an infinite number of identically possible alternatives! That is, Berkeley is in danger of falling into the trap that his ontology is only one of an infinite class of perfectly equivalent ontologies.

I am not sure that Berkeley, as a man of the cloth, had not prepared that trap himself. Because the only way out of it is to postulate "some other spirit" that can be there constantly vigilant, like an eternal beacon licensing the continuity of existence. This is of course God.

Although this argument is a necessary part of Berkeley's philosophy, it cannot in any way be considered as a convincing proof of the existence of God. (There is a clear circularity in creating a gap that must then be filled in: this is beautiful theology, but can hardly count as a philosophical argument.) If the existence of God were not to be invoked, then not only my tree would have existential hiccups, but also the Berkeley argument would fall into the trap of supporting an infinity of equivalent ontologies.

The reader may worry how far this bewildering notion is really possible. You might say that my mental life would automatically change if I were to use one of the alternative ontologies that I have mentioned because of having to contain such ideas in my mind. But nothing that I can perceive, and thus nothing that I can do, would change in such a case. I might of course be as foolish as to buy a bunch of flowers to place under the statue of Margaret Thatcher in my garden, but since I am a rational person, and I know that the moment I look at my garden the apple tree reappears, I would not waste good money that way. That you may not like the idea of my apple tree transvesting itself into a statue of Margaret Thatcher when I do not observe it, is neither here nor there. Philosophically, it is just as good as any other ontology.

Of course, this is nonsense. After all, the whole of our cultural history goes along with creating a good, clean, sensible ontology that saves the facts in the most parsimonious way possible. Pagans believed in satyrs and nymphs that populated their gardens. We have abolished them as parts of nature, and we have relegated them to the world of the imagination, of the purely mental, of ideas that are nice to have but that do not purport to map experience. After all, this work of ontological purging, so necessary for the healthy balance of our mental life, has largely been led by science, the ontological purifier par excellence.

So far, we have seen Berkeley as the propounder of idealistic monism, although very soon I shall argue that he was in fact some form of a dualist. The stronghold of dualism, however, arose from the investigation of what people call the *mind*; historically the strongest card held by those who believe that nature is not wholly physical.

THE INVENTION OF THE MIND

I am in my garden and show a friend a plant that I describe as a bush. My friend, looking me in the eye, says "I *know* it is a tree." Here we all understand why he uses the words "I know." Consider now a second scenario: I am in my study and look at my table, and think "I know this is a table." You will immediately realize that this is not usual language, that somebody is trying to make a meal out of the morsel of a very simple perception, that somebody is engaged in that unnatural activity that is called philosophy. The question is: is this good philosophy? And whatever answer you give, positive or negative, you are jumping into one or other philosophical pigeonhole.

Let me claim, to start the ball rolling, that there is absolutely nothing more in this case in saying "I know that this is a table," than in saying "This is a table." Try and find a difference. There isn't, unless some feature of the previous garden scenario remains, like somebody else's assertion that the object in question is not a table but is in fact a box. Yet, in the only significant, reduced, version that I suggested that we should keep, we have done a gentle form of murder: we have eliminated the "I." And it is here, in this very simple verbal maneuver, that the crux of the matter rests. If you think that there is no difference between the two forms that I have proposed, then you most probably would be prepared to believe the following: that, when you perceive a table, there is nothing more than just the photons reflecting from the table into your eye, and that they are causally efficacious in creating some sort of neurophysiological response in the brain. (The latter, in fact, is almost instantly propagated about the whole body, which, although not so evident when the perceived object is a table, is very clear when it is a lion.) If you insist, instead, that the "I know" is significant, then you are assuming that there is something, presumably in the brain, that is so to speak overseeing the perceptive process and you call that something the *mind*.[4]

You will notice that humans, or perhaps more precisely human bodies, engage in all sort of activities for which it never occurs to us to create an overlord: the body runs, but we do not postulate a "run controller," although in fact there must be some neurophysiological events in the brain that allow the body to adjust itself to a "running" state. The *mind* is unique in receiving a special status. People who believe in this special status of the mind, as some form of an organizer of our thoughts, often think that, whereas we do not have immediate perception of objects, because the senses must mediate, we do perceive our own minds by the wonderful faculty of *introspection*. This very moment the phrase floats in front of my *mind*: "I know that I am thinking." And this "I know" is, felicitously, unmediated. But when I say to myself that phrase, all that I am doing, *undoubtedly*, is: I am thinking! Introspection, thus, is no more than a belief that there is something in our inner lives that is split in two, the overseer or mind on one hand, and the thinker or perceiver on the other. At best, introspection is a doubtful tool to prove the existence of the mind.

The Oxford philosopher Gilbert Ryle (1900–1976) used a graphical expression for this belief in the mind as overseer: he called it "the ghost in the machine." Even if you do not believe in some sort of nonphysical overseer in the brain, you might still want to use the word *mind* as a convenient shorthand. Most people understand it without getting involved in ontological worries, as indicating some personal, private, activities of human bodies (and to some extent perhaps also of dogs, dolphins, etc.) that are communicated or expressed to others by "language," this being understood in its most general form, to include body language, blushing, winking, grimacing, and even touching.[5]

I have already indicated that the objective elements of Berkeley's ontology were ideas and not matter, and thus that he was an idealist-monist. Berkeley, in fact, was one of the great inventors of the mind: his world was made up of minds, my mind as well as those of others. If that were all, he would have been a straight monist. However, the contents, so to speak, of minds are ideas, and ideas cannot be caused by material objects external to the mind, since matter does not exist for Berkeley. So, he needed a cause for ideas to furnish the mind, and his chosen cause was the mind of God.[6] Thus, Berkeley is some sort of a dualist, since the infinite mind of God is of course a different "substance" from the human mind.

The relation between the mind and its contents, ideas, according to Berkeley, is very interesting, since, contrary to what you might expect, and as I shall discuss in a moment, he rejects introspection, which we have seen is one of the most formidable tools for manufacturing the mind. His fundamental postulate is the following: *Ideas are perceived, and the perceiver is the mind.* This appears to be unimpeachable. What we observe in the world are ideas and not objects; to be able to do that, we need some equipment, which is the mind. Let me try to paraphrase the italicized sentence with: *Food is eaten and the eater is the . . .* what? Is it the body? But then what part? If we say that it is the digestive system surely that is not right, because if your brainstem is dead, you cannot eat. One strongly suspects that we are caught in a tautology: "The eater is the eater," "The perceiver is the perceiver." Which is another way of saying that, probably, to define the mind as the perceiver, is to say nothing more than that there is perception.

It does not obviously follow from Berkeley's philosophy, therefore, that the existence of the mind as a separate entity is less dubious than that of matter. Of course, you might still believe that you can perceive your own mind by introspection, but even Berkeley agrees that minds do not perceive themselves but only their ideas because the mind is "a thing entirely distinct from them."[7] He thus falls into a double whammy: ideas, just as well as objects, should not be causally efficacious on the mind because of their distinct nature.

We finally run into a strange problem. I had so far given warnings that one should beware of philosophical principles that lead to an ontology that is not unique, but merely one out of a very large class of equivalent ontologies, that is, ontologies all of which are compatible with the same chosen principles. Here we face the opposite situation, namely that a serious pursual of the principles at hand will force our ontology to collapse into almost total annihilation. For, if with Berkeley we cannot perceive our own minds, even less can we perceive the minds of other people. How do we know that they exist? The result is an ontology that contains one and only one entity, *my* mind. No, dear reader, not *your* mind, because *you* do not exist. (And what is more profoundly dolorous, neither do my royalties!) Like in *La Vida es Sueño*, by Calderón de la Barca, life is a dream, *my* dream. This is *solipsism*, the purest and surest of all philosophies because it can never be disproved. And do not attempt to write to me in refutation: such letters as I receive are merely part of my dream (bills, incidentally, are nightmares). This is another example that irrefutability does not necessarily entail credibility.[8]

THE CASE AGAINST PHYSICALISM: DUALISM

Well before Locke and Berkeley, René Descartes (1596–1650) had created the preoccupation with mind and matter that was to inform philosophy for ever after his time. But he was also responsible for the important recognition that nothing in our understanding of nature and of ourselves should be taken for granted: a salutary skepticism and a rigorous analysis of one's suppositions were required. Although born in France, Descartes lived most of his life in Holland and died in Stockholm, where he had gone as tutor to Queen Christina. Descartes was perhaps the foremost introspectionist of all time: his dictum *I think, therefore I am* is one of the most famous phrases that philosophy has given to humanity. He did not, however, like Berkeley later, reject matter, which is extended substance, but averred that this is totally different from the mind or soul, which is a thinking substance without extension. The brain and the whole of the body was of course matter, but not only did he conceive of the possibility of a disembodied mind, but it seems sometimes as if he positively would have welcomed that rarefied state.

Given his assumption that mind and body are of entirely different kinds, Descartes bequeathed to us one of the most thorny problems in the philosophy of the mind: how can the mind, a substance without extension, act on the matter of the body, which is extended? In one of his later books, *Passions of the Soul*, published a year before his death, he suggests that the mind, though incorporeal, is active on the pineal gland. This is, of course, pretty silly neurophysiology, but it was not enough to put off Karl Popper and John Eccles, the latter one of the most eminent British neurophysiologists of his time, from broad support of Descartes's dualist view of the mind and body.[9]

Cartesian dualism was a very attractive philosophy at a time when most people believed in an immortal soul, but even in our time dualism of some sort is still very widespread, although in a highly modified form. The fundamental preoccupation in the latter case is a nagging feeling that the phenomenon that we roughly refer to under the name of *mind* cannot be fully explained in purely physical terms.

I should mention at this point that there are two slightly different forms of materialistic monism, often discussed in relation to the phenomenon of life. Even if you maintain that life is nothing more than some state of matter, you can either claim that this state can be entirely defined in terms of the laws of physics and chemistry (in which case you would be called a *physicalist*), or that life entails some natural laws, perhaps as yet unknown, distinct from those that hold for inert matter. Life and therefore the mind, require, so to speak, their own form of natural science. This latter position is often called *materialism,* although the distinction between these two views is sometimes blurred, or made in a different way. There is, nevertheless, an important distinction that makes the concept of *physicalism* more general and perhaps more acceptable than that of *materialism.* This is that physical laws are not only concerned with material objects but also with nonmaterial ones, such as forces and wave functions.

Leibniz (1646–1716), Locke's and Newton's contemporary, was as far as I know the first to propose an argument based on a thought experiment that allegedly demonstrates the insufficiency of a purely physical description of the mind; and because arguments of this type are considered by many to be valuable, I shall deal with such of them as have acquired some sort of a philosophical fan club. All these arguments are good examples of *reductionism,* in which phenomena are analyzed in minute physical elements and it is allegedly found that something is wanting.

Leibniz suggests in his *Monadology* that if he were reduced to microscopic size, and in this guise visited the brain, however exhaustively he would inspect the "machinery," he would never find a single thought or emotion. Neither would he if in the same guise he had visited the innards of a Pentium chip: and the reason is not that there is anything not physical in the chip, but rather that Leibniz would never have been able to decode the signals in it. And no one could possibly think that his neurophysiology would have been up to decoding such registers as the brain might possess.

In the early seventies the American philosopher Thomas Nagel (1937–) produced an example on Leibniz's lines that was grasped with great enthusiasm by the philosophical community, especially by the antimaterialists. Everyone knows that bats have an echolocation system that allows them to navigate at night never bumping into undesirable objects. Nagel argued that, if we were to examine in utmost microscopic detail what goes on in the bat's brain, then, even if we were able to understand its neurophysiology in its totality, we would never be able to reproduce the *firsthand* feelings experienced by the bat. *What Is It Like to Be a Bat?"* was the catching title of Nagel's paper, and his answer to this question was: we shall never know. As you can see, there is here an important shift from Leibniz: the emphasis is on the irreducible *privacy* of experience, privacy being then some *unique* property of minds not amenable to physical explanation.

To highlight the weak point of this argument let me ask a parallel if somewhat outrageous question: *What is it like to be a stone?* It is very tempting to dismiss this question out of hand as nonsense: even the most starry-eyed philosopher would not expect stones to have feelings or thoughts. But remember that the crux of Nagel's argument in my version is on the concept of privacy, and if we were to discover that it might be possible to find something *private* to the stone, then Nagel's argument would have to be looked at in an entirely different light. Imagine we do a visitation of the stone à la Leibniz, but now, as a difference with Leibniz, we assume that the microscopic visitor is a physicist who knows all about the laws of physics. Although he knows that the force of gravity acts on each particle of the stone, even if he measures it with all the instruments at his disposal, he would not be able to say that *he* can feel the forces that he is measuring, as these forces act, never on him, but on the particles of the stone. *The microscopic visitor is unable to feel the force of gravity that acts on the stone.* It is private to the stone, as *his* pain is private to *his* body. (Even if the stone falls on his head, what the hapless physicist experiences is not the force of gravity *on the stone* but its effects on *his* skull!)

Privacy, alas, is not unique to minds: everything in nature is accessible to us always at second remove. If one had a pessimistic disposition, one would talk, as the existentialists did, of the alienation of nature. In *Being and Nothingness*, Jean-Paul Sartre recognized that we ourselves experience this alienation as subjects of other people's perceptions (what he called *being-for-others*), and the impenetrability of the *being-in-itself* of material objects (not too alien to my stone example) is vividly illustrated by the visceral reactions in front of nature of Roquentin, the protagonist of his famous novel *Nausea*.

A couple of years before Nagel, two American philosophers, Ned Joel Block (1942–) and Jerry Alan Fodor (1935–), had produced an argument for the privacy of mental states which has recently been strongly supported by John Searle (of whom more in a moment). It goes as follows: a group of people have normal vision, except that what we see as red they see as green, and vice versa. They grow up among normal people, and they learn that, say, green is the color of grass and red that of fire engines. Thus, despite the color inversion of their perceptions, they correctly stop at traffic lights when their color is that of fire engines, and proceed when they have the color of grass (although they perceive green and red respectively in these cases). The argument here is that although *behaviorally* these people are indistinguishable from normal, their experiences are not the same. For all we know, you or I belong in fact to this group of people.

I shall gladly dispense with the discussion of ophthalmological experiments that would easily reveal the diversity of the color-inverted people, and which are easy to construct, since red and green have different diffraction indices, so that images of these colors can be produced that would be in focus at different lengths for the variant people: so much about the inaccessibility of their perceptions. I shall nevertheless assume that the details of the science are of no importance, and concede that Block and Fodor, and Searle after them, have here a case in favor of the privacy of mental events. But we have just seen that privacy is a currency of no exceptional value throughout the whole of nature, and that it is not unique to mental states. If mental states must be regarded as nonphysical, something else must be adduced.

The Australian philosopher Frank Jackson (1943–) produced in 1982 a famous argument against physicalism. Mary has been brought up in such a (cruel) way that she has not acquired color perception: she sees everything in black and white only. She is also a neurophysiologist of the highest standard: even more, she is omniscient in this respect. She does know, therefore, everything that physicalism relates to the experience of seeing red: but she does not know what seeing red is *like*. She now, in some way, acquires color vision, and Jackson concludes that "If her previous knowledge is defective, despite being all there is to know according to physicalism, then physicalism is false."

The keyword here is *defective*. My hearing is not *defective* because I am not able to perceive the ultrasonic frequencies that drive dogs demented. Surely, Mary, being an omniscient physicalist, would have known in her deprived sensorial condition that her neural networks had not been allowed to develop as those of the normal brains that she had studied, and she would have been entirely clear about the *physical* deficit she had. The important thing is not that I cannot experience what a dog experiences when he hears my whistle (or a bat when echolocating), but rather that I can confidently say that, if my perceptive and neural systems where like those of a dog or a bat, I would then perceive ultrasonic sounds. I could thus trace my current condition not to a lack of some nonmaterial "substance" in my brain but, instead, to some *physical* constraints. And this is precisely what Mary is able to say.

Let me discuss, however, the nature of Mary's deficit in a little more detail. There is no question that, even given that Mary's knowledge of vision is omniscient, she lacks the *experience* of seeing red. But when I say that I know what it *feels* like to see red, I mean that I can recall, for instance, the experience of having seen a fire engine. Even persons who have become blind as adults would say in this way that they know what it feels like to see red. So, there are two aspects to Mary's deficit. First, she has never had the experience of seeing red, just as most people have never experienced a frontal car collision. It would be wholly inappropriate in this latter case to conclude that, our physically omniscient observer not having had a given experience, entails that the "feeling" of that experience cannot be physical. Second, pace Jackson, Mary knows precisely the physical processes associated with other people seeing red: she can point out exactly to a given cluster of neurons in the brain that "fire" whenever the subject is exposed to light of the wavelength that she knows corresponds to red. Jackson's argument, however, cannot be that Mary's deficit is due to her lack of that neural cluster, or for that matter of any *physical* feature of the brain. Thus, if you claim that Mary has an experiential deficit, and that the existence of this deficit shows that Mary's omniscient physical understanding leaves in her a *physically inexpressible* experiential hole, you are presupposing the very duality that you are trying to prove.

In sum, the bat has states that I cannot experience, and Mary knows that there are states that she cannot perceive, not because of any peculiar property of their minds, but rather because the bat, and Mary, and all of us, share the property of having *bodies*, and things that happen to bodies, animate or inanimate, happen exclusively to *them* and not to other bodies. If something goes wrong with our bodies, the feeling of identity that is conjoined with that of privacy may be partially altered, as Oliver Saks vividly relates in *A Leg to Stand On*. As a result of a leg fracture he himself had a neurological condition whereby his leg became entirely alien to him. Not only the privacy with which anyone feels his legs was gone: the leg became an entirely foreign body. (So much so, that at one stage he tried to throw it out of his own bed!) And this was not a *mind* problem, but a clear neurophysiological condition.[10]

When you consider pain, hunger, and lust, you might call these *mental states*, as many people do. But: would you have them in a disembodied brain? Of course, such a brain might still experience similar states, just as an amputee has a pain in a missing leg, but this is still something bodily, probably as a remnant of nerve endings exciting neural networks structured before the amputation. *Mental states* is a shorthand for some body states in which the balance of determination of the state may have some localization in some part of the brain; but the state is still a bodily one. It is this *bodily* feeling that gives such states their privacy. This shows that Nagel's bat argument is far from conclusive: even if I understood in infinitesimal detail the neurophysiology of the bat, I still would not be able to *feel* the bodily reactions of *that* bat, not just because I happen to have a human body, but because I have a *different* body. *No other bat could possibly experience the sensations of the first one*, although, as Nagel put it, it could react sympathetically—which we can't—to the first bat's description of its own experience.

Straight dualism is not the only alternative to monism: wise men often like sitting on the fence. The University of Berkeley philosopher John Rogers Searle (1932–) is a *property dualist* but a *matter monist*. He accepts that the brain has states characterized by neurophysiological properties and that these are *physical properties*. But, he claims, the brain also has mental states that have *nonphysical properties*. In a way, Searle is a materialist, in the sense that he does not assume the mind to be a separate substance: it is the same matter, the brain, that has physical as well as nonphysical states, and it is the latter that are subjective, mental, and introspective. Searle stresses the importance of privacy or "first-person knowledge," and his approach, which is beautifully reasoned, stands or falls on the strength or weakness of this concept.

The Oxford mathematician Roger Penrose has an even more tenuous position than that of Searle. Some of his claims appear to be monist, insofar as he professes that there is nothing nonphysical. But the mind is so special, in accordance to him, that it cannot obey the ordinary laws of physics: thinking must be associated with some as yet largely undefined quantal phenomena occurring perhaps in some particular organelles of the brain. Penrose produces profound arguments to sustain his view, partly by using the results of the Gödel theorem which he, after Lucas, interprets as denying the possibility of human thinking being algorithmic. (As I have already mentioned on p. 437, this is in any case obvious in relation to creative mathematical thinking, as opposed to proof making.) It is because of this circumstance that he posits a quantal effect. Despite his professed show of monism, Penrose is a Platonist and claims that the truth of mathematical ideas is engraved in a Platonic world that is eternal, not created in any way by the human mind, and to which the human mind has access by *mysterious* means. So, it would be difficult to regard his monism as entirely kosher.[11]

I shall conclude this section by discussing one of the most serious objections to physicalism. If the mind were entirely physical, like a computer is, for example, then all that we may have are physically stored signals, which is in fact perfectly well established for the genetic code embedded in DNA. Philosophers, however, are very familiar with the fact that a string of physical signals can certainly entail a syntax but that it cannot per se encode any semantics.

Let me explain what these high-sounding words mean. That the physical signals entail a syntax, means that the signals can be read as a set of operational rules, that is, as a set of instructions about what to do. Thus, for instance, the genetic code. One talks here of a syntax, because in language a syntax will tell us, for instance, that the adjective must precede the noun, or that the subject and the corresponding verb must agree in plurality. But you can have a perfect syntax without any meaning at all, that is without any semantics. (As in the sentence "Hungry tables speak milk.")

I must now qualify a little the very important assertion I made about the semantic limitations of the physical, and I shall do this by quoting from the clearest philosophical discussion against physicalism that I know:

> nothing physical can ever possess semantic properties intrinsically.[12]

This appears to be the gravestone of physicalism, and it is indeed a very powerful point, as long as you look at it philosophically. Natural scientists, however, may weaken their rigor, because whatever they do must finally be accountable to nature, a prop that a Berkeleyan philosopher largely dispenses with when rejecting matter. Which remark gives us the first hint as to the significance of the above sentence, because the word "intrinsically" powerfully excludes at a stroke the possibility that the physical might, *through interaction with nature*, accumulate some sort of semantic experience, a point which I shall illustrate in a moment.

But before I do that, I have to say something about semantics, that is about *meaning*. If you mean by *meaning* something transcendental, some sort of Platonic understanding about what a word *means*, then there is nothing much that I can do, but there is no denying that the above-displayed sentence could then hardly entail a depreciation of physicalism, as it would already entail implicitly the denial of this position by the connotation of the words used in it. So, for the criticism to stick, a more neutral meaning of *meaning* must be given. I suppose that it would be perfectly adequate for a physicalist to claim, as Wittgenstein did, that the meaning of a word is its use: this is after all the principle that informs the construction of a good dictionary, like the admirable *Oxford English Dictionary*, in which the meaning of words is constructed from quotations of their use.

We can now start our work. Of course, you cannot have a string of physical signals being more than just signals. But imagine neural networks that hold signals, say associated with the word "table." Suppose your neural network is asked to accept or reject the sentence "Tables speak milk." Because the neural network will also have (empirical) relational probabilities stored (that is the frequencies with which one word follows another), it will "know" that the verb *to speak* cannot (normally) be used after "tables," and that "milk" is not known to be used after "speak." Thus, the neural network, which is entirely physical, would perfectly well declare that phrase to be *meaningless*. The antiphysicalists, of course, would claim that this does not entail that the neural network can cope with *meaning*, because they prefer to reserve the word *meaning* for something more exalted, more transcendental, something that would necessarily turn out to be unique to the human mind. If the latter is the premise they accept as evident truth, there is nothing to which one could object. But there is a long way from this to *proving* that premise by reasoned argument.

I hope that readers will realize at this point that any further discussion would be sterile, and that it is for them to make their choices; this does not mean that any one party is more right than the other but, if two *different languages* are used in a dialogue, final illumination is unlikely: natural scientists and idealist philosophers must agree to go friendly their own different ways.[13]

WHAT DR. JOHNSON SHOULD HAVE SAID: THE DEFENSE OF SCIENTIFIC REALISM

Scientific realism (a position which I shall try to define at the end of this section) has not a very good press these days; it is possible that Dr. Johnson's legendary kick has not done much good in this respect because as a philosophical argument it is extremely naive. Scientists, however, who almost as a professional necessity support some sort of materialism or realism, are often too much taken in by it.[14] So, in order to parallel the numerous conversations between Boswell and Dr. Johnson, I shall produce a short dialogue between a scientific realist, Mr. Fish, and an idealist philosopher, Sophia.

Sophia. Why are you swimming, Mr. Fish?
Mr. Fish. (*In a deep raucous voice.*) A fish has to do what a fish has to do.
Sophia. Then you do believe in the existence of water. Don't you?
Mr. Fish. Why shouldn't I?
Sophia. Because you know very well that you cannot *prove* the existence of matter.
Mr. Fish. Do not come to me with that. Not only have I gone to school, but I spend most of my time in one. And I know very well that Dr. Johnson was a mere *naive realist* when he thought that he could prove the existence of matter by just kicking it. He could just as well have eaten his stone, and yet would not have proved that it exists. Your request of a proof, however, gives me the grumbles. Only the other day, when we were together discussing God, you agreed that no philosopher has ever produced a satisfactory proof of His existence. Why should I be exposed to a higher standard of belief than the pope?
Sophia. Come, come, Mr. Fish, you say that because you are getting out of your depth. You have not yet persuaded me that Dr. Johnson perceived more that his own pain, and this, which is immaterial, as being within the mind, could not be caused by any material object.

Mr. Fish. Pain immaterial? My fin! Pain is a physical process, a neurophys-iological process. I can construct for you a perfectly sound causal chain from the stone to Dr. Johnson's mental state of pain. The force or action of Dr. Johnson on the stone causes, as you well know from classical mechanics, a force of reaction on his foot. This force excites nerve end-ings in the foot, and the signal propagates along the nervous system until it reaches a synapse in the brain. What happens at the synapse produces a state in the brain which you call a mental state of pain, but that is nothing else than a neurophysiological, that is, a physical, state.

Sophia. This is getting us into rather murky waters, Mr. Fish, because I do not agree with your view of the mind as being something that hap-pens solely in the material body, but if we go along that line, we shall never end. The question remains that you do not seem to be inclined to *prove* that water exists.

Mr. Fish. Not a bit. On the contrary, I shall tell you why I consider myself to be perfectly well entitled to *believe* that water exists. Do you agree that I have a buoyancy sack, or, as you call it, a swimming bladder?

Sophia. Yes of course, everyone knows that.

Mr. Fish. And do you broadly accept the theory of evolution?

Sophia. Naturally. I may be an idealist and a believer but I am also a modern woman.

Mr. Fish. And would you say that my buoyancy sack is there because it is a trait that has helped the survival of my species?

Sophia. That is evident.

Mr. Fish. So, I cannot prove logically anything, because logic cannot prove anything concerning nature. But I can try to reason in such a way that, not only does it not fall into contradiction, but that it also agrees with as much well-established knowledge as possible. (Only the other day at my school somebody was reading from some notes that called this procedure *entrenchment within the scientific mesh*.) So, I have measured my buoyancy sack, and from its volume, and from the den-sity of the water (which I can measure, even if I cannot prove that it exists), and from the laws of physics, I have worked out that it is the buoyancy sack that keeps me floating in the water. And because evo-lution took hundreds of thousands of years to favor my sack, which would be useless on dry land, I can safely assume that this would not have happened if water did not exist, and if I weren't surrounded by it. And although not a single step in my argument is capable of an ontological *proof*, the fact that everything in it fits perfectly together encourages me to posit that water exists. I believe in God, as you know, but I confess that I do not have such a good argument to sup-port this belief as to support me in the water.

Sophia. I am not sure that I agree with what you say, but I recognize the form of your argument to justify your inference that water exists. This is an example of what we philosophers call *inference to the best explanation.* As far as I can see, you are saying that the best explanation you can find of your own anatomy is that water exists. And that you consider that to be the best explanation because, in order to understand it, you have to sweep large tracks of the scientific mesh. To me, however, your argument is a bit fishy, if you don't mind my saying so, because on reducing everything to the physical you are throwing away the mind, and with it everything that matters most to me, like meaning and purpose.

Mr. Fish. It is you now who is getting into deep water. You wanted me to justify my belief in that liquid, and I have done my best to that effect. If you want to talk of meaning and purpose and of flying fish, we shall have to meet again because my school is calling me.

Mr. Fish had obviously learned his scientific realism from the American philosopher Wilfrid Sellars (1912–1989), who asserted that "To have good reason for espousing a theory is *ipso facto* to have good reason for saying that the entities postulated by the theory really exist."[15] All that Mr. Fish is doing is working out what he considers "good reasons" for holding that matter exists. We humans can go somewhat further than Mr. Fish, because the organ that we can present as evidence for the existence of matter is our own neural network, which could not have been a result of the evolutionary process unless appropriate external stimuli had been received by the organism. In this case, there is *laboratory evidence* that these stimuli are necessary for the development of the neural network in the brain (remember the mouse minus one whisker on p. 108). Moreover, the nature of this organ, the neural network, tells us something about the stimuli that the organisms had received during evolution, because the neural network is a learning system that could not have been created if the stimuli received and processed by the perceptive organs had not been repetitive. This remark gives the scientific realist good grounds for accepting the principle of regularity of nature as a good working tool *in dealing with the macroworld,* since the evidence for that principle comes entirely from within it.

WHERE HAS THE SPIRIT GONE?

Both scientific realism and materialism have for the general public connotations of philistinism, and science itself has been subject to pejoration under the name of *scientism*. You will never, alas, find a practitioner of this art: if you pop the question to anybody, "Do you engage in scientism?" they will almost certainly answer: "I don't, but I know a man who does." What scientism is, except it being a *bad thing*, I am not entirely sure. Possibly, the worst thing that a scientist could do is to claim that science can provide, now or perhaps in the future, answers to all the problems of humanity. If that is the accusation, I confess that I have never met any such miscreant. Curiously enough, and to my deepest despair, I have instead known many politicians, allegedly professionally concerned with humanity's predicaments, who defended to the death the view that *their* form of politics held the answer to all such problems. But I have never found a word derived from *politics* applied to this aberration.

Be it as it may, the fact is that many people regard with alarm the materialist stance that scientists, qua scientists, often have to adopt, and they feel that, by insisting purely on the material, the scientist denies that which is most dear to us, music, poetry, love, religion.[16] And that, even worse, all conception of meaning and purpose is thrown overboard by the swell of scientific rationalism. Patrick White's need for meaning expressed in the third epigraph to this chapter is neither unusual nor unimportant; and Salman Rushdie's poetic statement of the human condition could be read as a counsel of despair, when in fact it can be the beginning of a positive program. Let me try to discuss how such a program could evolve.

Over the centuries, humankind has tried to claim first a central and then a unique position in the evolutionary scale, often without much success. Yet, there is something unique to the human species: it is the only one that has a *recorded* history. Even if we have no guiding stars, as Rushdie says (and he is the last man who could be accused of scientism!), we do have our recorded history, and because of it we are not alone. Whatever we do has to be measured in relation to our past, to our culture, and to our language. If I lift my hand in a certain way, my movement will have a *meaning* for you, and if I continue with it to pick up a bottle to feed a baby, it will have a *purpose*.[17]

Many of my readers will think at this stage, "But this is not the meaning I want, this is not the purpose that I seek to understand. I want the meaning of life in toto; I want the meaning of the universe." You have seen, though, that both meaning and purpose, in the examples I have given, require a point of reference, and you cannot have a point of reference for the whole of life, or for the whole of the world, unless you posit something external to them, which is for many people the God they believe in. Science, however, is here out of its province. It would be imprudent for scientists to claim that their science either proves or disproves the existence of the deity, and thus to claim that the concepts of meaning and purpose, for the universe as a whole, receive or not a warrant.[18] Disbelieving scientists can at most say that they cannot conceive of a meaning for the use of the words "meaning" and "purpose" in any other way than as relative (contingent), and not as transcendental concepts. And when they say that life, or the evolutionary process, has no meaning or purpose, it is only in this, and not in the transcendental sense, that they speak; if only because a scientist qua scientist must necessarily say nothing at all about the transcendental.[19]

On the other hand, it is not scientism for scientists to fight their corner and produce such a vision of a self-organizing nature as it is within their duties. Let me briefly summarize what they can say in this respect. If we recognize the importance of having a recorded history, which is the result of having *language* and *writing*, we must admit that each of us shares in one way or another a pool of ideas, feelings and values, and that this pool has been subject to a process similar to the one of biological evolution, in which ideas evolve and fight to survive.[20] In order to describe such immaterial evolution, it is important to remember the possible modes of description of evolving systems. One possible mode, which was the classical theological one, was to describe the driving force for change in such systems as *teleological*. This means that all change is informed by a *purpose*, this being for instance that of creating the best possible species or the best possible idea. This *purpose* can best be described as Aristotle's *final cause*. Another mode of description is to use the principle of natural selection, which implicitly *negates* the metaphysical teleological explanation, because the acceptance of the latter contradicts the concept that the system is self-organizing, proceeding from one stage to the next by adaptive random changes. An important consequence of this approach is that the emerging structure is not now recognized as the best possible structure in all possible world histories. Rather, it is the structure best adapted to the circumstances holding during the period in which the structure evolved: in this type of process, the outcome at any one step depends on the *whole* of its previous history.

Because the acceptance of the principle of natural evolution implies a denial of metaphysical teleology, the latter might be disregarded by overzealous theorists who view such ideas as best forgotten cobwebs from the past. This we cannot do because, as we have seen, the evolutionary principle implies that the whole of the past is telescoped during any process of change. It is useless to consider possible but unrealized world histories. Thus, rational discourse cannot exist in a vacuum: we must reason from the *existing* pool of ideas. All of them, good or bad, are inputs which must be recognized as part of our condition, and the only way to be rational is to react to ideas such as they are, *not as they might have been*. Even the science mesh is not free from this constrain: we cannot argue from scientific concepts that *might have been*, but from scientific concepts *as we find them*. (Which makes the concept of the Kuhn's "paradigm" rather dubious, as I shall discuss a little later.)

Does this mean that we must accept everything as equally valid? No: if we were to do that, the great cultural evolutionary process would come to a standstill, which in evolution is a pathological situation. We must therefore make our own decisions and move in our own directions, because each of us is part of the great driving force for change. By this movement, whether we express the words or not, we are ostensibly defining our own meaning and our purpose, although they are not transcendental: they are contingent with respect to the whole of the history of our evolution. But, if in doing this we use a language divorced from our inheritance, and, because of excessive concern with a final result, we forget the condition we start from (always fundamental in an evolutionary process), and we are in danger of producing not change but upheaval.

Finally, is the acceptance of the principle of natural selection, even if regarded as meta-physical, incompatible with supernatural intervention? No, because the possibility cannot be denied of a creator who might have designed a self-regulated driving force for the evolutionary process of the world from the big bang on. But acceptance of the evolutionary principle requires acceptance that *intervention of the creator in this evolutionary process cannot be manifest*. Otherwise, we would fall back into a teleological view of evolution for which there is no evidence whatsoever, as will be discussed in the finale.

BACK TO MENTAL STATES

As I have remarked in the last chapter, especially in relation to the problem of quantum measurements, there has been a great deal of effort on the part of some scientists to establish what you might call a physical role for the mind in the description of nature. On the other hand, those who are concerned with the meaning of thinking, especially in mathematics, claim sometimes that the mind, even if considered as a physical phenomenon, cannot be described by the usual laws that regulate the latter. I have already shown that such views are by no means the only way in which you may go forward. The problem remains, though, that if mental states are basically regarded as neurophysiological, it is useful to have some idea of how, starting from mere matter, we eventually reach phenomena that appear to many of us as immaterial.

The first concept that is useful to formulate in this respect is that of *emer-*

gent properties. Take a molecular gas: we all understand what we mean by temperature, and yet, if we were to visit the gas à la Leibniz, you would see molecules moving with diverse speeds; but there will be nothing which you might recognize as a temperature. I had better qualify that statement: as you remember, a major problem with Leibniz was that he did not have the means to decode any signals that he might have been able to read. The same happens to our molecular visitation. Intelligent visitors, however, would go into the gas armed with some velocity-measuring device, would then measure the velocity of each of the molecules, and then from some physical law (the Maxwell-Boltzmann distribution), they would be able to produce a number which they would call the *temperature* of the gas. This is nothing else than a measure of the average kinetic energies of the molecules.

Imagine now a similar visitation but one instead that is totally possible. You go to a football field and you measure the velocity of each player with a gauge such as that used by the traffic police. You could then do precisely what I suggested in the last paragraph, and work out the "temperature" of the system of players. Far-fetched as this seems, it is perfectly possible, but nevertheless stupid, because the footballers form too small a system for their "temperature" (in the sense of some mean kinetic energy) to have any meaning. It is often the case, however, that as a system becomes more complex, properties *emerge* which, although entirely determined by the physics of the problem, are nevertheless adventitious to it. They are significant, and they have a physical meaning: although they are not attached to any single element of the system, they are useful in describing the whole, as temperature is for a molecular system.

The second concept that is important is the way in which properties *emerge* when the systems are very complex (as it is the case with the neural network in the brain). Consider a system in which there are lots of very simple units, all coupled in pairs. If you start with only a small number of such pairs, you will find that their interactions, which you could visualize as resulting in greater or less distances between the units, form no particular patterns. That is, the densities of the units, to follow on my example, are quite random and mean nothing. When you put together a very large number of precisely the same units, nevertheless, it follows from the mathematics (which is called *complexity theory*), and it can be demonstrated by computer simulations, that patterns appear in the system that are *global*, extending over the whole system, or at least over large parts of the system with respect to the size of the elementary units. Because these patterns entail information, that is, departure from the random distribution, this information *could* be understood (if you knew how to crack the code) as a message. It is not impossible that what we call thoughts are held in our brain in a way which is not too different from the patterns that I have just described.[21]

It might be useful to produce a simple example of how a highly significant global pattern would be lost from a reductionist point of view. Imagine a visitation of the sea à la Leibniz. You are now reduced to molecular size: would you we able to recognize a wave? No: all that you would see would be an apparently chaotic motion of the water molecules. Even if you were to measure most carefully their positions, since you will be doing this at different times, you would have to develop some pretty clever theory (assuming that you know nothing about waves) so as to finally identify the wave. (Because the water molecules are affected, not only by the wave, but also by thermal motion, this is not a minor task for our homunculus.) If you now reflect that a huge amount of the information that reaches us has done so through some form of wave transmission, at least during part of the process (even your printed newspaper started life as faxes, telephone calls, and computer links), you can see how a blunt reductionist approach to physicality may leave you with an entirely distorted view of what physicality might entail.

SCIENCE AND NATURE

Any treatment of physicalism must handle the question of what is the relation between science and nature, which is for us as always nature$_2$, the result of the interaction of our cognitive and rational system with whatever is "out there." I have argued in this book that there is an evolutionary process in science, whereby the scientific mesh is constantly revised, extended, and made as internally consistent as possible. Since the scientific mesh contains the whole of what we normally call scientific data (observations and results of experiments), the internal consistency of the mesh entails that its theoretical elements do fit the experimental ones. This is what one might call an optimist view of science (and likewise of physicalism), since it entails that science moves progressively to a better and more accurate picture of nature.

A strong claim for physicalism this is, but not the only one, alas, that is universally embraced. There are two separate strands to the history of this problem, the oldest one resulting from the interaction of science with theology, and the other, much more modern, arising from studies in the history and philosophy of science. Curiously enough, as I shall show, the tensions that for centuries have characterized the first strand, have been resolved very largely in favor of a view broadly in agreement with the role of science that I have proposed. It is historians and philosophers of science, instead, who have created a skeptical view about science's ability to describe nature; a view that has had far more popular influence than it deserves.

I shall deal in a moment with these two strands of the problem, but I should like to start with an observation that might help to get things into perspective. There is no doubt that science is a part of the evolutionary process that has informed the world of ideas that is the backbone of our culture. And when I say this, I immediately hear on my back: what about other cultures that do not share our scientific view? Of course, such cultures exist, and are entirely legitimate and rich in their own way, and *cultural relativity* is an important and delicate concept, so that let me postpone its consideration until we understand a little better what role science plays in our own Western (and indeed largely world) culture.

The beginnings of any culture must be based on irrationality, and please do not misunderstand me as propounding irrationality as the basis of culture. Not at all. But any advanced cognitive system must be booted up by some mechanism that is somewhat random: a baby moves his arms purposelessly, until he realizes that he can find his own mouth with his hand.[22]

Learning entails making hypotheses and trying them out. We are pretty good at present at making hypotheses, because we try to make them in relation to a reasonably substantial body of data to be explained, but this was not possible at the beginning of our cultural history, when hypotheses had to be made within an almost total lack of prior knowledge. And they were made: what is important is that such hypotheses, however ill founded, entail a desire to find explanations for physical phenomena. Thanks to the fact that we started by believing that thunder was the result of the action of a deity, Zeus, we booted up the line of thought that eventually led to the science of meteorology. The latter is a rational mode of thinking, in the sense that it is not based on arbitrary ad hoc hypotheses, but rather on huge well-known chunks of the scientific mesh. The structure of clouds, for instance, entails a great deal of the physico-chemical properties of water, of electrical theory, of fluid dynamics, and so on. The concept of Zeus as the thunder maker may be poetic, but it is irrational; and this is not a derogatory statement, but a description of the way in which hypotheses are sometimes generated.

If we look back at our knowledge of nature through the last millennium or two, it is easy to draw conclusions about science that are prejudiced by the fact that elements of irrationality persisted in it for a long time and resulted in false starts which, if taken seriously, create a precarious idea of the scientific enterprise. The evolutionary history of our species probably starts with amoebas, but it we want to understand *Homo sapiens*, it would be confusing to go so far back. Likewise, much can be made of alchemy and phlogiston theory to argue for the ephemeralness of scientific knowledge, but the question is whether those theories were science as *we understand it now*: they are not, because they do not satisfy even the most elementary modern principles of the scientific method.

Let me justify these assertions a little—although they should be fairly obvious from the work so far done in this book—for which purpose I shall look briefly at the *phlogiston* theory. The story starts with Aristotle and his belief in necessity: things must have reasons for being as they are, and it was his job to find them out. He did this in a grand scale, by producing classifications that in certain cases were based on what necessarily were superficial observations. One such was his "discovery" of the four elements—Earth, Water, Air, and Fire—as behind all properties of matter: although this is at the root of our civilization, it is hardly any nearer to being a scientific view of nature than Genesis is. To start with, the very *intention* behind such postulates is one that true science, from Galileo onward, has rejected. And the Aristotelian top-down method, starting from imagined concepts, in the hope of deducing from them the contingencies of nature, is the very opposite of our scientific practice.

Phlogiston was the intellectual heir of Aristotle's *fire*. In the seventeenth century combustion was regarded as the result of the escape of the *combustible principle* (very much a reincarnation of Aristotle's *fire*) from the burning body. It was the German physician George Ernest Stahl (1660–1734) who used the name *phlogiston* for that entirely hypothetical substance. The theory did not last long, however: it was observed that in many cases weight *increased* on combustion, so that this hypothetical substance had to have *negative weight*, a silly idea that the French chemist Antoine Laurent Lavoisier (1743–1794) soon discarded (see p. 228) by appropriate experiments. End of story: and this story is just a little more relevant to the significance of scientific ideas than Genesis is to cosmology. *Neanderthal* is perhaps nearer to *Homo sapiens* than the phlogiston dogma is to chemistry. Any treatment that ignores the fact that phlogiston and science are two different "species" in the evolutionary scale of ideas is bound to obscure fundamental issues.

Of course, those who embrace the view that everything that went on from Aristotle to Galileo to Einstein has the same right to be called *science* have to postulate great upheavals of the scientific world, resulting each in the total abandonment of previous conceptions. I have shown, however, that well-entrenched laws are normally not abandoned, but that their domains of application are instead circumscribed. Galileo's, I have argued, was not so much a scientific as a theological revolution, which brings me to the consideration of the first strand I had mentioned, involving the relations between science and theology.

Science and the Theologians

I have already said that it took a very long time for science and philosophy to be able to express themselves freely, without regard for the religious views of the time. One of the first theologian-philosophers defending the need to make these activities independent was the Italian Dominican Giordano Bruno (1548–1600), who argued for the separation of "philosophical reason in accordance with natural understanding and principles" from "the truth in accordance to the light of faith." He was made to recant this view by his inquisitor, the same Cardinal Bellarmino who later dealt with Galileo, but, as we have seen, it was his views on the Trinity that finally sent him to the stake.

Galileo had a double battle to fight, against both philosophers and theologians. First, he repudiated Aristotle's method of reasoning from first causes, which had to be guessed by pure thought in order to explain why things are as they *are*. It was Galileo who insisted that the job of science was instead that of finding laws or rules that describe how things *behave*. But, as Giordano Bruno before him, he bravely tried to assert science's independence from theology, he warned theologians:

> that in your desire to make matters of faith out of propositions relating to the fixity of sun and earth, you run the risk of eventually having to condemn as heretics those who would declare the earth to stand still and the sun to change position—eventually, I say, at such a time as it might be physically or logically proved that the earth moves and the sun stands still.[23]

Notice that in Galileo's time there was no cast-iron proof of the rotation of the Sun around Earth, but the alarm note raised by Galileo, should such a proof become eventually available, was not unappreciated by his inquisitor, Cardinal Bellarmino, a fact that does not appear to be well known. In such a case, Bellarmino wrote,

> it would be necessary to move with great care in explicating the Scriptures that appear in contradiction, and to rather say that we do not understand them, instead of saying that what had been demonstrated is false.[24]

This is quite remarkable, but it was much later that Galileo's contribution to theology became fully understood and accepted. As the truth of heliocentric theory became evident, Bellarmino's understanding of the implications of Galileo's position was reflected in various pontifical documents that hinted that the interpretation of the Scriptures could not be entirely literal. It was only at the Vatican Council II (1964–1966), however, that a document called *Dei Verbum* was issued, advising the church that a metaphorical reading of the Scriptures was appropriate when necessary. And, at last, the sciences were freed from their chains: in No. 59 of the conciliar constitution *Gaudium et Spes* "the legitimate autonomy of culture and especially of the sciences" was solemnly recognized.

Even more, in a speech read before the Pontifical Academy of Sciences on 10 November 1979, on the occasion of the first centennial of the birth of Einstein (which was extended also to a commemoration of Galileo), Pope John Paul II recognized that science is concerned with "the knowledge of the truth present in the mystery of the universe" and that the "search for truth is the task of basic science." Even more: "The Church willingly recognizes, moreover, that she has benefited from science." And the pope went on to make his a quotation from Monsignor Georges Lemaître: "Does the Church need science? Certainly not, the cross and the gospel are sufficient for her. But nothing human is alien to the Christian. How could the Church have failed to take an interest in the most noble of human occupations: the search for truth?"[25]

The implication of this is quite clear: even for the pope, science is engaged in "the most noble of human occupations," and in doing so it has helped religion to rid itself of the mental blinkers that considerably restricted its understanding of nature for centuries. Anyone who still thinks of science as a purely "materialistic" activity should seriously ponder over these remarks. And we can now move over to the second strand in our cultural history that affects the status of science vis-à-vis the description of nature.

Science and the Philosophers

Whereas, as we have seen, even the very church that condemned Galileo is now happy to acknowledge that the goal of pure science is freely to search for the truth about nature, many philosophers in the last half of the twentieth century have hotly denied, either that this is the case, or even that such a goal is possible. And any such who has read my last section, would be able to retort that reference to the Roman Catholic Church to license science's claims to pursue the truth are of very little value, since the doctrine of that church is based on the—for them dubious—principle that truth does exist.

The reader will have noticed that I have studiously avoided getting involved in the analysis of the concept of truth, but this is very thorny, and it, in fact, permits such criticisms to be legitimately raised. This, alas, does not mean that, in the particular case of science, one has to accept at face value the arguments that have been adduced to curtail its scope, and I shall try to review them briefly. But I shall attempt to place them first in relation to an important line of thought that appeared in Western culture toward the end of the nineteenth century, to which I shall roughly refer under the name of *cultural relativism*.

Since the days of the invasion of the Americas by Europeans, new cultures were discovered that were described at worst as savage, and at best as primitive: the lack of understanding of their richness was total, to wit the melting down of literally tons of priceless metal artifacts that took place. It was only toward the end of the nineteenth century that artists began to recognize the value of the work produced by some of these "savage" societies. Picasso, Gauguin, van Gogh, and Léger recognized that art schools from Benin to Khaligat had produced some of the most remarkable works of art known to humans. Naturally, at the same time, anthropologists like Bronislaw Malinowski and Claude Lévi-Strauss revalued our views of these societies, and the concept of *cultural relativity* became widely accepted. Let me try to describe it with a simple example: anyone who has seen a Byzantine mosaic will realize that it is very different from a Renaissance fresco, say. There is no play of light, the figures are largely two-dimensional, and so on. It would thus be tempting to conclude that the first is "primitive," compared to the "mature" Renaissance masterpiece. Yet, it would be wrong to say that one form of art is "better" than the other. Each has to be judged within its own terms of reference. Byzantine art was iconic and narrative, and the quality of a Byzantine mosaic has to be judged on how far these qualities have been achieved. Renaissance art is naturalistic and expressive; and likewise its quality depends on the fulfillment of these principles. As with art, also societies and cultures differ in their goals.

The doctrine of cultural relativism is an important one, and one that cannot lightly be disregarded. It is for cultures as important as equality is for individuals. And herein lies the trap that can easily mislead people in making the wrong claims for this doctrine. I am the equal of Einstein, no doubt: in law he (were he alive) and I have the same rights, pay the same taxes, and so on. But if I were to claim that I am also the equal of Einstein in the richness and scope of our minds, it would not just be silly, it would be positively demential. Cultural relativism, likewise, is right, but it is right only in the right context.

In order to illustrate this problem I shall skip numerous historical precedents and move on directly to the most influential historian of science of our generation, Thomas Samuel Kuhn (1922–1996) whose book *The Structure of Scientific Revolutions*, first published in 1962, changed the very language used by students of the humanities throughout the world. The following quotation will immediately connect the reader with my account of cultural relativism:

> The more carefully they [the historians of science] study, say, Aristotelian dynamics, phlogistic chemistry, or caloric thermodynamics, the more certain they feel that those once current views of nature were, as a whole, neither less scientific nor more the product of human idiosyncrasy than those current today. If these out-of-date beliefs are to be called myths, *then*

myths can be produced by the same sort of methods and held for the same sort of reasons that now lead to scientific knowledge (Kuhn 1962, p. 2, my italics).

I am afraid that I do not know of any other single sentence in the English language that has done so much damage to the public understanding of science as this one, although this might not have been Kuhn's intention. Let me compare its approach to different scientific cultures (say the Aristotelian and the modern one) with the treatment of Byzantine and Renaissance art under the concept of cultural relativism. We saw in this latter case that the two art cultures had different purposes, and that each work had to be understood in relation to its own goals. Modern science, instead, has one and only one goal, which is to understand and fit nature as closely as possible. As science got rid of its philosophical and theological fetters, this is the unique way in which the word science can now legitimately be used. If it is claimed that Aristotle's primary purpose was not that of fitting his theories to facts, but that he nevertheless produced, within his own terms of reference an *internally* consistent philosophy, then we are now comparing two entirely different activities: science stands or fall by saving the facts, and "internal" consistency means nothing, unless "internal" is meant to include a vast range of natural facts.

The reader must not be confused by the vagaries of language: a decoy duck is no more a duck than Aristotelian science is *science*.[26] And whoever is prepared to use the word science without careful analysis of the context, should look up 1 Tim. 6:20 in the King James version, where misunderstanding of the word *science* (therein meaning *gnosis*) would be disastrous. What Kuhn chooses to ignore is that the meaning of words is not cast in stone, and that what we now mean by "science" is the result of the struggle that started with Giordano Bruno and Galileo and still goes on. It is by praxis, by doing it, that we have created the current meaning of the word *science*, which has little to share with its own roots, the philosophical approach of Aristotle. To use the same word for the two activities would be almost as meaningless as not differentiating *Homo sapiens* from the higher primates, much as we share in the same genetic roots.

There is another crucial defect in Kuhn's argument. His work is based on historical research, and for his arguments to stand, he himself would have to be immune to the very criticisms that he adduces about science. Thus, on pp. 3–4 of the work just mentioned, he points out to "the insufficiency of methodological directives, by themselves, to dictate a unique substantive conclusion to many sorts of scientific questions. Instructed to examine electrical or chemical phenomena, the man who is ignorant of these fields but who knows what it is to be scientific may legitimately reach any one of a number of incompatible conclusions." It is absolutely amazing that a man or woman ignorant of the history of science, but who knows what it is to be a historian, must never disagree in his or her findings with anyone else in the same situation. And it is also amazing, contrariwise, that not a single European in 1820, who of course were all ignorant of the until then unknown relations between electricity and magnetism, did disagree with Ørsted when he found his extraordinary result. I have a nagging suspicion, in fact, that it is a great deal easier to find agreement in matters scientific than in matters historical.[27]

Kuhn, of course, would contest my claim that science has one and only one goal, the understanding and fitting of nature (remember, always nature$_2$). Thus, he writes, many philosophers of science wish:

> to compare theories as representations of nature, as statements about "what is really out there." Granted that neither theory of a historical pair is true, they nonetheless seek a sense in which the later is a better approximation to the truth. I believe that nothing of the sort can be found.[28]

Cultural relativism is taken here in entirely the wrong manner. The phlogiston practitioners were doing a good job by *their* standards, but their standards were not such as to allow them to fit even the simplest of chemical facts, as Lavoisier demonstrated with the simplest of means. The phlogiston culture and the Lavoisier culture may be equal, just as I am Einstein's equal, but Lavoisier's science is vastly richer because it fits a wider range of facts.

Which brings me to the more meaty argument implicit in Kuhn's quotation, which is the notion of "fact." To start with, very few people would claim that "facts" are just things that are "really out there." A scientist would have to be a very poor philosopher who asserts that, and you will have noticed that I have always kept prudently silent about the "out there." But well before Kuhn did his hatchet job on science (I wonder whether he had ever tried to persuade his gas company to allow him to pay his bills under the equally worthy phlogiston theory), a great deal of work had been going on about the complexity of "facts," a point about which I have remarked since the beginning of this book.

An influential figure in bringing forward the hybrid nature of "facts" was Benjamin Lee Whorf (1897–1941). After the war everybody talked about how Eskimos, who had very many words for snow, could distinguish more shades of white than other people. That is, language, it was claimed, affects perception, and thus what we call "facts." It was later discovered, alas, that Inuit do not have the vocabulary that Whorf claimed, and his views fell into disrepute, although it is recognized at present that they contain a grain of truth: it is prudent to accept that what we call "facts" contain in them an element of theoretical preconstruction. This was a point well elaborated by N. R. Hanson and very much exploited by Paul Feyerabend.[29]

I shall now try to discuss one or two examples which I hope will show that, though facts do indeed contain an element of theory, this does not in the least justify the view that science runs itself into circles of myths, as Kuhn claimed. To start with, facts do not come naked to us. At any one moment the world is full of billions of events, and you need a trained eye even to exploit the most serendipitous of discoveries, in which nature appears to offer someone a "fact" on a plate. It is quite possible that what Fleming noticed in his petri dish when discovering penicillin was previously ignored by others. But there is much more to it. I have already shown that even the simplest fact, such as "I perceive a stone falling," requires rational organization of perception; and of course the whole of science must do the same, because we would otherwise be swamped by trillions upon trillions of apparently unrelated events.

Take Newton's second law: we have seen that it crucially depends on the definition of the instantaneous velocity, and that the measurement of the latter (at least in Newton's time) depended not on a naked fact, but on an organized series of events combined with a theoretical construct. This happens throughout science; it would be impossible to sample nature blindly: even a fisherman fishing for trout has to know where to cast his line. We must organize ourselves, and it is because of this that we create the scientific mesh, in which facts and theories are not kept in watertight containers but are allowed to interact. We do this because we know that the internal consistency of the scientific mesh is not that of an unchecked closed system: its consistency will automatically entail events which inform facts and which must fit the natural world. The virtuous circles, by which we defined the timescale and physical states, do precisely this job of testing nature until a fit is obtained. Thus, the theoretical content of our "facts" is not left to an arbitrary whim of our feverish imagination, as Kuhn's "myths" seem to suggest, but they are so chosen that science can always proceed by inference to the best explanation, such inferences of course being subject to change as further knowledge becomes available.

Kuhn makes much of the problem that when an isolated fact springs up which does not fit the mesh (or what he calls the "paradigm"), scientists are reluctant to accept it. (He thus claims that scientists are so tied up to the accepted dogma that they would in principle refuse to countenance any fact or theory that contradicts it.) It is only natural, however, that scientists should not be ready to accept a change of outlook without the strictest of checks, and there is abundant evidence that they are always ready to do so. Let us consider two examples. Stomach ulcers were thought to have an organic cause until the then unknown Australian physician Barry Marshall thought that a bacterium, *H. pylori*, was their cause and suggested this at a meeting at Fremantle in 1983. His unorthodox views were disregarded by the science community until he caused himself to be infected by the bacterium and showed that he had produced an ulcer. Once this evidence was available, his theory of the disease gained worldwide recognition in a very short time.

Second, all nonorganic diseases were thought until fifteen or twenty years ago to be caused by either bacteria or viruses, that is, by some sort of life form, which always contains DNA or RNA. When the American Stanley B. Prusiner predicted that a straightforward protein molecule, which was called a *prion*, could be responsible for some forms of tissue degeneration, again this theory was regarded with great suspicion until evidence was accumulated, the theory accepted, and millions of cattle destroyed in the United Kingdom as a result. Prusiner the heretic was not burned at the stake, but was given the Nobel Prize in medicine for 1997. Where is Kuhn's evidence that scientists stick doggedly to their preconceptions? How do they compare, under such Kuhnian methods of analysis, for instance, with theologians? One dreads to think what absurdly crude, naive, and out-of-date results about religion would thus emerge.

That the scientific community is conservative should not be a great surprise. So are lawyers: they use precedent and stick to the interpretations of the law that they know, until pressure for change takes its effect. People who have to maintain an *objective standard*, that is, who are responsible for making sure that things *work*, as a difference with those who write books (including the present one) about the way other people are supposed to do their jobs, have to take endless precautions in performing their tasks.

And there is an even more important moral coming from my two examples. In Kuhn's views, what rebels like Marshall and Prusiner did was to substitute a new myth for an old one: would a Kuhnian philosopher eat a piece of prion-infected veal brain, even if deliciously cooked *au beurre noire*? Or, if suffering from ulcers, have his stomach surgically excised? And why should such a philosopher think twice in such cases, given that a myth is just as good as any other myth?

Let me go back to the question that the facts that science deals with are hybridized with theoretical constructs. Because of this, Kuhn believed that science, even in its modern form, is another variety of myth making, and that these myths are formed by the presuppositions that scientists use in their work. The whole body of principles that the scientific community is prepared to include in the mesh at any one period is what Kuhn called a *paradigm*, and he took the view that paradigms are largely arbitrary and disposable, the great purging operations being scientific revolutions.

The concept that the "paradigm" is arbitrary and disposable reveals a complete misunderstanding of the nature of a self-organizing evolutionary system, as the scientific mesh is. Of course, there is nothing *necessary* about the contents of the science mesh, just as there is nothing necessary, for instance, about the structure of the human eye. In an evolutionary system, things are as they are *because of the whole history of the evolutionary process*. Our eyes could see infrared and have a vision field of 180°, but to indulge in such counterfactuals may be profitable science fiction but abominable philosophy. And what is valid for our organs is also valid for the words and concepts that we laboriously tested and refined within the science mesh. To imagine that contemporary science is nothing more than another "paradigm," with no better status than that of the phlogiston era, is equivalent to imagining that there is no reason whatsoever why dinosaurs could not coexist with us: the evolutionary process has dealt with the latter in the same way as it has consigned phlogiston to the shelves of history.

It will be useful to consider an example that might appear to support Kuhn's contention that the "paradigm" is not cast in stone. As we have just seen, one of the drawbacks of Newton's theory is that the instantaneous velocity is not a naked fact, but has to be obtained as a combination of observation and theory. But, had Newton been clever, he would have borrowed a Doppler gauge from his local police station, and he would have measured velocities directly, just as directly as we measure lengths. The consequences of this approach for physics would have been remarkable. In our "paradigm" (to play Kuhn's game) we have two independent units, space and time, which can be directly measured with rulers and clocks, respectively; whereas velocity is a derivative unit, given as space (miles) divided by time (hours). With the Doppler gauge approach to velocities, they and space can be taken as independent units, and time *disappears* as an independent variable: it is determined by measuring lengths and dividing them by the velocities: no need for clocks!

Although this might be adduced to show the "flimsiness" of our "paradigm," it has a major flaw. Not only did this scenario not happen (which in evolutionary terms is a *final rejection* of it), but *it could not have happened* because in Newton's time Doppler gauges did not exist, and they could not exist because their invention would not have been possible without the *previous* knowledge of Newtonian mechanics. As in the construction of the causal timescale, humanity has to start from rough approximations and build on them. The fact that the "paradigm" is contingent is entirely insignificant: in an evolutionary process everything in it, every single step is contingent. That is, at every moment things could have been otherwise.

As for the existence of Kuhn's revolutions, insofar as they are genuine scientific events, rather than liberations from philosophical or theological cobwebs, it is important to understand what we are talking about. Up to the sixteenth century or so, we did not have any science mesh worth its name. Much is made of the Copernican revolution but, as I have argued, this was more than anything a theologico-philosophical revolution from which, by being liberated from such fetters, science as an *independent* activity emerged. Since then, theories have come and some, like phlogiston or ether, have gone, but in neither case has the science mesh been torn asunder. Even relativity theory did not destroy Newtonian mechanics, still alive and kicking in its rightful corner of the mesh.

History has to be applied to science with a lot of common sense. Numerous *technologies* existed in prehistoric times (like metal smelting and fabrication, astronomical observations, and so on), but no science existed, in the sense in which we now use this word, until not much more than two or three centuries ago. And to pretend that technologies are the protoelements of science is wrong: the whole concept of *enquiry* into the structure of nature is absent in technology, which is merely concerned with the use (if not the abuse) of nature. This is the reason why the very origins of the scientific quest (*as opposite to scientific praxis*) are rooted in philosophy and theology, rather than in technology.

I have now treated two cultural strands that have aligned themselves against science at one time or another. The cultural relativism of Kuhn has presented us with one more.

Social Constructivism

One of the more momentous discoveries of the social sciences is that the processes described by physics are constructed by means of social interactions entailed by teamwork, conferences, the internet, committees, and all that. I shall show in the next section (in which I shall comment on all matters raised in the present one) that the conclusion thereby derived that the entities of physics are merely complex social objects, is an unwarranted non sequitur. (The "social objects" thus engendered are of course not much better than Kuhn's myths.) Andrew Pickering, one of the advocates of social constructivism, observed that physical science is driven by more that just the search for truth. Anyone who has read my chapter 10 will recognize, in fact, that the impulses to pay your mortgage or find a mate, and so on, are legitimate driving forces of scientific activity, as they are indeed of any other actions taken by human beings distinct from going to bed for purposes of purely physiological sleep.

As a result of this observation, Pickering asserts that scientific statements are not straight claims about the natural world, and thus rejects scientific realism. Even more, within this mode of thinking, the analysis of "knowledge" or "reality" is contingent upon human relations, whence value-free knowledge just simply does not exist. All research is essentially biased, an argument which, if generally applicable, should therefore affect the very social scientists that propound it, and thus throw grave suspicions on its validity.[30] Be it as it may, because of this alleged bias, the distinctions between "knowledge" and "reality," and between "objective" and subjective," are considered to be seriously blurred.[31]

There are two forms of social constructivism that are deeply related to cultural relativism, and which concern problems of our world culture where the application of this latter concept is most legitimate. These are the questions of *feminism* and of the *emergent economies*. In both cases I am afraid, nevertheless, that the very important points that should be raised in these contexts are somewhat confused by the lack of distinction between the aims and values of *pure science*, on one hand, and both the *practice* of it and its *technological implications* on the other. Anyone who is familiar with the appalling treatment of Rosalind Franklin during the discovery of DNA in the early 1950s, or with the current exploitation of rural communities through cash crops and deforestation, cannot but feel very strongly that these are matters that must be at the forefront of our minds; but these are *political* matters that have nothing to do with the fundamental role of science in the understanding of nature.

It is quite legitimate of Sandra Harding, for instance, to make a plea to revise the social values and concerns that lead scientists to address certain questions rather than others, with the possible consequential neglect of women's interests. But it must be appreciated that whatever might emerge from this could affect women qua women only insofar as certain problems of technology might be given financial priority: and such decisions are *political* and not scientific. It is obvious that the understanding, for instance, of the way in which quantum mechanics and relativity may be made mutually compatible cannot have the slightest reference to the gender of those who might benefit from it or otherwise through some ensuing technology. What I consider a relevant preoccupation is whether funds that are now engaged in some expensive scientific research would be better employed to assist, for instance, mothers to pursue their careers, as protected from family problems as men were in the past: but again, this is a political, not a scientific problem. Of course, many scientists do engage in either open or covert politics, but, if they can do so for purposes that are not related to the aims of pure science, this is partly due to the comparative lack of understanding of what science is in the minds of politicians and of the general public. The understanding of such political problems is not in my view helped, alas, by describing Isaac Newton's *Principia* as a "rape manual."[32]

In what concerns *pure* science and mathematics one must admit, without diminishing them in any way, that the problems of the developing world do not exist. There are *political* problems as regards how much money is appropriate to divert to such activities, and about which *technologies* should be sponsored. And there are of course *pedagogical* problems about how best to teach these subjects to make them more interesting and relevant to the local communities. But the Lorentz equations are the Lorentz equations in the United Kingdom and in the Gobi desert: and if you think that they are irrelevant in this latter environment, I wonder how far development in such areas will be possible in the future without making use of the Global Positioning System, far cheaper than relying on expensive cartography and triangulation: and such a system could not work without the use of Einstein's relativity theory. Likewise, to talk of *ethnomathematics or ethnoscience* might not be the best of things, since it may obscure what the real political and social problems are that must be handled in order to further mathematical and scientific education and, when appropriate, research in such areas of the world. The negative aspect of such expressions is that they might stunt the progress of those societies with a form of neocolonialism, in which whatever is interesting, exciting, and perhaps eventually profitable is excluded from certain areas of the world, under the pretext that it disagrees with what is taken to be relevant for them. When after the war the Czech physicist Guido Beck, then at the Córdoba Observatory, transmitted to the department of physics of the University of Buenos Aires a request from his friend Werner Heisenberg to be given a chair there, it was precisely for those reasons that the professors of that department rejected the application, which if granted would probably have changed at a stroke the quality of South American physics. On the other hand, a healthy recognition that the cultural background of any given society must be sympathetically understood when planning mathematical teaching is important, as the book by Maria Ascher (1991) shows.[33]

THE PRESENT AND FUTURE OF SCIENCE

I shall deal in this section with some of the views about science presented in the last one. Of course, the pursual of science is a social enterprise, but it is plainly not true that its products are social objects, which presumably means that they have a value and meaning entirely dependent on the society that supports such enterprise. One of the major triumphs of science (and as always I include mathematics in this term) is that it has created a universal language. Newton's equations or the Lorentz equations are identically the same everywhere in the world, and sometimes even the script in which they are written is the same, irrespective of the local one. Scientific protocols, that is, the procedures by which the science community satisfies itself of the validity of a piece of work, are the same everywhere. But this universality of science is even deeper, and it is based in the fact that scientific statements must be made objective, that is, independent of the observer who reports them.

The reader might remember that one of the remarkable results of relativity theory is that lengths do not have an objective meaning, in the sense that different observers in different states of motion would not agree as to their measurements. It was thus concluded that lengths could not be used to describe in any form what one might expect to correspond to an "element of reality," and that they had to be replaced instead by *spacetime intervals*, for whose values all observers in whatever state of motion would agree. In the same way and in the same sense, all scientific statements in physics have to be scrutinized to make sure that they are *objective*. It is because of this objectiveness that science has produced a universal language, the only such since the Adamic one, which can be understood and used in exactly the same way in all corners of the world:

> A mathematical concept that is clear, a proposition that is true, a theory that is coherent, for an American mathematician is so for a Chinese or a Hindu mathematician and *vice versa*. That is a trite observation. There are individual differences of style, but no difference in substance, and none along national or racial lines.[34]

Thus Hermann Weyl, whose work ranged in physics, mathematics, and philosophy perhaps more widely than any other man of the century. Those who wish to pursue such things as ethnomathematics or ethnophysics would do well in pondering over these words and in making sure that they do not confuse technology with pure science or mathematics; and that they do not single out science as the one activity for which its praxis in the marketplace is confused with its essence. Politics, religion, economics, sociology, and art could hardly survive the same assault: the marketplace is there equally for all of us.

My last few paragraphs will be concerned with the claims of science as the only valid mode of enquiry to search for the truth about nature, leaving of course philosophy and religion to speculate about the nature "out there." This is a major claim, and one that illuminates the moral value of science, since the propagation of untruths has been, and still is, the moral scourge of humanity. That such a task has been conceded to science by the Vatican, thus wiping off centuries of obscurantism, will of course, as I averred, cut no ice with those philosophers who, following Kuhn, reject even the possibility of this claim.

As my readers will know, I have taken a very pragmatic view of "truth." I have looked at the body of science as a gigantic mesh, the elements of which are accepted or rejected as necessary, in order to bring consistencies with the challenges presented to that body by its interaction with nature. By definition of this evolutionary process, the more we move on, the wider the mesh we weave and the deeper we go in our understanding of nature. And I am perfectly happy to use the shorthand expression "the nearer we get to the truth about nature." Philosophers may object that this is far from being the Truth, but in this sublunar world all that we can achieve is to save the facts and to understand each other, and for that we must work together to build a consensual language: and this is the language that science helps to build, and which allows us to talk about "truth" as regards natural matters.

What I have said does not mean that at any one moment in this evolutionary process we can stand up and say: now we have reached the ultimate, now we understand everything. And there are two reasons for this. One is that the great scientific enterprise is open ended. If we were to achieve a situation where everything is known, that is, everything in the science mesh is in perfect internal consistency, we could say that we had achieved a remarkable advance (which of course contradicts Kuhn's conception that it is not possible to judge progress in the scientific evolution, but only change). What we could never say, though, is that this is the end of the road because no one could ever claim that no new phenomena will become known that will require more darning and weaving of the mesh.

The second reason is concerned with the nature of the evolution of any self-organizing system. The successful "mutations" in it (that is, the new facts and theories inserted at any one time in the mesh) ensure the best fitting available at that time *with respect to the state obtaining for the system when the "mutation" occurred.* Each time the eye evolved, it was not the best eye that was created; it was the best eye compatible with that initial state. Likewise, we weave and darn the science mesh, but we cannot make ourselves entirely independent of its origins,[35] and it is possible for some period of the scientific culture to make some progress, but only in some sort of cul-de-sac predetermined by the initial state. That we are or not in such cul-de-sac, only time and work can tell us: nature is always the supreme teacher.

FINALE. SCIENCE AND BELIEF

No treatment of the relation between science and nature could be complete without some reference to belief, since belief is the strongest motivation in positing a dual structure of nature. Because my own views on this delicate subject could interfere with my main job of trying to sort out the way in which one can legitimately argue about these matters, I leave the discussion in the capable hands, metaphorically speaking, of Mr. Fish and Ms. Sophia.

Sophia. Last time you argued for the existence of matter by means of an inference to the best explanation. But could I not likewise reason from the wonderful harmony of nature to the existence of God?

Mr. Fish. What harmony you talk about, when we have horrid earthquakes that generate the most disgusting tidal waves. And what about the people who come day after day to massacre us in the middle of our schools?

Sophia. You perceive disharmony because you ignore the global view of nature. What is bad to you may be good for others, and vice versa. This is the unavoidable condition of nature that the theologians describe with the metaphor of the fall. Thus, the grand design of nature, if you look at it transcendentally, that is, as if it were from outside nature, is harmonious and reveals the hand of God.

Mr. Fish. If you permit my saying so, Sophia, you are now talking through your goggles. As to the disharmony that you agree we experience: when I do my scientific thinking, I never find that my bad science is considered good by others. Science is harmonious, that is consistent, or it is not science at all. But there is something far more serious in your argument. I think that you must agree that the only way in which you recognize some transcendental harmony is if you accept the metaphor of the fall. Am I right?

Sophia. Yes, certainly, any theologian would tell you that.

Mr. Fish. But you must also agree that the concept of the fall comes to you through revelation and faith. And that you cannot have revelation and faith unless you believe in God.

Sophia. Yes, undoubtedly.

Mr. Fish. Therefore, you are disbarred from using the argument that the harmony of nature that you posit reveals to you the existence of God, since the first premise already involves the acceptance of the second one. I wish that you people once and for all renounced the use of dubious arguments about alleged harmonies and complexities. In the old days, preachers always tried to entice us with the beauty of the night sky. Now that this has worn thin, you come with the admirable design of nature, which, if you think about it, is nothing more than the result of the blind forces of evolution.

Sophia. I think it is you who is being blind, if you think that the vast complexity of the laws of nature, and of life, and of the mind, could have evolved without a purpose. Surely God is needed to guide the hand of evolution, in however hidden a manner you might claim.

Mr. Fish. This is very much like Paley and his clock: if you find a superb clock in the middle of a desert, you can be sure that some very able craftsman made it.[36]

Sophia. And he rightly argued then that the contemplation of the complexity of nature is a sure sign of a divine artificer responsible for it.

Mr. Fish. I am glad that Paley did not place his clock at the bottom of the sea, because his faith would have then crumbled, seeing how rusted and covered in barnacles the clock would have become. He could then appeal to the fall, but as I have said before, this would have revealed his argument as circular. This type of reasoning cannot but be discredited.

Sophia. I accept that Paley's argument is not convincing, but you must admit that there are things in nature that cannot possibly be understood unless you believe that a grand designer is behind them. Take the eye, for instance: how could you account for such a complex structure being the result of your blind evolutionary forces?

Mr. Fish. Of course, if the eye had emerged in evolution in one huge creative step you would (almost) be right. But it started by being a light-sensitive skin patch, and step by step over millions of years it evolved into the present organ.

Sophia. I agree, but the perfection of the result surely reveals a preordained design.

Mr. Fish. You'd better disabuse yourself, Sophia, of the idea that the design of the eye is of such perfection as to reveal a purpose in evolution, that of creating humans who share in some divine ideal. The eye, my friend, is a pretty blunt organ that clearly shows the robust utilitarianism of the evolutionary process. As soon as an eye was produced that was good enough for human survival at the time when its evolution took place, that is what you humans got. But at that time humans died well before their forties, and thus evolution produced a focusing system of such poor quality that it ceases to work just about at that age. Even your own father wears bifocals, doesn't he? If the eye had been designed to last longer in perfect condition, then you might have persuaded me that evolution responded to a kind guiding hand, rather than to the blind forces of the struggle for survival.[37]

Sophia. I think I got you there Mr. Fish, because in accordance to your argument our human hearts would conk out, if you pardon the expression, well before that age of forty that you mentioned, by which time humans were most unlikely to reproduce in the hostile environments they faced.

Mr. Fish. That is precisely the point. Humans at that time were subject to colossal strains. They had to be able to run away from predators, or behind their prey when hunting, and they had to face heavy tasks. Thus the heart had to evolve to become far more robust than most humans really need now (this is why they have to exercise them, to simulate the conditions under which the heart evolved), and this explains why it is so long-lasting. Now, if you agree, as you must, that the eye, one of the more extraordinary features of nature, is no evidence of preordained design, then all arguments about how amazing nature is, and how therefore we need a deity to explain it, must surely be consigned to the trash heap, once and for all.

Sophia. You are more skeptical than you confess, seeing that you allege that you are a believer. Do you then deny that nature has a meaning and a purpose?

Mr. Fish. I was just coming to that, because there is nothing that upsets my own school more than the sight of red herrings.

Sophia. Come on, Mr. Fish, surely, there is nothing wrong about recognizing the obvious existence of meaning and purpose in nature, which is a sure indication of the presence of the Divine Spirit.

Mr. Fish. I am as interested in the Paraclete as you are, but I would rather be hooked than invoke it on such flimsy logic. First, do you agree with me that, as a scientist, it is my job to study nature?

Sophia. Yes, of course.

Mr. Fish. And do you not agree that in order to do my job *as a scientist* I must not describe nature by postulating supranatural entities for which I could not obtain even inferential scientific evidence?

Sophia. I would accept that, as long as you claim this to be the case when you discuss nature purely as a scientist.

Mr. Fish. It follows therefore that as long as I argue qua scientist I must deal with nature as a self-organizing system; otherwise, I would have to postulate entities that are outside nature and organize it.

Sophia. This seems to me fair enough, but I do not see what it has to do with the question of meaning and purpose which, for me, is a self-evident feature of nature.

Mr. Fish. Don't rush me, Sophia, because at my age I can easily choke. First of all, it is only when I consider one *part* of nature in relation to *another* that I can meaningfully use the words "meaning" and "purpose." Thus, if the whole of creation consisted of nothing else than an automobile, I could identify the purpose of the engine as that of moving the wheels, but I would not be able to understand the purpose of the car itself, because without some point of reference its motion would not make any sense. In the same way, nature considered *as a whole* cannot have a purpose or a meaning, unless you introduce an external point of reference: purpose and meaning do not make sense for a *single* totally isolated entity.

Sophia. That is a very good idea, because it allows me to posit God as that point of reference that makes sense of the meaning and purpose of nature.

Mr. Fish. Now, if you choose to call God that point of reference, it is your own choice, and I have nothing to object. But you must come clean about what you are doing. You have to be very careful, even if you quite sensibly expect to produce not a fully logical argument, which is impossible, but at least a persuasive one. If you say that nature, *as a whole*, must have a meaning and purpose, and you think that that can thereby persuade you that God exists, then you are getting into another circular argument, because the first premise means absolutely nothing unless the (alleged) conclusion is valid. Like in most religious arguments, "reasons" often preimply belief, and they are not reasons at all. Such putative "reasons" are at best a poor substitute for good, clear, well-grounded faith.

Sophia. I see your point, Mr. Fish, but you are unrealistic. Theologians have a job to do; they cannot go about just telling people what to believe in without giving them some reasons. You almost seem to imply that theologians should stop writing books, and that scientists who believe should not appeal to what they *feel* to be true to persuade other people to follow the true path.

Mr. Fish. I don't deny anybody the right to their beliefs, and to propagate them in any manner they think fit. I only object to the cheapening of faith by bad argument: if we are now both agreed that water exists, let its clarity pervade our thoughts also. You must have noticed that I have shown repeatedly that you can make only one of two transcendental arguments. One is about whatever it is that suggests to you God's existence, and the other is that God exists. You cannot use the first to support your belief in the second because in every case the first is an entirely void proposition unless the second be true. What it all boils down to is that you should *affirm* the existence of God and then enjoy the fact that you are now entitled to talk about the meaning and purpose of nature, or its harmony, or the design of nature or whatever.

Sophia. You have now repeated that argument three times, I think, in one form or another, and yet you fail to see that your accusation of circularity, which is what your criticism entails, affects you own reasoning as well.

Mr. Fish. I do not see how.

Sophia. Of course, Mr. Fish. When we discussed the existence of matter, you agreed that you could not prove it, but that you could infer it from observations, measurements, and calculations, which demonstrated the consistency between the hypothesis that matter exists and those observations, and so forth. But you would not have made those observations, measurements, and calculations, had you not already postulated matter. In exactly the same way, I postulate purpose and meaning for the universe as a whole, and from this postulate I deduce the existence of God as a way of harmonizing the hypotheses I am positing.

Mr. Fish. That is a clever argument, Sophia, and one that I cannot answer in a purely philosophical way, but this does not worry me in the least, because philosophy on its own cannot deal fully with nature. I agree that when I demonstrated the existence of water, I already had in mind the principle of evolution, that is that nature is a self-organizing system. This principle, however, although it denies that the transcendental interacts with nature—which denial gives it metaphysical connotations—is a principle that can be used as an inference to the best explanation for a vast array of empirical observations. It is because of this that it can be regarded as meta-physical rather than metaphysical, since it postulates nothing except the nontranscendental.

Sophia. So do you concede that you have got as well into a circularity?

Mr. Fish. There might be a circularity in what I had argued, but I claim that mine is a virtuous and not a vicious circle. The difference between my postulates and yours is this. The meta-physical is for all practical purposes unique, since no other system better than the scientific mesh (which my meta-physical principle organizes) has been found that is equally consistent. The metaphysical, you must admit, is not unique. You can just as well postulate meaning as chaos, the deity as the devil. The fact that we are both circular is not as significant as you try to make it. My circularity is based on a carefully calculated method of successive approximations designed to achieve facticity, that is a fitting of nature. Your circularity, on the other hand, is such that it could not possibly fail to fit, because you have designed it from the beginning to suit exactly your purpose.

Sophia. Whatever you say, I do not need to pull my belief in God out of the blue, in the way that you suggest. You did mention believing scientists: I have a friend who is a famous nuclear physicist, who nevertheless believes in God as a necessary principle to understand nature. He actually maintains that the mathematical beauty of the laws behind the elementary particles of physics is revelatory of the deity.

Mr. Fish. When you challenged my belief in water, would you have been satisfied if I had told you that I held it because my friend the whale does? Famous as your nuclear friend may be, he is not free from the constraints that we have discovered in the various arguments that we rehearsed, of which his is but one example. Come, dear friend, it is time that you stopped using arguments from authority, that demonstrate nothing else than either the weakness of your case or the weakness of your faith. Choose which.

Sophia. You should not so lightly dismiss the opinion of my friend the physicist, because he understands cosmology better than I do, of course, and he claims that the big bang was an extraordinary event, not just because it was a big bang, but because it developed in such a way in the first tiny intervals of time, so as to fall within a very narrow range of physical parameters, that permitted the laws of nature to be as they are, and thus eventually led to life and consciousness. Surely, because the probability of this happening, of the universe being self-referential, is so remote, this cannot have been the result of blind chance and it is a sure indication of a divine purpose.

Mr. Fish. My dear Sophia, when are we going to learn from our past mistakes? Last century, the hand, the eye, and the stars were regarded as so wonderful as to be inconceivable that they could have been created without divine guidance. First evolution, and then cosmology, have taught us that those organs and those stars are the result, not of miraculous processes, but of perfectly understandable physical ones. And that what appears to be miraculous chance is an easily explainable event. I concede, though, that the dice, so to speak, were loaded at the big bang, and that without that bias we would not be here discoursing as part of this self-referential universe.

Sophia. So, do you concede that my friend was right?

Mr. Fish. Not at all. First, our knowledge of cosmology and of the big bang is still very raw, and it would be unscientifically arrogant to pretend that we know all about them. Further knowledge could perfectly well show that what you regard as a miraculous coincidence is, as it happened with all the arguments that I have mentioned, perfectly explainable. I do not know of course how it could be done, but I should like to make a suggestion, which I do not propose as a physical theory at all, but as an example of a physical chain of events that *could* explain the realization of your impossibly improbable result. Imagine that there have been not one big bang but billions upon billions of them, all of them, except the last one, fruitless, in the sense that they created universes with such crooked natural laws, that such universes not only never became self-referential but, even more, never became viable, rapidly decaying into nothingness. As you know very well, if you repeat billions upon billions of tosses of a coin, you will eventually find a sequence of a million successive heads, that for our own experience would be a miracle beyond comprehension. And this could have been "our" big bang. As I said, I do not claim for a moment that things have been thus, but I claim that the existence of this possibility enormously weakens your friend's argument, which I regard as no more than a modern version of "behold the beauty of the stars." So, you would do better, dear Sophia, listening to your own heart.[38]

Sophia. (*Covering her ears with her hands.*) Listen, Mr. Fish, you know full well that I need *reasons*, whereas science is incomplete and cannot answer *everything*. This incompleteness is rational proof that some additional hypothesis is necessary, and that is why it is rational to believe in God.

Mr. Fish. Let me go through what you have said bubble by bubble. First of all, when you say that science is incomplete, you do not say what you mean by *incomplete*. Be careful, you cannot accuse a sailor of not doing a complete job because he is not at the same time piloting a jet aircraft. Likewise, science deals with the whole of nature, nothing less but nothing more, and you cannot say that science is incomplete because it does not pose questions about nature that cannot be answered unless some supernatural agency is postulated. Even within the natural world science does not expect, and should not be expected, to answer *scientifically* questions such as what it is to be in love, or what goodness is. Materialist scientists will not be able to answer such questions qua scientists, but neither would they be as foolish as to deny that such questions are worth considering.

Sophia. You now puzzle me, because from what you said I thought that you would have considered discourse about God as impossible if not futile: "Doubt can exist only where a question exists, a question only where an answer exists, and an answer only where something *can be said*." I thought that you might have gone that way in order to wash your hands, sorry fins, of those questions.[39]

Mr. Fish. You are right to be puzzled, because I am myself puzzled, but I shall try to explain. Of course, when I am talking about the physical aspects of nature, I must be ruthlessly physical. But remember that I accept that the evolutionary process has so to speak two layers, of which the second and far more recent, our recorded history, supervenes on the first and embraces everything that humankind has ever done or thought. So, as a natural scientist, I have to try and comprehend this history, and because it is an undoubted fact that the existence of God has indeed exercised many worthy minds, I must find a way to *say* something about the question (despite the quotation with which you challenged me). That is, although I shall not speak now as a working scientist, I must meet your demand, because nothing is worse in human history than lack of dialogue between people with opposite ideas.

Sophia. Let me try you. I am perfectly sure that the answer to the question: does God exist? is *yes*. What do you have to say?

Mr. Fish. That *that* is not an answer but an *affirmation*. And it is so, because an answer is something that can be understood to be true or false by anyone, whereas it is an obvious fact that large numbers of people are agnostic, or atheistic, or believe in some god or gods that may have very little in common with the one you have in mind. The fact that yours is not an answer but an affirmation does not diminish in the slightest its value *to you*.

Sophia. I am not very clear about you distinction between a straight and honest answer and an affirmation. What do you mean?

Mr. Fish. To start with, there is nothing depreciatory about an affirmation, although it has a different status from that of an answer. Some of the most important statements in life are affirmations. Do you love this man? If you say *yes*, that is an affirmation: I can understand what you mean, but I cannot share in the meaning, in the way in which, when you say "I see a red table," I can share the perception with you. Had you answered *yes* to my question about the man, I would not have been able to agree or disagree with you, all that I could have done would have been to try and see what did you *mean* by saying that you love somebody.

Sophia. I do not understand from what you say how you can verify or falsify affirmations.

Mr. Fish. As you know, Sophia, when I was young, I spent some considerable time at the river Cherwell by Magdalen College at Oxford, and I once heard there a peripatetic tutorial by an Oxford don who talked about speech acts that he called *performatives*. I think that what I have called *affirmations* are really performatives, that is, utterances that can only be understood as a program of compliance with certain behavioral patterns expected by your community when such utterances are made. "I love you" means thus some promise of loyalty, companionship, and so on, and I believe that when people say "God exists," I must judge this utterance by their subsequent performance since, as you know, I cannot as a scientist deal with the transcendental. Of course, *performance* in this sense can at an external level be ritual compliance, but at what to me is a much deeper level, adherence to some code that I can easily associate with the faith that is stated. If I see that people adhere to the Beatitudes, for instance, I then understand what they mean by being Christians, and so on.[40]

Sophia. You cannot expect me to accept your very pallid version of theism. You know full well that when I posit a God, I do this in a clear ontological transcendental way. For me God exists in the same way as water exists for you.

Mr. Fish. Much as I love you in a nonperformative way, I confess that I find you idealistic philosophers somewhat inconsistent. In discussing the existence of matter you query the reliability of the senses, because

of the possibility of hallucinations, delirium tremens, and so on.
Surely, you must keep a very insalubrious company if you can doubt
your senses on the basis of such biased evidence.[41] On the other
hand, you appear to suspend such scruples when you posit God.

Sophia. I am not sure that you are being fair, Mr. Fish. Such company as
we keep is entirely irrelevant. I myself never met anyone who hallucinated, or who was in delirium tremens. What matters is the theoretical possibility of such events, even the possibility that we might
substitute a dream for reality.

Mr. Fish. All right, and yet, when you make transcendental arguments
about the Deity, you trust your reason, although you know that reason
is far more liable to go wrong than the senses. Not only have you para-
noid schizophrenics to worry about, but you know full well that perfectly balanced people often confuse the utterances of their own sub-
conscious minds for the revelations of the Paraclete. Even the great
prophet Muhammad was so tricked by the devil once, although he
soon discovered the truth, it is said. So, matter is subject to profound
critical doubt, but the transcendental is *known* to be what you make of
it. You yourself are prepared to consider it irrational to accept the
reality of matter, a concept that most humans are agreed upon, and yet
you claim that it is *rational* to postulate an invisible and omnipotent
God, about whom humans disagree even now to the death.

Sophia. It is you who is now irrational, Mr. Fish, because it would be
absurd to make the existence of God depend on consensus: you know
very well that free will does not allow you that.

Mr. Fish. Come, Sophia, you are now putting the propeller before the
boat. Of course, I do not expect consensus, I was only arguing as to
the entirely different standards by which you judge belief about
matter and belief about God.

Sophia. So, you agree that it is rational to believe in God.

Mr. Fish. You are pushing me too hard, Sophia. When I was in those
learned waters I told you about, I heard a great deal about a major
theologian who even claimed that the question of the existence of
God should not be asked, because you are in danger of trying to
answer it by using the word "existence" in the same sense for Him as
for us. If you want to explain whatever you transcendentally feel, you
must avoid, he claimed, doing so by creating another "object,"
another "thing," however immaterial, however "omnipotent." Because, if you do, you risk using for the deity meaningless words that
apply only within nature. That is why, as far as I understood him, this
theologian does not try to extrapolate imprudently from nature out
to something to be called God, and he prefers to find Him inside us,

within our inner experience, which is what faith is or, as he himself put it, in an *ultimate concern*, the need to experience in some form the *ground of being*. As a natural scientist, this is as near a statement of my own belief as I can find.[42]

Sophia. You are now too abstract for me, Mr. Fish; I need to figure out God as a person. Even if I cannot have direct evidence of this nonembodied person, surely to posit Him is not worse than postulating, as you scientists did, an unobserved planet to explain the orbit of Uranus.[43]

Mr. Fish. Are you not in danger, Sophia, of not seeing the sea for the water, you who could hardly accept that the latter existed? Of course, the situation as regards Uranus is *totally* different: Le Verrier hypothesized a planet, people trained their telescopes to the calculated position of it in the sky, and thus Neptune was *found*. Scientists can posit the existence of objects, because such hypotheses can always, at least eventually, be *tested* in some way. By postulating the nonembodied person that you call God you do nothing else than corroborate your preexisting faith in Him. Is it not that you must invoke *reasons* because you are unsure of your *faith*?

Sophia. I think I will have to read now a lot of books before I come back to discuss all this with you again. You have not convinced me, but it was fun arguing. Good-bye for now and good swimming.

Mr. Fish. Good-bye, dear Sophia. I also enjoyed our chat but remember: do not allow the flapping of my fin to delude you into thinking that the reverberation of one-hand clapping can also reach your ears.

NOTES

1. I have taken the Frege "laws" from Hacker (1972), p. 174.

2. There is, however, a much-disputed reference to the instant survival of the soul in Luke 23: 43. When at the crucifixion the good thief turned to Jesus for salvation, He responded: "To day shalt thou be with me in paradise." Luke was the only Gentile among the evangelists, and the most hellenized one; and none of the other Gospels contain this episode. The distinguished theologian Oscar Cullmann, professor of theology both at the Sorbonne and at Basel, demonstrated in a searching exegesis of this text that it cannot be read literally. I believe that not even early Christian sources did that. Thus, Jacopo da Varagine (1280), p. 258, in the most complete available compilation of early Christian legends, has Jesus meeting the thief, not in Paradise but during the Harrowing of Hell.

References to early Jewish tradition may be found in Grubbe (1996); Werblonsky et al. (1997), p. 21; and Vermes (1993, 2000). Cullmann (1973), p. 71, firmly placed the immortality of the soul outside Christianity.

3. Berkeley (1710), p. 66 for the last quotation. All the previous ones are from p. 73.

4. I have lifted the argument on the use of "I know" from Wittgenstein (1967b), p. 222. See also his paragraphs 413 and 587 on introspection. A comprehensive analysis of Wittgenstein's views on this subject, and on his general approach to the theory of mind appears in Hacker (1972).

5. Ryle (1949) is a classic, both for its robust approach to the mind and for its engaging style.

6. Berkeley (1710), p. 141.

7. Ibid., p. 66.

8. The view that solipsism is not refutable is pretty well universal: "There is only one philosophical theory which seems to be in a position to ignore physics, and that is solipsism. If you are willing to believe that nothing exists except what you directly experience, no other person can prove that you are wrong, and probably no valid arguments exist against your view" (Russell 1923). Wittgenstein, however, has produced the strongest argument so far presented against solipsism (and idealism) which is the so-called Private Language Argument. Wittgenstein denies that private language exists except in a trivial way, as a code of encryption of some public language. If somebody says "I am in pain," to assert that the latter is private is nothing more than to say: this is the way in which the phrase "I am in pain" is used. You cannot possibly have a language that has been created merely to speak to yourself: language is a game that requires at least two players. Therefore, if you were the one and only mind existing (solipsism) you would not have the language to say "Only my mind exists." Wittgenstein's argument is treated in Hacker (1972), chap. 7.

9. See Popper et al. (1977).

10. See also Sacks (1995).

11. I have somewhat shifted the emphasis of Nagel's example to the question of privacy, which he avoids. The major works referred to in the text are: Nagel (1974); Jackson (1982, 1986); Block et al. (1972); Searle (1992), especially pp. 36–37; and Searle (1999). The quotation from Jackson is from p. 295 of his 1986 paper. A criticism of Jackson appears in Stemmer (1989). Whether you agree with Searle or not, his books are a great read. Churchland (1995) contains an abundance of solid physicalist stuff, a strong criticism of introspection (p. 205), and on pp. 195–202 a strong rejection of the views of Nagel and Jackson. Boyd (1980) is a good review of physicalism. Essential reading for a critique of physicalism is Robinson (1993a), especially the Introduction, and Robinson (1993b), which contains a penetrating analysis of the Mary-type of argument. The main relevant books by Penrose are Penrose (1989, 1994). A clear and didactic review of the philosophy of mind at an easy level is Priest (1991). For other important references to introspection see note 4. I owe a great deal in this section to Daniel Altmann (1977).

12. Robinson (1993a), p. 7.

13. A very clear discussion of *meaning* in psycholinguistics from the point of view of the natural scientist may be found in chapter 9 of Gerry Altmann (1997).

14. See for instance Deutsch (1997), pp. 86–87. Edelman (1994), however, suggests an evolutionary answer to this problem.

15. Sellars (1962), p. 97. Another important reference is the book by the Australian philosopher John Jamieson Carswell Smart (1920–), Smart (1963). In one of the most important books on the philosophy of science of the last quarter of the century, the Princeton philosopher Bastiaan (Bas) C. van Fraassen (1941–) signaled his conversion to scientific realism (van Fraassen 1980, p. 204). The identity of the mind and body, which instantly dissolves Berkeley's rejection of the possibility of matter acting on the mind, was strongly defended by the English philosopher U. T. Place, in an important article published in 1956 (Place 1970). Thoughts, for him, are not caused by events in the brain; they *are* events in the brain. The dissolution of the mind-body problem as a pseudoproblem was suggested by the American philosopher Carl Gustav Hempel (1905–), who held that the word "mind" is a mere shorthand for some form of bodily behavior (Hempel 1980).

16. The British poet laureate Ted Hughes has asserted that "Scientific objectivity . . . has its own morality, which has nothing to do with human morality. . . . It is contemptuous of the 'human element' " (Hughes 1994). This is both historically and in scientific practice

absurd. The pursuit of truth, which is the supreme test of good science, is just as elevating of the human condition as the best of poems. There is nothing more harmful for humanity than the blind acceptance of ill-conceived notions about nature, and in thus improving the human condition, Galileo and Darwin must rank with the greatest benefactors of humanity.

The concept that monism or physicalism (and by implication science) must exclude the spiritual is utterly erroneous. The American theologian S. G. Post (1998) cogently argues that dualism is not a precondition of belief, favors physicalism, and opposes the concept of nonmaterial souls.

17. The Spanish philosopher José Ortega y Gasset (1883–1955) was perhaps the first (as far as I know) to emphasize the significance of man's history: "*el hombre no tiene natu-raleza, sino que tiene . . . historia. O, lo que es igual: lo que la naturaleza es a las cosas, es la historia—como res gestae—al hombre,*" Ortega y Gasset (1958), p. 51. [ellipes in the original.] In translation: "*Man has no Nature, but it has . . . history. Or, what is the same: what Nature is to things, history—as a royal treasure—is to man.*" This, however, is not entirely correct, because *every* species has a phylogenetic history. What is amazingly special for humanity is that our history is *recorded*, and that it has thus established a fairly self contained and self-organizing system, in which ideas are subject to the exigencies of the "struggle for life." The second part of Ortega y Gasset's proposition is important: *men are accountable to their history in the same way in which, through the material, they are accountable to nature.*

18. Of course, some scientists are prepared to make sweeping metaphysical statements. Thus Wolpert (1992), p. 144: "Religion and science are incompatible." Such a statement depends crucially on what "religion" is meant to be, and how it is supposed to deal with nature, points that will be addressed later in this chapter, especially in the Finale. See also the following note.

19. What I have said about the deity providing some sort of external point of reference, with respect to which "meaning" and "purpose" acquire significance, is necessarily dependent on the theological stance accepted. The Amsterdam philosopher of Jewish origin, Baruch Spinoza (1632–1677), whose views deeply inspired Einstein, identified God with nature, in a form of *pantheism* that makes nature partake of a divine meaning and purpose. Although this is not orthodox Christian doctrine, Keith Ward (1996), p. 297, appears to avoid the difficulty under discussion by negating God as a "reality 'outside' the universe." The fact remains that *meaning* and *purpose* for the *whole* of nature cannot be postulated independently of the postulation of the deity. This is a view implicitly held by Mr. Fish in the Finale of this chapter.

20. Unfortunately, cultural evolution is hundreds of times faster than the biological one, thus creating huge strains in our species. For scientific ideas as subject to an evolutionary struggle for survival see van Fraassen (1980), p. 40.

21. A simple account of complexity theory may be found in Lewin (1993). Popular expositions are Gell-Mann (1995) and Casti (1993). A detailed treatment may be found in Gell-Mann et al. (1990). I am indebted to Dan Altmann for discussion on this point.

22. I disregard prewired instinctive patterns, which necessarily play a much smaller part in the human species than in others: instinct in its strong form is not a suitable mechanism for learning, which is the most remarkable feature of humans. The moth *Sphex*, for instance, exhibits complex instinctive behavior, but, if its routine is slightly varied, it will restart from the beginning, and repeat the pattern however many times it is disturbed, until it virtually drops dead from exhaustion.

23. Galilei (1632), see Drake (1980), p. 62. For Giordano Bruno see Singer (1950).

24. Giacon (1966), p. 216. It was not until 1820, however, that the Holy Office, in a decree of Pius VII, permitted the teaching of heliocentric cosmology not purely as a scientific hypothesis. (See Brandmüller et al. 1992.)

25. John Paul II (1980), pp. 77 (first two quotations), 78, and 81, respectively. As a result

of the meeting mentioned in the text, the pope appointed a committee of the Pontifical Academy of Science to report on the church's position about Galileo, The committee presented its report on 31 October 1992, the year of the 350th anniversary of Galileo's death. In a speech he read that day, and in another on 28 December 1992, in an act of restitution for Galileo, the pope accepted that he had been unjustly condemned by the church for promoting a Copernican cosmology.

26. Even one of the earliest philosophers of science recognized the dangers of equating Aristotle's work with science. Thus Herschel (1830), p. 105, art. 97: "Previous to the publication of the Novum Organum of Bacon, natural philosophy, in any legitimate and extensive sense of the word, could hardly be said to exist." (Natural philosophy was in 1830 what we now refer to as science.) And he later relates (p. 110) how Galileo dismissed the Aristotelian style of reasoning as nonsense.

27. For a discussion of the scientific events in 1820 see Altmann (1992), pp. 13–14.

28. Kuhn (1980), p. 265. For a well-reasoned critique of Kuhn's ideas by a distinguished physicist see Steven Weinberg (1998).

29. See Whorf (1956) and Hanson (1958). For Feyerabend see for instance Feyerabend (1993). For further discussion about Whorf see note 13, p. 46.

30. I owe this remark to Leplin (1997), p. 4.

31. See Pickering (1992) and Berger et al. (1967). Bloor (1991) presents a consistent social constructivist model of science.

32. Sandra Harding (1986), but she later recanted. See also Harding (1996).

33. See also Hess (1995), p. 190.

34. Weyl (1968), pp. 556–57.

35. It is in this sense that something slightly similar to Kuhn's paradigm has a meaning, but it is not as he claims a sort of science fashion or arbitrary myth: it is a *necessary* part of the evolutionary process, *necessary* because the past cannot be changed.

36. William Paley, a British theologian, presented this famous argument in the opening paragraphs of his *Natural Theology*, published in 1806. A very well reasoned attack on this argument is given in Dawkins (1986).

37. Even then, it would have been necessary to prove that the apparently superfluous, from the point of evolution, long-endurance feature of the eye-focusing system did not respond to some hidden evolutionary challenge. (See an example on p. 379.)

38. The argument about the very low probability of the big-bang parameters is very common. See for example Ward (1996), p. 297.

39. Sophia is here teasing Mr. Fish with a quotation from Wittgenstein (1961), 6. 51.

40. The Oxford philosopher mentioned is John Langshaw Austin (1911–1960), whose most important works in this respect are Austin (1962a, 1962b). The performative could be considered superseded by Wittgenstein's view of language as a game, in which every utterance is challenged by some requirement of performance, in order to make its meaning ostensive to the game partner.

41. The distinguished Oxford psychiatrist Anthony Storr points out that "lack of trust in the senses [is] a characteristic familiar to psychiatrists who treat schizoid persons" (Storr 1990, pp. 90–91). Idealistic philosophers beware!

42. The theologian quoted is Paul Tillich (1886–1965) who, although born in Germany, spent the last thirty years of his life in the United States. The quotes come from Tillich (1965), pp. 1–2. See also Tillich (1964). Although Tillich was influential, he is not, of course, in the central tradition of theology, but he is nevertheless representative of an important school of modern "progressive" theologians. The bishop of Woolwich, whose book, Robinson (1963), was one of the most widely read books on religion of its time, was much influenced by Tillich.

43. See Swinburne (1996), pp. 35–37. In his inaugural lecture, Swinburne's predecessor

in the chair of the Christian religion at Oxford, gave a warning about the need to analyze the way in which the concept of the *personal* is used in theology. (See Ramsey 1952, especially p. 17, and also Ramsey 1965, pp. 75–76.) Most of Sophia's arguments in this section may be found in Swinburne (1996). See also the numerous books by Paul Davies, in particular, Davies (1983, 1992). Ward (1996) discusses the relation of the creation to the deity as understood by most of the major religions, and Pannenberg (1993) provides an extremely well documented treatment of the relations between theology and natural science.

When alternative connotations exist, I give only the one used in this book. The reader is advised to consult the index, where references to more detailed discussion, as well as examples of the use of these words, will be found. Cross-references are in *bold italic.*

Aggregate. A collection of objects, not to be confused with a *set*, which is an abstract mathematical concept.

Amplitude. This word is used in two slightly different meanings: loosely, as the numerical value of the perturbation of a wave at any point; and, correctly, as the maximum value of the above for the points along the line of propagation of the wave.

Analytic. p is q is analytic if p could not be not q.

Assemblage. In quantum mechanics, an *aggregate*, all the elements of which are physically prepared to be identical and in the same state. Not to be confused with an *ensemble*.

Causal relation. Given *precisely specified experimental conditions*, A is the cause of B if and only if: (1) whenever A holds B holds, (2) whenever A does not hold B does not hold, and (3) a continuous causal chain can be established from A to B.

Causal timescale. A scale for time measurement that is compatible with causality in the sense that physical laws stated in that scale are independent of the initial time at which the law is applied.

Ceteris paribus. Means "all other conditions being equal." A caveat required to state causal relations, effected in science by rigorous protocols.

Contingent. Whatever can or cannot be the case.

Deduction. A form of reasoning in which the truth of the proposition B is derived from that of A by the use of logical rules.

Empirical necessity. A nonlogical necessity *entrenched* by praxis in a self-organizing system. The *modality* of causal relations and physical laws.

Ensemble. A theoretical or imagined aggregate of a very large number of elements which although identical might be in different states.

Entrenchment. A procedure whereby propositions are validated by establishing their consistency with other propositions. It establishes a hierarchy of such propositions, those better entrenched (consistent with the larger number of propositions) providing support for others below them in the hierarchy.

Epistemology. The study of knowledge.

Explanandum. Whatever it is that has to be explained.

Extension. If a proposition applies to an object x, its extension is the *range* of x for which the proposition is valid.

Hamiltonian. A mathematical expression that gives the energy in classical mechanics and that is transformed into the energy operator in quantum mechanics.

Induction. A mode of reasoning from observed to unobserved regularities.

Inference. A mode of reasoning in which the truth of the proposition B is derived from that of A by *deduction* or *induction*.

Inference to the best explanation. Given alternative inferential arguments, the one that provides what is regarded as the best (simplest, wider ranging, etc.) explanation.

Instrumentalism. An approach to science in which physical theories are used purely as a device to predict facts without any ontological or explanatory commitment.

Intension. The meaning of the *referent* of a proposition.

Law of nature. A rule that relates two physical events or states, not necessarily simultaneous.

Mesh. The science mesh contains all physical laws and facts entrenched within it in varying levels of *entrenchment*. The degree of entrenchment for an element of the mesh is related to the extension of the mesh within which that element has been shown consistent. The mesh is organized by means of *meta-physical (normative) principles*.

Metaphysics. The ontological study of an ultimate (also called transcendental) reality that is beyond physical observation, either directly or by inference from physical principles.

Meta-physics. The study of nature by using certain principles (called meta-physical principles) that: (1) do not posit a reality external to nature, (2) cannot be directly *derived* from physical observation, and (3) can be *entrenched* from physical or evolutionary arguments. It is not explicitly *ontological*, except insofar as it ignores any reality external to nature.

Modality. The modality of a proposition is the necessity or contingency of its validity.

Nature. When used without qualification it is always nature$_2$, the nature that we perceive through the interaction of external stimuli with our perceptive and rational system.

Necessary and sufficient. See these two words. It is the same as *if and only if*.

Necessary. Philosophically: A is necessary if it is true in all possible worlds. Logically: A is a necessary condition for B if whenever not A, *then not B*.

Nomological principle. A normative principle that acts as a law of nature.

Normative principle. A prescriptive principle that establishes formal rules for reasoning without making any *epistemological* or *ontological* assumptions.

Ontogenesis. Pertaining to the development of the individual.

Ontology. The study of existence.

Phenomenological theory. A theory deprived of *ontological* or explanatory considerations that uses formulas or laws chosen purely to agree with the desired experimental results.

Phylogenesis. Pertaining to the development of the species.

Population. In statistics, a theoretical very large *aggregate*, the properties of which can only be inferred by studying actual subaggregates, called samples.

Prediction. A *projection* in which the untested property is in the future.

Projection. The use of *induction* to predict an untested property from observed regularities.

Range. The range of x with respect to a given property is the *aggregate* of objects x with that property.

Referent. The object for which a proposition is valid.

Set. Except when no confusion may arise this word is used to signify an abstract mathematical set, not an *aggregate*.

Sufficient. A is a sufficient condition for B if whenever A, then B.

Synthetic. p is q is synthetic if p could be not q.

Tautology. A proposition that is true by virtue of its form alone.

Teleology. Postulation and study of design, purpose , or ultimate causes.

BIBLIOGRAPHY

All the references in the text are given by author and date, as for instance Cassinello *et al.* (1996).

Abraham, R. and Marsden, J. E. (1978). *Foundations of mechanics: A mathematical exposition of classical mechanics with an introduction to the qualitative theory of dynamical systems and applications to the three-body problem*, (2nd edn). Benjamin/Cummings, Reading, Massachusetts.

Achinstein, P. (1968). The circularity of a self-supporting inductive argument. In *The philosophy of science*, (ed. P. H. Nidditch), pp. 144–48. Oxford University Press. [First published in *Analysis*, **22**, (1962), pp. 138–41.]

Achinstein, P. (1970). The problem of theoretical terms. In *Readings in the philosophy of science*, (ed. B. A. Brody), pp. 234–50. Prentice-Hall, Englewood Cliffs, New Jersey.

Agassi, J. (1959). How are facts discovered. *Impulse*, **3**, 2–4. [I was unable to check this reference.]

Agassi, J. (1963). *Towards an historiography of science*. Mouton, 's–Gravenhage.

Aharanov, Y., Bergmann, P. G., and Lebowitz, J. L. (1964). Time symmetry in the quantum process of measurement. *The Physical Review B*, **134**, 1410–16.

Aharonov, Y. and Albert, D. Z. (1981). Can we make sense out of the measurement process in relativistic quantum mechanics? *Physical Review D*, **24**, 359–70.

Aharonov, Y., Anandan, J., and Vaidman, L. (1993a). Meaning of the wave function. *Physical Review A*, **47**, 4616–26.

Aharonov, Y. and Vaidman, L. (1993b). Measurement of the Schrödinger wave function of a single particle. *Physics Letters A*, **178**, 38–42.

Albert, D. Z. (1992). *Quantum mechanics and experience*. Harvard University Press, Cambridge, Massachusetts.

Albert, D. Z. (1994). Bohm's alternative to quantum mechanics. *Scientific American*, **270**, (3), 32–39.

Aldhous, P. (1995). Doomsday bacteria thrive on radiation. *New Scientist*, **148**, (2007), 18.

Alexander, H. G. (ed.) (1956). *The Leibniz–Clarke correspondence*. Manchester University Press. [Introduction and notes by ed.]

Alighieri, Dante (*circa* 1314–21). *La Divina Commedia*. [Italian text from Lectura Dantis Scaligera (1968). *Paradiso*. Felice Le Monnier, Firenze.]

Altham, J. E. J. (1969). A note on Goodman's paradox. *The British Journal for the Philosophy of Science*, **19**, 257.

Altmann, D. (1977). *Physicalism and privacy.* Unpublished, Oxford. [Thesis submitted for the D. Phil. degree at the University of Oxford.]

Altmann, G. T. M. (1997). *The ascent of Babel: An exploration of language, mind, and understanding.* Oxford University Press.

Altmann, S. L. (1986). *Rotations, quaternions, and double groups.* Clarendon Press, Oxford.

Altmann, S. L. (1989). Hamilton, Rodrigues, and the quaternion scandal. *Mathematics Magazine,* **62**, 291–308.

Altmann, S. L. (1992). *Icons and symmetries.* Clarendon Press, Oxford.

Amaldi, E. (1979). Radioactivity, a pragmatic pillar of probabilistic conceptions. In *Problems in the foundations of physics,* (ed. G. Toraldo di Francia), pp. 1–28. North Holland, Amsterdam. [*Rendiconti della Scuola Internazionale di Fisica "Enrico Fermi",* Vol. **72**.]

Anandan, J. (1993). Protective measurement and quantum reality. *Foundations of Physics Letters,* **6**, 503–32.

Anandan, J. and Brown, H. R. (1995). On the reality of space-time geometry and the wave function. *Foundations of Physics,* **25**, 349–60.

Angel, R. B. (1980). *Relativity: The theory and its philosophy.* Pergamon Press, Oxford.

Apostol, T. M. (1957). *Mathematical analysis: A modern approach to advanced calculus.* Addison-Wesley, Reading, Massachusetts.

Apostol, T. M. (1962). *Calculus: Calculus of several variables with applications to probability and vector analysis,* Vol. 2. Blaisdell Publishing Company, New York.

Appel, K. and Haken, W. (1977). The solution of the four-color-map problem. *Scientific American,* **237**, (4), 108–21.

Ascher, M. (1991). *Ethnomathematics: A multicultural view of mathematical ideas.* Brooks/Cole, San Francisco.

Aspect, A. (1999). Bell's inequality test: more ideal than ever. *Nature,* **398**, 189–90.

Aspect, A., Grangier, P., and Roger, G. (1981). Experimental tests of realistic local theories via Bell's theorem. Physical Review Letters, 47, 460–3.

Aspect, A., Grangier, P., and Roger, G. (1982a). Experimental realization of Einstein–Podolky–Rosen–Bohm *Gedankenexperiment*: a new violation of Bell's inequalities. *Physical Review Letters,* **49**, 91–94.

Aspect, A., Dalibard, J., and Roger, G. (1982b). Experimental test of Bell's inequalities using time-varying analyzers. *Physical Review Letters,* **49**, 1804–7.

Attneave, F. (1959). *Applications of information theory to psychology: A summary of basic concepts, methods, and results.* Henry Holt, New York.

Aubineau, M. (1966). Les 318 serviteurs d'Abraham et le nombre des Pères au Concile de Nicée. *Revue des Hautes Études,* **61**, 5–43.

Auden, W. H. (1948). For T. S. Eliot. In *T. S. Eliot: A symposium,* (ed. R. Marsh and Tambimuttu), p. 43. Editions Poetry, London.

Auden, W. H. (1950). *Collected shorter poems 1930–1944.* Faber & Faber, London.

Augustine (397). *Confessions.* [Latin text from Augustine (1992).]

Augustine. (1992). *Confessions,* Vol. 1, (ed. J. J. O'Donnell). Clarendon Press, Oxford.

Austin, J. L. (1961). *Philosophical papers,* (ed. J. O. Urmson and G. J. Warnock). Clarendon Press, Oxford.

Austin, J. L. (1962a). *How to do things with words: The William James Lectures delivered at Harvard University in 1955,* (ed. and preface J. O. Urmson). Clarendon Press, Oxford.

Austin, J. L. (1962b). *Sense and sensibilia.* Clarendon Press, Oxford. [Reconstructed from the manuscript notes by G. J. Warnock.]

Auyang, S. Y. (1995). *How is quantum field theory possible?* Oxford University Press, New York.

Ayer, A. J. (ed.) (1959). *Logical positivism.* The Free Press, Glencoe, Illinois.

Ayer, A. J. (1972). *Probability and evidence.* Macmillan, London.

Baggott, J. (1992). *The meaning of quantum theory: A guide for students of chemistry and physics.* Oxford University Press.

Ballentine, L. E. (1970). The statistical interpretation of quantum mechanics. *Reviews of Modern Physics*, **42**, 358–81.

Barker, S. F. and Achinstein, P. (1960). On the new riddle of induction. *The Philosophical Review*, **69**, 511–22.

Barrow, J. D. and Tipler, F. J. (1986). *The anthropic cosmological principle.* Oxford University Press. [Corrected pbk 1994.]

Barrow, J. D. (1993). *Pi in the sky. Counting, thinking and being.* Penguin Books, London. [First published by Oxford University Press, 1992.]

Barton, R. A. (1998). Visual specialization and brain evolution in primates. *Proceedings of the Royal Society [London] B*, **265**, 1933–37.

Belinfante, F. J. (1975). *Measurements and time reversal in objective quantum theory.* Pergamon Press, Oxford.

Bell, J. S. (1964). On the Einstein–Podolsky–Rosen paradox. *Physics*, **1**, 195–200. [Reprinted in Bell (1987), pp. 14–21.]

Bell, J. S. (1966). On the problem of hidden variables in quantum mechanics. *Reviews of Modern Physics*, **38**, 447–52. [Reprinted in Bell (1987*b*), pp. 1–13.]

Bell, J. S. (1975). On wave packet reduction in the Coleman-Hepp model. *Helvetica Physica Acta*, **48**, 93–98.

Bell, J. S. (1987*a*). Are there quantum jumps? In Kilmister (1987). [Reprinted in Bell (1987*b*), pp. 201–12.]

Bell, J. S. (1987*b*). *Speakable and unspeakable in quantum mechanics.* Cambridge University Press.

Beller, M. and Fine, A. (1994). Bohr's response to EPR. In Faye and Folse (1994), pp. 1–31.

Benacerraf, P. (1967). God, the Devil, and Gödel. *The Monist*, **51**, 9–32.

Benacerraf, P. and Putnam, H. (ed.) (1983). *Philosophy of mathematics: Selected readings*, (2nd edn). Cambridge University Press.

Berenson, B. (1952). *The Italian painters of the renaissance.* Phaidon Press, London.

Berger, P. L. and Luckmann, T. (1967). *The social construction of reality: A treatise in the sociology of knowledge.* Allen Lane, Penguin Press, London.

Berkeley, George (1710). *A treatise concerning the principles of human knowledge.* Thomas Nelson, Edinburgh. [References from Berkeley (1962).]

Berkeley, George (1962). *Principles of human knowledge. Three dialogues between Hylas and Philonous*, (ed. G. J. Warnock). Collins/Fontana, London. [Introduction by ed.]

Bernays, P. (1935). Sur le Platonisme dans les mathématiques. *L'Enseignement Mathématique*, **34**, 52–69. [Reprinted in translation in Benacerraf and Putnam (1983), pp. 258–71.]

Berry, D. C. and Dienes, Z. (1993). *Implicit learning: Theoretical and empirical issues.* Lawrence Erlbaum, Hove.

Berry, M. V. and Robbins, J. M. (1997). Indistinguishability for quantum particles: spin, statistics and the geometric phase. *Proceedings of the Royal Society [London] A*, **453**, 1771–90.

Bertola, F. and Curi, U. (ed.) (1989). *The anthropic principle: Proceeedings of the second Venice conference on cosmology and philosophy.* Cambridge University Press.

Birkhoff, G. D. (1931). Proof of the ergodic theorem. *Proceedings of the National Academy of Sciences of the USA*, **17**, 656–60.

Bitbol, M. and Darrigol, O. (ed.) (1992). *Erwin Schrödinger: Philosophy and the birth of quantum mechanics. Philosophie et naissance de la mécanique quantique.* Editions Frontières, Gif-sur-Yvette.

Blackburn, S. (1973). *Reason and prediction.* Cambridge University Press.

Block, N. (ed.) (1980). *Readings in philosophy of psychology*, (2 Vols). Methuen, London.

Block, N. J. and Fodor, J. A. (1972). What psychological states are not. *The Philosophical Review*, **81**, 159–81.

Bloor, D. (1991). *Knowledge and social imagery*, (2nd edn). University of Chicago Press.

Bohm, D. (1951). *Quantum Theory*. Prentice-Hall, Englewood Cliffs. [The EPR treatment is reprinted in Wheeler and Zurek (1983), pp. 356–68.]

Bohm, D. (1952). A suggested interpretation of the quantum theory in terms of "hidden" variables, I and II. *The Physical Review*, **85**, 166–93. [Reprinted in Wheeler and Zurek (1983), pp. 369–96.]

Bohm, D. and Hiley, B. J. (1993). *The undivided universe: An ontological interpretation of quantum theory*. Routledge, London.

Bohr, N. (1928). The quantum postulate and the recent development of atomic theory. *Nature*, **121**, 580–90.

Bohr, N. (1935a). Quantum mechanics and physical reality. *Nature*, **136**, 65.

Bohr, N. (1935b). Can quantum-mechanical description of physical reality be considered complete? *The Physical Review*, **48**, 696–702. [Received 13 July 1935.]

Bohr, N. (1961). *Atomic physics and human knowledge*. Science Editions, New York. [First published 1958, Wiley, New York.]

Boltzmann, L. (1868). Studien über das Gleichgewicht der lebendigen Kraft zwischen bewegten materiellen Punkten. In Boltzmann (1909), Vol. 1, pp. 49–96.

Boltzmann, L. (1884). Über die Eigenschaften monozyklischer und anderer damit verwandter Systeme. In Boltzmann (1909), Vol. 3, pp. 122–52.

Boltzmann, L. (1887). Über die mechanischen Analogien des zweiten Hauptsatzes der Thermodynamik. In Boltzmann (1909), Vol. 3, pp. 258–71.

Boltzmann, L. (1909). *Wissenschaftliche Abhandlungen*, (3 Vols), (ed. F. Hasenhörl). J. A. Barth, Leipzig.

Borges, J. L. (1956). Funes el memorioso. In *Ficciones*, pp. 117–27. Emecé, Buenos Aires. [First published 1944 in the collection *Artificios*, Buenos Aires.]

Born, M. (1926). Zur Quantenmechanik der Stossvorgänge. *Zeitschrift für Physik*, **37**, 863–67. [Received 25 June 1926, four days after Schrödinger's fourth paper was sent to Annalen der Physik.]

Born, M. (1962). Die statistische Deutung der Quantenmechanik. In *Zur Begründung der Matrizenmechanik*, (ed. A. Hermann), pp. 1–12. Ernst Battenberg Verlag, Stuttgart. [Nobel lecture, 11 December 1954.]

Born, M. and Jordan, P. (1925). Zur Quantenmechanik. *Zeitschrift für Physik*, **34**, 858–88. [Received 27 September 1925.]

Born, M., Heisenberg, W., and Jordan, P. (1926). Zur Quantenmechanik II. *Zeitschrift für Physik*, **35**, 557–615. [Received 16 November 1925.]

Boswell, J. (1791). *Life of Johnson*, London. [Quotation from 1953 edn by R. W. Chapman, Oxford University Press, London.]

Bowie, G. L. (1982). Lucas's number is finally up. *Journal of Philosophical Logic*, **11**, 279–85.

Boyd, R. (1980). Materialism without reductionism: what physicalism does not entail. In Block (1980), Vol. 1, pp. 67–106.

Boyer, C. B. (1968). *A history of mathematics*. Wiley, New York.

Bradley, F. H. (1893). *Appearance and reality: A metaphysical essay*. Swan Sonnenschein, London. [2nd edn 1946, Oxford University Press.]

Braithwaite, R. B. (1953). *Scientific explanation: A study of the function of theory, probability and law in science*. Cambridge University Press.

Brandmüller, W. and Greipl, E. J. (1992). *Copernico, Galilei e la Chiesa: Fine della controversia (1821). Gli atti del Sant'Uffizio*, Firenze.

Brecht, B. (1955). *Leben des Galilei: Schauspiel*. Suhrkamp Verlag, Berlin. [First produced 1938/39.]

Bretschneider, K. G. (1828). *Handbuch der Dogmatik der evangelisch-lutherischen Kirche*, (3rd edn). [1st edn 1814.]

Broda, E. (1955). *Ludwig Boltzmann: Mensch, Physiker, Philosoph*. Franz Deuticke, Vienna.

Brown, H. I. (1979). *Perception, theory and commitment: The new philosophy of science*. University of Chicago Press.

Brown, H. R., Elby, A., and Weingard, R. (1996). Cause and effect in the pilot-wave interpretation of quantum mechanics. In *Bohmian mechanics and quantum theory: an appraisal*, (ed. J. T. Cushing, A. Fine, and S. Goldstein), pp. 309–19. Kluwer Academic, Dordrecht.

Brown, H. R. and Redhead, M. L. G. (1981). A critique of the disturbance theory of indeterminacy in quantum mechanics. *Foundations of Physics*, **11**, 1–20.

Brush, S. G. (1976). *The kind of motion we call heat: A history of the kinetic theory of gases in the 19th century. Book 1, Physics and the atomists. Book 2, Statistical physics and irreversible processes*, (2 Vols). North Holland, Amsterdam. [Reprinted 1986.]

Burgess, A. (1995). *Byrne: A novel*. Hutchinson, London.

Carlyle, T. (1831). *Sartor Resartus: The life and opinions of Herr Teufeldröckh*. Chapman & Hall, London. [Thomas Carlyle's collected works, Vol. 1.]

Carnap, R. (1936). Testability and meaning. *Philosophy of Science*, **3**, 419–71.

Carnap, R. (1937). Testability and meaning. *Philosophy of Science*, **4**, 1–40.

Carnap, R. (1947–8). On the application of inductive logic. *Philosophy and Phenomenological Research*, **8**, 133–48.

Carnap, R. (1956). *Meaning and necessity: A study in semantics and modal logic*, (2nd edn). The University of Chicago Press.

Carnap, R. (1966). *Philosophical foundations of physics: An introduction to the philosophy of science*, (ed. M. Gardner). Basic Books, New York.

Cartwright, N. (1983). *How the laws of physics lie*. Clarendon Press, Oxford.

Cartwright, N. (1989). *Nature's capacities and their measurements*. Clarendon Press, Oxford.

Cassinello, A. and Sánchez-Gómez, J. L. (1996). On the probabilistic postulate of quantum mechanics. *Foundations of Physics*, **26**, 1357–74.

Casti, J. L. (1993). *Searching for certainty: What the scientists know about the future*. Abacus, London.

Caveing, M. (1982). *Zénon d'Elée. Prolégomènes aux doctrines du continu. Étude historique et critique des fragments et témoignages*. J. Vrin, Paris.

Cercignani, C. (1998). *Ludwig Boltzmann: The man who trusted atoms*. Clarendon Press, Oxford.

Chaitin, G. J. (1982). Gödel's theorem and information. *International Journal of Theoretical Physics*, **21**, 941-54.

Chaitin, G. (1988). Randomness in arithmetic. *Scientific American*, **259**, (1), 52–57.

Chaitin, G. (1990). A random walk in arithmetic. *New Scientist*, **125**, (1709, March 24), 44–6.

Chaitin, G. J. (1995). Randomness in arithmetic and the decline and fall of reductionism in pure mathematics. In Cornwell (1995), pp. 26–44.

Chalmers, A. F. (1970). Curie's principle. *The British Journal for the Philosophy of Science*, **21**, 133–48.

Chalmers, D. J. (1996). *The conscious mind*. Oxford University Press.

Changeux, J.-P. and Connes, A. (1995). *Conversations on mind, matter, and mathematics*, (trans. and ed. M. B. DeBevoise). Princeton University Press.

Chart, D. (2000). Schulte and Goodman's Riddle. *British Journal for the Philosophy of Science*, **51**, 147–149.

Chihara, C. S. (1973). *Ontology and the vicious-circle principle*. Cornell University Press, Ithaca, New York.

Chihara, C. S. (1982). A Gödelian thesis regarding mathematical objects: Do they exist? And can we perceive them? The Philosophical Review, **91**, 211–27.

Chihara, C. S. (1990). *Constructibility and mathematical existence*. Clarendon Press, Oxford.

Church, A. (1940). On the concept of a random sequence. *Bulletin of the American Mathematical Society*, **46**, 130–35.

Churchland, P. M. (1995). *The engine of reason, the seat of the soul: A philosophical journey into the brain*. MIT Press, Cambridge, Massachusetts.

Churchland, P. M. and Churchland, P. S. (1990). Could a machine think? Scientific American, **262**, (1), 26-31.

Cini, M. and Lévy-Leblond, J.-M. (ed.) (1990). *Quantum theory without reduction*. Adam Hilger, Bristol.

Clauser, J. F., Horne, M. A., Shimony, A., and Holt, R. A. (1969). Proposed experiment to test local hidden-variables theories. *Physical Review Letters*, **23**, 880–84.

Clauser, J. F. and Shimony, A. (1978). Bell's theorem: experimental tests and implications. *Reports on Progress in Physics*, **41**, 1881–927.

Clausius, R. (1965). The nature of the motion which we call heat. In *Kinetic theory: The nature of gases and of heat*, Vol. 1, (ed. S. G. Brush), pp. 111–34. Pergamon Press, Oxford. [First published 1857.]

Cohen, L. J. (1989). *An introduction to the philosophy of induction and probability*. Clarendon-Press, Oxford.

Collingwood, R. G. (1938). On the so-called idea of causation. *Proceedings of the Aristotelian Society*, **38**, 85–112.

Cornwell, J. (ed). (1995). *Nature's imagination: The frontiers of scientific vision*. Oxford University Press.

Coveney, P. and Highfield, R. (1991). *The arrow of time: The quest to solve science's greatest mystery*. Flamingo, London.

Crombie, A. C. (1953). *Robert Grosseteste and the origins of experimental science, 1100–1700*. Clarendon Press, Oxford. [2nd augmented printing 1962.]

Cullmann, O. (1973). Immortality of the soul or resurrection of the dead? The witness of the New Testament. In *Immortality*, (ed. T. Penehulm), pp. 53–91. Wadsworth Publishing Co., Belmont, Calfornia.

Cushing, J. T. (1994). *Quantum mechanics. Historical contingency and the Copenhagen hegemony*. University of Chicago Press, Chicago.

Cushing, J. T. (1996). The causal quantum theory program. In *Bohmian mechanics and quantum theory: An appraisal*, (ed. J. T. Cushing, A. Fine, and S. Goldstein), pp. 1–19. Kluwer Academic, Dordrecht.

da Costa, N. C. A., Krause, D., and French, S. (1992). The Schrödinger problem. In Bitbol and Darrigol (1996), pp. 445–60.

d'Alembert, Jean le Ronde (1754). Croix ou pile (analyse des hasards). In *Encyclopédie, ou Dictionnaire raisonné des sciences, des arts et des métiers*, Vol. 4, (ed. Denis Diderot and Jean Le Rond d'Alembert), pp. 512–13. Briasson, David, Le Breton, Durand, Paris.

Dales, H. G. and Oliveri, G. (ed). (1998). *Truth in mathematics*. Clarendon Press, Oxford.

Dalla Chiara, M. L. (1985). Some foundational problems in mathematics suggested by physics. *Synthese*, **62**, 303–15.

Dalla Chiara, M. L. and Toraldo di Francia, G. (1985). Individuals, kinds and names in physics. *Versus. Quaderni di studi semiotici*, **40**, 29–50.

Dalla Chiara, M. L. and Toraldo di Francia, G. (1993). Individuals, kinds and names in physics. In *Bridging the gap: Philosophy, mathematics, physics. Lectures on the foundations of physics*, (ed. G. Corsi, M. L. Dalla Chiara, and G. C. Ghirardi), pp. 261–83. Kluwer Academic, Dordrecht. [*Boston Studies in the Philosophy of Science*, Vol. **140**.]

Darwin, Charles (1859). *On the origin of species by means of natural selection, or the preservation of favoured races in the struggle for life*. Murray, London.

Davidson, D. (1966). Emeroses by other names. *The Journal of Philosophy*, **63**, 778–80.

Davidson, D. (1980). *Essays on actions and events*. Clarendon Press, Oxford.

Davies, P. C. W. (1974). *The physics of time asymmetry*. University of California Press, Berkeley.

Davies, P. (1980). *Other worlds*. Dent & Sons, London.

Davies, P. C. W. (1981). Is thermodynamic gravity a route to quantum gravity? In *Quantum gravity 2: A second Oxford symposium*, (ed. C. J. Isham, R. Penrose, and D. W. Sciama), pp. 183–209. Clarendon Press, Oxford.

Davies, P. (1982). *The accidental universe*. Cambridge University Press.

Davies, P. (1983). *God and the new physics*. Dent & Sons, London. [Pbk edn 1990, Penguin Books, London.]

Davies, P. (1993). *The mind of God: Science and the search for ultimate meaning*. Penguin Books, London.

Davisson, C. and Kunsman, C. (1921). The scattering of electrons by nickel. *Science*, **64**, 522–24.

Davisson, C. and Kunsman, C. H. (1923). The scattering of low speed electrons by platinum and magnesium. *The Physical Review*, **22**, 242–58.

Davisson, C. and Germer, L. H. (1927). The scattering of electrons by a single crystal of nickel. *Nature*, **119**, 558–60.

Dawkins, R. (1986). *The blind watchmaker*. Longman, London. [Pbk edn 1988, Penguin Books, London.]

Dawson Jr, J. W. (1988a). Kurt Gödel in sharper focus. In Shanker (1988a), pp. 1–16.

Dawson Jr, J. W. (1988b). The reception of Gödel's incompleteness theorems. In Shanker (1998a), pp. 74–95.

de Broglie, L. (1923). Quanta de lumière, diffraction et interférences. *Comptes Rendus Hebdomanaires des Séances de l'Académie des Sciences*, **177**, 548–50.

de Finetti, B. (1990). *Theory of probability: A critical introductory treatment*, (2 Vols), (trans. A. Machi and A. Smith). Wiley, Chichester. [Italian first edn 1970.]

DeLong, H. (1971). *A profile of mathematical logic*. Addison–Wesley, Reading, Massachusetts.

Dennett, D. C. (1995). *Darwin's dangerous idea: Evolution and the meanings of life*. Simon and Schuster, New York.

de Regt, H. W. (1996). Philosophy of the kinetic theory of gases. *The British Journal for the Philosophy of Science*, **47**, 31–62.

d'Espagnat, B. (1976). *Conceptual foundations of quantum mechanics*, (2nd edn). W. A. Benjamin, Reading, Massachusetts.

d'Espagnat, B. (1983). *In search of reality*. Springer, New York.

d'Espagnat, B. (1989). *Reality and the physicist: Knowledge, duration and the quantum world*, (trans. J. C. Whitehouse and B. d'Espagnat). Cambridge University Press.

Deutsch, D. (1985a). Quantum theory as a universal physical theory. *International Journal of Theoretical Physics*, **24**, 1–41.

Deutsch, D. (1985b). Quantum theory, the Church–Turing principle and the universal quantum computer. *Proceedings of the Royal Society [London] A*, **400**, 97–117.

Deutsch, D. (1997). *The fabric of reality*. Allen Lane Penguin, London.

Deutsch, D. and Lockwood, M. (1994). The quantum physics of time travel. *Scientific American*, **270**, (3), 50–56.

Devlin, K. (1991). *Logic and information*. Cambridge University Press.

DeWitt, B. S. and Graham, R. N. (1973). *The many-worlds interpretation of quantum mechanics: A fundamental exposition*. Princeton University Press.

Dienes, Z. and McLeod, P. (1993). How to catch a cricket ball. *Perception*, **22**, 1427–39.

Dieudonné, J. (1971). Modern axiomatic methods and the foundations of mathematics. In *Great currents of mathematical thought*, Vol. 2, (trans. C. Pinter and H. Kline, ed. F. Le Lionnais), pp. 251–66. Dover, New York.

d'Inverno, R. (1992). *Introducing Einstein's relativity*. Clarendon Press, Oxford.

Drake, S. (1978). *Galileo at work: His scientific biography*. The University of Chicago Press. [Reprinted 1995, Dover, New York.]

Drake, S. (1980). *Galileo*. Oxford University Press. [Reissued 1996.]

Dreyfus, H. L. (1992). *What computers still can't do: A critique of artificial reason*, (3rd edn). MIT Press, Cambridge, Massachusetts.

Ducasse, C. J. (1966). Critique of Hume's conception of causality. *Journal of Philosophy*, **63**, 141–48.

Ducasse, C. J. (1968). *Truth, knowledge and causation*. Routledge & Kegan Paul, London.

Duhem, P. (1954). *The aim and structure of physical theory*, (trans. P. P. Wiener). Princeton University Press. [Trans. of 2nd French edn 1914.]

Dummett, M. (1954). Can an effect precede its cause? *Aristotelian Society Supplementary Volume*, **28**, 27–44.

Dummett, M. (1977). *Elements of intuitionism*. Oxford University Press. [With the assistance of Roberto Minio.]

Dummett, M. A. E. (1978). Is logic empirical? In *Truth and other enigmas*, pp. 269–89. Duckworth, London.

Dummett, M. (1986). Causal loops. In Flood and Lockwood (1986), pp. 135–169.

Duncan, D. E. (1998). *Calendar: Humanity's epic struggle to determine a true and accurate year*. Avon Books, New York.

Dürr, S., Nonn, T., and Rempe, G. (1998). Origin of quantum-mechanical complementarity probed by a "which-way" experiment in an atom interferometer. *Nature*. **395**, 33–37.

Earman, J. (1974). An attempt to add a little direction to "the problem of the direction of time". *Philosophy of Science*, **41**, 15–47.

Earman, J. and Rédei, M. (1996). Why ergodic theory does not explain the success of equilibrium statistical mechanics. *The British Journal for the Philosophy of Science*, **47**, 63–78.

Eccles, J. (1994). *How the self controls its brain*. Springer.

Eddington, A. S. (1928). *The nature of the physical world*. Cambridge University Press.

Eddington, A. S. (1939). *The philosophy of physical science*. Cambridge University Press. [Reprinted 1949 with corrections.]

Edelman, G. M. (1994). *Bright air, brilliant fire: On the matter of the mind*. Penguin, London.

Edgington, D. (1997). Mellor on chance and causation. *The British Journal for the Philosophy of Science*, **48**, 411–33.

Ehring, D. (1997). *Causation and persistence: A theory of causation*. Oxford University Press, New York.

Einstein, A. (1905*a*). Über einen die Erzeugung und Verwandlung des Lichtes betreffenden heuristischen Gesichtspunkt. *Annalen der Physik*, **17**, 132–48.

Einstein, A. (1905*b*). Zur Elektrodynamik bewegter Körper. *Annalen der Physik*, **17**, 891–921. [Translation in Sommerfeld (1923), pp. 35–65.]

Einstein, A. (1921). Prof. Einstein's lectures at King's College, London, and the University of Manchester. *Nature*, **107**, 504.

Einstein, A. (1922). *Sidelights on relativity*, (trans. G. B. Jeffery and W. Perrett). Methuen, London.

Einstein, A. (1936). Physik und Realität. *Journal of the Franklin Institute*, **221**, 313–47. [English version pp. 349–82.]

Einstein, A. (1950). *The meaning of relativity*, (4th edn), (trans. E. Plimpton, E. G. Straus, and S. Bargmann). Methuen, London. [1st edn 1922. The edition quoted is identical to the Princeton University Press 3rd edn.]

Einstein, A. (1954). *Relativity: The special and the general theory. A popular exposition by Albert Einstein*, (15th edn), (trans. R. W. Lawson). Methuen, London. [Enlarged reprint of the 1952 15th edn. 1st edn 1920.]

Einstein, A. (1959*a*). Autobiographical notes. In Schilpp (1959), Vol 1, pp. 2–95.

Einstein, A. (1959*b*). Reply to criticisms. In Schilpp (1959), Vol 2, pp. 663–88.

Einstein, A. (1971). *The Born–Einstein letters: Correspondence between Albert Einstein and Max and Hedwig Born from 1916 to 1955 with commentaries by Max Born*, (trans. I. Born). Macmillan, London.

Einstein, A., Podolsky, B., and Rosen, N. (1935). Can quantum-mechanical description of physical reality be considered complete? *The Physical Review*, **47**, 777–80. [Received 25 March 1935. Reprinted in Wheeler and Zurek (1983), pp. 138–41.]

Elsasser, W. (1925). Bemerkungen zur Quantenmechanik freier Elektronen. *Die Naturwissenschaften*, **13**, 711.

Enderton, H. B. (1972). *A mathematical introduction to logic*. Academic Press, New York.

Euler, L. (1942). Réflexions sur l'espace et le tems [*sic*]. In *Leonhardi Euleri opera omnia*, Series III, Vol. 2, (ed. E. Hoppe, K. Matter, and J. J. Burkhardt), pp. 376–83. Teubner, Leipzig.

Evans-Pritchard, E. E. (1937). *Witchcraft, oracles, and magic among the Azande*. Clarendon Press, Oxford. [The 1976 abridged edn is not suitable.]

Everett, H. (1957). "Relative state" formulation of quantum mechanics. *Reviews of Modern Physics*, **29**, 454–62.

Faye, J. and Folse, H. J. (ed.) (1994). *Niels Bohr and contemporary philosophy*. Kluwer Academic, Dordrecht. [Boston Studies in the Philosophy of Science, Vol. **153**.]

Feather, N. (1959). *An introduction to the physics of mass, length and time*. Aldine Publishing Company, Chicago. [Reprinted 1968 by Edinburgh University Press.]

Feferman, S. (1988). Kurt Gödel: conviction and caution. In Shanker (1988*a*), pp. 96–114.

Feller, W. (1968). *An introduction to probability theory and its applications*, (3rd edn). Wiley, New York.

Feyerabend, P. (1958). An attempt at a realistic interpretation of experience. *Proceedings of the Aristotelian Society*, **58**, 143–70.

Feyerabend, P. K. (1962). Explanation, reduction, and empiricism. In *Scientific explanation, space, and time*, (ed. H. Feigl and G. Maxwell), pp. 28–97. University of Minnesota Press, Minneapolis. [*Minnesota Studies in the Philosophy of Science*, Vol. **3**.]

Feyerabend, P. K. (1968). On a recent critique of complementarity: Part I. *Philosophy of Science*, **35**, 309–31.

Feyerabend, P. K. (1969). On a recent critique of complementarity: Part II. *Philosophy of Science*, **36**, 82–105.

Feyerabend, P. (1970). Philosophy of science: a subject with a great past. In *Historical and philosophical perspectives of science*, (ed. R. Stuewer), pp. 172–83. University of Minnesota Press, Minneapolis. [*Minnesota Studies in the Philosophy of Science*, Vol. **5**.]

Feyerabend, P. (1993). *Against method*, (3rd edn). Verso, London. [1st edn 1975, New Left Books, London.]

Feynman, R. P. (1992). *The character of physical law*. Penguin, London. [First published 1965, British Broadcasting Corporation, London.]

Fitzpatrick, P. J. (1966). To Gödel via Babel. *Mind*, **75**, 332–50.

Flood, R. and Lockwood, M. (ed.) (1986). *The nature of time*. Blackwell, Oxford.

Folse, H. J. (1994). Bohr's framework of complementarity and the realism debate. In Faye and Folse (1994), pp. 119–39.

Fontenelle, Bernard le Bovier de (1955). *Entretiens sur la pluralité des mondes – Digression sur les anciens et les modernes*, (ed. and Introduction R. Shackleton). Clarendon Press, Oxford. [1st edn 1686, C. Blageart, Paris.]

Foster, J. A. (1982). *The case for idealism*. Routledge & Kegan Paul, London.

Foster, J. (1993). The succinct case for idealism. In Robinson (1993*a*), pp. 293–313.

Frege, G. (1980). *Philosophical and mathematical correspondence*, (ed. G. Gabriel, H. Hermes, F. Kambartel, C. Thiel, and A. Veraart, trans. H. Kaal, abridged B. McGuinness). Blackwell, Oxford.

French, S. (1989*a*). Identity and individuality in classical and quantum physics. *Australasian Journal of Philosophy*, **67**, 432–46.

French, S. (1989*b*). Individuality, supervenience and Bell's theorem. *Philosophical Studies*, **55**, 1–22.

French, S. (1989c). Why the principle of the identity of indiscernibles is not contingently true either. *Synthese*, **78**, 141–66.

French, S. and Krause, D. (1995). Vague identity and quantum non-individuality. *Analysis*, **55**, 20–26.

French, S. and Redhead, M. (1988). Quantum physics and the identity of indiscernibles. *The British Journal of the Philosophy of Science*, **39**, 233–46.

Friedman, J. R., Patel, V., Chen, W., Tolpygo, S. K., and Lukens, J. E. (2000). Quantum superposition of distinct macroscopic states. *Nature*, **406**, 43–45.

Galilei, Galileo (1632). *Dialogo sopra i due massimi sistemi del mondo, Tolemaico e Copernicano.* [References from Galileo (1967).]

Galilei, Galileo. (1967). *Dialogue concerning the two chief world systems,* (2nd edn), (trans. S. Drake). University of California Press, Berkeley. [1st edn 1953.]

Galloway, D. (1992). Wynn on mathematical empiricism. *Mind and Language*, **7**, 333–58.

Gandy, R. (1996). Human versus mechanical intelligence. In Millican and Clark (1996), pp. 126–36.

Geach, P. and Black, M. (ed). (1970). *Translations from the philosophical writings of Gottlob Frege.* Blackwell, Oxford.

Gell–Mann, M. (1979). What are the building blocks of matter? In *The nature of the physical world, 1976 Nobel Conference* , (ed. D. Huff and O. Prewett), pp. 27–45. Wiley, New York.

Gell-Mann, M. (1995). *The quark and the jaguar: Adventures in the simple and the complex.* Abacus, London.

Gell-Mann, M. and Hartle, J. B. (1990). Quantum mechanics in the light of quantum cosmology. In *Complexity, entropy and the physics of information: The proceedings of the 1988 workshop on complexity, entropy and the physics of information, held May–June, 1989 in Santa fe, New Mexico,* (ed. W. H. Zurek), pp. 425–58. Addison-Wesley, Redwood City, California.

Gell-Mann, M. and Hartle, J. B. (1993). Classical equations for quantum systems. *Physical Review D*, **47**, 3345–82.

Gettier, E. L. (1963). Is justified true belief knowledge? *Analysis*, **23**, 121–23.

Ghirardi, G. C., Rimini, A., and Weber, T. (1986). Unified dynamics for microscopic and macroscopic systems. *Physical Review D*, **34**, 470–91.

Ghirardi, G. C., Rimini, A., and Weber, T. (1988). The puzzling entanglement of Schrödinger's wave function. *Foundations of physics*, **18**, 1–27.

Ghose, P. and Home, D. (1995). An analysis of the Aharanov-Anandan-Vaidman model. *Foundations of Physics*, **25**, 1105–9.

Giacon S. J., C. (1966). Bellarmino. In *Dizionario Letterario Bompiani degli autori di tutti i tempi e di tutte le letteratture,* Vol. 1, p. 216. Bompiani, Milano.

Gibbins, P. (1987). *Particles and paradoxes: The limits of quantum logic.* Cambridge University Press.

Giere, R. N. (1988). *Explaining science. A cognitive approach.* University of Chicago Press.

Giulini, D., Joos, E., Kiefer, C., Kupsch, J., Stamatescu, I. O., and Zeh, H. D. (1996). *Decoherence and the appearance of a classical world in quantum theory.* Springer, Berlin.

Glymour, C. (1980). *Theory and evidence.* Princeton University Press.

Glymour, C. (1996). The hierarchies of knowledge and the mathematics of discovery. In Millican and Clark (1996), pp. 265–91.

Gödel, K. (1983). What is Cantor's continuum problem? In *Philosophy of mathematics: Selected readings,* (2nd edn), (ed. P. Benacerraf and H. Putnam), pp. 470–85. Cambridge University Press. [Expanded version of an article written in 1947.]

Gold, T. (1962). The arrow of time. *American Journal of Physics*, **30**, 403–10.

Gombrich, E. H. (1962). *Art and illusion: A study in the psychology of pictorial representation,* (2nd edn). Phaidon Press, London.

Good, I. J. (1969). Gödel's theorem is a red herring. *The British Journal for the Philosophy of Science*, **19**, 357–58.

Goodman, N. (1946). A query on confirmation. *The Journal of Philosophy*, **43**, 383–85.

Goodman, N. (1947–8). On infirmities of confirmation theory. *Philosophy and Phenomenological Research*, **8**, 149–51.

Goodman, N. (1960). Positionality and pictures. *The Philosophical Review*, **69**, 523–25.

Goodman, N. (1965). *Fact, fiction, and forecast*, (2nd edn). Bobbs–Merrill, Indianapolis. [1st edn 1955.]

Goodman, N. (1967). Two replies. *The Journal of Philosophy*, **64**, 286–87.

Grabbe, L. L. (1996). *An introduction to first century Judaism: Jewish religion and history in the second temple period*. T. & T. Clark, Edinburgh.

Grünbaum, A. (1967). *Modern science and Zeno's paradoxes*. Wesleyan University Press, Middletown, Connecticut. [Revised edn 1968.]

Grünbaum, A. (1968). *Geometry and chronometry in philosophical perspective*. University of Minnesota Press, Minneapolis.

Grünbaum, A. (1969). Simultaneity by slow clock transport in the special theory of relativity. *Philosophy of Science*, **36**, 5–43.

Grünbaum, A. (1973). *Philosophical problems of space and time*, (2nd edn). Reidel, Dordrecht. [*Boston Studies in the Philosophy of Science*, Vol. **12**.]

Hacker, P. M. S. (1972). *Insight and illusion: Wittgenstein on philosophy and the metaphysics of experience*. Clarendon Press, Oxford.

Hafele, J. C. and Keating, R. E. (1972*a*). Around-the-world atomic clocks: predicted relativistic time gains. *Science*, **177**, 166–68.

Hafele, J. C. and Keating, R. E. (1972*b*). Around-the-world atomic clocks: observed relativistic time gains. *Science*, **177**, 168–70.

Haji-Hassan, T., Duncan, A. J., Perrie, W., Kleinpoppen, H., and Merzbacher, E. (1989). Polarization correlation analysis of the radiation from a two-photon deuterium source using three polarisers: a test of quantum mechanics versus local realism. *Physical Review Lettters*, **62**, 237–40.

Hamlyn, D. W. (1984). *Metaphysics*. Cambridge University Press.

Hampshire, S. N. (1963). Hume's place in philosophy. In *David Hume. A symposium*, (ed. D. F. Pears), pp. 1–10. Macmillan, London.

Hankel, H. (1874). *Zur Geschichte der Mathematik in Alterthum und Mittelalter*. Teubner, Leipzig.

Hanson, N. R. (1958). *Patterns of discovery: An inquiry into the conceptual foundations of science*. Cambridge University Press.

Harding, S. (1986). *The science question in feminism*. Cornell University Press, Ithaca, NY.

Harding, S. (1996). Rethinking standpoint epistemology: what is "strong objectivity'? In Keller and Longino (1996), pp. 235–48.

Hardy, G. H. (1992). *A mathematician's apology*. Cambridge University Press, Canto Edition. [First edn 1940, foreword by C. P. Snow 1967, pp. 9–58.]

Harré, R. (1964). Concepts and criteria. *Mind*, **73**, 353–63.

Harré, R. (1970). *The principles of scientific thinking*. Macmillan, London.

Harré, R. and Madden, E. H. (1975). *Causal Powers. A theory of natural necessity*. Blackwell, Oxford.

Hart, H. L. A. and Honoré, T. (1985). *Causation and the law*, (2nd edn). Oxford University Press.

Hartle, J. B. (1968). Quantum mechanics of individual systems. *American Journal of Physics*, **36**, 704–12.

Hartle, J. B. and Hawking, S. W. (1983). Wave function of the universe. *Physical Review D*, **28**, 2960–75.

Hawking, S. W. (1988). *A brief history of time: From the big bang to black holes*. Bantam Press, London.

Hawking, S. (1993). *Black holes and baby universes and other essays*. Bantam Press, London.

Healey, R. (1989). *The philosophy of quantum mechanics: An interactive interpretation*. Cambridge University Press.

Hebb, D. O. (1949). *The organization of human behaviour: A neuropsychological theory*. Wiley, New York.

Heisenberg, W. (1925). Über quantentheoretische Umdeutung kinematischer und mechanischer Beziehungen. *Zeitschrift für Physik*, **33**, 879–93. [Received 29 July 1925.]

Heisenberg, W. (1927). Über den anschaulichen Inhalt der quantentheoretischen Kinematik und Mechanik. *Zeitschrift für Physik*, **43**, 172–98. [Received 23 March 1927. Translated in Wheeler and Zurek (1983), pp. 62–84.]

Heisenberg, W. (1930). *The physical principles of the quantum theory*, (trans. C. Eckart and F. Hoyt). Dover, New York.

Heisenberg, W. (1955). The development of the interpretation of the quantum theory. In *Niels Bohr and the development of physics: Essays dedicated to Niels Bohr on the occasion of his seventieth birthday*, (ed. W. Pauli), pp. 12–29. Pergamon Press, Oxford.

Heisenberg, W. (1971). *Physics and beyond: Encounters and conversations*. Allen and Unwin, London.

Hellman, G. (1989). *Mathematics without numbers: Towards a modal-structural interpretation*. Clarendon Press, Oxford.

Hempel, C. G. (1965). *Aspects of scientific explanation and other essays in the philosophy of science*. The Free Press, New York.

Hempel, C. G. (1980). The logical analysis of psychology. In Block (1980), Vol 1, pp. 14–23.

Hepp, K. (1972). Quantum theory of measurement and macroscopic observables. *Helvetica Physica Acta*, **45**, 237–47.

Herschel, J. F. W. (1830). *A preliminary discourse on the study of natural philosophy*. Longman, Rees, Orme, Brown & Green, and John Taylor, London. [Reprinted 1996, Routledge/Thoemmes Press, London.]

Hersh, R. (1985). Some proposals for reviving the philosophy of mathematics. In Tymoczko (1986a), pp. 9–28.

Hersh, R. (1997). *What is mathematics, really?* Jonathan Cape, London.

Hertz, H. (1899). *The principles of mechanics presented in a new form*, (trans. D. E. Jones and J. T. Walley). Macmillan, London. [1st German edn 1894.]

Hess, D. J. (1995). *Science and technology in a multicultural world: The cultural politics of facts and artifacts*. Columbia University Press, New York.

Hesse, M. B. (1961). *Forces and fields: The concept of action at a distance in the history of physics*. Nelson, London.

Hiley, B. J. and Peat, F. D. (ed.) (1987). *Quantum implications: Essays in honour of David Bohm*. Routledge & Kegan Paul, New York.

Hinckfuss, I. (1975). *The existence of space and time*. Clarendon Press, Oxford.

Hirsch, E. (1982). *The concept of identity*. Oxford University Press, New York.

Hodges, A. (1983). *Alan Turing: The enigma*. Burnett Books, London.

Hofstadter, D. R. (1979). *Gödel, Escher, Bach: An eternal golden braid*. Basic Books, New York.

Hofstadter, D. R. (1998). *Fluid concepts and creative analogies: Computer models of the fundamental mechanisms of thought*. Penguin, London.

Holmes, B. (1994). Half a wing is better than none. *New Scientist*, **144**, (1950, 5 November), 18.

Holton, G. (1969). Einstein, Michelson, and the "crucial" experiment. *Isis*, **60**, 133–97.

Hopf, E. (1934). On causality, statistics and probability. *Journal of Mathematics and Physics (MIT)*, **13**, 51–102.

Hopf, E. (1937). *Ergodentheorie*. Springer, Berlin. [Reprinted 1948, Chelsea Publishing Co., New York.]

Horwich, P. (1982). *Probability and evidence*. Cambridge University Press.

Horwich, P. (1987). *Asymmetries in time: Problems in the philosophy of science*. MIT Press, Cambridge, Massachusetts.

Horwich, P. (1990). *Truth*. Blackwell, Oxford.

Howard, D. (1985). Einstein on locality and separability. *Studies in History and Philosophy of Science*, **16**, 171–201.

Howe, L. and Wain, A. (ed.) (1993). *Predicting the future*. Cambridge University Press.

Howson, C. (1995). Theories of probability. *The British Journal for the Philosophy of Science*, **46**, 1–32.

Hughes, R. (1980). *The shock of the new: Art and the century of change*. British Broadcasting Corporation, London.

Hughes, T. (1994). Myth and education. In *Winter pollen: Occasional prose*, pp. 136–53. Faber and Faber, London. [This article first published 1976.]

Hume, David (1739). *A treatise of human nature: Being an attempt to introduce the experimental method of reasoning into moral subjects*. John Noon, London. [References from Hume (1888).]

Hume, David (1748). *An enquiry concerning human understanding*. T. Cadell, London. [2nd, posthumous, edn 1777 as Vol. 2 of *Essays and treatises on several subjects*. First published 1748 as *Philosophical Essays*. Present title adopted in the 1758 edn. References from Hume (1902).]

Hume, David (1888). *A treatise of human nature*, (ed. L. A. Selby-Bigge). Clarendon Press, Oxford. [Reprinted from the original edition in three volumes, first published in 1739 (Vols 1 and 2) and 1740 (Vol. 3).]

Hume, David (1902). An enquiry concerning human understanding. In *Enquiries concerning the human understanding and concerning the principles of morals*, (2nd edn), (ed. L. A. Selby-Bigge), pp. 5–165. Clarendon Press, Oxford. [First published as posthumous edn of Hume (1748).]

Hunter, J. (1971). *Metalogic. An introduction to the metatheory of standard first order logic*. Macmillan, London.

Ismael, J. (1996). What chances could not be. *The British Journal for the Philosophy of Science*, **47**, 79–91.

Jackson, F. (1975). Grue. *The Journal of Philosophy*, **72**, 113–31.

Jackson, F. (1982). Epiphenomenal qualia. *Philosophical Quarterly*, **32**, 127–36.

Jackson, F. (1986). What Mary didn't know. *The Journal of Philosophy*, **83**, 291–95.

Jackson, F. and Pargetter, R. (1980). Confirmation and the nomological. *Canadian Journal of Philosophy*, **10**, 415–28.

Jackson, J. D. (1975). *Classical electrodynamics*, (2nd edn). Wiley, New York.

Jammer, M. (1957). *Concepts of force*. Harvard University Press, Cambridge, Massachusetts.

Jammer. (1966). *The conceptual development of quantum mechanics*. McGraw-Hill, New York

Jammer, M. (1970). *Concepts of space: The history of theories of space in physics*. Harvard University Press, Cambridge, Massachusetts.

Jammer, M. (1974). *The philosophy of quantum mechanics: The interpretations of quantum mechanics in historic perspective*. John Wiley & Sons, New York.

Jánossy, L. and Náray, Zs. (1958). Investigation into interference phenomena at extremely low light intensities by means of a large Michelson interferometer. *Supplemento al Nuovo Cimento*, **9**, 588–598.

Jauch, J. M. (1968). *Foundations of quantum mechanics*. Addison–Wesley, Reading, Massachusetts.

John Paul II. (1980). Speech of His Holiness John Paul II. In *Einstein: Galileo: Commemoration of Albert Einstein*, (ed. B. Bucciarelli), pp. 77–81. Pontificia Academia Scientarum, Libreria Editrice Vaticana, Vatican City.

Jones, S., Martin, R., and Pilbeam, D. (ed.) (1993). *The Cambridge encyclopedia of human evolution.* Cambridge University Press.

Kac, M. (1983). What is random? *American Scientist,* **71,** 405–406.

Kac, M. (1984). More on randomness. *American Scientist,* **72,** 282–83.

Kanigel, R. (1992). *The man who knew infinity: A life of the genius Ramanujan.* Abacus, London.

Kant, Immanuel (1781). *Kritik der reinen Vernunft.* J. F. Hartknoch, Riga. [2nd edn 1787. References from Kant (1929).]

Kant, Immanuel (1783). *Prolegomena zu einer jeden künftigen Metaphysik, die als Wissenschaft wird auftreten können.* J. F. Hartknoch, Riga. [References from Kant (1953).]

Kant, Immanuel (1929). *Immanuel Kant's Critique of Pure Reason,* (trans. Norman Kemp Smith). Macmillan, London. [Corrected edn 1933.]

Kant, Immanuel (1953). *Prolegomena to any future metaphysics that will be able to present itself as a science,* (trans. P. G. Lucas). Manchester University Press.

Keller, E. Fox and Longino, H. E. (ed). (1996). *Feminism and science.* Oxford University Press.

Kemble, E. C. (1937). *The fundamental principles of quantum mechanics: With elementary applications.* McGraw-Hill, New York.

Kennedy, A. L. (1998). *Looking for the possible dance.* Vintage, London.

Khintchine, A. (1933). Zur mathematischen Begründung der Statistischen Mechanik. *Zeitschrift für Angewandte Mathematik und Mechanik,* **13,** 101–103.

Kilmister, C. W. (ed.) (1987). *Schrödinger: Centenary celebration of a polymath,* Cambridge University Press.

Kirk, G. S., Raven, J. E., and Schofield, M. (1983). *The presocratic philosophers: A critical history with a selection of texts,* (2nd edn). Cambridge University Press.

Kitcher, P. (1984). *The nature of mathematical knowledge.* Oxford University Press, New York.

Kitcher, P. (1987). Mathematical naturalism. In *Essays in the history and philosophy of modern mathematics,* (ed. W. Aspray and P. Kitcher), pp. 293–325. University of Minnesota Press, Minneapolis. [Minnesota Studies in the Philosophy of Science, Vol. **11.**]

Kitcher, P. (1993). *The advancement of science.* Oxford University Press.

Klee, R. (1997). *Introduction to the philosophy of science: Cutting nature at its seams.* Oxford University Press, New York.

Klein, F. (1939). *Elementary mathematics from an advanced standpoint: Arithmetics, algebra, analysis,* (trans. E. R. Hedrick and C. A. Noble). Dover, New York. [From 3rd German edn 1924.]

Klir, G. and Folger, T. (1988). *Fuzzy sets, uncertainty, and information.* Prentice–Hall, Englewood Cliffs, New Jersey.

Kneale, W. C. (1949). *Probability and induction.* Clarendon Press, Oxford. [Reprinted with corrections 1952.]

Kolata, G. (1982). Does Godel's theorem matter to mathematics. *Science,* **218,** 779–80.

Kolmogorov, A. (1933). *Grundbegriffe der Wahrscheinlichkeitsrechnung.* Springer, Berlin. [Author's name transliterated as Kolmogoroff.]

Kolmogorov, A. (1963a). On tables of random numbers. *Sankhya. The Indian Journal of Statistics A,* **25,** 369–76.

Kolmogorov, A. (1963b). The theory of probability. In *Mathematics: Its contents, methods, and meaning,* Vol. 2, (ed. A.D. Aleksandrov, A. Kolmogorov, and M. Lavrent'ev), pp. 229–64. MIT Press, Cambridge, Massachusetts. [1st Russian edn 1956.]

Kolmogorov, A. N. (1965). Three approaches to the quantitative definition of information. *Problems of Information Transmission,* **1,** 1–7.

Kolmogorov, A. N. (1968). Logical basis for information theory and probability theory. *IEEE Transactions on Information Theory,* **14,** 662–64.

Krajewski, S. (1993). Did Gödel prove that we are not machines? (On philosophical consequences of Gödel's theorem). In *First international symposium on Gödel's theorems,* (ed. Z. W. Wolkowski), pp. 39–49. World Scientific, Singapore.

Kripke, S. (1971). Identity and necessity. In *Identity and individuation*, (ed. M. K. Munitz), pp. 135–64. New York University Press. [Reprinted in Schwartz (1977), pp. 66–101. Page references are to this version.]

Kripke, S. A. (1972). Naming and necessity. In *Semantics of natural language*, (2nd edn), (ed. D. Davidson and G. Harman), pp. 253–355. Reidel, Dordrecht. [See also *Addenda*, pp. 763–69.]

Kripke, S. A. (1980). *Naming and necessity*. Blackwell, Oxford. [1st published 1972, revised edn.]

Kuhn, T. A. (1962). *The structure of scientific revolutions*. University of Chicago Press. [References from 2nd edn 1970.]

Kuhn, T. S. (1970). Reflections on my critics. In *Criticism and the growth of knowledge: Proceedings of the International colloquium in the philosophy of science, London, 1965*, (ed. I. Lakatos and A. Musgrave), pp. 231–78. Cambridge University Press.

Lacey, H. and Joseph, G. (1968). What the Gödel formula says. *Mind*, **77**, 77–83.

Lakatos, I. (1978). *Mathematics, science and epistemology: Philosophical papers*, Vol. 2, (ed. J. Worrall and G. Currie). Cambridge University Press.

Lakatos, I. (1981). *Proofs and refutations: The logic of mathematical discovery*, (ed. J. Worral and E. Zahar). Cambridge University Press. [Corrected reprint of 1976 1st edn.]

Lechalas, G. (1896). *Études sur l'espace et les temps*. F. Alcan, Paris.

Leggett, A. J. (1980). Macroscopic quantum systems and the quantum theory of measurement. *Progress of Theoretical Physics (Supl.)*, **69**, 80–100.

Leggett, A. J. (1987). Quantum mechanics at the macroscopic level. In *Chance and matter: Proceedings of the 1986 les Houches Summer School*, (ed. J. Souletie, J. Vannimenus, and R. Stora), pp. 395–506. North Holland, Amsterdam.

Leggett, A. (1995). Time's arrow and the quantum measurement problem. In Savitt (1995*a*), pp. 97–106.

Leggett, A. J. and Garg, A. (1985). Quantum mechanics versus macroscopic realism: is the flux there when nobody looks? *Physical Review Letters*, **54**, 857–60.

Leibniz, G. W. (1696). On the principle of indiscernibles. In Parkison (1973), pp. 133–35. [Date given is approximate.]

Leibniz, G. W. (1697). On the ultimate origination of things. In *Leibniz. The Monadology and other philosophical writings*, 1898, (trans. and ed. R. Latta), pp. 337–51. Clarendon Press, Oxford.

Leibniz, G. W. (1714). *La Monadologie*, Vienna. [1st edn (German trans.) 1720. French original 1840. References from Leibniz (1991).]

Leibniz, G. W. (1951). *Theodicy: Essays on the goodness of God, the freedom of man, and the origins of evil*, (trans. E. M. Huggard, ed. A. Ferrrar). Routledge & Kegan Paul, London. [First published in 1710.]

Leibniz, G. W. (1991). *G. W. Leibniz's Monadology: An edition for students*, (ed. N. Rescher). Routledge, London.

Lemmon, E. J. (1965). *Beginning logic*. Nelson, London.

Leplin, J. (1997). *A novel defense of scientific realism*. Oxford University Press, New York.

Le Poidevin, R. and MacBeath, M. (ed.) (1993). *The philosophy of time*. Oxford University Press.

Levine, A. (1999). Letter to the Editor. *New York Review of Books*, **46**, (3, 18 February), 49.

Lévi-Strauss, C. (1962). *La pensée sauvage*. Plon, Paris.

Lewin, R. (1993). *Complexity: Life at the edge of chaos*. Phoenix, London. [First published 1993, J. M. Dent, London.]

Lewis, D. (1973). *Counterfactuals*. Blackwell, Oxford.

Lewis, D. (1979). Counterfactual dependence and time's arrow. *Noûs*, **13**, 455-76.

Lewis, D. (1986*a*). *On the plurality of worlds*. Blackwell, Oxford.

Lewis, D. (1986*b*). A subjectivist guide to objective chance. In *Philosophical papers*, Vol. 2, pp. 83–132. Oxford University Press, New York.

Lewis, G. N. (1926). The conservation of photons. *Nature*, **118**, 874–75.

Lindgren, B. W., McElrath, G. W., and Berry, D. A. (1978). *Introduction to probability and statistics*, (4th edn). Macmillan, New York.

Lindley, D. (1997). *Where does the weirdness go?: Why quantum mechanics is strange, but not as strange as you think*. Vintage, London.

Lines, M. E. (1994). *On the shoulders of giants*. Institute of Physics Publishing, Bristol.

Locke, J. (1690). *An essay concerning human understanding*. [References from Locke (1959).]

Locke, J. (1959). *An essay concerning human understanding*, Vol. 2, (ed. A. C. Fraser). Dover, New York.

Lockwood, M. (1989). *Mind, brain and the quantum: The compound 'I'*. Blackwell, Oxford. [Reprinted with corrections 1990.]

Lockwood, M. (1996). "Many minds" interpretation of quantum mechanics. *The British Journal for the Philosophy of Science*, **47**, 159–88.

London, F. and Bauer, E. (1939). *La théorie de l'observation en méchanique quantique*. Hermann, Paris.

Loschmidt, J. (1876). Über den Zustand des Wärmegleichgewichtes eines Systems von Körpern mit Rücksicht auf die Schwerkraft. *Sitzungsberichte der Kaiserlichen Akademie der Wissenschaften [Wien], Mathematisch-Naturwissenschaftlichen. Zweite Abtheilung*, **73**, 128–42.

Losee, J. (1993). *A historical introduction to the philosophy of science*, (3rd edn). Oxford University Press.

Lowe, E. J. (1994). Vague identity and quantum indeterminacy. *Analysis*, **54**, 110–14.

Lucas, J. R. (1961). Minds, machines and Gödel. *Philosophy*, **36**, (112–27)

Lucas, J. R. (1968). Satan stultified: a rejoinder to Paul Benacerraf. *The Monist*, **52**, 145–58.

Lucas, J. R. (1970a). *The concept of probability*. Clarendon Press, Oxford.

Lucas, J. R. (1970b). *The freedom of the will*. Clarendon Press, Oxford.

Lucas, J. R. (1996). Minds, machines, and Gödel: a retrospect. In Millican and Clark (1996), pp. 103–36.

Mach, E. (1896). *Principien der Wärmelehre: Historisch-kritisch entwickelt*. J. A. Barth, Leipzig.

Mach, E. (1960). *The science of mechanics: A critical and historical account of its development*, (6th edn), (trans. T. J. McCormack). The Open Court Publishing Company, Lasalle, Illinois. [Introduction by K. Menger. 1st German edn 1883, 1st American edn 1893.]

Mackie, J. L. (1963). The paradox of confirmation. *The British Journal for the Philosophy of Science*, **13**, 265–77.

Mackie, J. L. (1980). *The cement of the universe. A study of causation*. Clarendon Press, Oxford. [Pbk edn with addenda of first 1974 edn.]

McLaughlin, W. I. (1994). Resolving Zeno's paradoxes. *Scientific American*, **271**, (5), 66–71.

McLaughlin, W. I. and Miller, S. L. (1992). An epistemological use of nonstandard analysis to answer Zeno's objections against motion. *Synthese*, **92**, (3), 371–84.

McLeod, P. and Dienes, Z. (1996). Do fielders know where to go to catch the ball or only how to get there? *Journal of Experimental Psychology: Human Perception and Performance*, **22**, 531–43.

Maddy, P. (1981). Sets and numbers. *Noûs*, **15**, 494–511.

Maddy, P. (1990). *Realism in mathematics*. Clarendon Press, Oxford.

Maimonides, M. (1904). *The guide for the perplexed*, (trans., ed. M. Friedländer). Hebrew Publishing Co., New York.

Martín, A. (1940). *Poesías incompletas*. Editorial Laspataz, Sos del Rey Católico.

Mates, B. (1986). *The philosophy of Leibniz: Metaphysics and language*. Oxford University Press, New York.

Mehler, J. and Dupoux, E. (1990). *Naître humain*. Odile Jacob, Paris.

Melnyk, A. (1995). Two cheers for reductionism: or the dim prospects for non-reductive materialism. *Philosophy of Science*, **62**, 370–88.

Mermin, N. D. (1983). The great quantum muddle. *Philosophy of Science*, **50**, 651–56. [Reprinted in Mermin (1990b), pp. 190–97.]

Mermin, N. D. (1990a). Simple unified form for the major no-hidden variables theorems. *Physical Review Letters*, **65**, 3373–76.

Mermin, N. D. (1990b). *Boojums all the way through: Communicating science in a prosaic age.* Cambridge University Press.

Michotte, A. E. (1946). *La perception de la causalité.* Institut Supérieur de Philosophie, Louvain-la-Neuve. [Translated as *The perception of causality*, 1963, Methuen, London.]

Mill, J. S. (1843). *A system of logic ratiocinative and inductive: Being a connected view of the principles of evidence and the methods of scientific investigation. [Collected works of John Stuart Mill*, 1973, Vol VII, ed. J. M. Robson, University of Toronto Press.]

Millican, P. J. R. and Clark, A. (ed.) (1996). *Machines and thought: The legacy of Alan Turing,* Vol. 1. Clarendon Press, Oxford.

Millikan, R. A. (1916). Einstein's photoelectric equation and contact electromotive force. *The Physical Review*, **7**, 18–32.

Montaigne, Michel de (1588). *Essais*, (3 Vols), (5th edn), (ed. Pierre Michel). Gallimard, Paris. [This 1965 edn, which is the one used in the text, is based on the 1588 *exemplaire de Bordeaux* annotated by Montaigne.]

Moore, W. J. (1989). *A life of Erwin Schrödinger.* Cambridge University Press.

Morley, J. (1886). *Diderot and the Encyclopedists,* (2 Vols). Macmillan, London. [Reissued 1971.]

Morrison, P. (1966). Time's arrow and external perturbations. In *Preludes in theoretical physics: In honor of V. F. Weisskopf*, (ed. A. de Shalit, H. Feschbach, and L. van den Hove), pp. 347–51. North Holland, Amsterdam.

Mostowski, A. (1952). *Sentences undecidable in formalized arithmetic: An exposition of the theory of Kurt Gödel.* North Holland, Amsterdam.

Murdoch, D. (1994). The Bohr-Einstein dispute. In Faye and Folse (1994), pp. 303–24.

Murdoch, I. (1992). *Metaphysics as a guide to morals.* Chatto & Windus, London.

Nagel, E. (1955). Principles of the theory of probability. In *International encyclopedia of unified science*, Vol. 1, (ed. O. Neurath, R. Carnap, and C. Morris), pp. 341–422. University of Chicago Press. [Combined edn.]

Nagel, E. (1961). *The structure of science: Problems in the logic of scientific explanation.* Routledge & Kegan Paul, London.

Nagel, E. and Newman, J. R. (1958). *Gödel's proof.* Routledge & Kegan Paul, London.

Nagel, T. (1974). What is it like to be a bat? *The Philosophical Review*, **83**, 435–50.

Newton, I. (1686). *Sir Isaac Newton's mathematical principles of natural philosophy and his system of the world*, (trans. A. Motte, ed. F. Cajori). University of California Press, Berkeley. [This edn 1934, reprinted 1947.]

Newton–Smith, W. (1980). *The structure of time.* Routledge & Kegan Paul, London.

Nyhof, J. (1988). Philosophical objections to the kinetic theory. *The British Journal for the Philosophy of Science*, **39**, 81–109.

Omnès, R. (1994). *The interpretation of quantum mechanics.* Princeton University Press.

Omnès, R. (1999a). *Quantum Philosophy: Understanding and interpreting contemporary science,* (trans. A. Sangalli). Princeton University Press.

Omnès, R. (1999b). *Understanding quantum mechanics.* Princeton University Press.

Ornstein, R. (1991). *The evolution of consciousness: Of Darwin, Freud, and cranial fire: the origins of the way we think.* Prentice Hall, New York.

Ortega y Gasset, J. (1958). *Historia como sistema.* Revista de Occidente, Madrid. [First published 1936 in English translation in *Philosophy and history: Essays presented to Ernst Cassirer* (ed. R. Klibansky). Clarendon Press, Oxford.]

Owen, G. E. L. (1958). Zeno and the mathematicians. *Proceedings of the Aristotelian Society*, **58**, 199–222.

Owens, D. (1992). *Causes and coincidences*. Cambridge University Press.

Pais, A. (1982). *"Subtle is the Lord. . . .' The science and the life of Albert Einstein*. Oxford University Press.

Pais, A. (1986). *Inward bound. Of matter and forces in the physical world*. Clarendon Press, Oxford.

Pais, A. (1991). *Niels Bohr's times, in physics, philosophy, and polity*. Clarendon Press, Oxford.

Pannenberg, W. (1993). *Toward a theology of nature: Essays on science and faith*, (ed. Ted Peters). Westmister/John Knox Press, Louisville, Kentucky.

Papineau, D. (1996). Many minds are no worse than one. *The British Journal for the Philosophy of Science*, **47**, (2), 233–41. [This issue, pp 159–248, contains a useful symposium on the many-minds interpretation of quantum mechanics.]

Parkison, G. H. R. (ed). (1973). *Leibniz philosophical writings*, (trans. M. Morris and G. H. R. Parkison). J. M. Dent & Sons, London.

Pauli, W. (1925). Über den Einfluß der Geschwindigkeitsabhängigkeit der Elektronnenmasse auf den Zeemaneffekt. *Zeitschrift für Physik*, **31**, 373–85. [Received 2 December 1924.]

Peano, G. (1908). *Formulario Mathematico*, (5th edn). Bocca Editores, Turin. [Facsimile edn 1960, Edizioni Cremonese, Roma; introduction and notes by Ugo Cassina.]

Penrose, R. (1985). Quantum gravity and state-vector reduction. In Penrose and Isham (1986), pp. 129–46.

Penrose, R. (1986). Big bangs, black holes and "time's arrow." In *The nature of time*, (ed. R. Flood and M. Lockwood), pp. 37–62. Blackwell, Oxford.

Penrose, R. (1989). *The emperor's new mind: Concerning computers, minds, and the laws of physics*. Oxford University Press.

Penrose, R. (1990). Précis of The emperor's new mind: Concerning computers, minds, and the laws of physics (together with responses by critics and a reply by the author). *Behavioural and Brain Sciences*, **13**, 643–705.

Penrose, R. (1994). *Shadows of the mind: A search for the missing science of consciousness*. Oxford University Press.

Penrose, R. (1997). *The large, the small and the human mind*. Cambridge University Press. [With contributions by A. Shimony, N. Cartwright, and S. Hawking.]

Penrose, R. and Isham, C. J. (ed.) (1986). *Quantum concepts in space and time*. Clarendon Press. Oxford.

Peres, A. (1993). *Quantum theory: Concepts and methods*. Kluwer Academic, Dordrecht.

Pfau, T. (1996). Atom holography: not just an illusion. *Physics World*, **9**, (8), 20–21.

Pickering, A. (1992). *Science as practice and culture*. University of Chicago Press.

Pirandello, L. (1958). Sei personaggi in cerca d'autore. In *Tutto il teatro di Luigi Pirandello*, Vol. 1, (9th edn). Mondadori, Verona. [First published 1921.]

Place, U. T. (1970). Is consciousness a brain process? In *The mind-brain identity theory*, (ed. C. Borst), pp. 42–51. Macmillan, London. [First published in 1956.]

Planck, M. (1900). Zur Theorie des Gesetzes der Energieverteilung im Normalspectrum. *Verhandlungen der Deutschen Physikalischen Gesellschaft*, **2**, 237–45.

Poincaré, H. (1894). Sur la thórie cinétique des gaz. *Revue Générale des Sciences Pures et Appliquées*, **5**, 513–21.

Poincaré, H. (1905). *Science and hypothesis*, (trans. G. B. Halsted). Science Press, New York. [1st French edn 1902. References from the Dover 1952 reprint.]

Poincaré, H. (1912). *Calcul des probabilités*, (2nd edn). Gauthier–Villars, Paris.

Poincaré, H. (1929). *The value of science*, (trans. G. B. Halsted), New York.

Popper, K. R. (1959). *The logic of scientific discovery*, (trans. K. R. Popper, J. Freed, and L. Freed). Hutchinson, London. [First published as *Logik der Forschung* 1934 (1935 in the imprint), Vienna.]

Popper, K. R. (1967). Quantum mechanics without 'the observer' In *Quantum theory and reality*, (ed. M. Bunge), pp. 7–44. Springer, Berlin.

Popper, K. R. (1969). *Conjectures and refutations: The growth of scientific knowledge*, (3rd edn). Routledge & Kegan Paul, London.

Popper, K. R. (1972). *Objective knowledge: An evolutionary approach*. Clarendon Press, Oxford.

Popper, K. R. (1974). Replies to my critics. In *The philosophy of Karl Popper*, Vol. 2, (ed. P. A. Schilpp), pp. 961–1197. Open Court, La Salle, Illinois.

Popper, K. R. (1982). *Quantum theory and the schism in physics*, (ed. W. W. Bartley, III). Hutchinson, London.

Popper, K. R. and Eccles, J. C. (1977). *The self and its brain*. Springer, Berlin.

Post, S. G. (1998). A moral case for nonreductive physicalism. In *Whatever happened to the soul?: Scientific and theological portraits of human nature*, (ed. W. S. Brown, N. Murphy, and H. Newton Malony), pp. 195–212. Fortress Press, Minneapolis.

Poussin, N. (1964). *Lettres et propos sur l'art*, (ed. A. Blunt). Hermann, Paris.

Priest, S. (1991). *Theories of the mind*. Penguin Books, London. [Also published by Houghton Mifflin, Boston, 1991. References are to this edition.]

Primas, H. (1990). The measurement process in the individual interpretation of quantum mechanics. In Cini and Lévy Leblond (1990), pp. 49–68.

Pullum, G. K. (1991). *The great Eskimo vocabulary hoax, and other irreverent essays on the study of language*. University of Chicago Press.

Putnam, H. (1973). Meaning and reference. *The Journal of Philosophy*, **70**, 699–711. [Reprinted in Schwartz (1977), pp. 119–32. Page numbers refer to this version.]

Putnam, H. (1979). *Mathematics, matter and method. Philosophical papers.*, Vol. 1, (2nd edn). Cambridge University Press.

Quine, W. V. O. (1951). Two dogmas of empiricism. *The Philosophical Review*, **60**, 20–43. [References from reprint in Quine (1964), pp. 20–46.]

Quine, W. V. O. (1964). *From a logical point of view. 9 logico-philosophical essays*, (2nd edn). Harvard University Press, Cambridge, Massachusetts.

Quine, W. V. (1970). *Philosophy of logic*. Prentice–Hall, Englewood Cliffs, New Jersey.

Quinton, A. (1967). The *a priori* and the analytic. In *Philosophical logic*, (ed. P. F. Strawson), pp. 107–128. Oxford University Press.

Rae, A. (1989). Quantum mechanics *v.* local realism. *Physics World*, **2**, (April), 17.

Ramsey, I. T. (1952). *Miracles: An exercise in logical mapwork*. Clarendon Press, Oxford. [Inaugural lecture delivered before the University of Oxford on 7 December 1951.]

Ramsey, I. T. (1965). *Christian discourse: Some logical explorations*. Oxford University Press.

Redhead, M. (1987). *Incompleteness, non-locality and causality: A prolegomenon to the philosophy of quantum mechanics*. Cambridge University Press.

Redhead, M. and Teller, P. (1991). Particles, particle labels, and quanta: the toll of unacknowledged metaphysics. *Foundations of Physics*, **21**, 43–62.

Redhead, M. and Teller, P. (1992). Particle labels and the theory of indistinguishable particles in quantum mechanics. *The British Journal for the Philosophy of Science*, **43**, 201–18.

Reichenbach, H. (1938). *Experience and prediction. An analysis of the foundations and the structure of knowledge*. Chicago University Press. [1st Phoenix edn 1961.]

Reichenbach, H. (1946). *Philosophical foundations of quantum mechanics*. University of California Press, Berkeley. [2nd printing; 1st printing 1944.]

Reichenbach, H. (1956). *The direction of time*, (ed. M. Reichenbach). University of California Press, Berkeley.

Reichenbach, H. (1957). *The philosophy of space and time*, (trans. M. Reichenbach and J. Freund). Dover, New York.

Rescher, N. (1969). *Many-valued logic*. McGraw-Hill, New York.

Ridley, B. K. (1995). *Time, space and things*, (3rd edn). Cambridge University Press. [1st edn 1976, Penguin, London.]

Robinson, H. (ed.) (1993a). *Objections to physicalism*. Clarendon Press, Oxford. [Introduction by Howard Robinson, pp. 1–25.]

Robinson, H. (1993b). The anti–materialist strategy and the "knowledge" argument. In Robinson (1993a), pp. 159–83. Clarendon Press, Oxford.

Robinson, J. A. T. (1963). *Honest to God*. SCM Press, London.

Rojas, Fernando de (1499). *La Celestina: Tragicomedia de Calisto y Melibea*. Fadrique de Basilea, Burgos. [Quotation from the 1899 edn, after the 1500 Salamanca copy, by M. Menéndez y Pelayo, Eugenio Krapf, Vigo.]

Rose, S. (1998). *Lifelines: Biology, freedom, determinism*. Penguin Books, London. [First published 1997, Allen Lane The Penguin Press, London.]

Rosenfeld, L. (1955). On the foundations of statistical thermodynamics. *Acta Physica Polonica*, **14**, 9–39.

Rosser, B. (1936). Extension of some theorems of Gödel and Church. *Journal of Symbolic Logic*, **1**, 87–91.

Rosser, B. (1937). Gödel theorems for non-constructive logics. *Journal of Symbolic Logic*, **2**, 129–37.

Rucker, R. (1995). *Infinity and the mind: The science and philosophy of the infinite*. Princeton University Press.

Rushdie, S. (1995). *The moor's last sigh*. Jonathan Cape, London.

Russell, B. (1912). On the notion of cause. *Proceedings of the Aristotelian Society*, **13**, 1–26.

Russell, B. (1919). *Introduction to mathematical philosophy*. Allen & Unwin, London.

Russell, B. (1923). Vagueness. *The Australasian Journal of Psychology and Philosophy*, **1**, 84–92.

Russell, B. (1926). *Our knowledge of the external world as a field for scientific method in philosophy*. Allen & Unwin, London. [First published 1914.]

Russell, B. (1940). *An inquiry into meaning and truth*. Allen & Unwin, London.

Russell, B. (1946). *History of western philosophy and its connection with political and social circumstances from the earliest times to the present day*. Allen and Unwin, London.

Ryle, G. (1949). *The concept of mind*. Hutchinson, London.

Ryle, G. (1962). *Dilemmas. The Tarner Lectures 1953*. Cambridge University Press. [First printing 1954.]

Sacks, O. (1995). A new vision of the mind. In Cornwell (1995), pp. 101–21.

Sainsbury, R. M. (1988). *Paradoxes*. Cambridge University Press.

Salmon, W. C. (ed.) (1970). *Zeno's Paradoxes*. Bobbs–Merrill, Indianapolis.

Salmon, W. C. (1977). The philosophical significance of the one-way speed of light. *Noûs*, **11**, 253–92.

Sartre, J.-P. (1943). *L'être et le néant*. Gallimard, Paris.

Savitt, S. F. (ed.) (1995a). *Time's arrow today: Recent physical and philosophical work on the direction of time*. Cambridge University Press.

Savitt, S. F. (1995b). Introduction. In Savitt (1995a), pp. 1–19.

Scheffler, I. (1964). *The anatomy of inquiry*. Routledge & Kegan Paul, London.

Schilpp, P. A. (ed.) (1959). *Albert Einstein: Philosopher-scientist*, (2 Vols). Harper & Brothers, New York.

Schrödinger, E. (1926). Quantisierung als Eigenwertproblem. *Annalen der Physik*, **79**, 361–76. [Received 27 January 1926. Fourth part: **81** (1926), 109–39, received 21 June 1926.]

Schrödinger, E. (1935). Die gegenwärtige Situation in der Quantenmechanik. *Naturwissenschaften*, **23**, 807–12. [Parts II and III, *ibid.*, pp. 823–8, 844–9. Translated in Trimmer (1980).]

Schrödinger, E. (1953). What is matter? *Scientific American*, (September), 52–57.

Schrödinger, E. (1957a). *Science theory and man*. Dover, New York.

Schrödinger, E. (1957b). What is a law of nature? In Schrödinger (1957a), pp. 133–47.

Schrödinger, E. (1959). *Mind and matter*. Cambridge University Press.

Schulte, O. (1999). Means–Ends Epistemology. *British Journal for the Philosophy of Science*, **50**, 1–31.

Schulte, O. (2000). What To Believe and What To Take Seriously: A Reply to Davis Chart Concerning The Riddle of Induction. *British Journal for the Philosophy of Science*, **51**, 151–153.

Schwartz, R., Scheffler, I., and Goodman, N. (1970). An improvement in the theory of projectibility. *The Journal of Philosophy*, **67**, 605–8.

Schwartz, S. P. (ed.) (1977). *Naming, necessity, and natural kinds*. Cornell University Press, Ithaca. [Introduction by ed. pp. 13-41.]

Scriven, M. (1964). Review of E. Nagel's book *The structure of science. The Review of Metaphysics*, **17**, 403–24.

Scully, M. O., Englert, B.-G., and Walther, H. (1991). Quantum optical tests of complementarity. *Nature*, **351**, 111–16.

Searle, J. R. (1980). Minds, brains, and programs, with open peer commentaries. *Behavioural and brain sciences.*, **3**, (3), 417-58.

Searle, J. R. (1984). *Minds, brains and science*. British Broadcasting Corporation, London. [The 1984 Reith Lectures, first broadcast by the BBC.]

Searle, J. R. (1990). Is the brain's mind a computer program? *Scientific American*, **262**, (1), 20-25.

Searle, J. R. (1992). *The rediscovery of the mind*. MIT Press, Cambridge, Massachusetts.

Searle, J. R. (1995). The mystery of consciousness. *The New York Review of Books*, **42**, (17, November 2), 60-66.

Searle, J. R. (1999). *Mind, language, and society: Doing philosophy in the real world*. Weidenfeld and Nicolson, London.

Seeger, R. J. (1966). *Galileo Galilei, his life and his works*. Pergamon Press, Oxford.

Sellars, W. (1948). Concepts as involving laws and inconceivable without them. *Philosophy of Science*, **15**, 287–315.

Sellars, W. (1963). *Science, perception and reality*. Routledge & Kegan Paul, London.

Selleri, F. (1989). Wave-particle duality: recent proposals for the detection of empty waves. In *Quantum theory and pictures of reality: Foundations, interpretations, and new aspects*, (ed. W. Schommers), pp. 279–332. Springer, Berlin.

Selleri, F. (1990). *Quantum paradoxes and physical reality*, (ed. A. van der Merwe). Kluwer Academic, Dordrecht.

Shakespeare, W. (1600). *Much adoe about Nothing*. Andrew Wise and William Aspley, London. [*Shakespeare quarto facsimiles*, No. 15, 1971, Clarendon Press, Oxford. The through line number given is the one adopted in this edn.]

Shanker, S. G. (ed.) (1988a). *Gödel's theorem in focus*. Croom Helm, London.

Shanker, S. G. (1988b). Wittgenstein's remarks on the significance of Gödel's theorem. In Shanker (1988a), pp. 155–256.

Shanker, S. (1998). *Wittgenstein's remarks on the foundations of AI*. Routledge, London.

Shimony, A. (1984). Passion at a distance. In *Proceedings of the international symposium on the foundations of quantum mechanics*, (ed. S. Kamefuchi), pp. 225-30. Physical Society of Japan, Tokyo.

Shimony, A. (1988). The reality of the quantum world. *Scientific American*, **258**, (1), 36–43.

Simon, H. A. (1996). Machine as mind. In Millican and Clark (1996), pp. 82–102.

Singer, D. W. (1950). *Giordano Bruno. His life and thought with an annotated translation of his work On the infinite universe and worlds*. henry Schuman, New York.

Sklar, L. (1992). *Philosophy of physics*. Oxford University Press.

Sklar, L. (1993). *Physics and chance: Philosophical issues in the foundations of statistical mechanics*. Cambridge University Press.

Sklar, L. (1995). The elusive object of desire: in pursuit of the kinetic equations and the Second Law. In Savitt (1995a), pp. 191–216.

Smart, J. J. C. (1963). *Philosophy and scientific realism*. Routledge & Kegan Paul, London.

Smorynski, C. (1977). The incompleteness theorems. In *Handbook of mathematical logic*, (ed. J. Barwise), pp. 821–65. North Holland, Amsterdam.

Smullyan, R. (1987). *Forever undecided: A puzzle guide to Gödel*. Knopf, New York.

Snyder, A. W. and Mitchell, D. J. (1999). Is integer arithmetic fundamental to mental processing?: the mind's secret arithmetic. *Proceedings of the Royal Society [London] B*, **266**, 587–92.

Sommerfeld, A. (ed.) (1923). *The principle of relativity: A collection of original memoirs on the special and general theory of relativity*, (trans. W. Perrett and G. B. Jeffery). Methuen, London. [Reprinted by Dover, New York.]

Sorabji, R. (1983). *Time, creation and the continuum: Theories in antiquity and the early middle ages*. Duckworth, London.

Spinoza, B. (1670). *Tractatus Theologico-Politicus*, 1989,(trans. S. Shirley). E. J. Brill, Leiden.

Squires, E. (1986). *The mistery of the quantum world*. Adam Hilger, Bristol.

Squires, E. (1990). *Conscious mind in the physical world*. Adam Hilger, Bristol.

Stalker, D. (ed). (1994). *The new riddle of induction*. Open Court, Chicago.

Stamp, P. (1995). Time, decoherence, and "reversible" measurements. In Savitt (1995*a*), pp. 107–54.

Stanley, A. P. (1907). *Lectures on the history of the Eastern Church*, (3rd edn). Dent, London. [Quotation from 1924 Everyman's reprint.]

Stapp, H. P. (1972). The Copenhagen interpretation. *American Journal of Physics*, **40**, 1098–116.

Stapp, H. P. (1985). Consciousness and values in the quantum universe. *Foundations of Physics*, **15**, 35–47.

Stapp, H. P. (1993). *Mind, matter, and quantum mechanics*. Springer, Berlin.

Steiner, M. (1973). Platonism and the causal theory of knowledge. *Journal of Philosophy*, **70**, 57–66.

Stemmer, N. (1989). Physicalism and the argument from knowledge. *Australasian Journal of Philosophy*, **67**, 84–91.

Storr, A. (1990). *Churchill's black dog: And other phenomena of the human mind*. Fontana/ Collins, London.

Strawson, P. F. (1952). *Introduction to logical theory*. Methuen, London.

Strawson, P. F. (ed.) (1967). *Philosophical logic*. Oxford University Press.

Strawson, P. F. (1992). *Analysis and metaphysics. An introduction to philosophy*. Oxford University Press.

Strawson, P. F. (1997). *Entity and identity*. Clarendon Press, Oxford.

Sutherland, S. (1992). *Irrationality. The enemy within*. Constable, London.

Swinburne, R. G. (1968). Grue. *Analysis*, **28**, 123–28.

Swinburne, R. G. (1971). The paradoxes of confirmation—a survey. *American Philosophical Quarterly*, **8**, (318–29)

Swinburne, R. (1973). *An introduction to confirmation theory*. Methuen, London.

Swinburne, R. (1986). *The evolution of the soul*. Oxford University Press.

Swinburne, R. (1996). *Is there a God?* Oxford University Press.

Taylor, G. I. (1910). Interference fringes with feeble light. *Proceedings of the Cambridge Philosophical Society*, **15**, 114–15. [Read 25 January 1909.]

Tchehov, A. P. (1962). *Select tales of Tchehov*, Vol. 2, (trans. Constance Garnett). Chatto & Windus, London.

Temple, G. (1981). *100 years of mathematics*. Duckworth, London.

Thayer, H. S. (ed.) (1953). *Newton's philosophy of nature: Selections from his writings*. Hafner, New York.

Thom, R. (1975). *Structural stability and morphogenesis: An outline of a general theory of models*, (trans. D. H. Fowler). Benjamin/Cummings, Reading, Massachusetts.

Thomas, D. (1952). *Collected Poems 1934–1952.* London, Dent.

Thomson, W. and Tait, P. G. (1879). *Treatise on natural philosophy.* Cambridge University Press.

Thouless, R. H. (1931*a*). Phenomenal regression to the "real" object. I. *The British Journal of Psychology*, **21**, 339–59.

Thouless, R. H. (1931*b*). Phenomenal regression to the "real" object. II. *The British Journal of Psychology*, **22**, 1–30.

Tillich, P. (1964). *Theology of culture*, (ed. R. C. Kimball). Galaxy Book, New York. [First published in 1959 by Oxford University Press, New York.]

Tillich, P. (1965). *Ultimate concern: Tillich in dialogue*, (ed. D. Mackenzie Brown). SCM Press, London.

Toraldo di Francia, G. (1978). What is a physical object? *Scientia*, **113**, 57–65.

Toulmin, S. (1961). *Foresight and understanding: An enquiry into the aims of science.* Hutchinson, London.

Trimmer, J. D. (1980). The present situation in quantum mechanics: a translation of Schrödinger's "cat paradox" paper. *Proceedings of the American Philosophical Society*, **124**, 323–38.

Trollope, A. (1861). *The Bertrams: A novel*, (4th edn). Chapman & Hall, London. [1st edn 1859.]

Turing, A. M. (1950). Computing machinery and intelligence. *Mind*, **59**, 433–60.

Tymoczko, T. (ed.) (1986*a*). *New directions in the philosophy of mathematics: An anthology.* Birkhäuser, Boston.

Tymoczko, T. (1986*b*). Computers and mathematical practice: a case study. The four-color problem and its philosophical significance. In Tymoczko (1986*a*), pp. 243–66.

Unruh, W. (1995). Time, gravity, and quantum mechanics. In Savitt (1995*a*), pp. 23–65.

van der Waerden, B. L. (1974). *Science Awakening.* Noordhoff, Groningen.

van Fraassen, B. (1979). Foundations of probability: a modal interpretation. In *Problems in the foundations of physics*, (ed. G. Toraldo di Francia). North Holland, Amsterdam. [*Rendiconti della Scuola Internazionale di Fisica "Enrico Fermi"*, Vol. **72**.]

van Fraassen, B. C. (1980). *The scientific image.* Clarendon Press, Oxford.

van Fraassen, B. C. (1985). *An introduction to the philosophy of time and space*, (2nd edn). Columbia University Press, New York. [Reprint of 1st 1970 edn, Random House, New York, with new preface and postcript.]

van Fraassen, B. C. (1989). *Laws and symmetry.* Clarendon Press, Oxford.

van Fraassen, B. C. (1991). *Quantum mechanics: An empiricist view.* Clarendon Press, Oxford.

Varagine, Jacopo da (1280). *Leggenda Aurea*, (2 Vols). Libreria Editrice Fiorentina, Firenze. [The date given is conjectural. This edn, 1990.]

Vermes, G. (1993). *The religion of Jesus the Jew.* Fortress Press, Minneapolis.

Vermes, G. (2000). *The changing faces of Jesus.* Allen Lane, The Penguin Press, London.

von Mises, R. (1936). *Wahrscheinlichkeit, Statistik und Wahrheit: Einführung in die neue Wahrscheinlichkeitslehre und ihre Anwendung*, (2nd edn). Springer, Wien. [1st edn 1928.]

von Neumann, J. (1932*a*). *Mathematische Grundlagen der Quantenmechanik.* Springer, Berlin. [Reprinted 1943, Dover, New York. References from von Neumann (1955).]

von Neumann, J. (1932*b*). A proof of the quasi-ergodic hypothesis. *Proceedings of the National Academy of Sciences*, **18**, 70–82.

von Neumann, J. (1955). *Mathematical foundations of quantum mechanics*, (trans. R. T. Beyer). Princeton University Press.

von Plato, J. (1994). *Creating modern probability theory: Its mathematics, physics and philosophy in historical perspective.* Cambridge University Press.

von Wright, G. H. (1974). *Causality and determinism.* Columbia University Press, New York.

von Wright, G. H. (1993). On the logic and epistemology of the causal relation. In *Causation*, (ed. E. Sosa and M. Tooley), pp. 105–24. Oxford University Press.

Wang, H. (1987). *Reflections on Kurt Gödel.* MIT Press, Cambridge, Massachusetts.

Ward, K. (1996). *Religion and creation.* Clarendon Press, Oxford.

Warnock, G. J. (1963). Hume on causation. In *David Hume. A symposium,* (ed. D. F. Pears), pp. 55–66. Macmillan, London.

Webster, R. (1996). *Why Freud was wrong: Sin, science and psychoanalysis,* (Revised pbk edn). HarperCollins, London.

Weinberg, S. (1993). *The first three minutes: A modern view of the origin of the universe.* Flamingo, London. [First published 1977, Basic Books, New York.]

Weinberg, S. (1998). The revolution that didn't happen. *The New York Review of Books,* **65,** (15), 48–52.

Werblonsky, R. J. Z. and Wigoder, G. (ed.) (1997). *The Oxford dictionary of the Jewish religion.* Oxford University Press, New York.

Weyl, H. (1914). Sur une application de la théorie des nombres à la mécanique statistique et la théorie des perturbations. *L'Enseignement Mathématique,* **16,** 455–67.

Weyl, H. (1916). Ueber die Gleichverteilung von Zahlen mod. Eins. *Mathematische Annalen,* **77,** 313–52.

Weyl, H. (1949). *Philosophy of mathematics and natural science,* (trans. Olaf Helmer, revised and augmented by author). Princeton University Press. [First published in German in 1927.]

Weyl, H. (1968). *Gesammelte Abhandlungen,* Vol. 4, (ed. K. Chandrasekharan). Springer, Berlin.

Wheeler, J. A. (1983). Law without law. In Wheeler and Zurek (1983), pp. 182–213.

Wheeler, J. A. and Zurek, W. H. (ed.) (1983). *Quantum theory and measurement.* Princeton University Press.

Whewell, W. (1837). *History of the inductive sciences, from the earliest to the present time,* (3 Vols). J. W. Parker & Son, London.

Whewell, W. (1858–60). *The philosophy of the inductive sciences founded upon their history,* (3 Vols, 3rd edn). J. W. Parker & Son, London.

Whitaker, A. (1995). *Einstein, Bohr and the quantum dilemma.* Cambridge University Press.

Whitby, B. (1996). The Turing test: AI's biggest blind alley? In Millican and Clark (1996), pp. 53–62.

White, P. (1967). *The living and the dead,* (pbk edn). Penguin, London. [First published 1941.]

Whitehead, A. N. (1911). Mathematics. In *The Encyclopædia Britannica,* Vol. 17, (11th edn), pp. 878–83. Cambridge University Press.

Whitrow, G. J. (1956). Why physical space has three dimensions. *The British Journal for the Philosophy of Science,* **6,** 13–31.

Whitrow, G. J. (1961). *The natural philosophy of time.* Nelson, London.

Whittaker, E. T. (1951). *A history of the theories of aether and electricity: The classical theories,* (2nd edn). Nelson, London. [1st edn 1910.]

Whittaker, E. T. (1953). *A history of the theories of aether and electricity: The modern theories: 1900–1926.* Nelson, London.

Whorf, B. L. (1956). *Language, thought and reality: Selected writings of Benjamin Lee Whorf,* (ed. J. B. Carroll). MIT Press, Cambridge, Massachusetts.

Wigner, E. P. (1961). Remarks on the mind-body question. In *The scientist speculates,* (ed. I. J. Good). Heinemann, London. [Reprinted in Wigner (1967), pp 171–99.]

Wigner, E. P. (1963). The problem of measurement. *American Journal of Physics,* **31,** (1) [Reprinted in Wigner (1967), pp. 153-170.]

Wigner, E. P. (1964). Two kinds of reality. *The Monist,* **48,** 248–64.)

Wigner, E. P. (1967). *Symmetries and reflections: Scientific essays of Eugene P. Wigner.* Indiana University Press, Bloomington.

Wigner, E. P. (1983). Interpretation of quantum mechanics. In Wheeler and Zurek (1983), pp. 260–314. [Lectures delivered at the Physics Department, Princeton University, 1976; revised 1981.]

Will, C. M. (1995). *Was Einstein right? Putting general relativity to the test*, (2nd edn). Oxford University Press. [2nd edn, first published 1993, Basic Books, New York.]

Williams, L. P. (1965). *Michael Faraday.* Chapman & Hall, London.

Wittgenstein, L. (1961). *Tractatus Logico-Philosophicus*, (trans. D. F. Pears and B. F. McGuinness). Routledge & Kegan Paul, London. [Bilingual edition. First published 1921 as *Logisch-philosophische Abhandlung in Annalen der Naturphilosophie.* All references are given by section number, although page numbers may occasionaly also be used.]

Wittgenstein, L. (1967a). *Ludwig Wittgenstein und der Wiener Kreis: Shorthand notes recorded by F. Waismann*, (ed. B. F. McGuinness). Blackwell, Oxford.

Wittgenstein, L. (1967b). *Philosophical investigations*, (3rd edn), (trans. G. E. M. Anscombe). Blackwell, Oxford. [Bilingual edition. Page numbers with superscript "e" refer to the English version. Most references are given by section number alone (part I) or section number and page (part II). 1st edn 1953.]

Wittgenstein, L. (1974). *Philosophical grammar*, (trans, A. Kenny, ed. R. Rhees). Blackwell, Oxford.

Wittgenstein, L. (1975). *Philosophical remarks*, (trans. R. Hargreaves and R. M. White, ed. R. Rhees). Blackwell, Oxford.

Wittgenstein, L. (1978). *Remarks on the foundations of mathematics*, (3rd edn), (trans. G. E. M. Anscombe, ed. G. H. von Wright, R. Rhees, and G. E. M. Anscombe). Blackwell, Oxford. [All references are given by page and paragraph number.]

Wodehouse, P. G. (1958). *Uneasy money.* Penguin, Harmondsworth, Middlesex. [First published 1917 by Methuen.]

Wodehouse, P. G. (1979). *Psmith journalist.* Penguin Books, London. [First published 1915.]

Wolpert, L. (1992). *The unnatural nature of science.* Faber & Faber, London.

Wolter, A. B. (1990). *The philosophical theology of John Duns Scotus*, (ed. M. M. Adams). Cornell University Press, Ithaca, N. Y.

Yourgrau, P. (1991). *The disappearance of time. Kurt Gödel and the idealistic tradition in philosophy.* Cambridge University Press.

Zeh, H.-D. (1992). *The physical basis of the direction of time*, (2nd edn). Springer, Berlin.

Zeilinger, A. (1990). Experiment and quantum measurement theory. In Cini and Lévy-Leblond (1990), pp. 9–26. Adam Hilger, Bristol.

Zurek, W. H. (1982). Environment-induced superselection rules. *The Physical Review D*, **26**, 1862–80.

Zurek, W. H. (1991). Decoherence and the transition from quantum to classical. *Physics Today*, **44**, (10), 36–44. [See also correspondence in *Physics Today*, (1993), **46**, (4), pp. 13–15 and 81–90.]

Zylbersztajn, A. (1994). Newton's absolute space, Mach's principle and the possible reality of fictitious forces. *European Journal of Physics*, **15**, 1–8.

INDEX